Peptides and Protein Phosphorylation

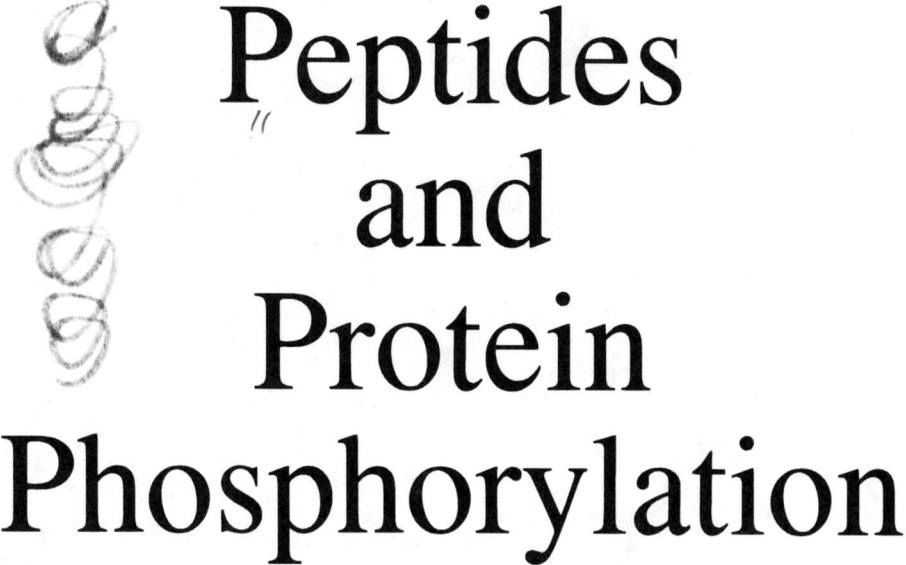

Editor

Bruce E. Kemp, Ph.D.
Deputy Director
St. Vincent's Institute of Medical Research
Melbourne, Australia

CRC Press, Inc.
Boca Raton, Florida

Library of Congress Cataloging-in-Publication Data

Peptides and protein phosphorylation/editor, Bruce E. Kemp.
    p.   cm.
    Includes bibliographical references.
    ISBN 0-8493-6530-9
    1. Phosphoproteins—Synthesis. 2. Protein kinase.
    3. Phosphorylation.   I. Kemp, Bruce E., 1946-
    [DNLM: 1. Chemistry, Organic. 2. Peptides. 3. Phosphoproteins.
    4. Protein Kinases.   QU 55 P4242]
    QP552.P5P46   1990
    612'.015754—dc20
    DNLM/DLC
    for Library of Congress   89-25145
                                                                                                                                           CIP

This book represents information obtained from authentic and highly regarded sources. Reprinted material is quoted with permission, and sources are indicated. A wide variety of references are listed. Every reasonable effort has been made to give reliable data and information, but the author and the publisher cannot assume responsibility for the validity of all materials or for the consequences of their use.

All rights reserved. This book, or any parts thereof, may not be reproduced in any form without written consent from the publisher.

Direct all inquiries to CRC Press, Inc., 2000 Corporate Blvd., N.W., Boca Raton, Florida 33431.

© 1990 by CRC Press, Inc.

International Standard Book Number 0-8493-6530-9

Library of Congress Card Number 89-25145
Printed in the United States

FOREWORD

Over the past 13 years, synthetic peptides have played a significant role in the explosion of knowledge about regulation by protein phosphorylation. This book highlights a number of important recent advances in our understanding of the enzymes involved, but is not intended to cover all of the protein kinases and protein phosphatases. The cAMP-dependent protein kinase has been a benchmark for progress in this area. It was the first protein kinase for which a model synthetic peptide substrate was developed. Since then structure/function studies by Taylor and her colleagues have resulted in advances in understanding the catalytic mechanism as well as the promise of an X-ray crystal structure (Chapter 1). This work has relevance for the entire family of protein kinases, which share homology in the catalytic domain. The development of synthetic peptide inhibitors also has been important and has provided a means of discriminating between protein kinase activities in crude extracts. Walsh and his colleagues have used synthetic peptides extensively to understand the primary structure/function relationships of the cAMP-dependent protein kinase inhibitor peptide as well as the solution structure determined by NMR. The advances here have paved the way for expression of synthetic genes for protein kinase inhibitors in living cells.

The early cAMP-dependent protein kinase peptide studies also provided a general basis for the subsequent development of the pseudosubstrate regulatory hypothesis (Chapter 4). It is now clear from studies with the myosin light chain kinase and protein kinase C that pseudosubstrate structures in the regulatory regions of these enzymes are responsible for maintaining them in inactive forms in the absence of their activators. The pseudosubstrate regulatory hypothesis provides a useful framework for exploring the DNA-derived amino acid sequences of protein kinases to locate potential regulatory regions. Moreover, by preparing synthetic peptide analogues of the pseudosubstrate region containing serine or threonine residues, it is possible to design substrates for these enzymes. With the impact of PCR in cloning protein kinases it is likely that many new protein kinases will be known first by their DNA-derived amino acid sequences before being isolated or their substrates identified. As more is learned about the substrate specificity and regulation it is hoped that inspection of the amino acid sequence alone will identify the pseudosubstrate regulatory domain and the protein substrate binding site.

Two chapters (3 and 10) have been devoted to tyrosine protein kinases. It is hard to overemphasize the impact of the discovery that a viral transforming protein, $pp60^{v-src}$, was a tyrosine protein kinase. Again, synthetic peptides have been used extensively in the measurement of these enzymes. The search for natural substrates for these enzymes has been difficult due to their low abundance. Antibodies to tyrosine phosphate are playing an increasingly important role in this work.

Chapters 11 and 12 are concerned with the preparation of antibodies to phosphorylation sites as well as the chemistry of preparing phosphorylated peptides. These techniques raise the possibility of being able to immunoprecipitate a putative phosphoprotein for which only the cDNA-derived amino acid sequence is known and a potential phosphorylation site postulated.

Given the large number of phosphoproteins, protein kinases, and protein phosphatases, as well as multisite phosphorylations and the possibility that some protein phosphorylation may be noise, the task of characterizing all of the regulatory pathways involving protein phosphorylation in a living cell would seem no less an endeavor than sequencing the entire human genome.

The completion of this book was seriously delayed, and any instances where particular chapters do not cover the very latest developments the responsibility rests with the Editor and does not reflect on the Contributor(s) or the Publisher.

Bruce E. Kemp
May 1989

THE EDITOR

Bruce E. Kemp, Ph.D., is Deputy Director of the St. Vincent's Institute of Medical Research in Melbourne, Australia, where Pehr Edman developed the first automated protein sequencer. Dr. Kemp received his Ph.D. at Flinders University in South Australia in 1975 under the supervision of Professor A. W. Murray. After doing postdoctoral work at the University of California at Davis with Dr. E. G. Krebs he returned to Australia, where he has been a National Health and Medical Research Fellow. He pioneered the development of synthetic peptide substrates and inhibitors for a number of protein kinases. The most widely known of these is the "Kemptide", a classic substrate for the cAMP-dependent protein kinase that has been used worldwide for the past 12 years. His major research interests are in the substrate specificity of protein kinases and their regulation by "pseudosubstrate structures". In 1988 he was awarded the Selwyn Smith Prize for Medical Research.

CONTRIBUTORS

Karen L. Angelos
Staff Research Associate
Department of Biological Chemistry
School of Medicine
University of California, Davis
Davis, California

Donald K. Blumenthal, Ph.D.
Associate Professor
Department of Biochemistry
University of Texas Health Center
at Tyler
Tyler, Texas

Joseph A. Buechler
Postgraduate Researcher
Department of Chemistry
University of California, San Diego
La Jolla, California

Jonathan A. Cooper, Ph.D.
Associate Member
Department of Cell Biology
Fred Hutchinson Cancer Research Center
Seattle, Washington

Lorentz Engström, M.D.
Professor
Department of Medical and Physiological
 Chemistry
University of Uppsala
Uppsala, Sweden

Lawrence P. Garetto
Assistant Professor
Indiana University School of Dentistry
Indianapolis, Indiana

Robert L. Geahlen, Ph.D.
Associate Professor
Department of Medicinal Chemistry
 and Pharmacognosy
Purdue University
West Lafayette, Indiana

David B. Glass, Ph.D.
Associate Professor
Departments of Pharmacology
 and Biochemistry
Emory University School of Medicine
Atlanta, Georgia

Marietta L. Harrison, Ph.D.
Assistant Professor
Department of Medicinal Chemistry
Purdue University
West Lafayette, Indiana

Tommy C. Hassell, M.Sc.
Staff Biochemist
Department of Biochemistry
MSDRL
West Point, Pennsylvania

Bruce E. Kemp, Ph.D.
Deputy Director
St. Vincent's Institute of
 Medical Research
Melbourne, Australia

Daniel R. Knighton, B.S.
Graduate Student
Department of Chemistry
University of California, San Diego
La Jolla, California

Fernando Marchiori, Ph.D.
Associate Professor
Department of Organic Chemistry
University of Padova
Padova, Italy

Ruthann A. Masaracchia, Ph.D.
Associate Professor
Department of Biochemistry
University of North Texas
Denton, Texas

Flavio Meggio, Ph.D.
Associate Professor
Department of Biological Chemistry
University of Padova
Padova, Italy

Fern E. Murdoch, Ph.D.
Postdoctoral Fellow
Department of Biochemistry
University of Wisconsin-Madison
Madison, Wisconsin

John W. Perich, Ph.D.
Postdoctoral Fellow
Department of Organic Chemistry
University of Melbourne
Parkville, Victoria
Australia

Lorenzo A. Pinna, Ph.D.
Professor and Director
Department of Biological Chemistry
University of Padova
Padova, Italy

Ulf Ragnarsson, Ph.D.
Docent
Department of Biochemistry
University of Uppsala
Uppsala, Sweden

James T. Stull, Ph.D.
Professor and Chairman
Department of Physiology
University of Texas Southwestern
 Medical Center at Dallas
Dallas, Texas

Susan S. Taylor, Ph.D.
Professor
Department of Chemistry
University of California, San Diego
La Jolla, California

Laura Anne Uphouse, Ph.D.
Research Associate
Department of Pharmacology
Emory University School of Medicine
Atlanta, Georgia

Scott M. Van Patten
Graduate Student
Department of Biological Chemistry
School of Medicine
University of California, Davis
Davis, California

Donal A. Walsh, Ph.D.
Professor
Department of Biological Chemistry
School of Medicine
University of California, Davis
Davis, California

Örjan Zetterqvist, M.D.
Lecturer
Department of Medical and
 Physiological Chemistry
University of Uppsala
Uppsala, Sweden

TABLE OF CONTENTS

Chapter 1
cAMP-Dependent Protein Kinase: Mechanism for ATP:Protein Phosphotransfer 1
Susan S. Taylor, Joseph A. Buechler, and Daniel R. Knighton

Chapter 2
The Inhibitor Protein of the cAMP-Dependent Protein Kinase 43
Donal A. Walsh, Karen L. Angelos, Scott M. Van Patten, David B. Glass, and Lawrence P. Garetto

Chapter 3
The *SRC* Family of Protein-Tyrosine Kinases 85
Jonathan A. Cooper

Chapter 4
Myosin Light Chain Kinases 115
Bruce E. Kemp and James T. Stull

Chapter 5
Use of Synthetic Peptides to Study Substrate Specificity of Protein
Phosphatases 135
Donald K. Blumenthal

Chapter 6
Type-2 Casein Kinases: General Properties and Substrate Specificity 145
Lorenzo A. Pinna, Flavio Meggio, and Fernando Marchiori

Chapter 7
Substrate Specificity of Cyclic AMP-Dependent Protein Kinase 171
Örjan Zetterqvist, Ulf Ragnarsson, and Lorentz Engström

Chapter 8
Unique Specificity Determinants for an S6/H4 Kinase and Protein Kinase C:
Phosphorylation of Synthetic Peptides Derived from the Smooth Muscle
Myosin Light Chain 189
Ruthann A. Masaracchia, Fern E. Murdoch, and Tommy C. Hassell

Chapter 9
Substrate Specificity of the Cyclic GMP-Dependent Protein Kinase 209
David B. Glass

Chapter 10
Protein-Tyrosine Kinases 239
Robert L. Geahlen and Marietta L. Harrison

Chapter 11
Antibodies against Synthetic Phosphorylation Site Peptides 255
David B. Glass and Laura Anne Uphouse

Chapter 12
Modern Methods of *O*-Phosphoserine- and *O*-Phosphotyrosine-Containing Peptide Synthesis 289
John W. Perich

Index 315

Chapter 1

cAMP-DEPENDENT PROTEIN KINASE: MECHANISM FOR ATP:PROTEIN PHOSPHOTRANSFER

Susan S. Taylor, Joseph A. Buechler, and Daniel R. Knighton

TABLE OF CONTENTS

I.	Introduction	2
II.	Kinases as Modular Structures	5
III.	ATP Binding Site	8
	A. Mapping with Analogues of ATP	8
	B. Fluorescent Analogues	10
	C. Affinity Labeling	11
	D. Nucleotide Fold	13
IV.	Peptide Binding Site	14
	A. Peptide Analogues	14
	B. Affinity Labeling	15
	C. Sulfhydryl Groups	16
	D. Carboxyl Groups	17
	E. NMR Analysis of Peptide Substrates	17
	F. Inhibitors	18
	1. Regulatory Subunit	18
	2. Heat-Stable Protein Kinase Inhibitor	20
V.	Other Active Site Residues	22
VI.	Enzyme Mechanism	23
VII.	Conformational Changes Associated with the C-Subunit	26
VIII.	Domain Structure	27
IX.	Future Directions	32
References		32

I. INTRODUCTION

From the classic studies on glycogen metabolism by the Coris[1] and their co-workers came the discovery that protein phosphorylation plays an important role in metabolic regulation. Their initial observations paved the way to the discovery of phosphorylase kinase[2] and to the eventual discovery of cyclic adenosine 3',5'-monophosphate (cAMP) by Sutherland.[3] Subsequent studies in the early 1960s elucidated the molecular basis for the action of cAMP involving a cascade mechanism initiated by activation of a protein kinase, originally called phosphorylase kinase kinase, but later named cAMP-dependent protein kinase.[4] This launched a new era in the saga of protein regulation. The number of new kinases that have been discovered in the ensuing decades continues to grow at an exponential rate. A new and exciting branch of the protein kinase family was discovered in 1978 when the transforming protein from Rous sarcoma virus, $pp60^{v-src}$, was found to have intrinsic protein kinase activity.[5] This was followed quickly by the discovery that protein kinase activity was inherent to many growth factor receptors, such as the epidermal growth factor (EGF) receptor[6] and the insulin receptor.[7] Most protein kinases that are classified in these latter two categories transfer phosphate from Mg·adenosine 5'-triphosphate (MgATP) to tyrosine instead of to the more frequently observed phosphate acceptors serine and threonine.[8] Although phosphorylations occurring on tyrosine residues account for a minute percentage of the total protein-bound phosphate, these kinases can be critical for growth regulation and mitogenesis and also can lead to transformation when they are genetically altered or rearranged.[9]

Historically, the classic role for protein phosphorylation initially was established for glycogen metabolism in the liver and subsequently was shown to affect both glycolysis and gluconeogenesis (Figure 1). An examination of these effects, mediated by the activation of a single protein kinase that subsequently acts on a wide range of substrate proteins, provides a prototype cascade for regulation by phosphorylation/dephosphorylation.[10-12] One of the primary mechanisms for elevating cAMP levels in the liver is mediated by the hormonal release of glucagon from the α-cells of the pancreas when blood glucose levels fall. It is the responsibility of the liver to replenish that blood glucose. Glucagon binding to the glucagon receptor leads to the activation of adenylate cyclase by a mechanism involving a guanylate nucleotide binding protein.[13] The major receptor for cAMP in eukaryotic cells is the regulatory (R) subunit of cAMP-dependent protein kinase. In the absence of cAMP, this R-subunit is part of a holoenzyme complex with a catalytic (C) subunit. This tetrameric holoenzyme, R_2C_2, is catalytically inactive. cAMP-binding to the R-subunit promotes dissociation of the complex into an R_2-dimer and two free C-subunits that are catalytically active.[14,15] Like the receptors that couple to adenylate cyclase, the protein substrates for this kinase also vary with each cell type, depending on the particular proteins that are expressed in any given cell. In the case of glycogen metabolism in the liver, two key target enzymes are glycogen synthase and phosphorylase kinase.[16-19] Phosphorylation of glycogen synthase by the C-subunit leads to its inactivation, thus slowing down glycogen synthesis.[16,17] Glycogen synthase is regulated by other kinases as well and demonstrates that phosphorylation at one site can enhance phosphorylation at other sites synergistically.[18] Thus, regulation by phosphorylation can have many subtleties and complexities. On the degradative pathway, phosphorylase *b* kinase also is a substrate for the C-subunit, and, in this case, phosphorylation converts the enzyme into its active *a* form.[19] Phosphorylase *a* kinase then phosphorylates and activates phosphorylase *b*, thereby leading to the breakdown of glycogen.[20,21] Thus, phosphorylations of multiple substrate proteins combine to yield the overall mobilization of glycogen. Two additional proteins are phosphorylated by the C-subunit, ensuring that glucose released from glycogen will not be utilized by the liver. These are pyruvate kinase, the terminal step in glycolysis, and phosphofructokinase-2/fructose-2,6-bisphosphatase. Phosphorylation of pyruvate kinase leads to its inactivation,[22] whereas phosphorylation of phos-

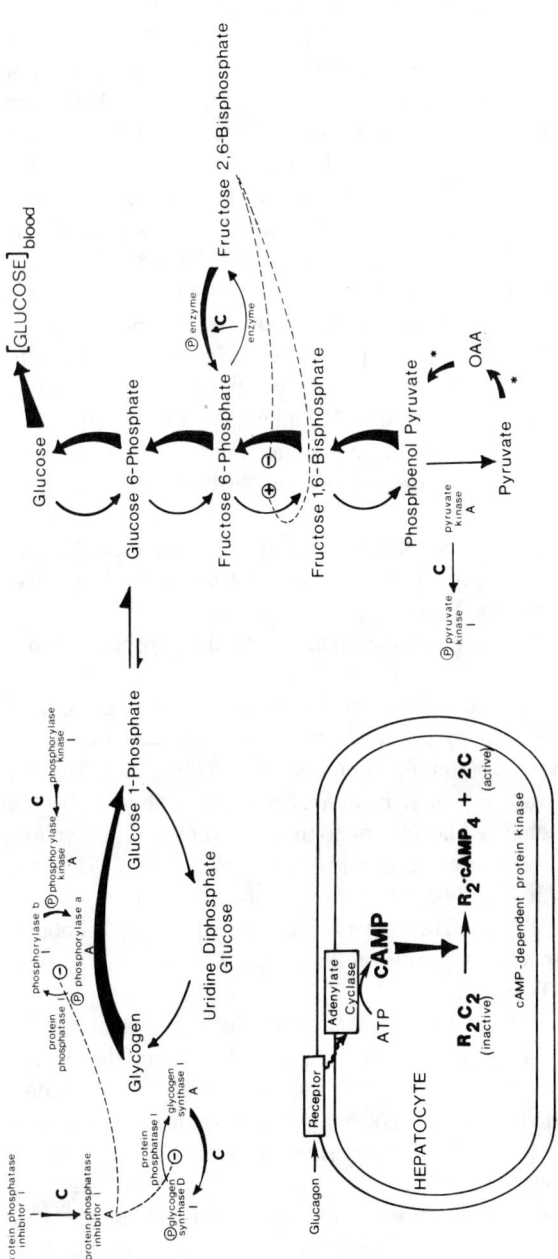

FIGURE 1. Hormonal regulation of glucose metabolism in the liver. The inset at the lower left summarizes the hormonal induction of cAMP and its action on cAMP-dependent protein kinase. The enzymes involved in glucose metabolism are indicated. Those enzymes or proteins known to be regulated by phosphorylation are designated, where I and A indicate the inactive and active forms, respectively. Regulation of glycogen synthase, phosphorylase kinase, phosphatase inhibitor, 6-phosphofructo-2-kinase/fructose-2,6-bisphosphatase, and pyruvate kinase is mediated directly by the C-subunit of cAMP-dependent protein kinase. In addition, PEPCK (*) and pyruvate carboxylase (*) are regulated at the transcriptional level by cAMP.

phofructokinase-2/fructose-2,6-bisphosphatase actually changes the reaction catalyzed by the enzyme.[23,24] In its dephosphorylated state, the enzyme functions as a kinase and converts fructose-6-phosphate into fructose-2,6-diphosphate, a potent activator of glycolysis and inhibitor of gluconeogenesis. The phosphorylated form of the enzyme, however, is a phosphatase that hydrolyzes fructose-2,6-diphosphate to fructose-6-phosphate, thus simultaneously removing the positive allosteric effects favoring glycolysis and the negative allosteric effects that inhibit gluconeogenesis. The supply of glucose that the liver provides to the blood from stored glycogen is further augmented by activation of the gluconeogenic pathway. Most of the gluconeogenic enzymes, such as phosphoenolpyruvate carboxykinase (PEPCK), are positively regulated at the transcriptional level by cAMP.[25] This ability of cAMP to induce transcription of cAMP-responsive genes has, until recently, been the least understood of the cAMP-mediated pathways. However, several recent findings indicate that the molecular basis for this regulation will be resolved soon. For example, cAMP-responsive elements have been identified preceding many cAMP-responsive genes, such as PEPCK,[25] somatostatin,[26] and the human glycoprotein hormone α-subunit.[27] Moreover, the transcriptional factors that bind to these cAMP-responsive elements have been purified.[28] Other transcriptional factors, such as the activator protein 2 (AP-2), appear to be activated by dual mechanisms — one involving phorbol esters and protein kinase C, and the other involving cAMP.[29] Grove et al.,[30] furthermore, demonstrated that the active C-subunit of cAMP-dependent protein kinase was necessary for cAMP stimulation of gene transcription, making it likely that many transcriptional factors may be sensitive to regulation by phosphorylation/dephosphorylation mechanisms.

The phosphate is removed from each protein substrate by specific phosphatases.[11] Maintaining all of the phosphoproteins in their phosphorylated state is enhanced by phosphorylation and activation of phosphatase inhibitor I. In its phosphorylated state, phosphatase inhibitor I forms a complex with phosphatase I, thereby maintaining the phosphatase in a relatively inactive state.[31]

This quick review of some of the effects of cAMP on glucose metabolism in the liver demonstrates several points. First, regulation by phosphorylation is frequently a complex and multifaceted process with many players combining to produce an overall effect. Second, phosphorylation can lead either to activation or inactivation of any given target substrate protein. Third, in some unusual cases, such as phosphofructokinase-2/fructose-2,6-bisphosphatase, phosphorylation actually can alter the reaction a given enzyme catalyzes. Fourth, phosphorylation can mediate regulation indirectly at the transcriptional level. In addition, if we compare the sequences of proteins that actually are phosphorylated by the C-subunit (Figure 1), it is clear that there is a conserved consensus sequence that the kinase recognizes, despite the diversity of the substrates (Table 1).

The examples cited above have provided a framework for establishing physiologically relevant phosphorylation events. For a given phosphoprotein, one must mimic the *in vivo* phosphorylation by *in vitro* modification at the same site with a specific kinase. Furthermore, it is essential to establish the functional consequences of the phosphorylation. These rules, originally established by Krebs and Beavo,[10] have been solid ones. In the case of the transforming protein kinases and the growth factor receptors, however, it has been extremely difficult to delineate similar cascade pathways that follow activation of the kinase.

Despite the wide diversity of the eukaryotic protein kinases, the available sequence information indicates that all protein kinases have evolved in part from a common origin.[36-38] Within this kinase family, the enzyme that is activated in response to cAMP is still the best understood biochemically. It is also one of the simplest and smallest protein kinases because, unlike the other kinases, the major regulatory elements are part of a separate subunit that readily dissociates in the presence of cAMP, leaving a catalytically active, monomeric enzyme. Because of the sequence similarities extending throughout most of this catalytic

TABLE 1
Alignment of Sequences Flanking Phosphorylation Sites that are Recognized by cAMP-Dependent Protein Kinase

Protein	Amino acid sequences	Ref.
Phosphorylase kinase, α-subunit	Phe-ARG-ARG-Leu-SER-Ile-Ser	33
Phosphorylase kinase, β-subunit	LYS-ARG-Ser-Gly-SER-Val-Tyr	33
Glycogen synthase, site 1a	Pro-ARG-ARG-Ala-SER-Cys-Thr	32
Glycogen synthase, site 1b	LYS-ARG-Ser-Asn-SER-Val-Asp	32
Glycogen synthase, site 2	Ser-ARG-Thr-Leu-SER-Val-Ser	32
Pyruvate kinase	Leu-ARG-ARG-Ala-SER-Leu-Ala	34
6-Phosphofructo-2-kinase/fructose-2,6-bis-phosphatase	Arg-ARG-ARG-Gly-SER-Ser-Ile	35
Protein phosphatase inhibitor I	Arg-ARG-ARG-Pro-THR-Pro-Ala	31

Note: These sequences correspond to proteins that are indicated in Figure 1. Arrows indicate location of phosphorylated residue. Phosphorylated residues and amino acids important for kinase recognition are capitalized.

subunit, much of the biochemical information gleaned from the C-subunit can be used as a framework for viewing the entire family of protein kinases.

II. KINASES AS MODULAR STRUCTURES

The protein kinases appear to have been assembled as modular structures, as summarized in Figure 2, presumably by a mechanism involving exon shuffling. The only module that is common to all and shares extensive sequence similarities is the one corresponding to the catalytic unit. The noncatalytic modules are important for regulation, for subcellular localization, and most likely for determining protein-protein interactions involving intracellular protein kinase complexes. In contrast to the conserved nature of the catalytic core, there is no conserved motif for the regulatory, noncatalytic modules. The flexibility and diversity of these modules is striking. They can lie either carboxy- or amino-terminal to the catalytic core, and there is no consistency in size. They can be intracellular, membrane bound, membrane spanning, or partially extracellular. Indeed, a wide variety of regulatory motifs appear to have been "stitched on" to the catalytic core in a somewhat arbitrary fashion. Many of the regulatory elements share a conserved function in that their role is to keep the enzyme in an inactive state when it is not required. These modules frequently contain a ligand binding site which, when occupied by the appropriate ligand, triggers conformational changes leading to activation of catalytic activity. Many of these regulatory regions also contain either a substrate binding site or a pseudosubstrate binding site that may occupy the peptide binding site and, thus, render the enzyme inactive toward external substrates.

The closest homologue to cAMP-dependent protein kinase is cyclic guanosine 3',5'-monophosphate (cGMP)-dependent protein kinase. This kinase shows extensive sequence similarities with both the R- and C-subunits, although cGMP-dependent protein kinase contains both the regulatory and catalytic components as part of a single, contiguous polypeptide chain.[42] The regulatory component is at the amino terminus, and, like the R^I-subunit, the two chains of the dimeric protein are linked by interchain disulfide bonds located close to the amino terminus.[55,56] The two cGMP binding domains are followed by the catalytic domain. In the absence of the activating ligand, cGMP, the enzyme is maintained in a catalytically inactive state. Binding of cGMP leads to conformational changes that result in activation. This ligand-induced conformational change is typical of most protein kinases

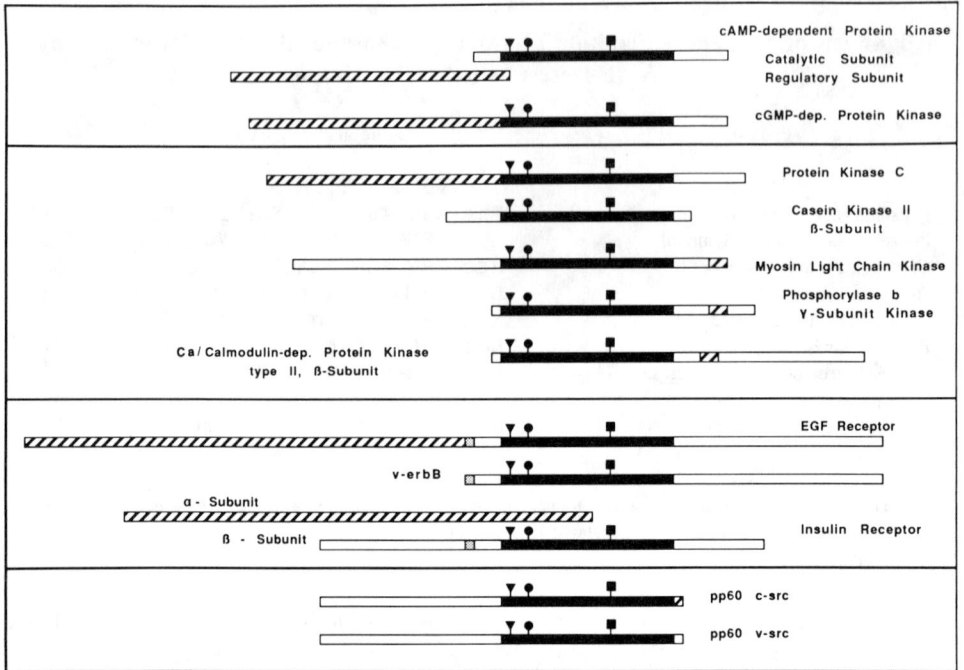

FIGURE 2. Sequence alignment of various protein kinases. Amino acid sequences were either determined directly or were deduced from DNA sequences: catalytic subunit of cAMP-dependent protein kinase;[39] regulatory subunits of cAMP-dependent protein kinase;[40,41] cGMP-dependent protein kinase;[42] protein kinase C;[43,44] myosin light chain kinase;[45] phosphorylase *b* kinase;[46] Ca^{2+}:calmodulin dependent protein kinase, type II, β-subunit;[47] casein kinase II, β-subunit;[48] EGF receptor;[49] v-erb B;[50] insulin receptor;[51,52] pp60^{c-src};[54] and pp60^{v-src}.[53] The conserved catalytic core is indicated in black, known regulatory regions are indicated by cross-hatching, and putative membrane-spanning segments are stippled. Conserved residues correspond to the glycine-rich segment Gly-50 to Gly-55 (▼), Lys-72 (●), and Asp-184 (■) in the C-subunit. (From Taylor, S. S., Buechler, J. A., Slice, L. W., Knighton, D. K., Dungerian, S., Ringheim, G. E., Neitzel, J. J., Yonemoto, W. M., Sowadski, J. M., and Dostmann, W., *Cold Spring Harbor Symp. Quant. Biol.*, 53, 121, 1988. With permission.)

with the exception of cAMP-dependent kinase, where ligand binding not only induces conformational changes, but also leads to the actual dissociation of the R- and C-subunits. This motif of fusing genes to assemble a complex, well-defined domain structure is repeated frequently in the family of kinases. Only the catalytic domain is conserved.

In some cases, the role played by the noncatalytic modules is well defined. For example, myosin light chain kinase (MLCK) is one of the protein kinases regulated by Ca^{2+}:calmodulin. This kinase has a major regulatory module following the catalytic domain.[45,57] This is a region that binds Ca^{2+}:calmodulin and, when occupied, leads to activation of the kinase. Limited proteolysis was used to identify the Ca^{2+}:calmodulin binding region.[58,59] Treatment of MLCK with a variety of proteases releases a small fragment from the carboxy terminus. This segment independently retains the ability to bind Ca^{2+}:calmodulin with a high affinity, while the amino-terminal domain retains the kinase activity. Following proteolysis, however, the kinase activity is independent of Ca^{2+}:calmodulin. Phosphorylase kinase and the Ca^{2+}:calmodulin-dependent protein kinase II also are activated by Ca^{2+}:calmodulin. Each of these kinases also has a Ca^{2+}:calmodulin recognition site that lies on the carboxy-terminal side of the catalytic core.[46,47,231]

In contrast to MLCK, protein kinase C contains its major extracatalytic modules in the region before the catalytic domain.[43,44,60] Protein kinase C is transiently associated with the plasma membrane, and the N-terminal domain preceding the catalytic region is essential for this association. The binding sites for Ca^{2+}, phosphatidyl serine, and diacylglycerol, the physiological activators of protein kinase C, also are contained in the modular units at the amino terminus. This region contains the phorbol ester binding site as well. Limited proteolysis cleaves protein kinase C into a fully active catalytic domain that is no longer dependent on Ca^{2+}, phosphatidyl serine, or diacylglycerol and no longer associates with the plasma membrane.[60]

The major regulatory regions of the EGF receptor are also on the N-terminal side of the catalytic core. The catalytic region of the EGF receptor is preceded almost immediately by a single membrane-spanning module that anchors the receptor in the plasma membrane. This membrane-spanning segment, in turn, connects the catalytic core with a large extracellular domain containing the binding site for the peptide hormone EGF.[49] One of the most intriguing questions now is to explain at the molecular level how the binding of EGF to the extracellular domain communicates with and activates the intracellular kinase domain and how kinase activation affects receptor internalization and processing.[61]

A close homologue of the EGF receptor is the insulin receptor.[62,63] Although this protein is synthesized as a single polypeptide chain, it is cleaved proteolytically into an α- and β-subunit posttranslationally.[64] The mature protein is an $\alpha_2\beta_2$ tetramer where all the subunits are cross-linked by disulfide bonds. The β-subunit contains the single membrane-spanning segment and is followed by the cytoplasmic domain, which contains the protein kinase activity.[51,52] The α-subunit, on the other hand, retains the insulin binding site. Like the EGF receptor, insulin binding to the extracellular domain leads to activation of the intracellular kinase domain.[51,52,62,63] The flexibility of these regulatory modules was demonstrated quite convincingly by Riedel et al.,[65] who constructed a chimeric protein that fused the extracellular insulin binding domain of the insulin receptor to the membrane-spanning and kinase segment at the carboxy terminus of the EGF receptor. This chimeric construct showed insulin-dependent protein kinase activity.

One major role that these extracatalytic modules play is in the regulation of kinase activity. Even though there is no conservation of structure, limited proteolysis has demonstrated repeatedly that these noncatalytic regions are essential for regulation. One can perhaps see this most clearly when the oncogenic transforming proteins are compared to their proto-oncogenic homologues. Regulation of the protein kinase activity may be mediated by several mechanisms, even within a single kinase. For instance, binding of EGF to the EGF receptor leads to activation of kinase activity. The next step, however, is autophosphorylation of three tyrosines. This autophosphorylation may be a necessary step in the activation process and a prerequisite for the EGF receptor to catalyze phosphorylations of other substrates.[66] A similar autocatalytic step has been invoked as a necessary step in the activation of the insulin receptor following insulin binding.[67,68] In spite of the fact that activation may involve several steps, the oncogenic transforming kinases all seem to share one common characteristic — the regulatory mechanism has been impaired. For example, the EGF receptor described above is most likely the proto-oncogenic form of the oncogenic transforming protein v-erb-B. v-erb-B lacks the entire EGF binding module, so activation of the kinase is decoupled from the binding of EGF.[50] There are many factors to consider with this truncated version of the EGF receptor such as internalization, rate of synthesis, and role of degradation; however, a dominant feature of the oncogenic version is that it has lost a major means of regulation and remains in a permanently active form.

The *src* family of kinases is another example where regulation is impaired when the oncogenic kinase is compared to its proto-oncogene counterpart.[53,54,69] In the case of *src*, it is primarily, but not exclusively, a small segment at the carboxy terminus that differs in v-*src* and c-*src*. This region in c-*src* contains a tyrosine that is phosphorylated,[70] while v-*src* lacks this tyrosine. This carboxy-terminal segment clearly is important for regulation, since replacement of Tyr-527 with a residue that cannot be phosphorylated is sufficient to convert c-*src* into a transforming protein.[71-73] There are other mutations in the amino-terminal region that also can mimic this phenotype,[69] but the evidence is striking that this small segment at the carboxy terminus of c-*src* is important in the overall regulation of the activity in the proto-oncogenic form of the enzyme.

Within the family of protein kinases, cAMP-dependent protein kinase is unique. It is the only kinase regulated by interaction with a distinct and dissociable subunit. Consequently, the catalytic subunit is one of the simplest protein kinases. Approximately 70% of the remaining catalytic subunit (residues 40 through 285) shares extensive sequence similarities with the catalytic core of every protein kinase. Since this conservation of sequence can be correlated with both structure and function, the C-subunit can serve as a prototype for the other kinases. For example, residues that can be identified as essential for MgATP binding and catalysis are likely to be invariant in the entire kinase family, whereas residues important for peptide recognition will differ. This review will deal primarily with the catalytic subunit of cAMP-dependent protein kinase and will focus in particular on residues and regions of the C-subunit that are thought to be essential for MgATP binding, catalysis, peptide recognition, R-subunit recognition, and secondary structure. It will not deal in depth with the configuration of the bound MgATP or the kinetic analysis of peptide analogues because these areas have been covered adequately in recent reviews.[56,74]

III. ATP BINDING SITE

A. MAPPING WITH ANALOGUES OF ATP

One of the first approaches for mapping the general features of the MgATP binding site in cAMP-dependent protein kinase utilized analogues of ATP.[75,76] By measuring the capacity of these analogues to compete with [^{32}P]ATP in a phosphotransfer assay, K_i values were determined and, on this basis, general requirements for nucleotide-protein interactions were predicted. Although these K_i values cannot be equated directly with either K_d or K_m values, they have indicated some general features of the MgATP binding site. For example, the initial studies carried out with the C-subunit of cAMP-dependent protein kinase showed that the enzyme was most sensitive to substitutions in the purine ring, particularly around the N_6 position, suggesting that this group might be hydrogen bonded to the protein.[76] The enzyme was least sensitive to perturbations in the ribose moiety, since the K_i values of 2'- and 3'-deoxyATP did not differ significantly from MgATP. On this basis it was concluded that no strong hydrogen bonding exists between the ribose and the protein. Substitutions in the triphosphate region were also sensitive to perturbations, although less so than the adenine ring. For example, Mg·adenosine 5'-diphosphate (MgADP) bound nearly as well as MgATP, whereas AMP bound very weakly. Adenosine, on the other hand, is a better inhibitor than AMP. Analogues of ATP that are inert in terms of phosphotransfer, such as adenosine 5'-(β,γ-methylene)-triphosphate (AMPPCP), do not bind as well as ATP, suggesting that the enzyme may be sensitive to the bridge between the β- and γ-phosphates. In contrast, substitution of the oxygen that bridges the α- and β-phosphates can be accommodated easily. Some notable features of these analogue binding studies are summarized in Table 2.

The same mapping was carried out with the type I holoenzyme form of cAMP-dependent protein kinase, which has a high-affinity binding site for MgATP.[75,76] Although the difference in affinity for MgATP between the free catalytic subunit and the holoenzyme is several orders of magnitude, the mapping of the ATP binding site with analogues of ATP showed very similar patterns with regard to substitutions in the adenine portion of the nucleotide. The major differences observed were associated with the triphosphate region of the nucleotide and, to a lesser extent, with the ribose moiety, leading to the conclusion that the adenine portion of the MgATP complex occupies the same site in both the holoenzyme and the free catalytic subunit.

TABLE 2
Ability of Various Analogues of ATP to Compete with ATP

	K_i(analogue) cAMP-dependent protein kinase		K_i(analogue)/K_m(ATP)			
Analogue	Catalytic subunit[77]	Holoenzyme[75]	Catalytic subunit[77]	cGMP-dependent protein kinase[77]	EGF receptor[79]	Ca:calmodulin-dependent protein kinase[78]	Phosphorylase b kinase[77]
ITP[a]	13,000	370	150	120	310	300	31
GTP[b]	2,100	105	150	110	1.7	160	31
N_6-Methyl ATP	180	2.5	27	37	50	6.7	—
N_6-Dimethyl ATP	2,600	95	—	—	>350	—	—
N_6-Benzyl ATP	>20,000	>1,000	—	120	>350	—	—
8-Thio-ATP	280	14	210	160	—	14	—
8-Br-ATP	150	4.1	—	—	>350	1.8	—
8-N_3-ATP	120	4.8	27	34	75	39	3.8
1,N^6-etheno-ATP	—	—	17	4	—	1.2	1.3
2'-Deoxy-ATP	38	4.5	4	4	0.9	0.15	0.76
3'-Deoxy-ATP	1.5	19.5	0.2	0.7	0.6	0.7	1.9
AMPCPP[c]	9	100	3	4	7.1	8.5	7.6
AMPPCP[d]	280	950	35	93	14	5.8	12.1
AMPPNP[e]	54	3	24	32	2.2	7.6	—
Adenosine	—	—	170	240	26	22	—
AMP	340	700	90	390	180	0.73	—
ADP	40	7	11	10	1.5		
K_m(MgATP)	—	0.1 μM	3.1 μM	7.1 μM	10 μM	33 μM	43.5 μM

Note: K_i values were determined using steady-state kinetic measurements with Mgγ[^{32}P]ATP and the following substrates: histone for cAMP-dependent protein kinase[75-78] and cGMP-dependent protein kinase,[77] phosphorylase b for phosphorylase b kinase,[77] angiotensin I for the EGF receptor,[79] and synapsin I for the Ca^{2+}: calmodulin-dependent protein kinase.[78]

[a] Inosine 5'-triphosphate.
[b] Guanosine 5'-triphosphate.
[c] Adenosine 5'-[α,β-methylene]-triphosphate.
[d] Adenosine 5'-[β,γ-methylene]-triphosphate.
[e] Adenosine 5'-[β,γ-imido]-triphosphate.

TABLE 3
Kinetic Parameters for Fluorescent Analogues of ATP for the Catalytic Subunit of cAMP-Dependent Protein Kinase[81]

Substrate	K_m (μM)	V_{max} (μmol/min/mg)	k_{cat} (min^{-1})
ATP	11.9 ± 2.7	24.9 ± 1.8	996 ± 72
lin-Benzo ATP	11.3 ± 3.1	5.0 ± 1.2	200 ± 48
ε-ATP	1800	1.1	44

Note: The experiments were carried out using C-subunit from bovine brain.

This kinetic mapping of the MgATP binding site subsequently was repeated with a number of protein kinases having very different overall structures.[77-79] The specific structural diversity of these enzymes was discussed in the previous section. Nevertheless, despite this diversity, the kinetic mapping was most striking because of the general similarities that were seen. As might be anticipated, cGMP-dependent protein kinase, the closest relative of the cAMP-dependent protein kinase, was very similar in its adenine nucleotide binding properties.[77] However, even more distantly related kinases, such as the type II calcium/calmodulin-dependent protein kinase,[78] phosphorylase *b* kinase,[77] and the EGF receptor,[79] showed similar profiles.

B. FLUORESCENT ANALOGUES

The MgATP binding site also was mapped with fluorescent analogues of ATP.[80,81] In this case, K_d values were measured directly by fluorescence displacement. In some instances, K_m values also were determined. Based on the general properties described above, it would be predicted that 1,N^6-*etheno*-ATP (ε-ATP) would not be a particularly good substrate for the C-subunit since it has a major substituent at the N_6 position of the adenine ring. As indicated in Table 3, ε-ATP is a very poor substrate for cAMP-dependent protein kinase and, thus, is not a good fluorescent probe for this enzyme.[80] In fact, in contrast to the earlier mapping that showed only a 4.2-fold difference between the K_d(ATP) and the K_i(ε-ATP), the direct measurement of the K_d(ε-ATP) by the fluorescence displacement method showed essentially no binding of the fluorescent analogue at concentrations up to 20 mM. This discrepancy could be explained by trace amounts of residual ATP in the earlier samples of ε-ATP. Because ε-ATP is such a poor substrate for the catalytic subunit, another fluorescent derivative of ATP, *lin*-benzo ATP (shown in Figure 3), was used for the fluorescence displacement studies.[80,81] In contrast to ε-ATP, this is a relatively good substrate for the C-subunit (Table 3). Hence, displacement of *lin*-benzo ATP was used to define the properties of the triphosphate region of the MgATP binding site more precisely. These K_d values were then compared with the K_is determined previously. With some notable exceptions, the K_d and K_i values corresponded reasonably well, as seen in Tables 2 and 4. In particular, the fluorescence titration confirmed the sensitivity of the MgATP binding site to perturbations between the β- and γ-phosphates. In addition, the similar binding properties of MgATP and MgADP were verified, and these K_d values contrasted strikingly to the very unfavorable binding of AMP. The fact that free adenosine bound well indicated that the poor binding of AMP may be due to the addition of an extra negative charge at this position. However, other analogues of AMP containing additional substituents at this 5′ position, as summarized in Table 4, indicated that it was not simply the presence of a negative charge that accounted for the poor binding of AMP. Hartl et al.[81] concluded that the portion of the nucleotide binding site recognizing the α-phosphate is hydrophobic in nature, since large hydrophobic substituents could be tolerated at the 5′-ribose position.

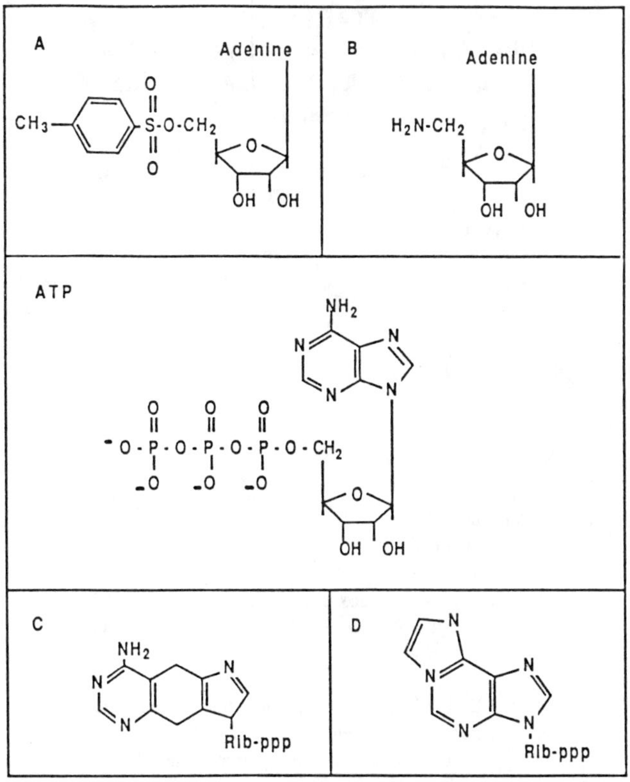

FIGURE 3. Nucleotide analogues that have been used to map the ATP binding site. (A) 5'-O-(p-toluene sulfonyl) adenosine, (B) 5'-amino-5'-deoxyadenosine, (C) 1,N^6-etheno-ATP, and (D) lin-benzo-ATP. ATP is shown in the center. Rib-ppp = ribose triphosphate.

lin-Benzo derivatives of ATP also were used to map the ATP binding site of cGMP-dependent protein kinase,[82] where similarities with the catalytic subunit again can be seen. In addition, Carlson and colleagues[83,84] recently reported a detailed kinetic analysis of the β-subunit of phosphorylase b kinase using the lin-benzo derivatives of ATP. In this case, the derivatives distinguished properties of the allosteric ADP binding site and the MgATP binding site in the active site. Although their results cannot be compared directly to those described here for the catalytic subunit and the cGMP-dependent protein kinase, they nevertheless emphasize the applicability of this fluorescent approach for probing the reactions catalyzed by these various ATP:protein phosphotransferases.[83,84]

In general, the results of the fluorescence titration were consistent with the earlier mapping and indicated that (1) the negative charges on the β- and γ-phosphates are important and most likely indicate an ionic interaction with the protein; (2) the extra negative charge on the α-phosphate of AMP is very unfavorable, suggesting a possible hydrophobic pocket in this region; (3) in contrast to AMP, adenosine binds reasonably well, emphasizing the dominant role the adenine moiety plays in binding the nucleotide; and (4) substitution of the oxygen between the β- and γ-phosphates leads to a significant decrease in affinity. When these displacement studies are compared with K_d values measured by equilibrium binding, the correlation is quite good.[85,86]

C. AFFINITY LABELING

Even for the catalytic subunit of cAMP-dependent protein kinase, our knowledge of specific amino acid residues that correlate with function is limited. The first clue was provided

TABLE 4
Dissociation Constants (K_d) for Various Analogues of ATP for the Catalytic Subunit

Analogue	K_d (μM)
8-Br-ATP	117
ITP	14,200
GTP	3,700
1,N^6-etheno-ATP	>20,000
2'-Deoxy-ATP	18
3'-Deoxy-ATP	2
AMPCPP	34
AMPPCP	46
AMPPNP	54
ADP	10
AMP	643
Adenosine	32
5'-O-(p-Toluenesulfonyl) adenosine	30
5'-Amino-5'-deoxyadenosine	506
5'-(Ethylamino)-5'-deoxyadenosine	403
5'-(Diethylamino)-5'-deoxyadenosine	284
5'-(Diallylamino)-5'-deoxyadenosine	102

Note: Determined by fluorescence polarization titration using *lin*-benzo-ATP.[80] The K_d(ATP) was 10 μM, as determined by steady-state kinetic measurements.

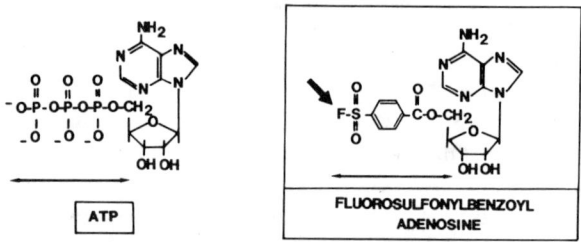

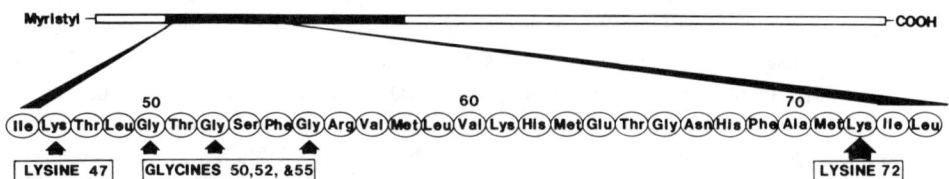

FIGURE 4. Localization of the ATP binding site. The structure of the affinity analogue FSBA in comparison to ATP is indicated above. The site of covalent modification, Lys-72, is localized in the polypeptide chain. The sequence flanking Lys-72, including the conserved glycines 50, 52, and 55, is shown below.

by affinity labeling with an analogue of ATP, p-fluorosulfonylbenzoyl-5'-adenosine (FSBA), when Zoller and Taylor[87] showed that FSBA was a specific inhibitor of the C-subunit and that MgATP protected against this inhibition. They later established that inhibition could be correlated with the covalent modification of a single residue, Lys-72.[88] As indicated in Figure 4, FSBA contains a reactive fluorosulfonyl group in a position that closely approximates that of the γ-phosphate of ATP. Bhatnagar et al.[89] used the *lin*-benzo analogues of ATP, ADP, AMP, and adenosine to measure nucleotide binding to the FSBA-modified protein. Modification by FSBA abolished binding of *lin*-benzo ATP and ADP, but had little effect on the binding of *lin*-benzo adenosine and AMP. Based on these results, it is likely that the ε-amino group of Lys-72 lies near the β- and/or γ-phosphates of ATP, even though in solution the fluorosulfonyl benzoyl group of FSBA appears to stack over the adenine ring.[90] The proximity of Lys-72 to the β-/γ-phosphate of ATP is also consistent with the analogue mapping studies.

Subsequent comparison of other protein kinase sequences has confirmed the initial prediction that this lysine plays an essential role in kinase function, since it is invariant in every protein kinase. The homologous lysine in pp60$^{v\text{-}src}$, Lys-296, also is modified by FSBA, and the resulting enzyme is devoid of kinase activity.[91] This confirmed that the conservation of sequence corresponds to a functional homology in these two protein kinases. Similar observations were made for the EGF receptor, where reaction with FSBA again led to inactivation of kinase activity. The site of modification subsequently was identified as Lys-721.[92]

The identification of the critical nature of this lysine residue has provided a framework for site-directed mutagenesis. For example, changing Lys-296 in pp60$^{v\text{-}src}$ to arginine or histine[93] or to methionine[94] by site-directed mutagenesis led to the synthesis of an inactive enzyme that had lost its ability to transform. Similar results subsequently have been observed with p130$^{gag\text{-}fps}$ and with p100$^{gag\text{-}mil}$, the translation products of the v-fps and v-mil oncogenes, respectively.[95,96]

Replacement of the corresponding lysine residue in the EGF-receptor, Lys-721, with methionine also led to the loss of kinase activity and, in addition, abolished the EGF-induced effects on cell proliferation, gene induction, and Ca^{2+} transport.[97] Some of the other processes the EGF receptor undergoes, such as clustering, endocytosis, cytoplasmic sorting, and recycling, also are altered as a consequence of the Lys-721 → Met replacement. EGF-induced dimerization is still observed, but ligand-induced endocytosis is blocked.[98] Honegger et al.[99] found that internalization per se was not impaired, but EGF-induced down regulation was abolished when Lys-721 was replaced by alanine. Replacement of the corresponding lysine in the insulin receptor likewise led to a loss of insulin-mediated metabolic responses.[100,101] Thus, modification of this residue has provided a very convenient and specific mechanism for establishing the fact that intrinsic protein kinase activity is essential for mediating the subsequent responses initiated by a wide variety of protein kinases.

D. NUCLEOTIDE FOLD

The identification of Lys-72 as being critical for ATP binding localized the ATP binding site in the amino-terminal region of the polypeptide chain. Other features of the primary structure are also consistent with this assignment. Precedence suggests such a nucleotide binding region will conform to the general rules that were first recognized by Rossmann and his co-workers[102] when they compared the crystal structure of glyceraldehyde-3-phosphate dehydrogenase with the crystal structure of lactate dehydrogenase. Despite the lack of any extensive sequence similarities, there clearly was a conservation of secondary and tertiary structure. One of the only features of amino acid sequence that is conserved in the nucleotide-fold motif predicted by Rossmann is a triad of glycines, Gly–X–Gly–X–X–Gly. Such a glycine-rich region in the C-subunit is located at residues 50 to 55. The identification of this conserved feature of the nucleotide-fold motif and its localization in the polypeptide chain is consistent with the prediction that the C-subunit contains a conventional nucleotide fold, and this fold constitutes the amino-terminal portion of the molecule. A detailed analysis of this nucleotide binding region will be made later in this review when the domain structure of the enzyme is discussed in greater depth (Section VIII).

TABLE 5
Synthetic Peptides that Serve as Substrates for the Catalytic Subunit of cAMP-Dependent Protein Kinase

	Peptide sequence	K_m app (μM)	V_{max} (μmol/min/mg)	Ref.
Kemptide	LEU-ARG-ARG-ALA-SER-LEU-GLY	16.0 ± 0.9	20.0 ± 0.6	104
Variations in flanking sequences	Leu-ALA-Arg-Ala-SER-Leu-Gly	4,900 ± 700	8.7 ± 0.6	104
	Leu-Arg-ALA-Ala-SER-Leu-Gly	6,300 ± 400	5.3 ± 0.2	104
	Leu-LYS-Arg-Ala-SER-Leu-Gly	1,400 ± 100	17.1 ± 0.4	104
	Leu-Arg-LYS-Ala-SER-Leu-Gly	260 ± 10	16.9 ± 0.3	104
	Leu-HIS-Arg-Ala-SER-Leu-Gly	415 ± 22	12.1 ± 0.3	104
	Leu-Arg-HIS-Ala-SER-Leu-Gly	1,340 ± 50	6.5 ± 0.1	104
	Leu-Arg-Arg-PRO-SER-Leu-Gly		20	107
Variations in phosphate acceptor	Leu-Arg-Arg-Ala-THR-Leu-Gly	590 ± 80	5.6 ± 0.3	104
	Leu-Arg-Arg-Ala-HYP-Leu-Gly	18,000	1	107
Inhibitor peptides	Leu-Arg-Arg-Ala-ALA-Leu-Gly	490 ± 10[a]		105
	Leu-Arg-Arg-Ala-HIS-Leu-Gly	2,150 ± 350[a]		105

Note: The phosphorylated residues as well as the residues that were replaced are capitalized.

[a] Value is K_i, not K_m.

IV. PEPTIDE BINDING SITE

The peptide binding site also was probed using different approaches. Synthetic peptide analogues were used to delineate specific requirements for peptide recognition. In contrast, affinity labeling with peptide analogues as well as group-specific labeling have localized portions of the peptide recognition site in the polypeptide chain. Finally, nuclear magnetic resonance (NMR) data have provided some structural information on the conformation of synthetic peptide substrates when they are bound to the active site of the C-subunit.

A. PEPTIDE ANALOGUES

As was seen in Table 1, many proteins are substrates for the C-subunit, and in most cases these phosphorylations have been shown to be physiological events correlating with functional changes in the protein. All of these protein substrates, however, contain a consensus sequence that is required for recognition by the C-subunit. Based on the identified phosphorylation sites in substrate proteins, synthetic peptides were used to map these requirements more precisely.[103-105] Since these results have been reviewed extensively,[74] they will only be summarized here. The peptides shown in Table 5 demonstrate some of the most essential requirements for peptide recognition. The parent peptide, referred to as Kemptide, is based on the sequence in pyruvate kinase that is phosphorylated by the C-subunit.[106] For example, the two arginines preceding the phosphorylation site are essential. Replacing either arginine, even with a residue that retains a positive charge, leads to a dramatic increase in K_m.[104] The spacing between the arginines and the serine is also critical.[103] The intervening residue between the arginines and the phosphorylation site as well as the residue immediately following the phosphorylation site are variable, and these positions usually are occupied by a small hydrophobic residue. Replacement of the serine with a threonine can be accommodated,[104] and even hydroxyproline can serve as an acceptor for the phosphate.[107] In both cases, however, the $K_{m,app}$ for the peptide is increased significantly, particularly with hydroxyproline. Replacement of the serine with a nonphosphorylatable residue such as alanine yields an inhibitory peptide with a K_i approximately two orders of magnitude greater than the K_m for the corresponding peptide containing serine.[105] Substitution of either a negatively

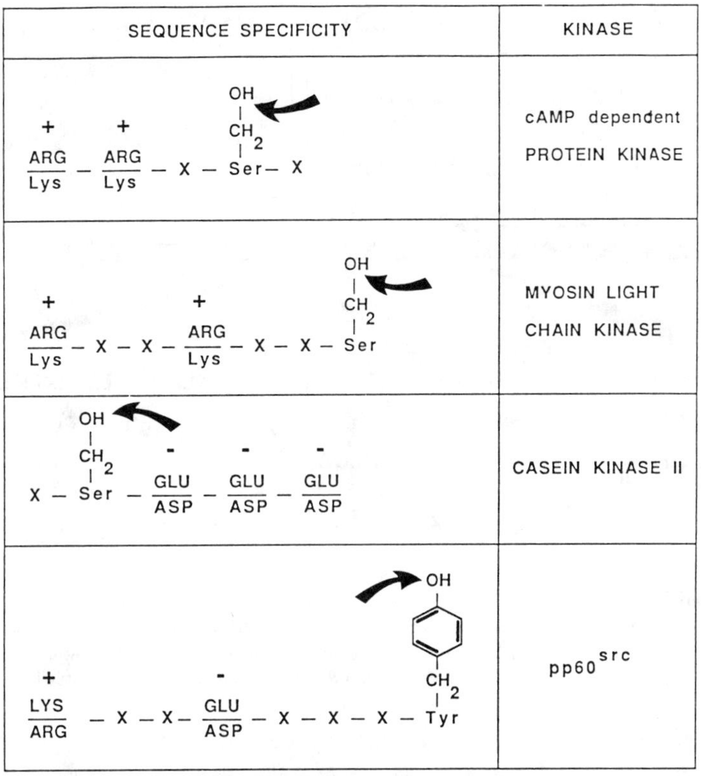

SEQUENCE SPECIFICITY	KINASE
$\overset{+}{\underset{Lys}{ARG}} - \overset{+}{\underset{Lys}{ARG}} - X - Ser - X$ (OH on Ser)	cAMP dependent PROTEIN KINASE
$\overset{+}{\underset{Lys}{ARG}} - X - X - \overset{+}{\underset{Lys}{ARG}} - X - X - Ser$ (OH on Ser)	MYOSIN LIGHT CHAIN KINASE
$X - Ser - \underset{ASP}{GLU} - \underset{ASP}{GLU} - \underset{ASP}{GLU}$ (OH on Ser)	CASEIN KINASE II
$\overset{+}{\underset{ARG}{LYS}} - X - X - \underset{ASP}{GLU} - X - X - X - Tyr$ (OH on Tyr)	pp60^{src}

FIGURE 5. Consensus sequences recognized by various protein kinases. Arrows indicate the site of phosphotransfer.

charged aspartate or a positive histidine at this position leads to a further increase in K_i. On this basis a general consensus sequence, Arg–Arg–X–Ser/Thr–X, was deduced. This sequence is, in fact, found in most of the physiological phosphorylations that have been identified and is a reliable predictor for a potential site that can be phosphorylated by the C-subunit.

Peptide mapping with synthetic substrates has been used for a number of different protein kinases, and they fall into several categories (Figure 5). Like the C-subunit, MLCK[108,109] and protein kinase C[110-112] require at least one basic residue preceding the phosphorylation site. In addition, phosphorylation by protein kinase C can be enhanced by a basic residue carboxy-terminal to the phosphorylation site. Casein kinase II, on the other hand, represents a family of kinases requiring acidic residues carboxy-terminal to the phosphorylation site.[113] The general features for peptide recognition by the transforming protein from Rous sarcoma virus, pp60^{v-src}, also are summarized in Figure 5.[114,115] These features seem to be consistent for most of the tyrosine-specific kinases and are quite distinct from those required by the C-subunit. In contrast to the synthetic peptides that are substrates for the C-subunit, which have K_m values in the micromole range, the synthetic peptides that mimic physiological tyrosine phosphorylation sites, in general, are relatively poor substrates for the tyrosine-specific protein kinases. The K_m values are typically in the millimole range, suggesting that more than the immediate amino acid residues flanking the phosphorylation site are required for recognition by this family of kinases.

B. AFFINITY LABELING

In order to localize the peptide binding site in the polypeptide chain, the affinity analogue

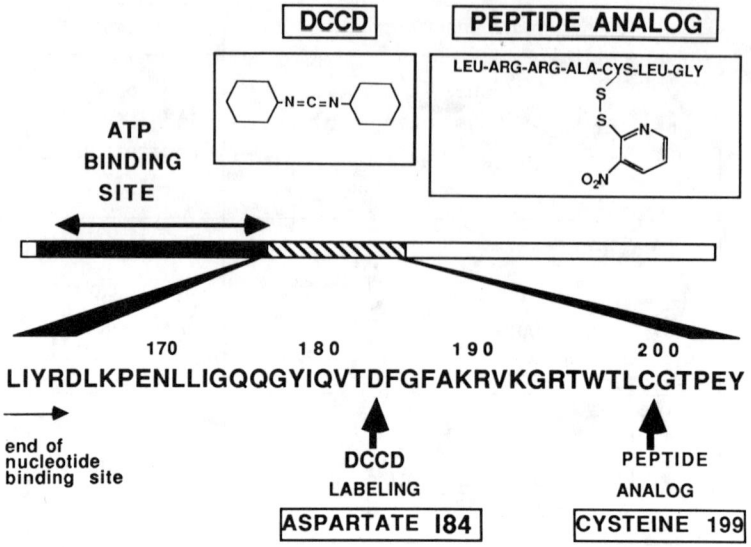

FIGURE 6. Localization of the catalytic and the peptide binding sites. The structures of dicyclohexylcarbodiimide (DCCD) and a peptide affinity analogue that served to localize these sites are indicated above. The amino acid sequence containing the amino acid residues that are labeled by DCCD and the peptide analogue are indicated below.

shown in Figure 6 was synthesized. In this peptide, the serine that typically would get phosphorylated at residue 5 was replaced by a cysteine, which was disulfide bonded to thionitropyridine. This peptide is an irreversible inhibitor of the C-subunit, although inactivation can be reversed in the presence of thiols. Inhibition by this analogue correlates with the covalent modification of Cys-199.[116] Thus, at least a portion of the peptide binding site has been localized to a region of the polypeptide chain lying carboxy-terminal to the proposed nucleotide binding region.

More recently, Miller and Kaiser[117] synthesized another peptide analogue that was used for affinity labeling. This peptide contains a photoactivatable residue, p-benzoyl-phenylalanine, in the position that is occupied by serine in the substrate peptide. Photolysis of the C-subunit in the presence of this peptide analogue correlates with the modification of Gly-126 and Met-128, which indicates another portion of the enzyme that may be close to the active site.

C. SULFHYDRYL GROUPS

Cys-199 is proposed to lie in close proximity to the active site of the enzyme, based on several criteria in addition to affinity labeling.[118] The C-subunit is sensitive to inhibition by a variety of sulfhydryl-specific reagents[119-129] (for review see Reference 74). There are two cysteines in the C-subunit, Cys-199 and Cys-343,[130] and both residues are modified by many sulfhydryl-specific reagents. Cys-199 is more reactive, and the loss of catalytic activity is associated exclusively with the covalent modification of Cys-199.[127,128] MgATP was able to protect the C-subunit against inhibition by these sulfhydryl-specific reagents.[118,121,123,124,126] Likewise, no inhibition was observed when the C-subunit was aggregated with the R-subunit in the holoenzyme complex.[122,124,126,127] This protection could be due, of course, to overall conformational changes in the enzyme and is not sufficient to localize Cys-199 at the active site. However, Jiménez et al.[125] demonstrated that MgADP and AMP did not protect against inhibition by 5,5'-dithio-bis(-2-nitrobenzoic acid) (DTNB) and, thus, concluded that the most reactive cysteine was in close proximity to the region of the protein that recognized the γ-phosphate of MgATP. This was confirmed by Bhatnagar et al.,[89] who showed that covalently modifying the C-subunit with DTNB prevented binding of the fluorescent ATP analogues *lin*-benzo ATP and *lin*-benzo ADP. On the other hand, *lin*-benzo adenosine bound with an affinity that was comparable to the unmodified C-subunit, and *lin*-benzo AMP bound to the modified enzyme with an efficiency of 25% compared to the unmodified C-subunit.[89] On this basis, Cys-199 could be assumed to be in close proximity to the region of the protein that recognizes the γ- and/or β-phosphate of ATP. Reaction of Cys-199 with a fluorescent analogue, *o*-phthalaldehyde, led to cross-linking with the ε-NH$_2$ group of a lysine that was predicted to be within 3 Å of Cys-199. The authors suggested this could correspond to Lys-72, but the specific residue involved in the cross-link was not identified.[129]

Despite the evidence placing Cys-199 near the γ-phosphate of ATP and close to the peptide binding site, there are several arguments making it unlikely that Cys-199 plays an essential role in catalysis. The initial studies of Peters et al.[118] demonstrated that the C-subunit retained 63% of its activity when the cysteines were percyanylated, even though modification with other alkylating agents led to complete loss of activity. In addition, since Cys-199 is not conserved in the larger family of protein kinases, it is not predicted to play an essential role in catalysis.

It is curious that Cys 199 immediately follows one of the two phosphorylation sites that are known to exist in the C-subunit, Thr-197 and Ser-338.[131,132] Since these phosphate groups cannot be removed readily by phosphatases, they are considered to be "silent" phosphorylations.[131,132] The fact that Thr-197 is preceded by a basic residue, Arg-195, suggests that its initial phosphorylation could be an autocatalytic event. The phosphate group is lost slowly from Ser 338 and can be rephosphorylated autocatalytically.[230] A comparison between the sequence of the C-subunit and pp60^{src} in this region reveals that (1) Cys-199 is not conserved; (2) both enzymes contain a phosphorylation site in this region, Thr-197 in the C-subunit and Tyr-416 in pp60^{src}; and (3) the sequence flanking this region has the characteristic requirements that are necessary for recognition by each kinase. In pp60^{v-src} this represents an autophosphorylation site.[133] The relationship of this segment to the active site of the enzyme can only be speculated at this time.

```
                                              (P)      199
Catalytic subunit:   Arg–Val–Lys–Gly–Arg–Thr–Trp–Thr–Leu–Cys–

                     410
pp60 src:            Arg–Leu–Ile – Glu–Asp–Asn–Glu–Tyr–Thr–Ala–
                                                   (P)
```

D. CARBOXYL GROUPS

Group-specific modification of the catalytic subunit with a water-soluble carbodimide, 1-ethyl-3(3-dimethylaminopropyl)carbodiimide (EDC), also identified three carboxylic acid residues that may interact with the peptide substrate. These amino acids, Glu-170, Glu-332, and Asp-328, were labeled with EDC and [^{14}C]glycine ethyl ester in the absence of substrates and with MgATP bound to the enzyme, but were protected from modification in the presence of MgATP and an inhibitor peptide.[232]

E. NMR ANALYSIS OF PEPTIDE SUBSTRATES

The conformations of synthetic peptides when bound to the active site of the C-subunit were characterized by NMR using β,γ-bidentate complexes of Co^{3+}(NH$_3$)$_4$AMPPCP. The inhibitory metal binding site was occupied by Mn^{2+}. The distances were measured from each backbone amide of a hexapeptide and a pentapeptide to Mn^{2+} when the peptides were bound to the C-subunit. These distances established that the bound peptide does not require either

TABLE 6
Apparent Constants for the Interaction of Regulatory and Catalytic Subunits of cAMP-Dependent Protein Kinase I and II[146]

Type of assay	App K (nM)	
	I	II
$R_2 + 2C \rightleftarrows R_2C_2$	0.36	0.22
$R_2C_2 \rightleftarrows R_2 + 2C$	0.23	0.27

Note: Binding constants were measured at 37°C in the presence of 1 mM Mg and 150 mM NaCl. The type I holoenzyme gave similar values in the presence of 10 mM NaCl, whereas the type II holoenzyme showed a 20-fold reduction in both app K in the presence of low salt.

a β-strand or an α-helical conformation.[134] Rosevear et al.[135] also ruled out a requirement for a β-turn based on the sequences of the peptides. By a process of elimination it was predicted that the peptide assumes an extended coil configuration when it is bound to the C-subunit. An additional series of depsi- and N-methylated peptide analogues were synthesized in order to determine whether any of the backbone amide residues were hydrogen bonded to the protein. With the possible exception of the serine amide, no hydrogen bonding to the main chain amides was observed.[136]

F. INHIBITORS

When the C-subunit is not required, it is maintained in an inactive state. There are two known mechanisms for inhibiting the C-subunit of cAMP-dependent protein kinase. The first mechanism is the reversible aggregation with an R-subunit in the absence of cAMP. The second mechanism involves aggregation with a heat-stable protein kinase inhibitor.[137] Unlike inhibition by the R-subunit, the inhibition with the heat-stable inhibitor protein is not thought to be readily reversible.

1. Regulatory Subunit

The R-subunit includes a family of unique gene products that are expressed differentially in each tissue.[138-143] They are antigenically distinct and can be resolved by ion exchange chromatography. In general, they can be classified into two basic categories based on their elution from anion exchange resins.[140] One general feature distinguishing the type I from the type II R-subunits is that the type II R-subunits have an autophosphorylation site, whereas the type I R-subunits do not.[138,139,144] The different forms of holoenzyme are also classified as type I and II based on the R-subunit.[145] In contrast to the type II holoenzymes, which can be autophosphorylated on the R-subunit, the type I holoenzymes have a high-affinity binding site for MgATP.[138] Both R-subunits exhibit high-affinity binding toward the C-subunit in the absence of cAMP[146] (Table 6).

Despite the differences seen between the R-subunits, each has a common and well-defined domain structure, as summarized in Figure 7. A highly conserved feature of this domain structure is the existence two tandem cAMP-binding domains that lie at the carboxy terminus, accounting for approximately two thirds of the protein. These gene-duplicated segments share extensive sequence similarities in all of the R-subunits, as well as with the

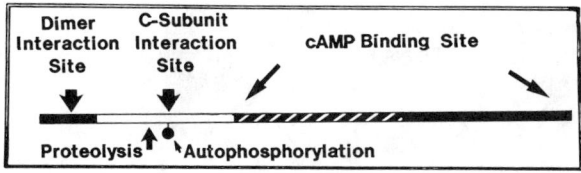

FIGURE 7. General domain structure of the regulatory subunits of cAMP-dependent protein kinase. The two tandem cAMP-binding sites at the carboxy terminus account for two thirds of the protein. Site A (▨) is followed by site B (▬). The point of contact between the protomers in the dimer is shown in black at the amino terminus. The "hinge" region includes the major site of proteolysis in the free R-subunit and the autophosphorylation site (●) and, in addition, is the major site of interaction with the C-subunit.

cAMP-binding domain of the catabolite gene activator protein, the major cAMP-binding protein in *Escherichia coli*.[147,148] The amino-terminal portion of the R-subunit, on the other hand, is a highly variable region. All of the antigenic determinants for the R-subunits and the holoenzymes appear to be localized at this amino-terminal segment.[149] Within this variable region is a so-called "hinge" region that is highly susceptible to limited proteolysis in the dissociated R-subunit.[150-154] This hinge region contains the autophosphorylation site in the case of the R^{II}-subunits[154] (Figure 8 and Table 7). The R^I-subunits, on the other hand, contain a "pseudophosphorylation" site in this region; the two essential arginines are retained, but the serine serving as a phosphate acceptor is missing. Typically it is replaced with alanine. The catalytic subunit can catalyze the autophosphorylation of the R^{II}-subunit in the absence of dissociation;[155] in addition, phosphorylation of the R^{II}-subunit affects dissociation and reassociation of holoenzyme.[156] More recently, recombinant techniques were used to introduce an autophosphorylation site into the hinge region of the R^I-subunit by replacing Ala-97 with serine. Holoenzyme formed with this mutant R-subunit also can catalyze the autophosphorylation of Ser-97 as an intramolecular event that does not require dissociation of the subunits.[158] Thus, it is probable that both the autophosphorylation site in the R^{II}-subunits and the pseudophosphorylation site in the R^I-subunits occupy the substrate recognition site of the C-subunit in the holoenzyme complex. Consistent with this conclusion is the finding that the R-subunit is a competitive inhibitor with regard to peptide substrates and the observation that butane-dione, an arginine-specific reagent, prevents the R-subunit from binding to the C-subunit.[158,159] The R^{II}- and C-subunits also can be cross-linked by a specific disulfide bond between Cys-97 in the R^{II}-subunit and Cys-199 in the C-subunit, further supporting the hypothesis that the autophosphorylation site of the R^{II}-subunit lies near the peptide binding site of the C-subunit when the two proteins are part of the holoenzyme complex.[160] Limited proteolysis provides a final piece of evidence indicating that the hinge region of the type II R-subunit, and in particular the pair of arginines preceding the autophosphorylation site, is essential for recognition by the C-subunit. Limited proteolysis with a variety of proteases yields a carboxy-terminal fragment that retains two functional cAMP-binding sites. If cleavage occurs after residue 90, leaving the two arginines with the carboxy-terminal fragment, a monomeric R-subunit is generated that retains the capacity to interact with the C-subunit (Figure 8). If, on the other hand, cleavage occurs after Arg-93, the carboxy-terminal fragment no longer is capable of forming a stable complex with the C-subunit.[161] This hinge region, therefore, is essential for recognition by the C-subunit, and this region most likely occupies the peptide binding site, thus preventing other substrates from binding.

cAMP-dependent protein kinase also is found in *Saccharomyces cerevisiae*, and genetic evidence has identified the threonine phosphorylation site in the C-subunit[162] and the autophosphorylation site in the R-subunit as important regions for subunit interaction.[163] For

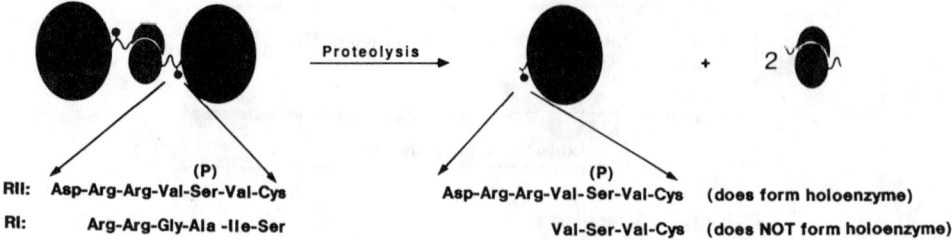

FIGURE 8. Schematic representation of the overall domain structure of the R-subunits of cAMP-dependent protein kinase. The subunits composing the asymmetric dimer interact at the amino terminus. This dimer interaction site is linked by a "hinge" region to the cAMP-binding domains. The hinge region of the R^{II}-subunit contains an autophosphorylation site, while that of the R^{I}-subunit contains a pseudophosphorylation site. The dissociated R-subunits are very susceptible to limited proteolysis at the hinge region. Retention of the two arginine residues in the hinge region of the R^{II}-subunit is essential for the remaining monomeric carboxy-terminal fragment to retain the ability to interact with the C-subunit.

example, a mutant form of the yeast C-subunit was isolated that was no longer sensitive to regulation by the R-subunit. This mutant C-subunit contained a single point mutation replacing Thr-241 with alanine. Thr-241 in the yeast C-subunit corresponds to Thr-197 in the mammalian C-subunit. This single change led to a 100-fold reduction in the affinity of the R-subunit for the C-subunit. In contrast, the kinetic properties of the mutant C-subunit were unchanged.[162] Synthetic oligonucleotides then were used to introduce selective mutations into the hinge region of the R-subunit. These point mutations also had a significant effect on holoenzyme formation. The yeast R-subunit, which is the product of the BCY1 gene, contains an autophosphorylation site, Ser-145, in the hinge region. Replacement of Ser-145 with alanine or glycine converted the R-subunit into a more potent inhibitor of the C-subunit, whereas replacement with glutamic acid, lysine, threonine, or aspartic acid reduced its inhibitory properties.[163]

2. Heat-Stable Protein Kinase Inhibitor

The heat-stable protein kinase inhibitor (PKI) is a 10-kDa protein that also forms a stable complex with the C-subunit.[164,165] The high-affinity binding of PKI to the C-subunit (K_d = 0.2 nM) is comparable to the binding affinities observed for the R-subunits. All of the inhibitory properties of PKI are localized at the amino terminus.[166-170] The sequences of several inhibitory peptides from this region of the protein are shown in Table 5. These synthetic peptides are nearly as potent in their inhibitory properties as the intact PKI. Although the inhibitory region of PKI contains regions distinct from the R-subunit and most likely interacts with some different regions of the C-subunit, it shares a common sequence that includes two tandem arginines. This corresponds to a pseudophosphorylation site. This segment presumably occupies the peptide binding site of the C-subunit, since PKI is a competitive inhibitor with respect to peptide substrates. Replacement of either arginine severely interferes with the ability of the inhibitor peptide to form a high-affinity complex with the C-subunit.[167,169,170] Circular dichroism was used to predict a tentative structure for this peptide in solution. The predicted structure has an amphipathic helix at the amino terminus, and the pseudophosphorylation site is predicted to lie within a β-turn and random coil configuration.[171]

The mechanisms of inhibition by PKI and the R-subunits share common properties, although the two inhibitory proteins are quite different overall. Despite this diversity, both appear to function as inhibitors by occupying the peptide recognition site of the C-subunit, thus rendering that site inaccessible to other potential substrates. This occupancy of the peptide recognition site by an inhibitory pseudosubstrate site or by a phosphorylation site that is part of a regulatory element appears to be a common mechanism for regulation of

TABLE 7
Comparison of Peptides that Correspond to the Inhibitory Regions of the Heat-Stable Protein Kinase Inhibitor and of the Regulatory Subunits of cAMP-Dependent Protein Kinase

Peptide	Amino Acid Sequence	K_m or K_i	Ref.
Synthetic Peptides			
Substrate	Leu ARG ARG Ala SER Leu Gly	4.7 μM	86
Inhibitor	Leu ARG ARG Ala Ala Leu Gly	320 μM	86
PKI Inhibitory Peptides			
I	Thr Thr Tyr Ala Asp Phe Ile Ala Ser Gly Arg Thr Gly ARG ARG Asn Ala Ile His Asp	0.3 nM	166
II	Thr Thr Tyr Ala Asp Phe Ile Ala Ser Gly Arg Thr Gly ARG ARG Asn Ala Ile His Asp NH2	2.3nM	167
III	Gly Arg Thr Gly ARG ARG Asn Ala Ile His Asp NH2	57 nM	167
IV	Gly Lys Thr Gly ARG ARG Asn Ala Ile His Asp NH2	0.37 μM	167
V	Gly Arg Thr Gly Lys ARG Asn Ala Ile His Asp NH2	36 μM	167
VI	Gly Arg Thr Gly ARG Lys Asn Ala Ile His Asp NH2	4.2 μM	167
VII	Tyr Ala Asp Phe Ile Ala Ser Gly Arg Thr Gly ARG ARG Asn Ala Ile His Asp Ile Leu Val Ser Ser Ala	240 nM	169
VIII	Tyr Ala Asp Phe Ile Ala Ser Gly Arg Thr Gly ARG ARG Asn Ala Ile His Asp Ile Leu Val Ser Ser Ala	120 μM	169
IX	Tyr Ala Asp Phe Ile Ala Ser Gly Arg Thr Gly ARG ARG Asn Ala Ile His Asp Ile Leu Val Ser Ser Ala	20 μM	169
Hinge Regions			
R^I subunit (α)	Ser Pro Pro Asn Pro Val Val Lys Gly Arg Arg ARG ARG Gly Ala Ile Ser Ala		40
R^{II} subunit (α)	Glu Glu Asp Leu Asp Val Pro Ile Pro Gly Arg Phe Asp ARG ARG Val SER Val Cys Ala		41
R^{II} subunit (β)	Ala Gly Ala Phe Asn Ala Pro Val Phe Asn Arg Thr ARG ARG Ala SER Val Cys Ala		142

Note: The phosphorylation (or pseudophosphorylation) sites that occur in each sequence are underlined. R^I (α) = bovine skeletal muscle; R^{II} (α) = bovine heart; R^{II} (β) = rat ovarian.

protein kinases. For example, a similar mechanism has been proposed for protein kinase C,[172,173] MLCK,[174] the EGF receptor,[66] and the insulin receptor.[67] Furthermore, all protein kinases, with the notable exception of the oncogenic transforming proteins, characteristically are tightly regulated and maintained in an inactive state in the absence of the activating ligand.

V. OTHER ACTIVE SITE RESIDUES

Until recently, the only residues that had been identified at the active site of the C-subunit were Lys-72 and Cys-199. Subsequently, two carboxyl groups have been identified. If one considers the C-subunit in particular and the protein kinases in general, there are several roles that carboxyl groups could play. For example, carboxyl groups on the protein may contribute to the recognition of the basic residues preceding the phosphorylation site in the substrate. Inhibition of the C-subunit with a water-soluble carbodiimide has led to the prediction that at least one glutamic acid contributes to this recognition site.[175] There are at least two additional roles for carboxyl groups. One is a general base catalyst, which would interact with the hydroxyl group on the substrate, thus making the oxygen more nucleophilic and a better attacking group. The other role is a ligand of Mg^{2+} in the MgATP complex or in the inhibitory Mg^{2+} binding site. There is precedence for both roles in other ATP-phosphotransferases that act on small molecules. In an effort to identify a carboxyl group in the C-subunit that may contribute in one of the latter two roles, the C-subunit was treated with a hydrophobic carbodiimide, dicyclohexylcarbodiimide (DCCD). DCCD specifically partitions into hydrophobic regions,[176,177] typically associated with nucleotide binding sites.

DCCD was a good inhibitor of the C-subunit. Inhibition followed first order kinetics (K_i = 60 μM), and both MgATP and R-subunit were capable of protecting against inactivation. Attempts to covalently label the enzyme and to identify an essential carboxyl group with [^{14}C]DCCD, or to trap a reactive carboxyl group with an amine such as [^{14}C]glycine ethyl ester or [^{3}H]aniline, were initially unsuccessful. However, peptide mapping did allow Toner-Webb and Taylor[178] to identify a peptide, residues 165 to 189, that was consistently lost in conjunction with inhibition by DCCD. This peptide contained four carboxyl groups. It was concluded that one of these carboxyl groups reacted with DCCD, but the reactive intermediate covalently cross-linked to a nearby nucleophile on the protein. Of the four carboxyl groups contained within this peptide, two of them, Asp 166 and Asp 184, are invariant in the larger family of protein kinases and, thus, were good candidates for playing an essential role. In order to identify the specific carboxyl group that was labeled by DCCD, the C-subunit was first modified with acetic anhydride to block potential reactive lysine residues. Under these conditions DCCD, in the presence of [^{14}C]glycine ethyl ester, led to the labeling of one major residue, Asp-184. Since MgATP protects against this modification, Asp-184 is most likely close to the nucleotide binding site and, thus, in a position to contribute in an essential way to catalysis (Figure 6). A second residue that reacted to a lesser extent, but also was protected by MgATP, was Glu-91.[179] Both Asp-184 and Glu-91 are invariant in all protein kinases.

Several lines of evidence indicate that the C-subunit has a requirement for a general base in the catalytic mechanism. The most convincing evidence comes from the pH dependency of the reaction. Yoon and Cook[180] showed that the V/K$_{ser-pep}$ had a biphasic pH dependency requiring that one group with a pK_a of 6.2 be in its deprotonated form and a second group with a pK_a of 8.5 be in its protonated form. The residue with the pK_a of 6.2 was presumed to be a general base. This was consistent with the earlier pH study of Bramson et al.[74] Although the pK_a of the β-carboxyl group of aspartic acid is typically about 3.9, Asp-184 is located in a hydrophobic environment based on its reactivity with DCCD. This hydrophobic environment would make ionization less favorable by excluding the aqueous

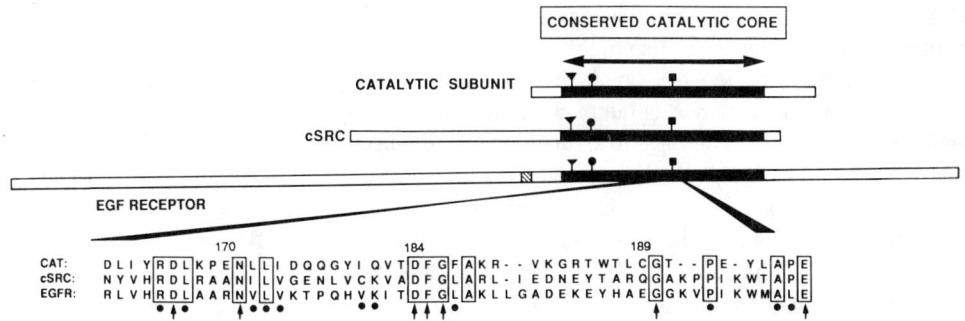

FIGURE 9. Comparison of kinase structures. The structure of the C-subunit (CAT) is compared to v-*src* (cSRC) and the EGF receptor (EGFR). The conserved catalytic core is designated in black, and the specific conservation of sequence in the region flanking the catalytic site is shown below. Residues that are invariant in all known protein kinases are indicated by arrows, whereas residues that are highly conserved are designated by filled circles. Invariant residues (the glycine-rich loop, ▼; Lys 72, ●; and Asp 184, ■) are indicated. (From Taylor, S. S., Buechler, J. A., Slice, L. W., Knighton, D. K., Durgerian, S., Ringheim, G. E., Neitzel, J. J., Yonemoto, W. M., Sowadski, J. M., and Dostmann, W., *Cold Spring Harbor Symp. Quant. Biol.*, 53, 121, 1988. With permission.)

solvent. Thus, Asp-184 could easily have an elevated pK_a, similar to other carboxyl groups known to be in active-site regions. For example, several small-molecule kinases, such as yeast hexokinase[181,182] and bacterial phosphofructokinase,[183] are predicted to have an aspartic acid at the active site that functions as a general base. In the case of hexokinase, the pH dependency of the reaction has established that the residue participating as a general base has a pK_a of 6.2.[184]

The available data cannot rule out the possibility that Asp-184 is a ligand of Mg^{2+}, although this role is not as probable based on the following results. Metal binding studies with the ligand substitution-inert complex $Co^{3+}(NH_3)_4ATP$ led to the conclusion that the metal binds as an enzyme-ATP-metal complex[185] and that residues in the C-subunit do not directly ligand the metal in the metal-ATP complex. The fact that $Co^{3+}(NH_3)_4ATP$ can serve as a substrate for the C-subunit supports this conclusion.[186] These results, however, do not preclude that H_2O, which could replace the NH_3 ligands in the active MgATP complex, bridges an enzyme residue to the Mg^{2+}, similar to what has been proposed for adenylate kinase.[187] A protein residue also could contribute to the inhibitory metal binding site first described by Armstrong et al.[185]

The conserved nature of Asp-184 is further evidence that it plays an essential role. Like Lys-72, this aspartic acid is a conserved feature of every protein kinase. The region flanking Asp-184 is also highly conserved in the entire family of protein kinases. Finally, the localization of Asp-184 in the linear sequence is consistent with its playing a role as a general base catalyst. Asp-184, unlike Glu-91, is probably not part of the nucleotide-fold motif, as would be predicted for residues contributing specifically to MgATP binding.

The second carboxylic acid residue that was labeled by DCCD and protected by MgATP is Glu-91. Based on its location in the polypeptide chain, Glu-91 probably is an integral part of the nucleotide binding domain. Its potential role in the nucleotide binding region will be discussed in Section VIII when the domain structure is described more extensively.

VI. ENZYME MECHANISM

Chemical and kinetic evidence both contribute to our understanding of the enzymatic mechanism for the reaction catalyzed by the C-subunit. For example, several specific amino acid residues have been identified that may participate directly in catalysis and substrate binding. These include Lys-72, Asp-184, and Cys-199. Of these, only Lys-72 and Asp-184 are shared by all protein kinases and, as such, are likely to play a common, universal role.

The sequence conservation in the region flanking Asp-184 is seen in Figure 9, while the sequence flanking Lys-72 is also highly conserved. It has been demonstrated repeatedly that Lys-72 shares a common function within the kinase family. Thus, in this case, sequence identity can be correlated accurately with functional homology. Conserved functions are anticipated in the ATP binding site and in the catalytic mechanism. In these regions, residues that are deemed essential should, by definition, be invariant in all related protein kinases. Cys-199 is common to some of the protein serine kinases, but for reasons outlined earlier (Section IV.C) is not thought to be essential. In general, this region flanking Cys-199, which most likely is associated with peptide binding, differs predictably in each kinase.

Treatment of the C-subunit with DCCD led to irreversible inactivation due to the generation of an intramolecular cross-link.[178] Buechler and Taylor[188] subsequently demonstrated that treatment of the apoenzyme with DCCD resulted in the cross-linking of Asp-184 to Lys-72. Identification of this cross-link localizes these two residues, which are both thought to be important for kinase function, close to each other at the active site. Based on the proximity of these two residues and their known location at the active site of the enzyme, a mechanism indicated in Figure 10 can be proposed. The first step in catalysis, according to this scheme, is the binding of MgATP. This potentially weakens any interaction between the side chains of Lys-72 and Asp-184 and makes the side chain of Asp-184 available for peptide binding in the next step. Asp-184 then hydrogen bonds to the proton on the serine hydroxyl group, making the oxygen more nucleophilic for attacking the γ-phosphate of ATP. An alternative mechanism where Asp-184 is a ligand for the metal in the MgATP complex is also possible. This interaction would occur either through direct contact between the carboxyl group of Asp-184 and Mg^{2+} or with an H_2O molecule bridging Asp-184 with Mg^{2+}. The interaction between Asp-184 and Mg^{2+} potentially would help orient the γ-phosphate of ATP for transfer to the peptide substrate.

The proposed mechanism is consistent with the pH dependency of the kinase reaction.[180] It is also consistent with kinetic evidence based on steady-state kinetics and equilibrium binding results.[85,86] Feramisco and Krebs[105] were among the first to demonstrate that the K_d for Kemptide, based on equilibrium dialysis, was nearly two orders of magnitude greater than the K_m, leading them to postulate an ordered mechanism of binding. A preferred ordered pathway also was predicted by Bolen et al.[189] and is consistent with the early studies of Pomerantz and colleagues.[190] Whitehouse and co-workers[86] subsequently confirmed that the C-subunit followed a preferred sequential pathway, with MgATP binding first. The binding of MgATP then provided a higher affinity binding site for the peptide substrate. The peptide substrate (Kemptide) had a K_d of 180 μM in the absence of MgATP, analogous to an inhibitor peptide where the phosphorylatable serine in Kemptide was replaced by a substitution-inert alanine (K_d = 230 μM). In the presence of MgATP, however, the observed K_m for the substrate peptide was 3 μM. Their kinetic analysis of the catalytic mechanism also was based on the ATP dependency of the interaction between the C-subunit and the heat-stable protein kinase inhibitor. PKI was a competitive inhibitor for the peptide substrate, but was an uncompetitive inhibitor with regard to MgATP. MgATP was required for high-affinity binding of PKI to the C-subunit.[86,191,192] In addition, the binding of MgATP was enhanced by nearly three orders of magnitude in the presence of PKI. This MgATP dependency for complex formation led them to predict a preferred ordered pathway for binding, with MgATP binding first. The preferred ordered pathway also is consistent with the high ATPase activity associated with the C-subunit.[193] Unlike hexokinase, where glucose binds prior to MgATP,[194] the C-subunit must be capable of existing as a binary complex, where water has access to the active site of the enzyme containing bound MgATP in a configuration that is poised for phosphate transfer. In contrast to the preferred ordered pathway, Cook et al.[85] suggest a steady-state random kinetic mechanism based on initial velocity studies carried out in the presence and absence of MgADP and dead-end inhibitors. They also emphasize the importance of free Mg^{2+} in kinetic measurements.

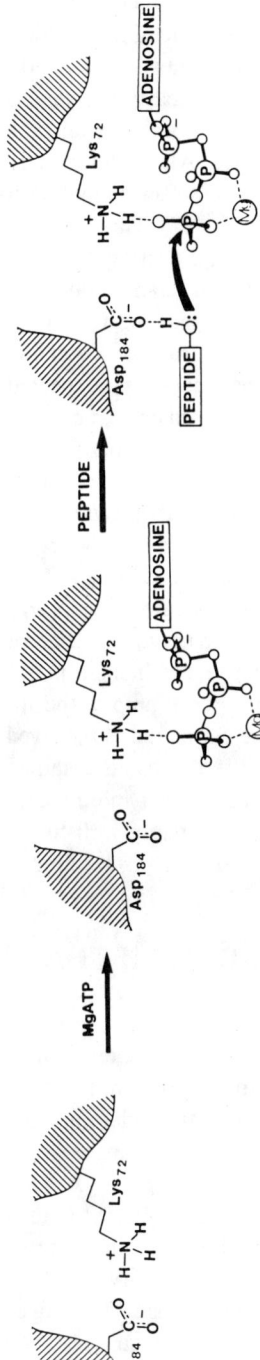

FIGURE 10. A proposed mechanism for the reaction that is catalyzed by the C-subunit of cAMP-dependent protein kinase. (From Taylor, S. S., Buechler, J. A., Slice, L. W., Knighton, D. K., Durgerian, S., Ringheim, G. E., Neitzel, J. J., Yonemoto, W. M., Sowadski, J. M., and Dostmann, W., *Cold Spring Harbor Symp. Quant. Biol.*, 53, 121, 1988. With permission.)

Granot and Mildvan[186] used substitution-inert metal·ATP complexes to establish the conformation of the nucleotide when it is bound to the active site of the C-subunit. Their results established that the conformation serving as a substrate for the enzyme is the Δ-isomer of β,γ-bidentate $Co(NH_3)_4ATP$. Although the activity was only 0.1% compared to the normal substrate, the substitution-inert complex, nevertheless, was capable of transferring the γ-phosphate to an appropriate phosphoryl acceptor peptide. These studies subsequently were extended using the nonhydrolyzable form of ATP, AMPPCP, to measure distances between the active (Co) and inhibitory (Mn) metal sites and between each of the metal sites and the site of phosphoryl transfer on a synthetic peptide, Leu–Arg–Arg–Ala–Ser–Leu–Gly (for review see Reference 74). From these values, the distance between the serine hydroxyl oxygen and the γ-phosphate of ATP was deduced to be 5.3 ± 0.7 Å.[74,196] This distance was used to support a dissociative mechanism for the enzyme, involving a metaphosphate intermediate. However, an associative mechanism could not be ruled out unambiguously.

Several lines of evidence subsequently have led to the conclusion that the C-subunit follows an associative mechanism. First are the kinetic, pH dependency results of Yoon and Cook[180] which predicted a general base-catalyzed reaction. A general base catalyst is expected in the associative mechanism, but is not required in the dissociative mechanism. A potential candidate for the general base catalyst is Asp-184, as was discussed in Section V. The stereochemistry for this reaction recently was confirmed by Ho et al.,[197] who used chiral analogues of ATP to distinguish whether the reaction occurred by a double displacement or a single displacement mechanism. They showed an inversion of configuration of the transferred phosphate group, unambiguously ruling out a phosphoenzyme intermediate and confirming a single displacement mechanism. The primary criterion for the dissociative mechanism, based on the NMR studies of Granot et al.,[196] was the intersubunit distance of 5.3 ± 0.7 Å, where the maximum distance allowed for the associative mechanism is 4.9 Å. The authors point out several qualifications in their conclusions. For example, the intersubstrate distance is likely to differ somewhat in the transition state compared to the fixed phosphotransfer-inert complex that was characterized. More importantly, the results assumed that AMPPCP was analogous to ATP since the K_d values were similar. Other more recent results, however, have demonstrated that although AMPPCP binds reasonably well to the catalytic subunit, it is unable to induce the necessary conformational changes leading to efficient peptide binding.[86]

VII. CONFORMATIONAL CHANGES ASSOCIATED WITH THE C-SUBUNIT

Substrate binding induces significant conformational changes in the C-subunit. Such conformational changes have been confirmed in several ways. Circular dichroism (CD), in particular, was used to demonstrate that peptide binding induces discrete and sequential conformational changes in the enzyme. MgATP also induces conformational changes in the enzyme.

The initial observations of conformational changes in the C-subunit were made by Reed and co-workers.[198-200] By measuring changes in intrinsic UV circular dichroism in the presence and absence of a peptide substrate, they found substantial changes in secondary structure. The magnitude of these changes may be exaggerated, and attributing the intrinsic CD changes to a specific tyrosine residue as the authors did, particularly one that is not even conserved in other C-subunits, is still speculative. However, the results do establish convincingly that the enzyme undergoes significant conformational changes as a consequence of peptide binding. The initial results measured CD changes induced by peptide binding to C-subunit that had the MgATP binding site occupied by Blue Dextran-Sepharose®.[199] The ability of peptides to induce conformational changes was explored further using several peptides, including a peptide substrate (Leu–Arg–Arg–Ala–Ser–Leu–Gly) and its inhibitory analogue (Leu–Arg–Arg–Ala–Ala–Leu–Gly). These later studies allowed Reed et al. to detect several stages in peptide binding. The first step was assumed to be an ionic interaction between the guanidinium groups on the peptide and negatively charged groups on the protein. Both the substrates and the inhibitor peptides were capable of mediating this change in conformation. The next step presumably involved the peptide assuming a random coil conformation. The inhibitor Ala-peptide lacks the serine hydroxyl group and, thus, was not capable of inducing the induced conformational changes. Reed et al.[200] also used another inhibitor peptide containing D-serine instead of L-serine. This peptide showed intermediate CD changes, leading to the conclusion that the hydroxyl group needed to be in a precise orientation, and the inability to assume a random coil conformation was detrimental to productive peptide binding and the prerequisite conformational changes.[200]

Whitehouse and Walsh[191] were the first to demonstrate a difference between ATP and adenosine 5′-[β,γ-imido]-triphosphate (AMPPNP). As indicated earlier, MgATP enhanced the binding of peptide substrate to C-subunit by nearly two orders of magnitude. In contrast, when peptide binding was measured in the presence of MgAMPPNP, the K_d (peptide) remained at 210 μM, nearly identical to the equilibrium binding of peptide alone. On this basis it was concluded that MgAMPPNP was not capable of inducing the conformation of the C-subunit that has a high-affinity binding site for peptide. This subsequently was confirmed by binding the heat-stable inhibitor protein (PKI) to the catalytic subunit. High-affinity binding of PKI requires MgATP; MgAMPPNP was not able to substitute.[191,201]

Kochevar et al.[202] characterized the capacity of the C-subunit, in the presence of MgATP, to enhance the dissociation of bound cAMP from the R-subunit. The enhanced dissociation of cAMP did not occur when MgAMPPCP was substituted for MgATP, providing additional evidence that AMPPCP cannot substitute completely for ATP. The basis for this distinction is not apparent; however, it is clear that the group bridging the β- and γ-phosphates in ATP is extremely important for inducing conformational changes in this enzyme. The oxygens on the γ-phosphate of AMPPCP are not as acidic as ATP, and, in addition, the methylene and amino groups that bridge the β- and γ-phosphates may impose certain steric constraints.

First and Taylor[128] demonstrated conformational changes indirectly by characterizing the distance between the two sulfhydryl groups in the C-subunit. Based on criteria described earlier, Cys-199 is thought to be in close proximity to the active site of the enzyme. This cysteine is readily alkylated in the apoenzyme, but is protected by MgATP, and loss of activity following alkylation is due to the modification of Cys-199.[127,128] However, MgATP also protects Cys-343 against alkylation.[123] Either Cys-343 is protected because it, too, participates directly in ATP binding, or it is protected because of conformational changes extending beyond the immediate environment of the active site. To resolve these two possibilities, First and Taylor attached fluorescent probes to both cysteine residues. One cysteine was modified by a fluorescence donor, whereas the other was modified by an appropriate fluorescence acceptor. Fluorescence energy transfer subsequently established that Cys-343 was approximately 35 ± 5 Å from Cys-199. Since the maximum distance that MgATP itself can extend is 16 Å, it was concluded that Cys-343 is protected by conformational changes that extend over a substantial distance.[203]

VIII. DOMAIN STRUCTURE

Several structures for the nucleotide binding site of the C-subunit have been predicted.[204-206] These proposed structures consider some of the chemical information that is available, but are primarily based on comparisons with other adenine nucleotide binding proteins where crystal structures are available. One must be cautious about such predictions, and each of the structures predicted so far has inconsistencies when considered in the context of all of the chemical information.

The evidence that sheds some light on the specific domain structure for this region of the C-subunit can be summarized as follows. First is the identification of Lys-72 as an essential residue.[88] Second is the placing of the side chain of Lys-72 in close proximity to the β- and/or γ-phosphate of MgATP.[89,125] Third is the NMR evidence predicting the specific conformation for the bound metal-ATP complex.[74,195] Fourth is the mapping of the MgATP binding site with various analogues of MgATP, providing a general framework that must conform with any model.[76,77] Fifth is the modification of lysine residues with acetic anhydride in the presence and absence of MgATP, which gives some indication of residues or regions that are reactive in the apoenzyme and that are protected when MgATP is bound.[207] Finally, the cross-linking with DCCD means that Lys-72 must be positioned so that it can be close to Asp-184.[188]

The basic nucleotide-fold motif that was first recognized by Rossmann et al.[102] had a sheet of six β-strands with two α-helices above and below the plane of β-strands (Figure 11). They predicted that this general structure would be conserved in most adenine nucleotide binding proteins. The carboxy-terminal end of the β-sheet and a crevice between two of the β-strands constitute a major part of the adenine nucleotide binding site.[208,209] Although the order of the β-strands and the size of the helices vary, this overall structure has been reproduced in nearly every adenine and guanine nucleotide binding protein where crystallographic data are available.[208-214] One of the few exceptions is the cAMP binding site in the catabolite gene activator protein[215] and, based on sequence similarities, in the R-subunit. This cAMP binding site has a different motif best described as a β-roll structure.

Although the secondary and tertiary structures in the nucleotide-fold motif are conserved, the primary sequence generally is not, with the notable exception of a triad of glycines. These glycines lie at a bend between the first β-strand and the subsequent α-helix, and they appear to place a clear constraint on secondary structure. The bound nucleotide, particularly the phosphate region, frequently is close to this region of the polypeptide chain containing the triad of glycines, although the orientation of the nucleotide base varies in each structure. The dipole of the helix is also thought to be a significant feature of this structural motif.[209]

Now let us consider other nucleotide binding proteins — emphasizing not only their similarities, but also their differences. Since the elongation factor Tu (EF-Tu) and *ras* p21 are homologous proteins belonging to the same family, the crystal structure of EF-Tu was used as a general framework for modeling the structure of *ras* p21, in addition to the entire family of guanosine 5'-triphosphate (GTP)-binding proteins.[212,216] The recent solution of the crystal structure of *ras* p21 at 2.7 Å confirmed the validity of this approach.[213,217] Most of the overall structural homologies are conserved as predicted. Initially it was thought that the order of the β-strands in the region of the glycine-rich loop differed between EF-Tu[212] and *ras* p21.[217]. However, the earlier *ras* p21 structure recently was corrected,[218] and the *ras* p21 structure also was solved independently.[214] Based on these results the order of the β-strands in the two proteins is identical. Hopefully, the crystal structure of one protein kinase likewise can serve as a framework for the catalytic core of the entire family, but it is important to recognize the limitations of such modeling.

Nearly all of the adenine nucleotide binding proteins have followed the general α/β motif that was first predicted by Rossmann et al.[102] A comparison of lactate dehydrogenase, adenylate kinase, phosphoglycerate mutase, and hexokinase, as well as c-H-*ras* p21 and EF-Tu, is sufficient to demonstrate that general features of secondary structure are conserved (Figure 11). At the same time, this comparison exemplifies the diversity of these structures. Although each maintains the same general α/β motif, they differ in several important regards. First, the order of the β-strands forming the β-sheet differs. Second, the β-strands differ with respect to parallel vs. antiparallel orientation. Finally, each protein differs with regard to essential residues that contribute to nucleotide binding. Each of these structures has five to six β-strands extending throughout the center of the fold to form a β-sheet (Figure 11).

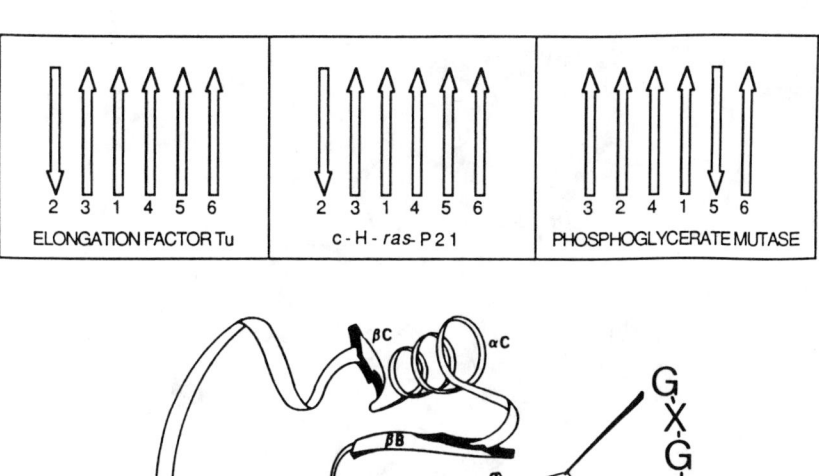

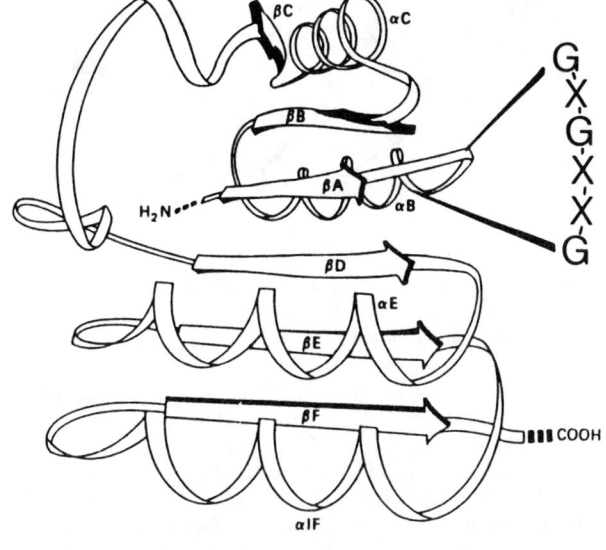

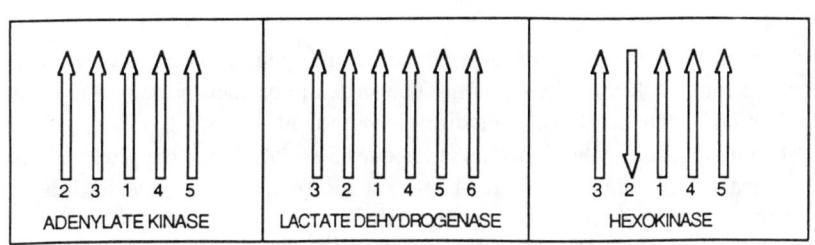

FIGURE 11. Schematic representation of the nucleotide folds for various adenine nucleotide binding proteins. The helix and β-strand organization of lactate dehydrogenase, shown in the center, is based on the crystallographic coordinates.[210] The numbers refer to the order of the strands, and the arrows go from the amino-terminal to the carboxyl-terminal end. The alignments of the β-strands in the other proteins also were taken from the crystal structures: EF-Tu,[212] ras-p21,[213,217-219] phosphoglycerate mutase,[211] adenylate kinase,[187] and hexokinase.[214] It should be noted that the order of the β-strands in ras p21, originally reported to be 213456,[217] is actually 231456,[218,219] similar to Ef-Tu. (From Taylor, S. S., Buechler, J. A., Slice, L. W., Knighton, D. K., Durgerian, S., Ringheim, G. E., Neitzel, J. J., Yonemoto, W. M., Sowadski, J. M., and Dostmann, W., *Cold Spring Harbor Symp. Quant. Biol.*, 53, 121, 1988. With permission.)

As Brändén[208] pointed out, a comparison of the order of these strands and their orientation displays one difference immediately. The β-strands in lactate dehydrogenase (LDH), for example, have a 321 456 pattern,[210] whereas adenylate kinase has a 23145 pattern.[187] Elongation factor Tu and p21-*ras* also have a 231 456 motif, with one major difference — strand 2 is antiparallel with respect to the others.[212,218,219] Hexokinase also has one antiparallel strand,[214] while LDH and adenylate kinase have only parallel strands.[187,210]

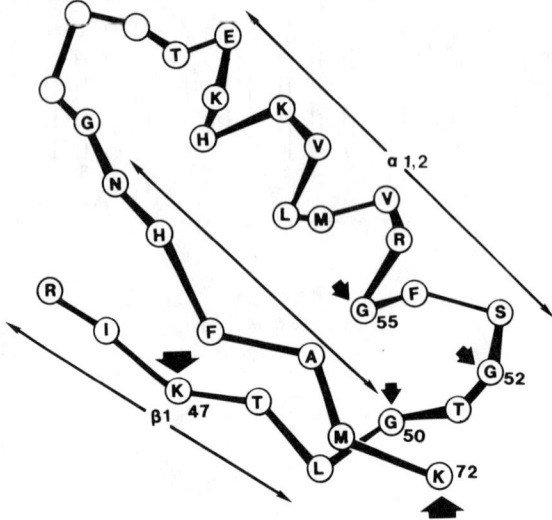

FIGURE 12. A proposed model for the glycine-rich loop region of the catalytic subunit. The backbone coordinates were taken from LDH[210] based on the modeling of Wierenga et al.[206] The corresponding sequences of the C-subunit were then substituted into these coordinates, as indicated by the single-letter code for each amino acid. The proposed positions of the two residues that are protected by MgATP, Lys-72 and Lys-47, are indicated by arrows. (From Taylor, S. S., Buechler, J. A., Slice, L. W., Knighton, D. K., Durgerian, S., Ringheim, G. E., Neitzel, J. J., Yonemoto, W. M., Sowadski, J. M., and Dostmann, W., *Cold Spring Harbor Symp. Quant. Biol.*, 53, 121, 1988. With permission.)

In conjunction with the chemical evidence, some general conclusions can be drawn about the nucleotide fold in the C-subunit. The structure of each protein with a nucleotide-fold motif is consistent in that the glycine-rich segment follows β-strand 1. This highly conserved region in the C-subunit occurs at residues 50 to 55. Since the sequence similarities with other kinases in the catalytic domain begin at approximately residue 40, it is reasonable to predict that the first β-strand is contained in this region (residues 40 to 48). The next problem is locating Lys-72 with respect to this secondary structure. The distance from Gly-55 to Lys-72 is most consistent with this segment lying at the end of the first β-sheet, which would be linked by an α-helical segment (α1,2) to β-strand 2. The model for this region, shown in Figure 12, is based on the crystallographic coordinates of LDH. The amino acid sequence of the C-subunit has been fitted to the corresponding portion of the LDH structure in order to determine whether the location of Lys-72, based on this model, would place them near the proposed nucleotide binding site. Three residues have been deleted between the α-helix that links β-strand 1 and β-strand 2, and insertions and deletions that occur in the overall family of protein kinases indicate that the length of such a loop may vary slightly with each enzyme. The sequence requirements posed by Wierenga et al.[206] are fulfilled reasonably well in this model.

In *ras* p21, the loop joining β-strand 1 with an α-helix comes in direct contact with the β-phosphate of guanosine 5′-diphosphate (GDP).[213,218,219] Like the C-subunit, *ras* p21 also has hydrolytic activity. It is a GTPase, just as the C-subunit can function as an ATPase in the absence of protein or peptide substrate. The glycine-rich loop in adenylate kinase also is predicted to lie near the phosphates of ATP, although an actual crystal structure of a binary complex containing enzyme and MgATP has not been solved.[187,220] Recently, a binary complex of adenylate kinase and an inhibitory ATP analogue, P^1,P^5-di(adenosine-5'-)pentaphosphate, was crystallized and confirmed this relationship of the glycine-rich loop to the phosphate region of the analogue.[221] It is reasonable to predict that the C-subunit has an analogous glycine-rich loop and that it will lie close to the phosphates of ATP.

Some additional insights into the folding of this nucleotide binding region have emerged from the labeling of lysine residues with [³H]acetic anhydride.[207] This has allowed accessible and reactive lysines to be labeled and residues protected by substrates to be identified. Several residues were protected by MgATP, and all were located in the amino-terminal region of the protein, again consistent with the prediction that this region constitutes the nucleotide binding site. Lys-72 is labeled readily in the apoprotein, but is protected from modification in the presence of MgATP. This was predicted based on its location in the MgATP binding site and also on its potential to form a salt bridge with Asp-184, which may sequester that side chain and make the ϵ-NH$_2$ group of Lys-72 less reactive toward acetic anhydride. However, in addition to Lys-72, MgATP afforded substantial protection to Lys-47. These results are consistent with Lys-47 being near the carboxy-terminal end of the first β-strand and Lys-72 being at the carboxy-terminal end of β-strand 2, since the nucleotide typically binds at the carboxy end of the sheet. Lys-76 is very unreactive in the apoprotein, but is totally protected by MgATP, whereas Lys-78, Lys-81, and Lys-83 are all relatively reactive and show only slight protection. Thus, the reactivity of certain lysines in the presence and absence of MgATP gives an indication of regions in the C-subunit that are sensitive to MgATP binding.

The protection of Lys-72 by MgATP is consistent with its labeling by FSBA, and the model shown in Figure 12 places Lys-72 in an ideal position to interact with the bound nucleotide. This position in LDH is occupied by an essential amino acid, Asp-32, which participates in NAD binding.[210] The pattern of accessibility observed for the region following Lys-72 also is consistent with a helical motif and can be fitted easily to an amphipathic helix with Lys-78, Lys-81, and Lys-83 being exposed on the surface and the hydrophobic surface facing inward and making contact with the protein. The other lysine that was protected by MgATP is Lys-47, which we predict to be in the first β-strand.

The model proposed by Sternberg places the triad of glycines at the end of β-strand 1 and also places Lys-72 near the β- and γ-phosphates of ATP, which agrees with the protein chemistry.[204] Whether Lys-72 ends β-strand 2 or begins the α-helix will be resolved when the crystal structure is solved. However, Sternberg's subsequent placing of Asp-184 at a position corresponding to the carboxy-terminal end of β-strand 3 is not plausible. Although Asp-184 is close to Lys-72, it most likely lies outside the nucleotide-fold motif.

Fry et al.[205,222] also have proposed a model for this nucleotide-fold region of the protein based on potential sequence similarities with adenylate kinase. Their model is based on NMR studies of metal-ATP complexes binding to adenylate kinase.[223] The putative ATP binding site was deduced previously by Pai et al.[220] from the crystal structure of the apoenzyme. In addition, Fry et al. used a synthetic peptide corresponding to residues 1 to 45. Hamada et al.[224] had demonstrated previously that this fragment bound ϵ-ATP with an affinity that was comparable to the native enzyme. The NMR studies were carried out with rabbit muscle adenylate kinase and with the synthetic peptide, and on this basis residues were identified that might interact with the nucleotide. Three segments of particular significance were identified that were related to other ATP-binding proteins, including the C-subunit. Segment 1 corresponds to a glycine-rich loop following β-strand 1 and terminates with Lys-21. Segment 2 is an α-helix containing Lys-27, which is predicted to interact with the γ-phosphate of ATP. Segment 3 is a hydrophobic strand terminated by Asp-119 that may accept a hydrogen bond from a water molecule liganded to Mg^{2+}. Subsequent crystallographic analysis of yeast adenylate kinase is not consistent with some of these assignments.[209,221] Thus, even though both enzymes show a similar in-line mechanism with inversion of configuration,[225] it is not clear what the actual structural homologies will be when the tertiary structures of the catalytic subunit and adenylate kinase are compared. A glycine-rich loop clearly is an important feature of both ATP binding sites. The importance of this region has been emphasized again with the recent crystal structure of p21-*ras*.[213,217-19] Lys-72 in the C-subunit is likely to be close to the γ-phosphate of ATP, but convincing evidence for a similar role for Lys-27 in adenylate kinase is lacking. In addition, Lys-27 is not conserved in all adenylate kinases, arguing against its playing an essential role.[226] Asp-119 in rabbit muscle adenylate kinase also is not conserved, and the recent crystallographic model predicts that Asp-93, which is invariant, interacts with Mg^{2+}.[187,221] Having chemical evidence to support predicted is obviously extremely important.

IX. FUTURE DIRECTIONS

Structural information can be deduced indirectly from affinity labeling and from group-specific labeling; however, the detailed understanding of any protein molecule eventually must be based on high-resolution analysis, either X-ray crystallography or two-dimensional NMR. In order to fully understand a protein the size of the C-subunit, it is essential to have a high-resolution crystal structure, not only of the apoenzyme, but also of a complex containing bound substrates. Even for many of the kinases that act on small molecules, such as hexokinase, this information is not available. For example, no crystals of hexokinase have been characterized containing bound nucleotide. In addition to having a high-resolution crystal structure, it is necessary to have chemical information that allows the structure to be interpreted in terms of function. Not only the amino acid sequence, but also the identification of functional residues at the active site is essential. Combining all of this information can provide a framework for viewing and interpreting related molecules. Elucidation of the crystal structure of the C-subunit will be the next step forward in understanding how protein kinases function.[227-229]

REFERENCES

1. **Cori, G. T. and Cori, C. F.**, The enzymatic conversions of phosphorylase *a* to *b*, *J. Biol. Chem.*, 158, 321, 1945.
2. **Fischer, E. H. and Krebs, E. G.**, Conversion of phosphorylase *b* to phosphorylase *a* in muscle extracts, *J. Biol. Chem.*, 216, 121, 1955.
3. **Sutherland, E. W. and Rall, T. W.**, Fractionation and characterization of a cyclic adenine ribonucleotide formed by tissue particles, *J. Biol. Chem.*, 232, 1077, 1958.
4. **Walsh, D. A., Perkins, J. P., and Krebs, E. G.**, An adenosine 3′,5′-monophosphate-dependent protein kinase from rabbit skeletal muscle, *J. Biol. Chem.*, 243, 3763, 1968.
5. **Collett, M. S. and Erickson, R. L.**, Protein kinase activity associated with the avian sarcoma virus *src* gene product, *Proc. Natl. Acad. Sci. U.S.A.*, 75, 2021, 1978.
6. **Staros, J. V., Cohen, S., and Russo, M. W.**, Epidermal growth factor receptor: characterization of its protein kinase activity, in *Molecular Mechanisms of Transmembrane Signaling*, Cohen, P., Ed., Elsevier, Amsterdam, 1985, 253.
7. **Roth, A. R. and Cassell, D. J.**, Insulin receptor: evidence that it is a protein kinase, *Science*, 219, 299, 1983.
8. **Hunter, T. and Sefton, B. M.**, Transforming gene product of Rous sarcoma virus phosphorylates tyrosine, *Proc. Natl. Acad. Sci. U.S.A.*, 77, 1311, 1980.
9. **Sefton, B. M., Hunter, T., Beemon, K., and Eckhart, W.**, Evidence that the phosphorylation of tyrosine is essential for cellular transformation by Rous sarcoma virus, *Cell*, 20, 807, 1980.
10. **Krebs, E. G. and Beavo, J. A.**, Phosphorylation-dephosphorylation of enzymes, *Annu. Rev. Biochem.*, 48, 923, 1979.
11. **Cohen, P.**, The coordinated control of metabolic pathways by broad-specificity protein kinases and phosphatases, *Curr. Top. Cell. Regul.*, 17, 23, 1985.
12. **Shacter, E., Chock, P. B., Rhe, S. G., and Stadtman, E. R.**, Cyclic cascades and metabolic regulation, in *The Enzymes*, Vol. 17, Boyer, P. D. and Krebs, E. G., Eds., Academic Press, New York, 1986, 21.

13. **Gilman, A. G.,** G proteins: transducers of receptor-generated signals, *Annu. Rev. Biochem.,* 56, 615, 1987.
14. **Brostom, C. O., Corbin, J. D., King, C. A., and Krebs, E. G.,** Interaction of the subunits of adenosine 3',5'-monophosphate-dependent protein kinase of muscle, *Proc. Natl. Acad. Sci. U.S.A.,* 68, 2444, 1971.
15. **Krebs, E. G.,** Protein kinases, *Curr. Top. Cell. Regul.,* 5, 99, 1972.
16. **Cohen, P.,** Muscle glycogen synthase, in *The Enzymes: Control by Phosphorylation Part A,* Vol. 17, Boyer, P. D. and Krebs, E. G., Eds., Academic Press, New York, 1986, 462.
17. **Roach, P. J.,** Liver glycogen synthase, in *The Enzymes: Control by Phosphorylation Part A,* Vol. 17, Boyer, P. D. and Krebs, E. G., Eds., Academic Press, New York, 1986, 500.
18. **Fiol, C. J., Mahrenholz, A. M., Wang, Y., Roeske, R. W., and Roach, P. J.,** Formation of protein kinase recognition sites by covalent modification of the substrate, *J. Biol. Chem.,* 262, 14042, 1987.
19. **Pickett-Gies, C. A. and Walsh, D. A.,** Phosphorylase kinase, in *The Enzymes: Control by Phosphorylation Part A,* Vol. 17, Boyer, P. D. and Krebs, E. G., Eds., Academic Press, New York, 1986, 396.
20. **Madsen, N. B.,** Glycogen phosphorylase, in *The Enzymes: Control by Phosphorylation Part A,* Vol. 17, Boyer, P. D. and Krebs, E. G., Eds., Academic Press, New York, 1986, 366.
21. **Sprang, S., Goldsmith, E., and Fletterick, R.,** Structure of the nucleotide activation switch in glycogen phosphorylase a, *Science,* 237, 1012, 1987.
22. **Engström, L.,** The regulation of liver pyruvate kinase by phosphorylation-dephosphorylation, *Curr. Top. Cell. Regul.,* 13, 29, 1978.
23. **Pilkis, S. J., Regen, D. M., Stewart, B. H., Chrisman, S. T., Pilkis, J., Kountz, P., Pate, T., McGrane, M., El-Maghrabi, M. R., and Claus, T. H.,** Rat liver 6-phosphofructo-2-kinase/fructose-2,6-bisphosphatase: a unique bifunctional enzyme regulated by cyclic AMP-dependent phosphorylation, in *Enzyme Regulation by Reversible Phosphorylation,* Vol. 3., Cohen, P., Ed., Elsevier, Amsterdam, 1984, 95.
24. **Lively, M. O., El-Maghrabi, M. R., Pilkis, J., D'Angelo, G., Colosia, A. D., Ciavola, J.-A., Fraser, B. A., and Pilkis, S. J.,** Complete amino acid sequence of rat liver 6-phosphofructo-2-kinase/fructose-2,6-bisphosphatase, *J. Biol. Chem.,* 263, 839, 1988.
25. **Loose, D. S., Wynshaw-Boris, A., Meisner, H. M., Hod, Y., and Hanson, R. W.,** Hormonal regulation of phosphoenolpyruvate carboxy kinase gene expression, in *Molecular Basis for Insulin Action,* Dzech, M. P., Ed., Plenum Press, New York, 1985, 347.
26. **Motminy, M. R., Sevarino, K. A., Wagner, J. A., Mandel, G., and Goodman, R. H.,** Identification of a cAMP responsive element within the rat somatostatin gene, *Proc. Natl. Acad. Sci. U.S.A.,* 83, 6682, 1986.
27. **Silver, B. J., Bokar, J. A., Virgin, J. B., Vallen, E. A., Milsted, A., and Nilson, J. H.,** Cyclic AMP regulation of the human glycoprotein hormone α-subunit gene is mediated by an 18-base-pair element, *Proc. Natl. Acad. Sci. U.S.A.,* 84, 2198, 1987.
28. **Gonzalez, G. A., Yamamoto, K. K., Fischer, W. H., Karr, D., Menzel, P., Biggs, W., III, Vale, W. W., and Montminy, M. R.,** A cluster of phosphorylation sites on the cyclic AMP-regulated nuclear factor CREB predicted by its sequence, *Nature (London),* 337, 749, 1989.
29. **Imagawa, M., Chiu, R., and Karin, M.,** Transcription factor AP-2 mediates induction by two different signal-transduction pathways: protein kinase C and cAMP, *Cell,* 51, 251, 1987.
30. **Grove, J. R., Price, D. J., Goodman, H. M., and Avruch, J.,** Recombinant fragment of protein kinase inhibitor blocks cyclic AMP-dependent gene transcription, *Science,* 238, 530, 1987.
31. **Cohen, P., Rylatt, D. B., and Nimmo, G. A.,** The hormonal control of glycogen metabolism: the amino acid sequence at the phosphorylation site of protein phosphatase inhibitor-1, *FEBS Lett.,* 76, 182, 1977.
32. **Embi, N., Parker, P. J., and Cohen, P.,** A reinvestigation of the phosphorylation of rabbit skeletal-muscle glycogen synthase by cyclic-AMP-dependent protein kinase. Identification of the third site of phosphorylation as serine-7, *Eur. J. Biochem.,* 115, 405, 1981.
33. **Yeaman, S. J., Cohen, P., Watson, D. C., and Dixon, G. H.,** The specificity of adenosine 3',5'-cyclic monophosphate-dependent protein kinase, *Biochem. Soc. Trans.,* 4, 1027, 1976.
34. **Hjelmquist, G., Andersson, J., Edlund, B., and Engström, L.,** Amino acid sequence of a [^{32}P]phosphopeptide from pig liver pyruvate kinase phosphorylated by cyclic 3',5'-AMP-stimulated protein kinase and [^{32}P]ATP, *Biochem. Biophys. Res. Commun.,* 61, 559, 1974.
35. **Murray, K. J., El-Maghrabi, M. R., Kountz, P. D., Lukas, T. J., Soderling, T. R., and Pilkis, S. J.,** Amino acid sequence of the phosphorylation site of rat liver 6-phosphofructo-2-kinase/fructose-2,6-bisphosphatase, *J. Biol. Chem.,* 259, 7673, 1984.
36. **Barker, W. C. and Dayhoff, M. O.,** Viral *src* gene products are related to the catalytic chain of mammalian cAMP-dependent protein kinase, *Proc. Natl. Acad. Sci. U.S.A.,* 79, 2836, 1982.
37. **Hunter, T.,** One thousand and one kinases, *Cell,* 50, 823, 1987.
38. **Hanks, S. K., Quinn, A. M., and Hunter, T.,** The protein kinase family: conserved features and deduced phylogeny of the catalytic domains, *Science,* 241, 42, 1988.

39. Shoji, S., Ericsson, L. H., Walsh, K. A., Fischer, E. H., and Titani, K., Amino acid sequence of the catalytic subunit of bovine type II adenosine cyclic 3′,5′-phosphate dependent protein kinase, *Biochemistry*, 22, 3702, 1983.
40. Titani, K., Sasagawa, T., Ericsson, L. H., Kumar, S., Smith, S. B., Krebs, E. G., and Walsh, K. A., Amino acid sequence of the regulatory subunit of bovine type I adenosine 3′,5′-phosphate dependent protein kinase, *Biochemistry*, 23, 4193, 1984.
41. Takio, K., Smith, S. B., Krebs, E. G., Walsh, K. A., and Titani, K., Amino acid sequence of the regulatory subunit of bovine type II adenosine cyclic 3′,5′-phosphate dependent protein kinase, *Biochemistry*, 23, 4200, 1984.
42. Takio, K., Wade, R. D., Smith, S. B., Krebs, E. G., Walsh, K. A., and Titani, K., Guanosine cyclic 3′,5′-phosphate dependent protein kinase, a chimeric protein homologous with two separate protein families, *Biochemistry*, 23, 4207, 1984.
43. Parker, P. J., Coussens, L., Tatty, N., Rhee, L., and Ulrich, A., The complete primary structure of protein kinase C — the major phorbol ester receptor, *Science*, 233, 853, 1986.
44. Knopf, J. L., Lee, M.-H., Sultzman, L. A., Kriz, R. W., Loomis, C. R., Hewick, R. M., and Bell, R. M., Cloning and expression of multiple protein kinase C cDNAs, *Cell*, 46, 491, 1986.
45. Takio, K., Blumenthal, D. K., Walsh, K. A., Titani, K., and Krebs, E. G., Amino acid sequence of rabbit skeletal muscle myosin light chain kinase, *Biochemistry*, 25, 8049, 1986.
46. Reimann, E. M., Titani, K., Ericsson, L. H., Wade, R. D., Fischer, E. H., and Walsh, K. A., Homology of the subunit of phosphorylase b kinase with cAMP-dependent protein kinase, *Biochemistry*, 23, 4185, 1984.
47. Bennett, M. K. and Kennedy, M. B., Deduced primary structure of the β-subunit of brain type II Ca^{2+}/calmodulin-dependent protein kinase determined by molecular cloning, *Proc. Natl. Acad. Sci. U.S.A.*, 84, 1794, 1987.
48. Saxena, A., Padmanatha, R., and Glover, C. V. C., Isolation and sequencing of cDNA clones encoding the alpha and beta subunits of *Drosophila melanogaster* casein kinase II, *Mol. Cell. Biol.*, 7, 3409, 1987.
49. Ullrich, A., Coussens, L., Hayflick, J. A., Dull, T. J., Gray, A., Tam, A. W., Lee, Y., Yarden, Y., Libermann, T. A., Schlessinger, J., Downward, J., Mayes, E. L. V., Whittle, N., Waterfield, M., and Seeburg, P. H., Human epidermal growth factor receptor of cDNA sequence and aberrant expression of the amplified gene in S431 epidermal carcinoma cells, *Nature (London)*, 309, 418, 1984.
50. Yamamoto, T., Nishida, T., Miyajima, N., Kawai, S., Oi, T., and Toyoshima, K., The erbB gene of avian erythroblastosis virus is a member of the *src* gene family, *Cell*, 35, 71, 1983.
51. Ullrich, A., Bell, J. R., Chen, E. Y., Herrera, R., Petruzzelli, L. M., Dull, T. J., Gray, A., Coussens, L., Liao, Y.-C., Tsubokawa, M., Mason, A., Seeburg, P. H., Grunfeld, C., Rosen, O. M., and Ramachandran, J., Human insulin receptor and its relationship to the tyrosine kinase family of oncogenes, *Nature (London)*, 313, 756, 1985.
52. Ebina, Y., Ellis, L., Jarnagin, K., Edery, M., Graf, L., Clauser, J.-H. O., Masiarz, F., Kan, Y. W., Goldfine, I. D., Roth, R. A., and Rutter, W. J., The human insulin receptor cDNA: the structural basis for hormone-activated transmembrane signalling, *Cell*, 40, 747, 1985.
53. Van Beveren, C., Galleshaw, J. A., Jonas, V., Berns, A. J. M., Doolittle, R. F., Donoghue, D. J., and Verma, I. M., Nucleotide sequence and formation of the transforming gene of a mouse sarcoma virus, *Nature (London)*, 289, 258, 1981.
54. Cooper, J. A., Gould, K. L., Cartwright, C. A., and Hunter, T., Tyr 527 is phosphorylated in p60[c-src]: implications for regulation, *Science*, 231, 1431, 1986.
55. Monken, C. E. and Gill, G. H., A comparison of the cyclic nucleotide-dependent protein kinases using chemical cleavage at tryptophan and cysteine, *Arch. Biochem. Biophys.*, 240, 888, 1985.
56. Flockhart, D. A. and Corbin, J. D., Regulatory mechanisms in the control of protein kinases, *Crit. Rev. Biochem.*, 12, 133, 1982.
57. Takio, K., Blumenthal, D. K., Edelman, S. M., Walsh, K. A., Krebs, E. G., and Titani, K., Amino acid sequence of an active fragment of rabbit skeletal muscle myosin light chain kinase, *Biochemistry*, 24, 6028, 1985.
58. Mayr, G. W. and Heilmeyer, L. M. G., Jr., Shape and substructure of skeletal muscle myosin light chain kinase, *Biochemistry*, 22, 4316, 1983.
59. Edelman, A. M., Takio, K., Blumenthal, D. K., Hansen, R. S., Walsh, K. A., Titani, K., and Krebs, E. G., Characterization of the calmodulin-binding and catalytic domains in skeletal muscle myosin light chain kinase, *J. Biol. Chem.*, 260, 11275, 1985.
60. Kikkawa, U. and Nishizuka, Y., Protein kinase C, in *The Enzymes: Control by Phosphorylation Part A*, Vol. 17, Boyer, P. D. and Krebs, E. G., Eds., Academic Press, New York, 1986, 167.
61. Gill, G. N., Chen, W. S., Lazar, C. S., Clenney, J. R., Jr., Wiley, H. S., Ingraham, H. A., and Rosenfeld, M. G., Role of intrinsic protein tyrosine kinase in function and metabolism of the epidermal growth factor receptor, *Cold Spring Harbor Symp. Quant. Biol.*, 53, 467, 1988.

62. **White, M. F. and Kahn, C. R.,** The insulin receptor and tyrosine phosphorylation, in *The Enzymes: Control by Phosphorylation Part A,* Vol. 17, Boyer, P. D. and Krebs, E. G., Eds., Academic Press, New York, 1986, 248.
63. **Rosen, O. M.,** After insulin binds, *Science,* 237, 1452, 1987.
64. **Ronnett, G. V., Knutson, V. P., Kohanski, R. A., Simpson, T. L., and Lane, M. D.,** Role of glycosylation in processing of newly translated insulin proreceptor in 3T3-L1 adipocytes, *J. Biol. Chem.,* 259, 4566, 1984.
65. **Riedel, H., Dull, D. J., Schlesinger, J., and Ullrich, A.,** A chimeric receptor allows insulin to stimulate tyrosine kinase activity of epidermal growth factor receptor, *Nature (London),* 324, 68, 1986.
66. **Bertics, P. J. and Gill, G. N.,** Self-phosphorylation enhances the protein-tyrosine kinase activity of the epidermal growth factor receptor, *J. Biol. Chem.,* 260, 14642, 1985.
67. **White, M. F., Maron, R., and Kahn, C. R.,** Insulin rapidly stimulates tyrosine phosphorylation of a M_r-185,000 protein in intact cells, *Nature (London),* 318, 183, 1985.
68. **Ellis, L., Clauser, E., Morgan, D. O., Edery, M., Roth, R. A., and Rutter, W. J.,** Replacement of insulin receptor tyrosine residues 1162 and 1163 compromises insulin-stimulated kinase activity and uptake of 2-deoxyglucose, *Cell,* 45, 721, 1986.
69. **Hunter, T.,** A tail of two *src*'s: mutatis mutandis, *Cell,* 49, 1, 1987.
70. **Cooper, J. A., Gould, K. L., Cartwright, C. A., and Hunter, T.,** Tyr^{527} is phosphorylated in $pp60^{c-src}$: implications for regulation, *Science,* 231, 1431, 1986.
71. **Kmiecik, T. E. and Shalloway, D.,** Activation and suppression of $pp60^{c-src}$ transforming ability by mutation of its primary sites of tyrosine phosphorylation, *Cell,* 49, 65, 1987.
72. **Piwnica-Worms, H., Saunder, K. B., Roberts, T. M., Smith, A. E., and Cheng, S. H.,** Tyrosine phosphorylation regulates the biochemical and biological properties of $pp60^{c-src}$, *Cell,* 49, 75, 1987.
73. **Cartwright, C. A., Eckhart, W., Simon, S., and Kaplan, P. L.,** Cell transformation by $pp60^{c-src}$ mutated in the carboxy-terminal regulatory domain, *Cell,* 49, 83, 1987.
74. **Bramson, H. N., Mildvan, A. S., and Kaiser, E. T.,** Mechanistic studies of cAMP-dependent protein kinase action, *Crit. Rev. Biochem.,* 15, 93, 1984.
75. **Hoppe, J., Marutsky, R., and Freist, W.,** Mechanism of activation of the protein kinase I from rabbit skeletal muscle, *Eur. J. Biochem.,* 80, 369, 1977.
76. **Hoppe, J., Freist, W., Marutzky, R., and Shaltiel, S.,** Mapping the ATP binding site in the catalytic subunit of cAMP-dependent protein kinase: spatial relationships with the ATP binding site of the undissociated enzyme, *Eur. J. Biochem.,* 90, 427, 1978.
77. **Flockhart, D. A., Freist, W., Hoppe, J., Lincoln, T. M., and Corbin, J. D.,** ATP analog specificity of cAMP-dependent protein kinase, cGMP-dependent protein kinase, and phosphorylase kinase, *Eur. J. Biochem.,* 140, 289, 1985.
78. **Kwiatkowski, A. and King, M. M.,** Mapping of the adenosine 5'-triphosphate binding site of type II calmodulin-dependent protein kinase, *Biochemistry,* 26, 7636, 1987.
79. **Vogel, S., Freist, W., and Hoppe, J.,** Assignment of conserved amino acid residues to the ATP site in the protein kinase domain of the receptor for epidermal growth factor, *Eur. J. Biochem.,* 154, 529, 1986.
80. **Bhatnagar, D., Roskoski, R., Jr., Rosendahl, M. S., and Leonard, N. J.,** Adenosine cyclic 3',5'-monophosphate dependent kinase: a new fluorescence displacement titration technique for characterizing the nucleotide binding site on catalytic subunit, *Biochemistry,* 22, 6310, 1983.
81. **Hartl, F. T., Roskoski, R., Jr., Rosendahl, M. S., and Leonard, N. J.,** Adenosine cyclic 3',5'-monophosphate dependent protein kinase: interaction of the catalytic subunit and holoenzyme with *lin*-benzoadenine nucleotides, *Biochemistry,* 22, 2347, 1983.
82. **Bhatnagar, D., Glass, D. B., Roskoski, R., Jr., Lessor, R. A., and Leonard, N. J.,** Interaction of guanosine cyclic 3',5'-phosphate dependent protein kinase with *lin*-benzoadenine nucleotides, *Biochemistry,* 24, 1122, 1985.
83. **Cheng, A., Fitzgerald, T. J., Bhatnagar, D., Roskoski, R., Jr., and Carlson, G. M.,** Allosteric nucleotide specificity of phosphorylase kinase: correlation of binding, conformational transitions, and activation, *J. Biol. Chem.,* 263, 5534, 1988.
84. **Cheng, A. and Carlson, G. M.,** Competition between nucleoside diphosphates and triphosphates at the catalytic and allosteric sites of phosphorylase kinase, *J. Biol. Chem.,* 263, 5543, 1988.
85. **Cook, P. F., Neville, M. E., Jr., Vrana, K. E., Hartl, F. T., and Roskoski, R. R., Jr.,** Adenosine cyclic 3',5'-monophosphate dependent protein kinase: kinetic mechanism for the bovine skeletal muscle catalytic subunit, *Biochemistry,* 21, 5794, 1982.
86. **Whitehouse, S., Framisco, J. R., Casnellic, J. E., Krebs, E. G., and Walsh, D. A.,** Studies on the kinetic mechanism of the catalytic subunit of the cAMP-dependent protein kinase, *J. Biol. Chem.,* 258, 3693, 1983.
87. **Zoller, M. N. and Taylor, S. S.,** Affinity labeling of the nucleotide binding site of the catalytic subunit of cAMP-dependent protein kinase using *p*-fluorosulfonyl-[^{14}C]benzoyl 5'-adenosine. Identification of a modified lysine residue, *J. Biol. Chem.,* 254, 8363, 1979.

88. **Zoller, M. N., Nelson, N. C., and Taylor, S. S.**, Affinity labeling of cAMP-dependent protein kinase with p-fluorosulfonylbenzoyl adenosine, *J. Biol. Chem.*, 256, 10387, 1981.
89. **Bhatnagar, D., Hartl, F. T., Roskoski, R., Jr., Lessor, R. A., and Leonard, N. J.**, Adenosine cyclic 3′,5′-monophosphate dependent protein kinase: nucleotide binding to chemically modified catalytic subunit, *Biochemistry*, 23, 4350, 1984.
90. **Jacobson, M. A. and Colman, R. F.**, Evaluation of the intramolecular stacking of the fluorosulfonylbenzoyl derivatives of 1,N^6-ethenoadenosine, adenosine, and guanosine, *J. Biol. Chem.*, 259, 1454, 1984.
91. **Kamps, M. P., Taylor, S. S., and Sefton, B. M.**, Oncogenic tyrosine kinases and cAMP-dependent protein kinases have homologous ATP binding sites, *Nature (London)*, 310, 589, 1984.
92. **Staros, J. V., Cohen, S., and Russo, M. W.**, Epidermal growth factor receptor: characterization of its protein kinase activity, in *Molecular Mechanisms of Transmembrane Signaling*, Cohen, P., Ed., Elsevier, Amsterdam, 1985, 253.
93. **Kamps, M. P. and Sefton, B. M.**, Neither arginine nor histidine can carry out the function of lys 295 in the ATP binding site of p60^{src}, *Mol. Cell. Biol.*, 6, 751, 1986.
94. **Snyder, M., Bishop, J. M., McGrath, J., and Levinson, A.**, A mutation at the ATP-binding site of pp60^{v-src} abolishes kinase activity, transformation, and tumorigenicity, *Mol. Cell. Biol.*, 5, 1772, 1985.
95. **Weinmaster, G., Zoller, M. J., and Pawson, T.**, A lysine in the ATP-binding site of P130gag-fps is essential for protein-tyrosine kinase activity, *EMBO J.*, 5, 69, 1986.
96. **Denhez, F., Heimann, B., d'Auriol, L., Graf, T., Coquillaud, M., Coll, J., Galibert, F., Moeling, K., Stehenlin, D., and Ghysdael, J.**, Replacement of lys 622 in the ATP binding domain of P100$^{gag-mil}$ abolishes the in vitro autophosphorylation of the protein and the biological properties of the *v-mil* oncogene of MH2 virus, *EMBO J.*, 7, 541, 1988.
97. **Chen, W. S., Lazar, C. S., Poenie, M., Tsien, R. Y., Gill, G. N., and Rosenfeld, M. G.**, Requirement for intrinsic protein tyrosine kinase in the immediate and late actions of the EGF receptor, *Nature (London)*, 328, 820, 1987.
98. **Glenney, J. R., Jr., Che, W. S., Lazar, C. S., Walton, G. M., and Gill, G. N.**, Ligand-induced endocytosis of the EGF receptor is blocked by mutational inactivation and by microinjection of anti-phosphotyrosine antibodies, *Cell*, 52, 675, 1988.
99. **Honegger, A. M., Dull, T. J., Felder, S., Van Obberghen, E., Bellot, F., Szapary, D., Schmidt, A., Ullrich, A., and Schlessinger, J.**, Point mutation at the ATP binding site of EGF receptor abolishes protein-tyrosine kinase activity and alters cellular routing, *Cell*, 51, 199, 1987.
100. **Chou, C. K., Dull, T. J., Russell, D. S., Gherzi, R., Lobwohl, D., Ullrich, A., and Rosen, O.**, Human insulin receptor mutated at the ATP binding site lacks protein tyrosine-kinase activity and fails to mediate post-receptor effects of insulin, *J. Biol. Chem.*, 262, 1842, 1987.
101. **Ebina, Y., Araki, E., Taira, M., Shimada, F., Mori, M., Craik, C. S., Sidle, K., Pierce, S. B., Roth, R. A., and Rutter, W. J.**, Replacement of lysine residue 1030 in the putative ATP-binding region of the insulin receptor abolishes insulin- and antibody-stimulated glucose uptake and receptor kinase activity, *Proc. Natl. Acad. Sci. U.S.A.*, 84, 704, 1987.
102. **Rossmann, M. G., Moras, D., and Olsen, K.**, Chemical and biological evolution of nucleotide binding proteins, *Nature (London)*, 250, 194, 1974.
103. **Feramisco, J. R., Glass, D. B., and Krebs, E. G.**, Optimal spatial requirements for the location of basic residues in peptide substrates for the cyclic AMP-dependent protein kinase, *J. Biol. Chem.*, 255, 4240, 1980.
104. **Kemp, B. E., Graves, D. J., Benjamini, E., and Krebs, E. G.**, Role of multiple basic residues in determining the substrate specificity of cyclic AMP-dependent protein kinase, *J. Biol. Chem.*, 252, 4888, 1977.
105. **Feramisco, J. R. and Krebs, E. G.**, Inhibition of cAMP-dependent protein kinase by analogs of a synthetic peptide substrate, *J. Biol. Chem.*, 253, 8968, 1978.
106. **Zetterqvist, Ö., Ragnarsson, U., Humble, E., Berglund, L., and Engström, L.**, The minimum substrate of cyclic AMP-stimulated protein kinase, as studied by synthetic peptides representing the phosphorylatable site of pyruvate kinase (type L) of rat liver, *Biochem. Biophys. Res. Commun.*, 70, 696, 1976.
107. **Feramisco, J. R., Kemp, B. E., and Krebs, E. G.**, Phosphorylation of hydroxyproline in a synthetic peptide catalyzed by cyclic AMP-dependent protein kinase, *J. Biol. Chem.*, 254, 6987, 1979.
108. **Kemp, B. E. and Pearson, R. B.**, Spatial requirements for location of basic residues in peptide substrates for smooth muscle myosin light chain kinase, *J. Biol. Chem.*, 260, 3355, 1985.
109. **Michnoff, C. H., Kemp, B. E., and Stull, J. T.**, Phosphorylation of synthetic peptides by skeletal muscle myosin light chain kinases, *J. Biol. Chem.*, 261, 8320, 1986.
110. **Turner, R. S., Kemp, B. E., Su, H.-D., and Kuo, J. F.**, Substrate specificity of phospholipid/Ca^{2+}-dependent protein kinase as probed with synthetic peptide fragments of the bovine myelin basic protein, *J. Biol. Chem.*, 260, 11503, 1985.
111. **House, C., Wettenhall, R. E. H., and Kemp, B. E.**, The influence of basic residues on the substrate specificity of protein kinase C, *J. Biol. Chem.*, 262, 772, 1987.

112. **Ferrari, S., Marchiori, F., Borin, G., and Pinna, L. A.**, Distinct structural requirements of Ca^{2+} / phospholipid-dependent protein kinase (protein kinase C) and cAMP-dependent protein kinase as evidenced by synthetic peptide substrates, *FEBS Lett.*, 184, 72, 1985.
113. **Meggio, F., Marchiori, F., Borin, G., Chessa, G., and Pinna, L. A.**, Synthetic peptides including acidic clusters as substrates and inhibitors of rat liver casein kinase TS (type-2), *J. Biol. Chem.*, 259, 14576, 1984.
114. **Patschinsky, T., Hunter, T., Esch, F. S., Cooper, J. A., and Sefton, B. M.**, Analysis of the sequence of amino acids surrounding sites of tyrosine phosphorylation, *Biochemistry*, 79, 973, 1982.
115. **Hunter, T.**, Synthetic peptide substrates for a tyrosine protein kinase, *J. Biol. Chem.*, 257, 4843, 1982.
116. **Bramson, H. N., Thomas, N., Matsueda, R., Nelson, N. C., Taylor, S. S., and Kaiser, E. T.**, Modification of the catalytic subunit of bovine heart cAMP-dependent protein kinase with affinity labels related to peptide substrates, *J. Biol. Chem.*, 257, 10575, 1983.
117. **Miller, W. T. and Kaiser, E. T.**, Probing the peptide binding site of the cAMP-dependent protein kinase using a novel photoaffinity label, *Proc. Natl. Acad. Sci. U.S.A.*, 85, 5429, 1988.
118. **Peters, K. A., Demaille, J. G., and Fischer, E. H.**, Adenosine 3':5'-monophosphate-dependent protein kinase from bovine heart: characterization of the catalytic subunit, *Biochemistry*, 16, 5691, 1977.
119. **Severin, E. S., Sushchenko, L. P., Kochetkov, S. N., and Karochkin, S. N.**, Structure and function of protein kinase from pig brain, *Adv. Cyclic Nucleotide Res.*, 9, 171, 1978.
120. **Bechtel, P. J., Beavo, J. A., and Krebs, E. G.**, Purification and characterization of catalytic subunit of skeletal muscle adenosine 3':5'-monophosphate-dependent protein kinase, *J. Biol. Chem.*, 252, 2691, 1977.
121. **Sugden, P. H., Holladay, L. A., Reimann, E. H., and Corbin, J. D.**, Purification and characterization of the catalytic subunit of adenosine 3':5'-cyclic monophosphate-dependent protein kinase, *Biochem. J.*, 159, 409, 1976.
122. **Armstrong, R. N. and Kaiser, E. T.**, Sulfhydryl group reactivity of adenosine 3',5'-monophosphate-dependent protein kinase from bovine heart: characterization of the catalytic subunit, *Biochemistry*, 17, 2840, 1978.
123. **Nelson, N. C. and Taylor, S. S.**, Differential labeling and identification of the cysteine-containing tryptic peptides of catalytic subunit from porcine heart cAMP-dependent protein kinase, *J. Biol. Chem.*, 256, 3743, 1981.
124. **Kupfer, A., Jiménez, J. S., Gottlieb, P., and Shaltiel, S.**, On the protein accommodating site of the catalytic subunit of adenosine cyclic 3',5'-phosphate dependent protein kinase, *Biochemistry*, 21, 1631, 1982.
125. **Jiménez, J. S., Kupfer, A., Gani, V., and Shaltiel, S.**, Salt-induced conformational changes in the catalytic subunit of adenosine cyclic 3',5'-phosphate dependent protein kinase. Use for establishing a connection between one sulfhydryl group and the γ-P subsite in the ATP site of this subunit, *Biochemistry*, 21, 1623, 1982.
126. **Hartl, F. T. and Roskowski, R., Jr.**, Adenosine 3',5'-cyclic monophosphate protein kinase from bovine brain: inactivation of the catalytic subunit and holoenzyme by 7-chloro-4-nitro-2,1,3-benzoxadiazole, *Biochemistry*, 21, 5175, 1982.
127. **Nelson, N. C. and Taylor, S. S.**, Selective protection of sulfhydryl groups in cAMP-dependent protein kinase II, *J. Biol. Chem.*, 258, 10981, 1983.
128. **First, E. A. and Taylor, S. S.**, Selective modification of the catalytic subunit of cAMP-dependent protein kinase with sulfhydryl-specific fluorescent probes, *Biochemistry*, 28, 3589, 1989.
129. **Rajinder, N. P., Bhatnagar, D., and Roskoski, R., Jr.**, Adenosine cyclic 3',5'-monophosphate dependent protein kinase: fluorescent affinity labeling of the catalytic subunit from bovine skeletal muscle with o-phthalaldehyde, *Biochemistry*, 24, 6499, 1985.
130. **Shoji, S., Titani, K., Demaille, J. G., and Fischer, E. H.**, Sequence of two phosphorylated sites in the catalytic subunit of bovine cardiac muscle adenosine 3':5'-monophosphate-dependent protein kinase, *J. Biol. Chem.*, 254, 6211, 1979.
131. **Shoji, S., Parmelee, D. C., Wade, R. D., Kumar, S., Ericsson, L. H., Walsh, K. A., Neurath, H., Long, G. L., Demaille, J. G., Fischer, E. H., and Titani, K.**, Complete amino acid sequence of the catalytic subunit of bovine cardiac muscle cyclic AMP-dependent protein kinase, *Proc. Natl. Acad. Sci. U.S.A.*, 78, 848, 1981.
132. **Chiu, Y. S. and Tao, M.**, Autophosphorylation of rabbit skeletal muscle cyclic AMP-dependent protein kinase I catalytic subunit, *J. Biol. Chem.*, 253, 7145, 1978.
133. **Cartwright, C. A., Kaplan, P. L., Cooper, J. A., Hunter, T., and Eckhart, W.**, Altered sites of tyrosine phosphorylation in pp60^{c-src} associated with polyoma middle tumor antigen, *Mol. Cell. Biol.*, 6, 1562, 1986.
134. **Granot, J., Mildvan, A. S., Bramson, H. N., Thomas, N., and Kaiser, E. T.**, Nuclear magnetic resonance studies of the conformation and kinetics of the peptide-substrate at the active site of bovine heart protein kinase, *Biochemistry*, 20, 602, 1981.

135. **Rosevear, P. R., Fry, D. C., Mildvan, A. S., Doughty, M., O'Brian, C., and Kaiser, E. T.**, NMR studies of the backbone protons and secondary structure of pentapeptide and heptapeptide substrates bound to bovine heart protein kinase, *Biochemistry*, 23, 3161, 1984.
136. **Bramson, H. N., Thomas, N. E., and Kaiser, E. T.**, The use of N-methylated peptides and depsipeptides to probe the binding of heptapeptide substrates to cAMP-dependent protein kinase, *J. Biol. Chem.*, 260, 15452, 1985.
137. **Walsh, D. A., Ashby, C. D., Gonzalez, C., Calkins, D., Fischer, E. H., and Krebs, E. G.**, Purification and characterization of a protein inhibitor of adenosine 3',5'-monophosphate-dependent protein kinases, *J. Biol. Chem.*, 246, 1977, 1971.
138. **Hofmann, F., Beavo, J. A., Bechtel, P. J., and Krebs, E. G.**, Comparison of adenosine 3':5'-monophosphate-dependent protein kinases from rabbit skeletal and bovine heart muscle, *J. Biol. Chem.*, 250, 7795, 1975.
139. **Rosen, O. M. and Erlichman, J.**, Reversible autophosphorylation of cyclic 3':5'-AMP-dependent protein kinase from bovine cardiac muscle, *J. Biol. Chem.*, 250, 7788, 1975.
140. **Corbin, J. D. and Keely, S. L.**, Characterization and regulation of heart adenosine 3':5'-monophosphate-dependent protein kinase isozymes, *J. Biol. Chem.*, 252, 910, 1977.
141. **Lee, D. C., Carmichael, D. F., Krebs, E. G., and McKnight, G. S.**, Isolation of a cDNA clone for the type I regulatory subunit of bovine cAMP-dependent protein kinase, *Proc. Natl. Acad. Sci. U.S.A.*, 80, 3608, 1983.
142. **Jahsen, T., Hedin, L., Kidd, V. J., Beattie, W. G., Lohmann, S. M., Walter, V., Durica, J., Schulz, T. Z., Schlitz, E., Browner, M., Lawrence, C. B., Goldman, D., Ratoosh, S. L., and Richards, J. S.**, Molecular cloning, cDNA structure, and regulation of the regulatory subunit of type II cAMP-dependent protein kinase from rat ovarian granulosa cells, *J. Biol. Chem.*, 261, 12352, 1986.
143. **Scott, J. D., Glaccum, M. B., Zoller, M. J., Uhler, M. D., Helfman, D. M., McKnight, G. S., and Krebs, E. G.**, The molecular cloning of a type II regulatory subunit of the cAMP-dependent protein kinase from rat skeletal muscle and mouse brain, *Proc. Natl. Acad. Sci. U.S.A.*, 84, 5192, 1987.
144. **Erlichman, J., Rosenfeld, R., and Rosen, O. M.**, Phosphorylation of a cyclic adenosine 3':5'-monophosphate-dependent protein kinase from bovine cardiac muscle, *J. Biol. Chem.*, 249, 5000, 1974.
145. **Zoller, M. J., Kerlavage, A. R., and Taylor, S. S.**, Structural comparisons of cAMP-dependent protein kinases I and II from porcine skeletal muscle, *J. Biol. Chem.*, 254, 2408, 1979.
146. **Hofmann, F.**, Apparent constants for the interaction of regulatory and catalytic subunit of cAMP-dependent protein kinase I and II, *J. Biol. Chem.*, 255, 1559, 1980.
147. **Weber, I. T., Steitz, T. A., Bubis, J., and Taylor, S. S.**, Predicted structures of the cAMP-dependent protein kinase, *Biochemistry*, 26, 343, 1987.
148. **Weber, I. T., Takio, K., Titani, K., and Steitz, T. A.**, The cAMP-binding domains of the regulatory subunit of cAMP-dependent protein kinase and the catabolite gene activator protein are homologous, *Proc. Natl. Acad. Sci. U.S.A.*, 79, 7679, 1982.
149. **Mumby, M. C., Weldon, S. L., Scott, C. W., and Taylor, S. S.**, Monoclonal antibodies as probes of structure, function and isoenzyme forms of the type II regulatory subunit of cyclic AMP-dependent protein kinase, *Pharm. Ther.*, 28, 367, 1985.
150. **Potter, R. L. and Taylor, S. S.**, Relationships between structural domains and function in the regulatory subunit of cAMP-dependent protein kinases I and II from porcine skeletal muscle, *J. Biol. Chem.*, 254, 9000, 1979.
151. **Taylor, S. S., Kerlavage, A. R., Zoller, M. J., Nelson, N. C., and Potter, R. L.**, Nucleotide-binding sites and structural domains of cAMP-dependent protein kinases, in *Cold Spring Harbor Symp. Protein Phosphorylation*, Book A, Vol. 8, Cold Spring Harbor Laboratory, Cold Spring Harbor, NY, 1981, 3.
152. **Rannels, S. R. and Corbin, J. D.**, Studies of functional domains of the regulatory subunit from cAMP-dependent protein kinase isozyme I, *J. Cyclic Nucleotide Res.*, 6, 201, 1980.
153. **Takio, K., Walsh, K. A., Neurath, H., Smith, S. B., Krebs, E. G., and Titani, K.**, The amino acid sequence of a hinge region in the regulatory subunit of bovine cardiac muscle cyclic AMP-dependent protein kinase II, *FEBS Lett.*, 114, 83, 1980.
154. **Potter, R. L. and Taylor, S. S.**, Correlation of the cAMP binding domain with a site of autophosphorylation on the regulatory subunit of cAMP-dependent protein kinase II from porcine skeletal muscle, *J. Biol. Chem.*, 254, 9000, 1979.
155. **Rangel-Aldao, R. and Rosen, O. M.**, Mechanism of self-phosphorylation of adenosine 3':5'-monophosphate-dependent protein kinase from bovine cardiac muscle, *J. Biol. Chem.*, 251, 7526, 1976.
156. **Rangel-Aldao, R. and Rosen, O. M.**, Dissociation and reassociation of the phosphorylated and non-phosphorylated forms of adenosine 3':5'-monophosphate-dependent protein kinase from bovine cardiac muscle, *J. Biol. Chem.*, 251, 3375, 1976.
157. **Rangel-Aldao, R. and Rosen, O. M.**, Effect of cAMP and ATP on the reassociation of phosphorylated and nonphosphorylated subunits of the cAMP-dependent protein kinase from bovine cardiac muscle, *J. Biol. Chem.*, 252, 7140, 1977.

158. **Durgerian, S. and Taylor, S. S.**, The consequences of introducing an autophosphorylation site into the type I regulatory subunit of cAMP-dependent protein kinase, *J. Biol. Chem.*, 264, 9807, 1989.
159. **Beebe, S. J. and Corbin, J. D.**, Cyclic nucleotide-dependent protein kinases, in *The Enzymes*, Vol. 17, Boyer, P. D. and Krebs, E. G., Eds., Academic Press, San Diego, 1986, 44.
160. **First, E. A., Bubis, J., and Taylor, S. S.**, Subunit interaction sites between the regulatory and catalytic subunits of cAMP-dependent protein kinase. Identification of a specific interchain disulfide bond, *J. Biol. Chem.*, 263, 5176, 1988.
161. **Weldon, S. L. and Taylor, S. S.**, Monoclonal antibodies as probes for functional domains in cAMP-dependent protein kinase II, *J. Biol. Chem.*, 260, 4203, 1985.
162. **Levin, L. R., Kuret, J., Johnson, K. E., Powers, S., Cameron, S., Michaeli, T., Wigler, M., and Zoller, M. J.**, A mutation in the catalytic subunit of cAMP-dependent protein kinase that disrupts regulation, *Science*, 240, 68, 1988.
163. **Kuret, J., Johnson, K. E., Nicolette, C., and Zoller, M. J.**, Mutagenesis of the regulatory subunit of yeast cAMP-dependent protein kinase. Isolation of site-directed mutants with altered binding affinity for catalytic subunit, *J. Biol. Chem.*, 263, 9149, 1988.
164. **Demaile, J. G., Peters, K. A., and Fischer, E. H.**, Isolation and properties of the rabbit skeletal muscle protein inhibitor of adenosine 3',5'-monophosphate dependent protein kinases, *Biochemistry*, 16, 3080, 1977.
165. **Scott, J. D., Fischer, E. H., Takio, K., Demaille, J. G., and Krebs, E. G.**, Amino acid sequence of the heat-stable inhibitor of the cAMP-dependent protein kinase from rabbit skeletal muscle, *Proc. Natl. Acad. Sci. U.S.A.*, 82, 5732, 1985.
166. **Cheng, H.-C., VanPatten, S. M., Smith, A. J., and Walsh, D. A.**, An active twenty-amino-acid-residue peptide derived from the inhibitor protein of the cyclic AMP-dependent protein kinase, *Biochem. J.*, 231, 655, 1985.
167. **Cheng, H.-C., Kemp, B. E., Pearson, R. B., Smith, A. J., Misconi, L., VanPatten, S. M., and Walsh, D. A.**, A potent synthetic peptide inhibitor of the cAMP-dependent protein kinase, *J. Biol. Chem.*, 261, 989, 1986.
168. **Scott, J. D., Fischer, E. H., Demaille, J. G., and Krebs, E. G.**, Identification of an inhibitory region of the heat-stable protein inhibitor of the cAMP-dependent protein kinase, *Proc. Natl. Acad. Sci. U.S.A.*, 82, 4379, 1985.
169. **Scott, J. D., Glaccum, M. B., Fischer, E. H., and Krebs, E. G.**, Primary-structure requirements for inhibition by the heat-stable inhibitor of the cAMP-dependent protein kinase, *Proc. Natl. Acad. Sci. U.S.A.*, 83, 1613, 1986.
170. **Glass, D. B., Cheng, H.-C., Mende-Mueller, L., Reed, J., and Walsh, D. A.**, Primary structural determinants essential for potent inhibition of cAMP-dependent protein kinase by inhibitory peptides corresponding to the active portion of the heat-stable inhibitor protein, *J. Biol. Chem.*, 264, 8802, 1989.
171. **Reed, J., Kinzel, V., Cheng, H.-C., and Walsh, D. A.**, Circular dichroic investigations of secondary structure in synthetic peptide inhibitors of cAMP-dependent protein kinase: a model for inhibitory potential, *Biochemistry*, 26, 7641, 1987.
172. **House, C. and Kemp, B. E.**, Protein kinase C contains a pseudosubstrate prototype in its regulatory domain, *Science*, 238, 1726, 1987.
173. **Lee, M.-H. and Bell, R. M.**, The lipid binding, regulatory domain of protein kinase C, *J. Biol. Chem.*, 261, 14867, 1986.
174. **Kemp, B. E., Pearson, R. B., Guerriero, V., Jr., Bagchi, I. C., and Means, A. R.**, The calmodulin binding domain of chicken smooth muscle myosin light chain kinase contains a pseudosubstrate sequence, *J. Biol. Chem.*, 262, 2542, 1987.
175. **Matsuo, M., Huang, C.-H., and Huang, L. C.**, Modification and identification of glutamate residues at the arginine-recognition site in the catalytic subunit of adenosine 3':5'-cyclic monophosphate-dependent protein kinase of rabbit skeletal muscle, *Biochem. J.*, 187, 371, 1980.
176. **Solioz, M.**, Dicyclohexylcarbodiimide as a probe for proton translocating enzymes, *Trends Biochem. Sci.*, 103, 309, 1984.
177. **Satre, M., Lunardi, J., Pougeois, R., and Vignais, P. V.**, Inactivation of *Escherichia coli* BF_1-ATPase by dicyclohexylcarbodiimide. Chemical modification of the β subunit, *Biochemistry*, 18, 3134, 1979.
178. **Toner-Webb, J. and Taylor, S. S.**, Inhibition of the catalytic subunit of cAMP-dependent protein kinase by dicyclohexylcarbodiimide, *Biochemistry*, 26, 7371, 1987.
179. **Buechler, J. A. and Taylor, S. S.**, Identification of aspartate 184 as an essential residue in the catalytic subunit of cAMP-dependent protein kinase, *Biochemistry*, 27, 7356, 1988.
180. **Yoon, M. Y. and Cook, P. F.**, Chemical mechanism of the adenosine cyclic 3',5'-monophosphate dependent protein kinase from pH studies, *Biochemistry*, 26, 4118, 1987.
181. **Anderson, C. M., Stenkamp, R. E., McDonald, R. C., and Steitz, T. A.**, A refined model of the sugar binding site of yeast hexokinase B, *J. Mol. Biol.*, 123, 207, 1978.

182. **Stachelek, C., Stachelek, J., Swain, J., Botstein, D., and Konigsbeng, W.**, Identification, cloning, and sequence determination of the genes specifying hexokinase A and B from yeast, *Nucleic Acids Res.*, 14, 945, 1986.
183. **Shirakihara, Y. and Evans, P. R.**, Crystal structure of the complex of phosphofructokinase from *Escherichia coli* with its reaction products, *J. Mol. Biol.*, 204, 973, 1988.
184. **Viola, R. E. and Cleland, W. W.**, Use of pH studies to elucidate the chemical mechanism of yeast hexokinase, *Biochemistry*, 17, 4111, 1978.
185. **Armstrong, R. N., Kondo, J., Granot, J., Kaiser, E. T., and Mildvan, A. S.**, Magnetic resonance and kinetic studies of the manganous (II) ion and substrate complexes of the catalytic subunit of adenosine 3′,5′-monophosphate dependent protein kinase from bovine heart, *Biochemistry*, 18, 1230, 1979.
186. **Granot, J. and Mildvan, A. S.**, Specificity of bovine heart protein kinase for the Δ-stereoisomer of the metal-ATP complex, *FEBS Lett.*, 103, 265, 1979.
187. **Dreusicke, D., Karplus, R. A., and Schulz, G. E.**, Refined structure of porcine cytosolic adenylate kinase at 2.1 Å resolution, *J. Mol. Biol.*, 199, 359, 1988.
188. **Buechler, J. A. and Taylor, S. S.**, Dicyclohexylcarbodiimide crosslinks two essential residues at the active site of the catalytic subunit of cAMP-dependent protein kinase: Asp 184 and Lys 72, *Biochemistry*, 28, 2065, 1988.
189. **Bolen, D. W., Stingelin, J., Bramson, H. N., and Kaiser, E. T.**, Stereochemical and kinetic studies on the action of the catalytic subunit of bovine cardiac muscle adenosine 3′,5′-monophosphate dependent protein kinase using metal ion complexes of ATP βS, *Biochemistry*, 19, 1176, 1980.
190. **Pomerantz, A. H., Allfrey, V. G., Merrifield, R. D., and Johnson, E. M.**, Studies on the mechanism of phosphorylation of synthetic polypeptides by a calf thymus cAMP-dependent protein kinase, *Proc. Natl. Acad. Sci. U.S.A.*, 74, 4261, 1977.
191. **Whitehouse, S. and Walsh, D. A.**, MgATP^{-2}-dependent interaction of the inhibitor protein of the cAMP-dependent protein kinase with the catalytic subunit, *J. Biol. Chem.*, 258, 3682, 1983.
192. **Van Patten, S. M. V., Fletcher, W. H., and Walsh, D. A.**, The inhibitor protein of the cAMP-dependent protein kinase-catalytic subunit interaction, *J. Biol. Chem.*, 261, 5514, 1986.
193. **Armstrong, R. N., Kondo, H., and Kaiser, E. T.**, Cyclic AMP-dependent ATPase activity of bovine heart protein kinase, *Proc. Natl. Acad. Sci. U.S.A.*, 76, 722, 1979.
194. **Bennett, W. S. and Steitz, T. A.**, Glucose-induced conformational change in yeast hexokinase, *Proc. Natl. Acad. Sci. U.S.A.*, 75, 4848, 1978.
195. **Rosevear, P. R., Bramson, H. N., O'Brian, C., Kaiser, E. T., and Mildvan, E. T.**, Nuclear overhauser effect studies on the conformations of tetraamminecobalt(III)-adenosine 5′-triphosphate free and bound to bovine heart protein kinase, *Biochemistry*, 22, 3439, 1983.
196. **Granot, J., Mildvan, A. S., Bramson, H. N., and Kaiser, E. T.**, Magnetic resonance measurements of intersubstrate distances at the active site of protein kinase using substitution-inert cobalt(III) and chromium (III) complexes of adenosine 5′-(β-methylenetriphosphate), *Biochemistry*, 19, 3537, 1980.
197. **Ho, M.-F., Bramson, H. N., Hansen, D. E., Knowles, J. R., and Kaiser, E. T.**, Stereochemical course of the phospho group transfer catalyzed by cAMP-dependent protein kinase, *J. Am. Chem. Soc.*, 110, 2680, 1988.
198. **Reed, J. and Kinzel, V.**, Near- and far-ultraviolet circular dichroism of the catalytic subunit of adenosine cyclic 5′-monophosphate dependent protein kinase, *Biochemistry*, 23, 1357, 1984.
199. **Reed, J. and Kinzel, V.**, Ligand binding site interaction in adenosine cyclic 3′,5′-monophosphate dependent protein kinase catalytic subunit: circular dichroic evidence for intramolecular transmission of conformational change, *Biochemistry*, 23, 968, 1984.
200. **Reed, J., Kinzel, V., Kemp, B. E., Cheng, H.-C., and Walsh, D. A.**, Circular dichroic evidence for an ordered sequence of ligand/binding site interactions in the catalytic reaction of the cAMP-dependent protein kinase, *Biochemistry*, 24, 2967, 1985.
201. **Van Patten, S. M., Fletcher, W. H., and Walsh, D. A.**, The inhibitor protein of the cAMP-dependent protein kinase-catalytic subunit interaction. Parameters of complex formation, *J. Biol. Chem.*, 261, 5514, 1986.
202. **Kochevar, L. E., Huang, L. C., and Huang, C.-H.**, The function of Mg-ATP in interactions between the regulatory and catalytic subunits of type I cAMP-dependent protein kinase from rabbit skeletal muscle, *Int. J. Biochem.*, 18, 519, 1986.
203. **First, E. A., Johnson, D., and Taylor, S. S.**, Fluorescence energy transfer between cysteine 199 and cysteine 343: evidence for MgATP dependent conformational change in the catalytic subunit of cAMP-dependent protein kinase, *Biochemistry*, 28, 3606, 1988.
204. **Sternberg, M. J. E. and Taylor, W. R.**, Modelling the ATP-binding site of oncogene products, the epidermal growth factor receptor and related proteins, *FEBS Lett.*, 175, 1821, 1984.
205. **Fry, D. C., Stephen, A. K., and Mildvan, A. S.**, ATP-binding site of adenylate kinase: mechanistic implications of its homology with *ras*-encoded p21, F_1-ATPase, and other nucleotide-binding proteins, *Proc. Natl. Acad. Sci. U.S.A.*, 83, 907, 1986.

206. **Wierenga, R. K., Terpstra, P., and Hol, W. G. J.**, Prediction of the occurrence of the ADP-binding βαβ-fold in proteins, using an amino acid sequence fingerprint, *J. Mol. Biol.*, 187, 101, 1986.
207. **Buechler, J. A., Vedvick, T. A., and Taylor, S. S.**, Differential labeling of the catalytic subunit of cAMP-dependent protein kinase with acetic anhydride, *Biochemistry*, 28, 3018, 1989.
208. **Brändén, C.-I.**, Relation between structure and function of α/β-proteins, *Q. Rev. Biophys.*, 13, 317, 1980.
209. **Hol, W. G. J., Van Duijnen, P. T., and Berendsen, H. J. C.**, The α-helix dipole and the properties of proteins, *Nature (London)*, 273, 443, 1978.
210. **Abad-Zapatero, C., Griffith, J. P., Sussman, J. L., and Rossmann, M. G.**, Refined crystal structure of dogfish M^4 apo-lactate dehydrogenase, *J. Mol. Biol.*, 198, 445, 1987.
211. **Winn, S. I., Watson, H. C., Fothergill, L. A., and Harkins, R. N.**, The active site of yeast phosphoglycerate mutase, *Biochem. Soc. Trans.*, 5, 657, 1977.
212. **Jurnak, F.**, Structure of the GDP domain of EF-Tu and location of the amino acids homologous to *ras* oncogene proteins, *Science*, 230, 32, 1985.
213. **deVos, A. M., Tong, L., Milburn, M. V., Matias, P. M., Jancarik, J., Noguchi, S., Nishimura, S., Miura, K., Ohtsuka, E., and Kim, S.-H.**, Three-dimensional structure of an oncogene protein: catalytic domain of human c-H-*ras* p21, *Science*, 239, 888, 1988.
214. **Steitz, T. A., Fletterick, R. J., Anderson, W. F., and Anderson, C. M.**, High resolution X-ray structure of yeast hexokinase, an allosteric protein exhibiting a non-symmetric arrangement of subunits, *J. Mol. Biol.*, 104, 367, 1976.
215. **Steitz, T. A. and Weber, I. T.**, Structure of catabolite gene activator protein, in *Biological Macromolecules and Assemblies*, Vol. 2, Jurnak, F. and McPherson, A., Eds., John Wiley & Sons, New York, 1985, 290.
216. **Masters, S. B., Stroud, R. M., and Bourne, H. R.**, Family of G protein α chains: amphipathic analysis and predicted structure of functional domains, *Protein Eng.*, 1, 47, 1986.
217. **Tong, L., de Vos, A. M., Milburn, M. V., Jancarik, J., Noguchi, S., Nishimura, S., Miura, K., Ohtsuka, E., and Kim, S.-H.**, Structural differences between a *ras* oncogene protein and the normal protein, *Nature (London)*, 337, 90, 1989.
218. **Tong, L., Milburn, M. V., de Vos, A. M., and Kim, S.-H.**, Structure of *ras* protein, *Science*, 245, 244, 1989.
219. **Pai, E. F., Kabsch, W., Krengel, U., Homes, K. C., John, J., and Wittinghofer, A.**, Structure of the guanine-nucleotide-binding domain of the Ha-*ras* oncogene product p21 in the triphosphate, *Nature (London)*, 341, 209, 1989.
220. **Pai, E. F., Sachsenheimer, W., Schirmer, R. H., and Schulz, G. E.**, Substrate positions and induced-fit in crystalline adenylate kinase, *J. Mol. Biol.*, 114, 37, 1977.
221. **Egner, U., Tomasselli, A. G., and Schulz, G. E.**, Structure of the complex of yeast adenylate kinase with the inhibitor P^1,P^5-di(adenosine-5'-)pentaphosphate at 2.6 Å resolution, *J. Mol. Biol.*, 195, 649, 1987.
222. **Mildvan, A. S. and Fry, D. C.**, NMR studies of the mechanism of enzyme action, *Adv. Enzymol.*, 59, 242, 1987.
223. **Fry, D. C., Kuby, S. A., and Mildvan, A. S.**, NMR studies of the MgATP binding site of adenylate kinase and of a 45-residue peptide fragment of the enzyme, *Biochemistry*, 24, 4680, 1985.
224. **Hamada, M., Palmieri, R. H., Russell, G. A., and Kuby, S. A.**, Studies on adenylate triphosphate transphosphorylase. XIV. Equilibrium binding properties of the crystalline rabbit and calf muscle ATP-AMP transphosphorylase (adenylate kinase) and derived peptide fragments, *Arch. Biochem. Biophys.*, 195, 155, 1979.
225. **Richard, J. P. and Frey, P. A.**, Stereochemical course of thiophosphoryl group transfer catalyzed by adenylate kinase, *J. Am. Chem. Soc.*, 100, 7757, 1978.
226. **Schulz, G. E., Schiltz, E., Tomasselli, A. G., Frank, R., Brune, M., Wittinghofer, A., and Schirmer, R. H.**, Structural relationships in the adenylate kinase family, *Eur. J. Biochem.*, 161, 127, 1986.
227. **Sowadski, J. M., Xuong, N.-H., Anderson, D., and Taylor, S. S.**, Crystallization studies of cAMP-dependent protein kinase: crystals of catalytic subunit suitable for high resolution structure determination, *J. Mol. Biol.*, 182, 617, 1985.
228. **Taylor, S. S., Bubis, J., Toner-Webb, J., Saraswat, L. D., First, E. A., Buechler, J. A., Knighton, E. R., and Sowadski, J.**, cAMP-dependent protein kinase: prototype for a family of enzymes, *FASEB J.*, 2, 2677, 1988.
229. **Taylor, S. S., Buechler, J. A., Slice, L. W., Knighton, D. K., Burgerian, S., Ringheim, G. E., Neitzel, J. J., Yonemoto, W. M., Sowadski, J. M., and Dostmann, W.**, cAMP-dependent protein kinase: a framework for a diverse family enzymes, *Cold Spring Harbor Symp. Quant. Biol.*, 53, 121, 1988.
230. **Toner, J. and Taylor, S. S.**, unpublished data.
231. **Payne, M. E., Fong, Y.-L., Ono, T., Colbran, R. J., Kemp. B. E., Soderling, T. R., and Means, A. R.**, Calcium Calmodulin-dependent protein kinase II. Characterization of distinct calmodulin binding and inhibitory domains, *J. Biol. Chem.*, 263, 7190, 1988.
232. **Buechler, J. A. and Taylor, S. S.**, Differential labeling of the catalytic subunit of cAMP-dependent protein kinase with a water soluble carbodiimide: identification of carboxyl groups protected by MgATP and inhibitor peptides, *Biochemistry*, in press, 1989.

Chapter 2

THE INHIBITOR PROTEIN OF THE cAMP-DEPENDENT PROTEIN KINASE

Donal A. Walsh, Karen L. Angelos, Scott M. Van Patten, David B. Glass, and Lawrence P. Garetto

TABLE OF CONTENTS

I. Introduction 44

II. The cAMP-Dependent Protein Kinase 44

III. Purification of the Inhibitor Protein and Its Physicochemical Properties 49
    A. Purification 49
    B. Physicochemical Characteristics 50
    C. Multiple Molecular Forms 51
    D. Relationship of Other Protein Kinase Inhibitors to the Inhibitor Protein 52
        1. Inhibitor Protein from Testis 52
        2. Protein Kinase Modulator 53
        3. Type I and Type II Cyclic Nucleotide-Dependent Protein Kinase Inhibitors 53
        4. Other Protein Kinase Inhibitors 54

IV. Mechanism of Interaction of the Inhibitor Protein with the cAMP-Dependent Protein Kinase 54
    A. Nature of the Inhibitor Protein-Catalytic Subunit Complex 54
    B. Recognition Determinants of the Inhibitor Protein — Studies with Active Peptide Fragments 56
    C. PKI Peptides as a Model for Substrates of the cAMP-Dependent Protein Kinase 64
    D. Three-Dimensional Structure of PKI Peptides 66

V. The Use of Inhibitor Protein as a Tool to Study cAMP Action 71

VI. The Physiological Function of the Inhibitor Protein 72
    A. Experimental Approaches 72
    B. Potential Physiological Functions 74

VII. Appendix 76

Acknowledgments 77

References 78

I. INTRODUCTION

This is the first comprehensive review on the inhibitor protein of the cAMP-dependent protein kinase since its discovery in 1968. One past brief review[1] centered on the results of a single laboratory; the inhibitor protein has been discussed briefly in reviews of the cAMP-dependent protein kinase.[2-5]

The first indications of the inhibitor protein arose in experiments of Posner et al.[6] in 1964, in which it was observed that muscle extracts contained a trypsin-labile factor which interfered with an assay of cAMP based upon the cAMP-dependent activation of phosphorylase kinase. This phenomenon was then investigated by Gonzalez.[7] The cAMP-dependent activation and phosphorylation of "pure" phosphorylase kinase was indeed inhibited by a protein in cell extracts, but since both activation and phosphorylation were only partially cAMP dependent and since the mechanism by which cAMP worked was not known, the resulting complex kinetics proved difficult to analyze. With partial purification of this trypsin-labile factor, it was then shown by Appleman and colleagues[8] to also inhibit cAMP-dependent glycogen synthase inactivation, thus pointing to a commonality of regulation of both the synthesis and degradation of glycogen by cAMP. Further investigation of the phosphorylase kinase activation mechanism, with dissection of cAMP-dependent- and autophosphorylation,[9] then led to the codiscovery of both the cAMP-dependent protein kinase and the inhibitor protein.[10,11] The experiment in which this occurred is depicted in Figure 1, with both the inhibitor protein and the cAMP-dependent protein kinase being provided by treated crude muscle extracts. This experiment showed that the cAMP-dependent activation of phosphorylase kinase, as catalyzed by "endogenous" contaminating cAMP-dependent protein kinase, was inhibited by a crude fraction of inhibitor protein, as was activation catalyzed by exogenous protein kinase provided by the extract. Depending on which was in excess, cAMP-dependent protein kinase or inhibitor protein, the result was either enhanced phosphorylation or full inhibition, respectively. Thus, the inhibition by inhibitor protein showed that the catalyst of endogenous activation was identical to that provoked by exogenously added protein kinase (i.e., the cAMP-dependent protein kinase). Since this initial experiment, the widespread pivotal role of the cAMP-dependent protein kinase in mammalian regulatory mechanisms has been well recognized. In many of the subsequent studies the inhibitor protein has been used as a means to examine cAMP-dependent phosphorylation, both *in vitro*[12] and in intact cells via microinjection,[13-19] liposome fusion,[20] or, recently, by transfection with a pseudo partial cDNA.[21] The particular usefulness of the inhibitor protein in each of these is to distinguish cAMP-mediated phosphorylation from that catalyzed by the myriad of other protein kinases now known to also regulate cell function. The inhibitor protein so far has been found to be absolutely specific for the cAMP-dependent protein kinase.

Two elusive goals that have existed since the original time of its discovery remain to be achieved. The first is to have a form of the inhibitor that could readily be taken up into populations of cells, or even better into tissues in an intact animal, in a directed manner. Perhaps the design for this is close to becoming approachable, both using molecular biological techniques and by direct synthesis (see Section VI.A). The second important question that still remains enigmatic is an understanding of the role of the inhibitor protein in the cell. Direct evidence that the inhibitor protein is a *physiological* modulator of cAMP-mediated function has yet to be obtained. This review provides a progress report of our knowledge of the inhibitor protein, suggestions as to its role, and details of our understanding of its mechanism of interaction with the protein kinase.

II. THE cAMP-DEPENDENT PROTEIN KINASE

Several reviews,[2-4,22] the latest two being in this volume,[23,23a] negate any need here for a detailed description of the protein kinase; however, some selected points directly related

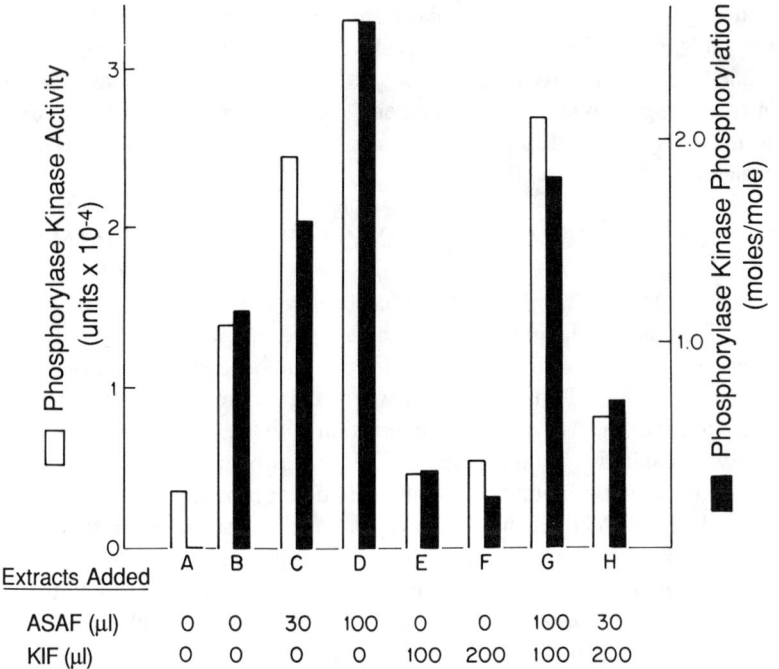

FIGURE 1. Codiscovery of the inhibitor protein and the cAMP-dependent protein kinase — activation, and inhibition of activation, of phosphorylase kinase by cellular extracts containing the inhibitor protein (designated as KIF) and cAMP-dependent protein kinase (designated as ASAF).[9] Full conditions of this experiment have been presented in Reference 9. (A) The activity of nonactivated phosphorylase kinase. All activations were for 10 min. Shown is the cAMP-dependent activation and phosphorylation of phosphorylase kinase by endogenous contaminating protein kinase (B), which is further increased (C, D) by exogenous cAMP-dependent protein kinase provided by a cellular extract (Acid Supernatant Activating Factor) from which phosphorylase kinase had been removed by pH 6.0 precipitation. Both endogenous activation (E, F) and exogenous activation (H) can be inhibited by a crude extract containing inhibitor protein prepared by boiling to remove both cAMP-dependent protein kinase and phosphorylase kinase. High levels of exogenous cAMP-dependent protein kinase, however, overcome the inhibitory action of the inhibitor protein (G).

to inhibitor protein action will be summarized briefly. The structure and mode of activation of the kinase are summarized by the equation

$$R_2C_2 + 4cAMP \leftrightarrow R_2cAMP_4 + 2C \tag{1}$$

where R and C refer to the regulatory and catalytic subunits of the protein kinase, respectively.

Two primary types of regulatory subunit occur, designated R^I and R^{II}, that have distinctive properties. These two forms differ in their interaction with the catalytic subunit in that the reassociation of holoenzyme with R^I is markedly enhanced by the presence of ATP;[24-26] in contrast, with R^{II} autophosphorylation the rate of holoenzyme formation is diminished.[27] The $K_{a,app}$ for ATP with the R^I-catalytic subunit holoenzyme is approximately 35 nM,[24] two orders of magnitude lower than the nucleotide concentration required for catalysis (K_m = 7.6 μM, K_d = 4.4 μM^{28}). Since free R^I does not bind ATP, it has been suggested from these data, and partially supported by ATP analogue studies,[29] that the binding of ATP to the R^I-holoenzyme might be to a binding site partially composed of the adenine binding site of the catalytic site on the catalytic subunit, together with a binding pocket on the regulatory subunit for the ribose and triphosphate moieties.[30] While this remains a possibility, it is in fact not necessary to invoke the presence of such a high-affinity binding site (i.e., one of higher affinity than that required for catalysis). Alternatively, the apparent high-affinity binding of ATP to the R^I-holoenzyme could simply reflect the direct binding of ATP to the catalytic site, followed by the consequential displacement of the equilibrium by the formation of an $R^I \cdot C \cdot ATP$ complex: i.e.,

$$C + ATP \leftrightarrow C \cdot ATP + R^I_2 \cdot cAMP_4 \leftrightarrow R^I_2 \cdot (C \cdot ATP)_2 + 4cAMP \qquad (2)$$

As discussed below, a similar proposal has been made to account for the apparent high affinity of ATP binding to the inhibitor protein-catalytic subunit complex. If such a proposed reaction does indeed account for ATP-stimulated R^I-holoenzyme formation, then the magnitude of stimulation by ATP is such[24] that when ATP is present the reassociation reaction would occur almost exclusively via an R^I interaction with the catalytic subunit-ATP complex (rather than with catalytic subunit alone). As discussed below, this has analogies with the interactions of the catalytic subunit with both the inhibitor protein and with protein substrates. With R^{II} no effect of ATP on holoenzyme formation is seen, other than the effects of autophosphorylation, although the consequences of autophosphorylation possibly would make its detection less apparent. These data suggest that with R^{II} (in contrast to the interactions of R^I) either holoenzyme formation occurs primarily (if not exclusively) via interaction with free catalytic subunit (rather than with the catalytic subunit-ATP complex) or R^{II} interaction is with a catalytic subunit-ATP complex, but the presence of ATP makes no difference to the binding interaction.

There are at least two species of catalytic subunit, designated C_α and C_β, that have been identified by molecular cloning approaches.[31-33] Actual data exist only on the interaction of C_α with the inhibitor protein, since C_α appears to be the predominant (if not sole) form in most tissues,[32] particularly in bovine heart, the primary tissue source from which most investigators have obtained the pure enzyme. However, given that the inhibitor protein fully inhibits total brain cAMP-dependent protein kinase,[12] where C_β would be most prevalent, it is probable that if any differences exist in the interaction of the inhibitor protein with C_α and C_β, at most they are only subtle. This does merit exploration, which will be possible shortly with the establishment of an expression vector containing C_β. Other forms of catalytic subunit exist that can be distinguished by charge and separated by CM-cellulose chromatography.[34-36] The two most prominent have been termed C_A and C_B[37] and appear distinct from C_α and C_β.[38] Whether the difference between them is physiological remains to be determined, and there is no observed difference in their interaction with the inhibitor protein.[38] A form of catalytic subunit termed "mute" has been suggested[39] which is detected following incubation with a partially purified preparation of the inhibitor protein; this entity was not detected with a homogeneous inhibitor preparation.

Detailed kinetic investigations of the cAMP-dependent protein kinase catalytic mechanism have shown that the reaction is steady-state ordered Bi-Bi.[28] This was studied with the peptide substrate Kemptide (LRRASLG), and it is possible, as discussed in that work,[28] that this might not fully apply to all substrates or to the same set of substrates under different experimental conditions. The established reaction mechanism with Kemptide is illustrated in Figure 2. Several points in this mechanism are pertinent to a consideration of inhibitor protein interaction with the protein kinase. As illustrated, the primary enzyme form to which substrate-protein or peptide binds is an enzyme-MgATP complex. From atomic coordinate measurements obtained using nuclear magnetic resonance (NMR), Granot et al.[40,41] proposed that ATP might initially exist within the catalytic site as a metaphosphate intermediate. Since Kemptide binding to the protein kinase is not facilitated by 5'-adenylimidodiphosphate (AMP·PNP),[28] and since AMP·PNP cannot form such a metaphosphate intermediate, this

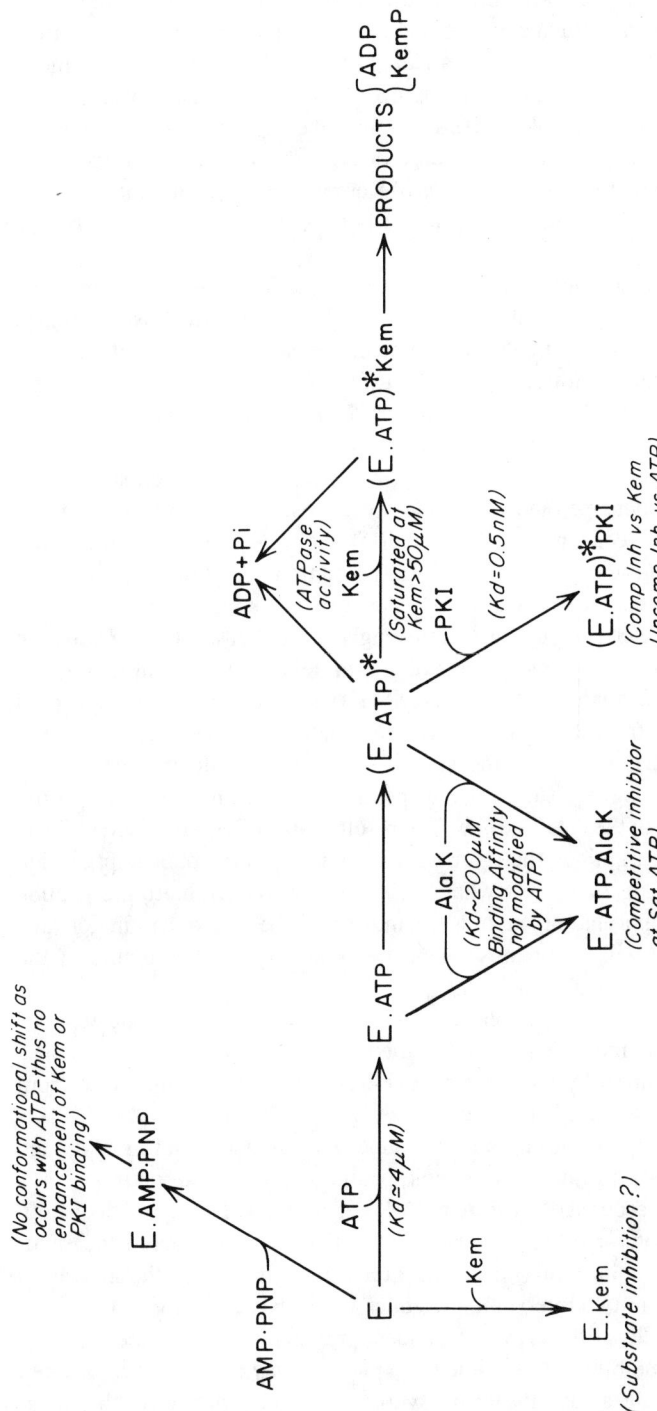

FIGURE 2. Kinetic mechanism of the cAMP-dependent protein kinase. The productive catalytic reaction proceeds via initial ATP binding, followed by the proposed formation of a metaphosphate intermediate (E·ATP)*, peptide substrate binding, and then phosphorylation. Direct Kemptide binding to the enzyme is likely the mode for substrate inhibition. Abbreviations used: Kem, Kemptide; AlaK, Ala-Kemptide; PKI, inhibitor protein. Experimental data and rationale for this mechanism are given in Reference 28. (From Whitehouse, S., Feramisco, J. R., Casnellie, J. E., Krebs, E. G., and Walsh, D. A., *J. Biol. Chem.* 258, 3693, 1983. With permission.)

has led to the suggestion[28] that Kemptide (or protein) primarily recognizes a metaphosphate intermediate form of ATP, illustrated in Figure 2 as (E·ATP)*. Recognition of this form by Kemptide was suggested to be due to the presence of the seryl hydroxyl, since Ala-Kemptide (LRRAALG), the inhibitory peptide in which this hydroxyl is not present, does not exhibit ATP-dependent binding when measured at equilibrium, even though kinetically, at saturating ATP, it is a competitive inhibitor.[28] However, Ala-Kemptide is not a very potent inhibitor, potentially as a consequence of its inability to recognize the metaphosphate form of ATP in the catalytic site. Its K_i is almost 100-fold higher than the K_m of its close analogue Kemptide, and it binds to the protein kinase with more than five orders of magnitude lower affinity than the inhibitor protein. High concentrations of Kemptide itself are inhibitory,[28,42,43] due perhaps to a secondary binding that, as discussed below, might be mirrored by the inhibitor protein.

Productive catalytic competency of the protein kinase requires Mg^{2+} as the divalent ion; Mn^{2+} at low concentrations can support catalysis, but at high concentrations is inhibitory, even in the presence of excess Mg^{2+}. Physical parameter measurements that utilize ions other than Mg^{2+} therefore must be interpreted cautiously. Within this caveat, however, Granot et al.,[44] using $Co(NH_3)_4ATP-Mn^{2+}$ and CrAMP·PCP, have concluded that Kemptide is present in the catalytic site of the protein kinase as a random coil (i.e., does not contain either α-helix, β-sheet, or β-bend conformation). This view is supported by their subsequent studies using N-methylated or depsipeptide Kemptide analogues.[43,45] A contrasting, albeit more indirect, observation has been provided by Small et al.[46] and Matsuo et al.,[47] who have suggested, based on Chou and Fasman structural predictions of phosphorylation sequences, that the phosphoryl acceptor serine is likely to be located as part of (or juxtaposed to) a β-bend (in the substrate). D-Ser-Kemptide, interestingly, is neither a substrate nor even a weak inhibitor.[48] However, if the peptide chain of the protein kinase substrate did not need to be in an ordered conformation, the hydroxyl of D-Ser-Kemptide actually could assume a quite similar distance from the required basic recognition site provided by the pair of arginines (see Section IV.B). However, the Kaiser laboratory,[45] with their studies of methylated or depsipeptide analogues, suggests that peptides do assume a rather specific random coil arrangement when in the catalytic site. The results with D-Ser-Kemptide indicate that in addition to the pair of arginines there are several other features of Kemptide that contribute to its efficacy as a substrate. Just what, if any, secondary structure the peptide region encompassing the phosphorylation site must be in, or must be able to assume, awaits fuller resolution; this probably will occur only when the crystallographic structure of the cAMP-dependent protein kinase has been solved.[49]

Studies of the sequences of phosphorylation sites of target proteins and synthetic peptides based upon these sequences have identified several of the peptide constituents (i.e., residues or structure) needed for recognition by the protein kinase catalytic site. Summarized from the works of Kemp, Engstrom, Krebs, Feramisco, Glass, and their colleagues,[50-52] Table 1 illustrates the essential role for a basic amino acid subsite located NH_2-terminal to the phosphoryl acceptor serine. As illustrated with simple peptides, this subsite optimally is a pair of arginines located in the sequence –Arg–Arg–X–Ser–, with lysine only inadequately replacing either arginine. In natural substrates such a specificity is not adhered to, either with respect to the type of basic amino acid (i.e., lysine or arginine) or whether the separation from the seryl residue is by one or two intervening residues.[53,54] A basic subsite is, however, a universal feature of all substrates of the cAMP-dependent protein kinase. In some proteins, of which protein phosphatase inhibitor I is a prime example,[55] it may be composed of three or more consecutive basic amino acids. Based on synthetic peptides mirroring the phosphorylase kinase β-subunit phosphorylation site, Zetterqvist and Ragnarsson[56] have proposed that a general sequence of –Arg–X–X–Arg–X–X–Ser– constitutes an alternative possible array as the target recognition sequence for the protein kinase. Were this sequence in an α-

TABLE 1
Specific Amino Acid Residue Requirements in Substrates of the cAMP-Dependent Protein Kinase

Peptide	K_m (μM)	Ref.
Leu-Arg-Arg-Ala-Ser-Leu-Gly	16	50
Leu-Ala-Arg-Ala-Ser-Leu-Gly	4900	50
Leu-Arg-Ala-Ala-Ser-Leu-Gly	6300	50
Leu-Lys-Arg-Ala-Ser-Leu-Gly	1400	50
Leu-Arg-Lys-Ala-Ser-Leu-Gly	260	50
Gly-Gly-Gly-Gly-Gly-Gly-Gly-Arg-Arg-Ser-Leu-Gly	1080	52
Gly-Gly-Gly-Gly-Gly-Gly-Arg-Arg-Ala-Ser-Leu-Gly	2.4	52
Gly-Gly-Gly-Gly-Gly-Arg-Arg-Gly-Ala-Ser-Leu-Gly	400	52
Gly-Gly-Gly-Gly-Arg-Arg-Gly-Gly-Ala-Ser-Leu-Gly	4070	52
Gly-Gly-Gly-Arg-Arg-Gly-Gly-Gly-Ala-Ser-Leu-Gly	290	52
Gly-Arg-Arg-Gly-Gly-Gly-Gly-Gly-Ala-Ser-Leu-Gly	960	52

Peptide	Relative rate of phosphorylation (%)	Ref.
Arg-Arg-Ala-Ser-Val	100	51
Arg-Arg-Ala-Ser-Phe	137	51
Arg-Arg-Ala-Ser-Gly	7	51

helix, the two guanidino groups could fairly readily be positioned into the same acidic pocket of the catalytic site as likely occurs with peptide substrates with the sequence –Arg–Arg–X–Ser–. With the inhibitor protein, as discussed below, arginines in all three locations are important for maximum potency (i.e., Arg–X–X–Arg–Arg–X–"Ser"–X).

The nature of the "X" residues in any of the above peptides has not been studied systematically. Earlier data from Zetterqvist et al.[51] on the pyruvate kinase Kemptide sequence (summarized briefly in Table 1) and later studies by Meggio et al.[57] with the protamine phosphorylation sequences have suggested that a hydrophobic residue on the COOH-terminal side of the phosphoacceptor serine may enhance substrate recognition. This feature has not been seen consistently in all target proteins,[58] but, as discussed below, has been found to be crucial for maximum inhibitory potency of the inhibitor protein.

III. PURIFICATION OF THE INHIBITOR PROTEIN AND ITS PHYSICOCHEMICAL PROPERTIES

For the purposes of this chapter the term "inhibitor protein" has been used to refer to the protein isolated from rabbit skeletal muscle and its equivalent in other tissues and species. It is this entity which is the primary subject of this review. However, as described below, several studies have examined inhibition of protein kinase activity by other entities, and the exact relationships of these to the "inhibitor protein" are not always apparent. A comparison of inhibitor protein with these other "species" of inhibitor and a discussion of similarities and dissimilarities are given in Section III.D of this chapter.

A. PURIFICATION

The inhibitor protein has been purified to homogeneity from rabbit skeletal muscle[59,60] and bovine brain[61] and has been partially purified from bovine heart[62] and thyroid,[63] rat brain,[64] and chicken liver.[65] Presented in Section VII is our recommended procedure for the purification of the rabbit skeletal muscle inhibitor protein to homogeneity. This protocol is derived from the procedures described by Walsh et al.,[10] Demaille et al.,[59] Whitehouse and Walsh,[60] and Van Patten et al.[66] and, in particular, employs as the final step a protein kinase catalytic subunit affinity column, as modified from the original procedure of Demaille et al.[59] Several comments on this purification protocol provide some necessary insight into the current level of knowledge.

Following tissue extraction, the next two steps of purification are heat denaturation and acid precipitation by 5% trichloroacetic acid (TCA). The harshness of these treatments raises questions as to whether the purified inhibitor protein may not be in its physiological form and/or may exist in the cell as a component of a larger protein. Neither seems to be the case. Inhibitor protein activity, with gel filtration and ion-exchange chromatographic characteristics identical to those of the pure protein, can be identified in crude cell extracts.[62,63,67] The inhibitor protein has not been purified to homogeneity, however, without the use of denaturing conditions.

Although the stability of inhibitor protein to denaturation might tend to suggest that the protein has little secondary or tertiary structure, the data discussed below on the inhibitor peptide PKI(5—22)amide suggest either that the inhibitor protein has a structure that is stable to the denaturing conditions used for purification or that the conformation readily reassembles upon return to more physiological buffer conditions. By NMR studies, PKI(5—22)amide structure itself has been found to be quite resistant to denaturation,[68] suggesting that the inhibitor protein might behave similarly.

Multiple-charge isomers of inhibitor protein (I-1, I-2, I-3) have been demonstrated,[61,67] with I-2 being the form that has been purified to homogeneity (Section VII). Since the relationship between the isomers is not yet known and may be artifactual rather than physiological[61,67] (caused, for example, by deamidation [see Section III.C]), it remains a possibility that the form of inhibitor protein purified so far might not fully represent the physiological species. If so, it may be relevant that, when nondenaturing extraction conditions are used,[67] a slightly higher proportion of I-1 to I-2 is observed than is obtained with denaturing extraction. Were deamidation to have occurred during purification, I-1 would represent the form with a higher number of amidated residues.

The rate of addition of TCA to the crude cell extract in the purification is of critical importance for obtaining the maximum yield of inhibitor protein.[69] Slow addition of TCA, and in consequence a slow transition of pH across the mid-acidic range, causes a marked decrease in final yield.

Some trouble has been experienced in obtaining suitable protein kinase catalytic subunit affinity resins. The protocol outlined in Section VII optimizes maintenance of catalytic subunit activity with a slightly diminished coupling efficiency. With this protocol, 20 mg of catalytic subunit can be suitably coupled to provide an affinity column sufficient for the purification of 400 μg of inhibitor protein, and the column can be used repeatedly. High pressure liquid chromatography (HPLC) has also been used as an alternate purification mode.[70]

B. PHYSICOCHEMICAL CHARACTERISTICS

In Table 2 is presented a summary of the physical parameters of the physiological I form (see Section III.C) of the inhibitor protein isolated from rabbit skeletal muscle, and Figure 3 depicts its amino acid sequence. The protein is distinctive due to the lack of tryptophan, cysteine, proline, and methionine, and it contains only single residues of tyrosine, phenylalanine, and histidine; the latter is compatible with its low UV absorption. As isolated it contains no phosphate[59] or carbohydrate,[59,67] and based upon activity determinations it is insensitive to digestion by neuraminidase, DNase, or RNase.[10,62] The blocking group at the NH_2-terminus remains to be determined; it is not thought to be myristyl, and data on peptide

TABLE 2
Physicochemical Characteristics of the Inhibitor Protein

	Skeletal muscle inhibitor protein	Ref.	Testis inhibitor protein	Ref.
Molecular weight (Da) Method				
Amino acid composition	7829	71	19,700	76
Gel filtration	22,000[a]	62, 69	21,000	76
Nondenaturing gel electrophoresis (Hedrick and Smith)	10,000	69		
SDS-gel electrophoresis	10—11,000[a]	67	26,000	76
Sedimentation coefficient ($S_{20,w}$)	1.5	94	1.2	76
Stokes radius	21 Å	69	20.8 Å	76
Frictional coefficient	1.3	69	1.17	76
Axial ratio	4	69		
Isoelectric point	4.24	59, 67	4.4	76
Absorption coefficient (276 nm/1 cm/1%)	1.3	59		
Phosphate content	<0.1 mol/mol	59, 61		
Specific activity	3—6 × 10^5 U/mg	60		
K_i	0.2 nM	60, 103	10 nM	76

[a] Data for I form; see Table 3 for comparison of I and I'.

FIGURE 3. The amino acid sequence of inhibitor protein. (From Scott, J. D., Fischer, E. H., Takio, K., Demaille, J. G., and Krebs, E. G., *Proc. Natl. Acad. Sci. U.S.A.*, 82, 572, 1985. With permission.)

fragments suggest that it is most likely either formyl or acetyl.[71] One of the most distinctive features of this form of the protein is its apparent asymmetry as suggested by the disparities in size determinations by various techniques and the calculated axial ratio. The data would be compatible with a protein containing a central core of structure around the active site (see discussion on PKI(5—22)amide in Section IV.3), with the remainder of the molecule (i.e., the COOH-terminal half) having little ordered structure.

C. MULTIPLE MOLECULAR FORMS

Three charge isomers of the inhibitor protein have been detected in skeletal and cardiac muscle[59,67,72] and in the brain;[61] they are separated by DEAE chromatography during purification[69] or with the final pure protein[59] and also can be distinguished by nondenaturing gel electrophoresis.[67] These have been designated I-1, I-2, and I-3, in order of DEAE elution, with the first two being the most abundant in the tissue sources examined so far. Each of the three has the same size characteristics as presented for the I-2 form in Table 2 and, within experimental error, an identical isoelectric point. I-1 and I-2 appear to have slightly higher inhibitory activities than I-3.[72] The relationship between these three forms is unknown, and they may arise artifactually during isolation, such as by deamidation. All three are detectable using either denaturing or nondenaturing tissue extraction,[67] and none contain covalently bound phosphate.[59]

In addition to the possible existence of charge isomers, the inhibitor protein has been

TABLE 3
Comparison of the Apparent Molecular Weights (Da) of the I and I' Forms of Inhibitor Protein

Method	I	I'
G-75 exclusion chromatography	22,000	11,000
Hedrick and Smith analysis	10,000	10,000
SDS-gel electrophoresis	10,500	7,750

Data from Reference 60.

shown to exist in two molecular species that have been designated I and I'. The different physical characteristics of these two species are presented in Table 3. The I form is that present in initial cell extracts, using either denaturing or nondenaturing isolation conditions,[60,67,69] and is therefore believed to be the primary (if not sole) physiological form.[60,69] Whether I' can occur physiologically is not known, and it currently appears most likely that it is actually an artifact produced by some manipulation occurring during purification.[60,69] Adherence to the isolation protocol presented in Section VII, with all steps performed rapidly, has consistently (>90%) resulted in a preparation containing >95% I.[66] The exact relationship between I and I' has not yet been elucidated and, thus, as a consequence, neither have the reasons for the occasional spontaneous I → I' transitions that arise during purification. Because the I' form is sometimes produced during purification, it is essential to characterize the final product. This can be suitably accomplished by either gel filtration or SDS gel electrophoresis.[60] Both initial descriptions of the purification of the inhibitor protein to homogeneity in fact produced the I' form.[59,69]

As noted, knowledge as to the exact relationship between the I and I' forms is currently lacking and, importantly, an I' → I conversion has never been observed. Because I and I' were first distinguished by gel filtration, with elutions equivalent to apparent molecular weights of 22 kDa and 11 kDa, respectively, it initially was supposed that they might simply represent a dimer and monomer (respectively) of the same protein. This clearly is not so, since both exhibit an apparent molecular weight of ~11,000 by the nondenaturing gel electrophoretic procedure of Hedrick and Smith,[145] albeit they migrate separately; they also migrate separately on SDS gel electrophoresis. Some of the differences in the apparent molecular weights between the different procedures are clearly a consequence of the inaccuracy of these methods for a protein of this small size. In this molecular weight range none of the methods yield linear molecular weight relationships, and the measurements are adventitiously dependent upon the standards chosen. Nevertheless, this does not compromise the observation that the two forms can be separated by gel filtration or by nondenaturing or denaturing gel electrophoresis,[60,66,67,69] and these methods do give some indication of differences in shape and/or charge of the two forms. The current, most reasonable explanation is that I and I' represent distinct stable conformers of the same polypeptide chain.[69] How these different conformers arise and what might contribute to their separate stabilities can only be speculated upon this time. I' does not appear to be a proteolytic product derived from I, although this still remains a possibility. Both I and I' inhibit the protein kinase in an identical manner with equivalent affinities.[73] All three charge isomers from either cardiac or skeletal muscle exhibit I and I' species.[67]

D. RELATIONSHIP OF OTHER PROTEIN KINASE INHIBITORS TO THE INHIBITOR PROTEIN
1. Inhibitor Protein from Testis

A form of inhibitor protein has been isolated from testis[74-76] whose relationship to the protein present in rabbit skeletal muscle or bovine brain still remains to be determined. The protein, purified from testis to apparent homogeneity,[76] is clearly a high-affinity inhibitor of the cAMP-dependent protein kinase. As evaluated by a titration assay, the testis inhibitor exhibits a K_i of 10 nM, which is clearly comparable to the high affinity exhibited by the skeletal muscle inhibitor protein. (A direct comparison, however, has not been reported.)

Physical characteristics of the testis protein preparation are presented in Table 2 for comparison. Like the skeletal muscle protein, the testis inhibitor is an acidic protein (pI = 4.4) of low molecular weight, has similar heat stability and DEAE elution characteristics, and also exhibits multiple size species,[77] possibly analogous to the I and I' forms. (The precise size determined for the testis protein has depended upon the method of determination and has varied somewhat between reports.[76,77]) The testis inhibitor protein, however, appears distinct from the skeletal muscle inhibitor protein in several properties. The testis protein contains an abundance of proline (13% of the total protein vs. none) and 3 mol of tryptophan (vs. none), but no tyrosine (vs. one). The testis inhibitor also was reported to inhibit the cAMP phosphodiesterase with high affinity (K_i = 20 nM), an observation of renewed interest with the recent report[78] that peptides derived from the skeletal muscle inhibitor protein (but not the inhibitor protein itself) can bind calmodulin with high affinity.

The exact relationship of the testis inhibitor to the inhibitor protein found in other tissues is still in question. In particular, the presence of high levels of proline suggests that it either is a distinct protein or contains an additional segment of peptide chain. Further study is needed, however, to assess whether the criteria used to examine the homogeneity of the testis protein preparation were sufficient and, thus, whether some of the apparent differences in physical characteristics and properties could be attributable to the presence of contaminants. The testis inhibitor protein is of particular interest given the findings of marked changes in its amount with a variety of physiological manipulations.[74,75]

2. Protein Kinase Modulator

In initial studies by Donnelly et al.[79,80] it was reported that lobster tail muscle contained a heat-stable protein that inhibited the cAMP-dependent protein kinase in a manner similar to the inhibitor protein from rabbit skeletal muscle, but also activated the cGMP-dependent enzyme. Such an entity was subsequently examined in a range of mammalian tissues,[81,82] and while it is clear that many tissues contain a cGMP-dependent protein kinase stimulator, this stimulator is not the cAMP-dependent protein kinase inhibitor protein. Pure preparations of skeletal muscle inhibitor protein neither activate nor inhibit the cGMP-dependent protein kinase, even when tested at a concentration five orders of magnitude higher than that necessary to fully inhibit the cAMP-dependent enzyme.[83] Purified modulator protein from mammalian sources is likewise fully inert toward the cAMP-dependent protein kinase.[81,82] While it remains a possibility that lobster muscle contains a protein that has both cAMP-inhibitory and cGMP-stimulatory properties as initially reported,[79] this early observation may have been an artifact of insufficient purification.

3. Type I and Type II Cyclic Nucleotide-Dependent Protein Kinase Inhibitors

Szmigielski et al.[1,64] have extensively investigated the potential physiological manipulation of two species of protein kinase inhibitor, designated Types I and II. Type I is clearly identical to the skeletal muscle inhibitor protein, whereas Type II appears to be a distinct entity. Type II is of lower molecular weight (15,000 Da by gel filtration), can be separated from Type I by gel filtration, and blocks the phosphorylation of cAMP-dependent, cGMP-dependent, and cyclic nucleotide-independent protein kinases as a competitive inhibitor of substrate protein.[64] It has usually been assayed at low concentrations of protein substrate;[84] it has been identified in several tissues[64] and has been shown to change with physiological manipulation.[85,86] Like the Type I inhibitor, it is heat stable and is believed to be a protein since its activity is destroyed by trypsin or chymotrypsin, but it has yet to be characterized in any detail. Although similar in size, it definitely appears not to be the I′ form of the inhibitor protein since the latter has no effect on the cGMP-dependent protein kinase.[69] Given its lack of protein kinase specificity, the significance of this Type II inhibitor in physiological regulation appears equivocal. It is probably unrelated to the skeletal muscle inhibitor protein.

4. Other Protein Kinase Inhibitors

Several inhibitors of various protein kinases have been described,[87-91] including one of considerable interest directed to protein kinase C.[92] None of these appear to be related to the skeletal muscle inhibitor protein.

IV. MECHANISM OF INTERACTION OF THE INHIBITOR PROTEIN WITH THE cAMP-DEPENDENT PROTEIN KINASE

A. NATURE OF THE INHIBITOR PROTEIN-CATALYTIC SUBUNIT COMPLEX

The initial studies of the inhibitor protein demonstrated that it acted by direct interaction with the cAMP-dependent protein kinase catalytic subunit,[93] did not act as an ATPase or phosphatase,[10] and, as a tightly bound inhibitor, was competitive vs. protein substrate.[59,69] By a variety of approaches it was shown that it does not interact with either the holo (i.e., undissociated) protein kinase[5,93,94] or with free regulatory subunit[93] and that the inhibitor protein and the regulatory subunit compete for binding and share, at least in part, the same binding site on the catalytic subunit.[93,95] Binding of the regulatory subunit to the catalytic subunit appears to be approximately two- to fivefold higher in affinity than for the inhibitor protein,[73,96,97] although a comparison has not been performed using identical conditions. Under such circumstances it would be expected that, at equimolar concentrations of all constituents (i.e., inhibitor protein, catalytic subunit, and regulatory subunit monomer), catalytic subunit would be present predominantly as holoenzyme, with little to no formation of a catalytic subunit-inhibitor protein complex. Several predicted consequences thus arise. Addition of regulatory subunit (R_2cAMP_4) to a catalytic subunit-inhibitor protein complex would result in holoenzyme formation with the consequent release of cAMP.[93] In contrast, when inhibitor protein was added to protein kinase holoenzyme, the catalytic subunit-inhibitor protein complex would be formed only if the inhibitor protein was present in some excess. Excess inhibitor can, however, result in a less-than-saturating concentration of cAMP, promoting a greater degree of holoenzyme dissociation.[26,94] However, this is unlikely to be meaningful in any physiological situation because of the low cellular concentrations of inhibitor protein. Even though the affinity of inhibitor protein for the catalytic subunit is very high, formation of the complex is rapidly reversed, as readily demonstrated by simple dilution.[94]

The interaction of the inhibitor protein with the catalytic subunit has been extensively examined kinetically.[28,59,69,73] Using an extension of the Henderson analysis to two-substrate reactions,[73] the inhibitor protein has been demonstrated to be competitive vs. protein substrate and uncompetitive vs. MgATP^{2-}. Kinetically, therefore, the inhibitor protein does not bind directly to the protein kinase, but rather to an enzyme-MgATP complex (Figure 2). This binding, and its requirement for both Mg^{2+} and ATP, have been examined further by an extension of the kinetic experiments.[73] It was shown that the binding of the inhibitor protein to the catalytic subunit is increased by preincubation of the two in the absence of protein substrate. MgATP is required for this increase in effectiveness at an apparent half-maximal nucleotide concentration of 60 nM. Such a concentration of ATP is two orders of magnitude lower than the K_m (and K_d) required to support protein substrate phosphorylation (K_m = 7.6 μM, K_d = 4.4 μM^{28}). Most likely, however, such a difference in the required nucleotide concentration does not imply that a high-affinity binding site for ATP exists on the catalytic subunit (in addition to that at the catalytic site) or on the catalytic subunit-inhibitor protein complex. More simply, such a high apparent affinity of ATP for complex formation is most probably a reflection of the formation of a catalytic subunit-MgATP-inhibitor protein complex with a resultant displacement of the equilibrium of ATP binding to the protein kinase; i.e.,

$$C + MgATP \leftrightarrow C \cdot MgATP + I \leftrightarrow C \cdot MgATP \cdot I \qquad (3)$$

A similar mechanism is suggested to occur with the ATP-dependent interaction of R^I with the catalytic subunit (see Equation 2). The nucleotide requirement for the formation of the catalytic subunit-inhibitor protein complex cannot be met by ADP, AMP, cAMP, adenosine, or AMP·PNP. Similarly, analysis of multiple inhibition of the protein kinase by AMP·PNP plus inhibitor protein[73] also demonstrates that the inhibitor protein does not bind to an enzyme·Mg·AMP·PNP complex. Data examining the efficacy of inhibitor protein interaction by these kinetic approaches further showed that Mg^{2+} could be substituted for by Mn^{2+} as the cation required for complex formation. Such a finding is likely correlated with the observation that Mn^{2+} only poorly supports catalysis[73,98] and, thus, clearly interacts with the protein kinase in a somewhat different manner.

The nature of the catalytic subunit-inhibitor protein complex and the requirements for its formation have also been explored using either CM-cellulose column chromatography[95] or nondenaturing gel electrophoresis[38,66] to separate the complex from its free constituents. In the studies employing CM-cellulose chromatography,[95] the extent of binding was assessed as the amount of inhibitor retained on the CM-cellulose in the presence of catalytic subunit (plus Mg^{2+}), since free inhibitor protein, being anionic, does not bind. With this method, not only ATP, but (in contrast to the kinetic data discussed above) also ADP, dATP, AMP·PNP, and the αβ (but not the βγ) methylene analogue of ATP appeared to enhance the inhibitor protein-protein kinase interaction. Inhibitor protein-catalytic subunit complex formation has also been examined using nondenaturing gel electrophoresis, with which the CI complex is readily separated from the free constituent proteins.[66] This method has given results more closely analogous to the results obtained kinetically. In the presence of Mg^{2+}, complex formation was specific for ATP, and this requirement could not be substituted for by AMP, ADP, GTP, ITP, or AMP·PNP.[66] Interestingly, when Mg^{2+} was replaced by Mn^{2+}, complex formation occurred with any of the triphosphate nucleotides listed. This difference in the binding obtained with the two cations most likely represents differences between the exact conformations of Mn^{2+}-nucleotide and Mg^{2+}-nucleotide in the protein kinase catalytic site, as is also compatible with the inhibition of Mg^{2+}-supported catalysis by Mn^{2+}.

The nondenaturing gel electrophoretic method for examining inhibitor protein-catalytic subunit complex formation has provided further information on the nature of the complex. Using this approach (both by measurement of protein and by tagging the catalytic subunit with a fluorescent probe), the molar stoichiometry of the inhibitor protein-catalytic subunit complex was demonstrated to be 1:1. Furthermore, the procedure allowed separation of complexes composed of either the C_A or C_B form of catalytic subunit and showed that complex formation is ATP dependent for both.[38]

Taken all together, these results indicate that the inhibitor protein's interaction with the protein kinase occurs by the mechanism illustrated in Figure 4. As shown, there appears to be a close similarity in the interaction of the catalytic subunit with inhibitor protein, R^I-regulatory subunit, and protein kinase substrate. With all three, the preferred (and likely physiological) reaction is with the catalytic subunit-MgATP complex rather than with free catalytic subunit alone. This is of interest mechanistically, but physiologically, given its cellular concentration, it is likely that ATP is always intrinsically resident at the catalytic

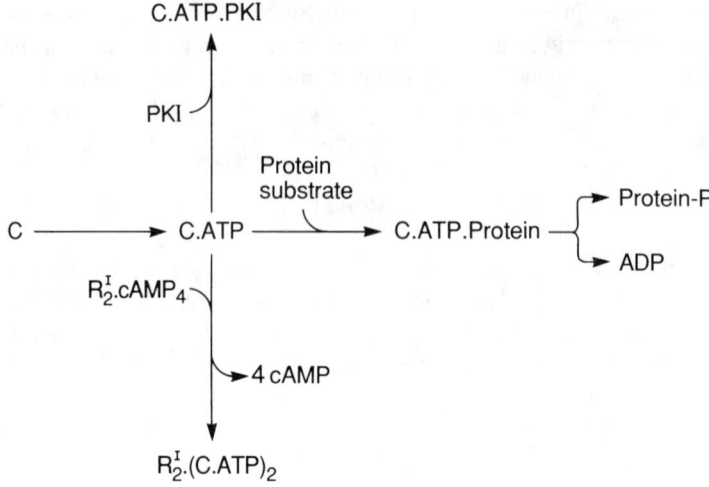

FIGURE 4. Comparison of the kinetic mechanism of the protein kinase catalytic subunit with its mode of interaction with regulatory subunit (R^I) and inhibitor protein (PKI). (See text for rationale.)

site (except, of course, immediately following substrate phosphorylation). As discussed above, interaction of R^{II} may be distinct, either occurring primarily with the catalytic subunit alone or with ATP not affecting the rate of reassociation.

Although it does seem clear that the major route of binding of inhibitor protein is to a catalytic subunit-MgATP complex, ATP-independent binding, while probably not physiological, does occur also. ATP-independent binding was seen in early studies at high protein concentrations, when binding was examined by sucrose density-gradient ultracentrifugation,[94] but in those experiments the extent of complex formation was far less than the stoichiometric level that might have been expected. It is possible that ATP-independent binding of protein substrates to the protein kinase also occurs. As with the inhibitor protein, the kinetic data (as summarized in Figure 2 and discussed above) show that the interaction of the peptide substrate Kemptide with the protein kinase in the productive catalytic reaction essentially occurs exclusively via the binding of peptide to enzyme-ATP. Direct binding studies of the rather poor inhibitory peptide Ala-Kemptide[28] do not show ATP dependence, although kinetically it is clearly a competitive inhibitor vs. Kemptide (thus indicating that, under the conditions of the kinetic assay, binding of Ala-Kemptide is to an enzyme-ATP form). Some kinetic studies with other substrates of the protein kinase (histone or ATPβS "A" diastereomer) also have suggested the possibility that not all protein substrate binding is ATP dependent.[99-101] Furthermore, high levels of Kemptide are inhibitory;[28,42,43] this binding may be analogous to that of high levels of Ala-Kemptide or the inhibitor protein and also may be ATP independent.

B. RECOGNITION DETERMINANTS OF THE INHIBITOR PROTEIN — STUDIES WITH ACTIVE PEPTIDE FRAGMENTS

The observations that all protein kinase substrates contained an essential pair of basic amino acids on the NH_2-terminal side of the phosphorylated serine and that the inhibitor protein was competitive vs. protein substrate and clearly shared with protein substrates many analogous interactions with the catalytic subunit made it highly likely that a similar basic amino acid recognition site would exist in the inhibitor protein. Initial clues that this was indeed so arose from observations that the inhibitor protein was inactivated by cyclohexanedione[59] and also that inhibitor protein-catalytic subunit complex formation was inhibited by guanethidine.[95] Other tentative clues as to possible recognition factors of the

TABLE 4
Effect of Length of PKI Peptides on Inhibitory Potency

Peptide number	Peptide sequence	K_i (nM)	Ref.
P1	T T Y A D F I A S G R T G R R N A I[a]	3.1	105
10	T Y A D F I A S G R T G R R N A I[a]	1.7	58
11	[c]T Y A D F I A S G R T G R R N A I[a]	2.5	58
12	Y A D F I A S G R T G R R N A I[a]	28	58
13	[c]Y A D F I A S G R T G R R N A I[a]	34	58
14	A D F I A S G R T G R R N A I[a]	90	58
15	D F I A S G R T G R R N A I[a]	97	58
P4	G R T G R R N A I[a]	36	58
P2*	T T Y A D F I A S G R T G R R N A I H D[a]	2.3	103, 105
16	F I A S G R T G R R N A I H D[a]	73	105
17	I A S G R T G R R N A I H D[a]	115	105
18	A S G R T G R R N A I H D[a]	119	58
P5	G R T G R R N A I H D[a]	57	105
19	R T G R R N A I H D[a]	250	58
20	T G R R N A I H D[a]	1,250	58
P3	Y A D F I A S G R T G R R N A I H D I L V S S A	240	104
P6	I A S G R T G R R N A I H D I L V S S A	800	70
21	G R T G R R N A I H D I L V S S A	75,000	70
22	R R N A I H D I L V S S A	150,000	70
23	N A I H D I L V S S A	nd	70
24	H D I L V S S A	nd	70
25	I L V S S A	nd	104
P2*	T T Y A D F I A S G R T G R R N A I H D	8.8	104
26	T D V E T T Y A D F I A S G R T G R R N A I H D	4.8	104
27	[c]T D V E T T Y A D F I A S G R T G R R N A I H D[a]	35	104

Note: [a] = COOH-terminal amide; [c] = NH$_2$-terminal acetyl; nd = no inhibitory activity detected.

* The two values for P2 were derived from different laboratories, so comparisons should be made within the groups.

inhibitor protein also could be drawn from the observation that either iodination or acetylation (possibly of the single tyrosine) diminished inhibitory potency, as did maleylation or reaction with Bolton and Hunter reagent (possibly modifying lysine). Inhibitory activity also was found to be destroyed rapidly by trypsin, at a lesser rate by chymotrypsin, and partially by *Staphylococcus aureus* protease digestion.[102] (As seen from the data presented below, however, some of these effects now appear to be more peripherally related to the actual active site of the inhibitor protein.)

Prompted by the initial observations of Demaille et al.,[102] two groups, ours[103] and Scott et al.,[70] successfully isolated from the inhibitor protein two distinct, but overlapping, 20-amino acid active peptide fragments. These were obtained by digestion with either *S. aureus*[103] or Mast cell[70] proteases to yield peptides PKI(5—24) or PKI(11—30), respectively. (Nomenclature is based upon the sequence of the native protein as determined by Scott et al.;[71] see Figure 3.) These two peptides are listed in Table 4 as P2 and P6, respectively. Of these two peptides, PKI(5—24) was the most active and exhibited approximately 20% of the activity of the native inhibitor protein (Figure 5). These original peptides have served to extensively characterize the essential features of the inhibitor protein that provide for its high-affinity interaction with the protein kinase.[9,58,68,104-106] This has been approached by the synthesis of a range of peptides varying in length and/or with substituted residues.

Table 4 addresses the question of what length of peptide is necessary for full inhibitory potency. The peptides PKI(5—22), PKI(5—24), and PKI(7—30) have served as the parent peptides with which to examine the effects of serial deletion from the NH$_2$-terminus; they

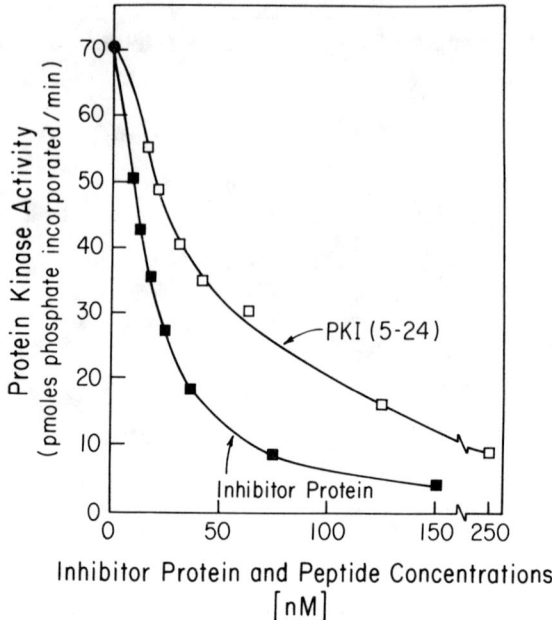

FIGURE 5. Comparison of the titration of the cAMP-dependent protein kinase by inhibitor protein and PKI(5—24). Data taken from Reference 103.

are designated P1, P2, and P3, respectively. Comparisons between these three parent peptides (and their derivatives) provide insight into the requirements at the COOH-terminus. The shortest peptide found so far that retained the highest inhibitory potency was PKI(6—22)amide (Peptide 10). Extension on the NH_2-terminal side to the full sequence of the native inhibitor protein (Peptide 26) did not increase activity significantly, and the activity of this extended peptide was actually diminished when the NH_2-terminus was then acetylated (peptide 27), even though this acetylation may, in fact, mimic the blocked NH_2-terminus of the native protein. Extension of the COOH-terminal end of PKI(5—22)amide with the addition of residues histidine-23 and aspartate-24 (peptide P2) did not increase or decrease inhibitory activity. Further extension of the COOH-terminus by the addition of residues 25 to 30, however, appeared to markedly diminish inhibitory activity (10- to 1000-fold), as seen by comparison of peptides P3 and 12, P6 and 17, and 21 and P5.* This detrimental effect of the presence of residues 25 to 30 on inhibitory activity, as seen with these peptides, most likely does not occur with the native inhibitor protein (given the latter's very high potency). In these extended peptides, residues 25 to 30 are possibly in (or can assume) a conformation and/or an interaction with the remainder of the peptide which is not allowed when the full inhibitor protein sequence is present. (As reported by Grove et al.,[21] PKI(1—31) exhibits a K_i of 4 nM, which would indicate that the decrease in inhibitory potency provoked by the presence of residues 25 to 30 is not seen when the full NH_2-terminus is present.*)

Examination of peptides in which the NH_2-terminal amino acids of the most active peptide PKI(6—22) have been serially deleted shows several interesting features. The first five residues of PKI(6—22) clearly make an important contribution to inhibitory potency [as also seen with peptides based upon PKI(5—24) and PKI(7—30)]. The serial removal of

---

* This comparison involves values obtained by different laboratories, although under quite similar assay conditions. In addition to the values presented in Tables 4 to 7, Grove et al.[21] report a K_i for PKI(1—31) of 4 nM; however, no assay conditions were stated.

TABLE 5
Effects of Substitution of NH$_2$-Terminal Amino Acids of PKI(5—22)amide on Inhibitory Potency

Peptide number	Peptide sequence	K$_i$ (nM)	Ref.
P1	T T Y A D F I A S G R T G R R N A I[a]	3.1	105
28	**S^r** T Y A D F I A S G R T G R R N A I[a]	5.7	58
29	T **A** Y A D F I A S G R T G R R N A I[a]	7.1	58
30	T T **A** A D F I A S G R T G R R N A I[a]	14	58
31	T T Y **L** D F I A S G R T G R R N A I[a]	8.9	58
32	T T Y A **A** F I A S G R T G R R N A I[a]	8.2	58
33	T T Y A D **A** I A S G R T G R R N A I[a]	270	58
34	T T Y A D F **A** A S G R T G R R N A I[a]	8.1	58

Note: Sr represents sarcosine. [a] = COOH-terminal amide.

threonine-6 and then tyrosine-7 decreased potency by tenfold (peptide 12) and a further threefold (peptide 14), respectively. Subsequent removal of alanine-8 and aspartate-9 produced no further effect, but the deletion of phenylalanine-10 then increased the K$_i$ by a further 50%. Overall, the removal of the first five residues of PKI(6—22) decreased inhibitory potency by 50-fold. Interestingly, the full removal of this region, together with the next three residues (isoleucine-11, alanine-12, and serine-13, giving peptides P4 and P5), caused some recovery of activity. As will be discussed in Section IV.D below, there is now good evidence that the NH$_2$-terminal region of PKI(6—22) is present as an α-helix. Thus, it appears that partial removal of this helical region (as in peptides 14, 15, 16, 17, and 18) results in peptides where the residual residues of the NH$_2$-terminal region (that were originally present as an α-helix) now, in some manner, interfere with binding of the remainder of the inhibitory peptide to the protein kinase. Serial deletion of NH$_2$-terminal residues beyond glycine-14 caused large losses of potency, not surprisingly associated with the elimination of the critical arginines, eventually resulting in totally inactive fragments.

As an alternate approach to examine the role of the NH$_2$-terminal region of PKI(5—22) amide, a series of substitution analogues have also been explored (Table 5). The substitutions made were either alanine for threonine-6, tyrosine-7, aspartate-9, phenylalanine-10, or isoleucine-11, sarcosine for threonine-5, or leucine for alanine-8. With the exception of the latter, such substitutions were considered "nondisruptive". The alanine or sarcosine would not introduce a bulky hydrophobic or charged hydrophilic group, nor would they either disrupt structure (as would proline, for example) or add the kind of flexibility that would be introduced by a glycine. The results of these substitutions have provided interesting insight into the requirements for maximum inhibitory potency provided by this region. In apparent contrast to what was observed by the deletion studies discussed above, substitution of threonine-6, alanine-8, aspartate-9, or isoleucine-11 at most only slightly diminished inhibitory potency, and substitution of tyrosine-7 produced a four- to fivefold effect. Interestingly, whereas deletion of the phenylalanine caused only a slight change, its substitution diminished inhibitory potency 100-fold. A graphic comparison of the effects of substitution vs. deletion in this area of the peptide is provided by Figure 6. The apparent discrepancies have a simple explanation. As we now know, this part of the peptide is structured. What occurs with deletion is that this structure is destroyed, whereas with the single substitutions (tested so far) the structure does not appear to be disrupted. The structural integrity of PKI(6—22) thus requires the presence of suitable residues at positions 6 and 7, but does not require that they be threonine and tyrosine, respectively. (This tyrosine in the intact inhibitor protein, however, can be phosphorylated, leading to a diminished inhibitory ac-

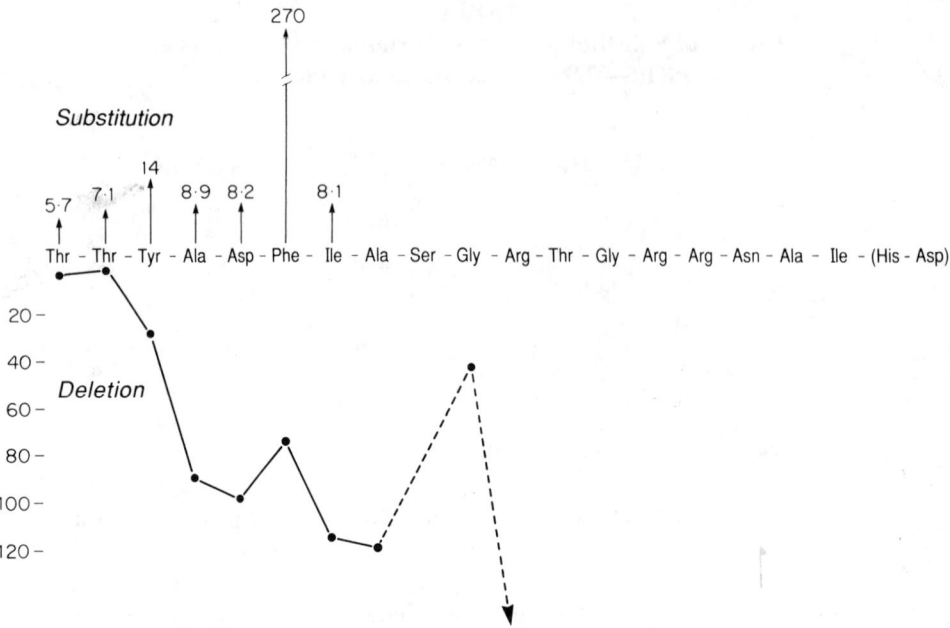

FIGURE 6. Comparison of the inhibitory potency of NH_2-terminal deletion and substitution PKI peptides. Listed above the peptide sequence are the K_i values of peptide derivatives of the parent peptide PKI(5—22)amide with single substitutions of the indicated amino acid. In comparison, shown below the peptide sequence are the K_i values of derivatives of the parent peptide PKI(5—22)amide with serial deletions from the NH_2-terminus. Values are beneath the residue corresponding to the new NH_2-terminus. All values are nanomolar. The illustration, for convenience, has combined the data for deletion peptides derived from both PKI(5—22)amide and PKI(5—24)amide.

tivity.) The situation with phenylalanine is clearly different. At the stage where phenylalanine is removed as the NH_2-terminus (i.e., transition from peptide 16 to 17), inhibitory potency already has been compromised greatly. If the structure existent in residues 5 to 12 is essential for the correct orientation of the phenylalanine, then it is likely that even though phenylalanine is present in peptide 16 the critical orientation structure for it is not. What peptides 28 through 32 and 34 show is that the mild substitutions so far examined appear not to have disrupted the structure of this portion of the peptide (as is also evident from NMR studies;[68] also see Section IV.B below). With peptide number 33, in which phenylalanine-10 has been substituted, there are two apparent explanations for the decreased potency: either phenylalanine is essential for the integrity of the structure, or the structure is still retained but the peptide no longer possesses the recognition signal provided by the phenylalanine side chain. We would favor the latter possibility; a series of additional substitutions for phenylalanine recently tested[146] has provided support for this conclusion and has shown a precise need for the orientation of the phenylalanine aromatic ring with reference to the peptide backbone.

A range of substitutions, both single and multiple, have also been used to examine the role of critical residues in the COOH-terminal portion of PKI peptides. These results are summarized in Table 6. For these studies, several different parent peptides have been used, namely P1 to P6. (Comparisons between the activities of these parent peptides are provided by Table 4; for convenience their inhibitory activities have been repeated in Table 6 with each of the sets of substitutions.) Not surprisingly, a critical role for a basic subsite was readily apparent (Table 6). This subsite contains not only arginine-18 and arginine-19, however, but also arginine-15. With both of the two parent peptides examined, the substitution of either arginine-18 or arginine-19 produced 100- to 1000-fold decreases in activity (peptides 36, 37, 39, and 40), and substitution of arginine-15 caused a 10- to 30-fold change

TABLE 6
Effects of Substitution of Residues 13 to 24 in PKI Peptide on Inhibitory Potency

Peptide number	Peptide sequence	K_i (nM)	Ref.
Substitution of arginine-15, -18, and -19			
P5	G R T G R R N A I H D[a]	57	105
35	G **K** T G R R N A I H D[a]	370	105
36	G R T G **K** R N A I H D[a]	36,000	105
37	G R T G R **K** N A I H D[a]	4,200	105
P3	Y A D F I A S G R T G R R N A I H D I L V S S A	240	104
38	Y A D F I A S G **T** G R R N A I H D I L V S S A	6,600	104
39	Y A D F I A S G R T **G** R R N A I H D I L V S S A	20,000	104
40	Y A D F I A S G R T G R **G** N A I H D I L V S S A	120,000	104
Substitution of threonine-16			
P1	T T Y A D F I A S G R T G R R N A I[a]	3.1	105
41	T T Y A D F I A S G R **U** G R R N A I[a]	2.3	58
P4	G R T G R R N A I[a]	36	58
42	G R **A** G R R N A I[a]	140	58
Substitution of glycine-17 and asparagine-20			
P1	T T Y A D F I A S G R T G R R N A I[a]	3.1	105
43	T T Y A D F I A S G R T G R R **A** A I[a]	21	58
44	T T Y A D F I A S G R T **L** R R N A I[a]	26	58
P4	G R T G R R N A I[a]	36	58
45	G R T G R R **A** A I[a]	550	58
46	G R T G R **G** A I[a]	110	58
47	G R T **L** R R N A I[a]	390	58
48	G R T **L** R R **A** A I[a]	4,000	58
Substitution of alanine-21			
P6	I A S G R T G R R N A I H D I L V S S A	800	104
49	I A S G R T G R R N **S**[b] I H D I L V S S A	96,000	104
50	I A S G R T G R R N **U** I H D I L V S S A	19,000	104

TABLE 6 (continued)
Effects of Substitution of Residues 13 to 24 in PKI Peptide on Inhibitory Potency

Peptide number	Peptide sequence	K_i (nM)	Ref.
Substitution of isoleucine-22			
P1	T T Y A D F I A S G R T G R R N A I[a]	3.1	105
51	T T Y A D F I A S G R T G R R N A **L**[a]	11	58
52	T T Y A D F I A S G R T G R R N A **G**[a]	470	58
P4	G R T G R R N A I[a]	36	58
53	G R T G R R N A **L**[a]	180	58
54	G R T G R R N A **G**[a]	1,400	58
Substitution of histidine-23			
P3	Y A D F I A S G R T G R R N A I H D I L V S S A	240	104
55	Y A D F I A S G R T G R R N A I **G** D I L V S S A	1,200	104
Substitution of aspartate-24			
56	Y A D F I A S G R T G R R N A I H **E**	240	104
Multiple substitutions			
57	T T Y A D F I A **L I** R T G R R N A I[a]	130	58
58	I **A A** G R T G R R N A I H **E** I L V S S A	780	104
59	I **A A** G R T G R R N A I H E	800	104
60	Y A D F I **A A** G R **U** G R R N A I H D I L V **A A A**	12,000	104

Note: [a] = COOH-terminal amide; S^p = phosphoserine; U = aminobutyrate.

(peptides 35 and 38). Even though lysine is seen in some protein substrates in lieu of arginine, the effect of this substitution on inhibitory potency in all three positions of the PKI peptide was quite marked (peptides 35 to 37) and, although tested with different parent peptides, the changes were of the same order of magnitude as seen with a glycine substitution. Some apparent differences in the relative degree of change were seen in the substitution of arginine-18 vs. arginine-19 by lysine (peptides 36 and 37) compared to glycine (peptides 39 and 40). Whether this is due to the residue that was substituted or the parent peptide into which it was substituted has not yet been evaluated. The sequence of arginines in the inhibitor protein is directly comparable to both of the two proposed basic subsite sequences of protein kinase substrates (–Arg–Arg–X–Ser or –Arg–X–X–Arg–X–X–Ser–; see discussion in Section II on the protein kinase). Thus, it appears that for full inhibitory potency (and likely also for optimal substrate effectiveness) the full complement of all three arginines is required to maximize interaction with the protein kinase catalytic site. In the inhibitor peptides, the amino acid in the position equivalent to the phosphorylatable serine (in substrates) is alanine. Since in Ala-Kemptide the sequence –Arg–Arg–X–Ala– is far from sufficient to produce an effective inhibitor of the protein kinase (i.e., K_i of Ala-Kemptide = 348 μM), and D-Ser-Kemptide is totally ineffective as either a substrate or an inhibitor,[48] these data alone indicate that in addition to this arginine basic subsite there must be one or more other critical components of the PKI peptides (and of the inhibitor protein itself) that are necessary to create a high-affinity inhibitor. However, as illustrated by the large effects created by substitution of any one of these three arginines (and especially arginine-18 and arginine-19), a fully constituted basic subsite is critical for high potency, as seen with protein kinase peptide substrates.

In addition to the requirement for the three arginines, several other residues in the COOH-terminal half of PKI(6—22)amide are clearly also of considerable importance. So far all of the residues from arginine-15 to isoleucine-22 have been tested by at least one type of substitution in at least one (and for several residues two) parent peptides (Table 6). Surrounding arginine-18 and arginine-19, the substitution of either asparagine-20 with alanine (peptides 43 and 45) or glycine-17 with leucine (peptides 44 and 47) results in a 7- to 15-fold decrease in inhibitory potency, with substitution of both (peptide 48) producing a synergistic effect. As discussed below, this region may contain a β-bend (as has also been suggested for some peptide substrates of the protein kinase[46,47]), and it is possible that these substitutions disrupt this structure. Compatible with this was the observation that substitution of asparagine-20 by glycine (peptide 46) did not produce as large a change as occurred with alanine (peptide 45). Further information is required to elucidate what is indeed occurring with these substitutions of asparagine-20 and glycine-17. It is not yet certain that this region does contain a β-bend,[58,68] and for peptide substrates the presence of a β-bend structure in an equivalent region is also quite questionable (see discussion in Section II and References 44 and 45). Whether this region does or does not contain secondary structure, it is nevertheless clear that both asparagine-20 and glycine-17 make important contributions to inhibitory potency.

Another residue of considerable importance to inhibitory activity is isoleucine-22. Since its substitution by glycine (peptides 52 and 54) greatly decreased inhibitory potency, but substitution by leucine (peptides 51 and 53) produced a much more modest decrease, it appears likely that it is the bulky hydrophobicity of isoleucine-22 that is important. This is similar to what has been suggested for peptide substrates of the protein kinase (see Section II and Table 1). Substitution of alanine-21 by either phosphoserine (peptide 49) or α-aminobutyric acid (peptide 50) also caused considerable decreases in inhibitory potency. Given that with Kemptide the substitution of an equivalent serine by alanine resulted in the conversion of a good substrate (K_m = 5 μM) into a poor inhibitor (K_i = 348 μM), this observation was somewhat surprising. Possibly alanine-21 itself makes a minimal contribution to inhibitory activity, but the additional charge of phosphoserine or the bulk of aminobutyrate cannot be tolerated in the binding interaction with the protein kinase. Substitutions for threonine-16 so far have produced ambiguous results. Aminobutyrate (peptide 41) was without effect, whereas in a shorter peptide substitution by alanine diminished inhibitory activity fourfold (peptide 42). This difference may be due to the substitution residue tested or type (i.e., length) of parent peptide used for the test. This has yet to be resolved.

Of all the residues of PKI(6—22)amide, only serine-13 and glycine-14 have not yet been examined by a single amino acid substitution. Some partial insight can be obtained, however, from the construct in which both were substituted by the bulky hydrophobic groups of leucine and isoleucine, respectively (peptide 57). This caused a 40-fold increase in the K_i (relative to the parent peptide P1), pointing to a critical role for one or both of these residues, either in the structure of PKI(6—22)amide or as recognition signals. Which is the case remains to be evaluated, but it may be of importance that this is a region of the peptide where a β-bend is a likelihood.[58,106]

Although PKI(6—22)amide is the most fully active peptide, substitutions outside of these 17 residues also have caused changes in inhibitory potency. Such effects, however, are likely the result of introducing a residue whose presence is not allowed for full inhibitory potency rather than reflecting a positive contribution of that residue in the native protein. An example of this is the substitution of histidine-23 by glycine (peptide 55). This resulted in a fivefold increase in the K_i, even though the elimination of histidine-23 (and aspartate-24) did not affect inhibitory potency (i.e., peptide P2 vs. P1, Table 4). One possible reason in this case is that the added flexibility in the peptide chain which would occur with glycine present in peptide 55 allowed the negative effects of residues 25 to 30 to be more pronounced. Substitution of aspartate-24 by glutamate (peptide 56) also appeared to result in a decrease in potency (when compared with the activity of peptide 12; Table 4), but the correct parent peptide for this comparison has not yet been synthesized. Aspartate-24 substitution also has been examined in two doubly substituted peptides (58 and 59), but without equivalent single substitutions the individual contributions of the two different residues are not clear.

C. PKI PEPTIDES AS A MODEL FOR SUBSTRATES OF THE cAMP-DEPENDENT PROTEIN KINASE

Considerable insight has been obtained by comparison of PKI peptides with Kemptide and its inhibitory analogue, Ala-Kemptide. As noted in a previous discussion in this chapter (Section II), Kemptide, the peptide derived from the phosphorylation site of hepatic pyruvate kinase, is a highly effective substrate of the cAMP-dependent protein kinase and exhibits K_m and V_{max} values essentially identical to those obtained with the parent protein.[107] Thus, Kemptide appears to contain all the required "substrate features" of at least one physiological substrate for the kinase. Ala-Kemptide, however, is fairly impotent as an inhibitor. Given these observations, we have examined those features which distinguish Ala-Kemptide from some of the short PKI peptides of high inhibitory potency. We also have asked whether Kemptide would produce an even better substrate if those features present in the PKI peptides were provided. As noted in Section IV.B above, the nonapeptide P4 is a highly effective inhibitor (Tables 4 and 7). When it is modified by the combined substitution of glycine-17, asparagine-20, and isoleucine-22 (by leucine, alanine, and leucine, respectively; peptide 62), its inhibitory potency is diminished quite substantially, as would be expected from what occurs when each residue is substituted singly (peptides 47, 45, and 53). Peptide 62 is of interest, however, because it differs from Ala-Kemptide only by the substitution of the COOH-terminal amide by glycine and the addition of Gly[14]–Arg[15]–Thr[16]; peptide 62 is 30-fold more potent than Ala-Kemptide. One or more of the additional residues thus has a major impact on the effectiveness of peptide 62 as an inhibitor. Part or all of this may be

TABLE 7
Comparison of PKI- and Kemptide-Derived Inhibitors and Substrate Peptides of the cAMP-Dependent Protein Kinase

Peptide number	Peptide sequence	K_i or K_m (nM)	Ref.
Inhibitors			
P4	G R T G R R N A I[a]	36	58
61	**A A A** G R R N A L[a]	3,800	58
62	G R T **L** R R **A A** L[a]	11,000	58
Ala-Kemptide	L R R A A L G	348,000	28
Substrates			
Kemptide	L R R A S L G	4,700	28
63	G R T G R R N S I[a]	110	58
64	I A S G R T G R R N S I H D I L V S S A	280,000	104

Note: [a] = COOH-terminal amide.

due to the presence of arginine-15, since it is clear from the data with peptides 35 and 38 that this arginine contributes to the binding of the peptide to the protein kinase. However, because of the magnitude of the difference in inhibitory potency between peptide 62 and Ala-Kemptide, it appears that glycine-14 and/or threonine-16 also may contribute to binding affinity. This needs clarification. Another construct, peptide 61, is also of considerable interest in comparison to Ala-Kemptide. It is nearly 100-fold more potent than Ala-Kemptide, but differs from it by the presence of glycine-17 and asparagine-20 and three alanines at the NH$_2$-terminus. It might be presumed that these three additional alanines are inert. As noted above, however, glycine-17 and asparagine-20 make an important contribution to the inhibitory potency of PKI-derived peptides and therefore presumably are critical factors in the still substantial inhibitory potency of peptide 61.

The studies with Ala-Kemptide and PKI peptides, as discussed above, strongly suggested that, although Kemptide is a good substrate for the protein kinase, it may not be as good as a short peptide might possibly be. When this was initially investigated by Scott et al.[104] using the construct (Ser^{21})PKI(11—30), disappointing results were obtained, since it exhibited a K_m of only 280 µM. This result, however, can now be seen to underrepresent the full potential of PKI peptides to serve as substrates. (Ser^{21})PKI(11—30) is compromised by the presence of Ile^{11}–Ala^{12}–Ser^{13}– at the amino terminus and residues 25 to 30 at the COOH-terminus, both of which diminish potency in PKI inhibitory peptides. Thus, in illustration, peptide P5 has a greater activity than peptide 17, and peptides P3, P6, and 21 are markedly less active than peptides 12, 17, and P5, respectively. We now believe that the activity of (Ser^{21})PKI(11—30) as a substrate is severely limited because it contains these regions that reduce effective protein kinase binding. This conclusion prompted a reexamination of whether the PKI peptides might serve as models for protein kinase substrates and especially whether the features of these peptides that allow high-affinity binding as an inhibitor would also promote high efficacy as a substrate. The construct (Ser^{21})PKI(14—22)amide was thus examined, and it proved to be an excellent substrate for the protein kinase, with a K_m 40-fold lower (Table 7) and a V_{max} 2-fold higher[58] than those exhibited by Kemptide. Other peptide substrates are now being examined, but it is clear from this information alone that those features of PKI(14—22)amide that contribute to its efficacy as an inhibitor (in addition to arginine-18 and arginine-19) also contribute to the efficiency of (Ser^{21})PKI(14—22)amide as a substrate.

Interestingly, although Kemptide mirrors the requirements of a physiological substrate of the cAMP-dependent protein kinase and is identical in this regard to hepatic pyruvate kinase,[107] nevertheless it apparently has not "evolved" to take advantage of all of the possibilities that could make it bind maximally to the protein kinase. Perhaps, however, there are good reasons for this. One distinction between the inhibitor protein and protein kinase substrates is that, whereas substrates must not only bind, but also (once phosphorylated) must be released efficiently, the inhibitor protein, once bound, might remain in the catalytic site for a substantially longer period. This could readily be one reason why the inhibitor protein is bound more efficiently and tightly to the protein kinase. This alone, however, does not fully explain why natural substrates do not appear to have "evolved" to optimize binding to the protein kinase. (Ser^{21})PKI(14—22)amide, as noted, is an efficient substrate with not only a high affinity but also a high V_{max}, and in this regard is the best substrate found so far for the protein kinase, artificial or otherwise. Thus, it is possible to have a protein substrate for the protein kinase that both binds with much higher affinity than pyruvate kinase (and Kemptide) and is also efficiently released upon being phosphorylated. The question that therefore must be addressed is whether or not there is a physiological reason why protein kinase substrates apparently have not "evolved" to be "better" substrates, given that there is clearly a potential for this to occur, when judged by the parameters of K_m and V_{max}. Perhaps this is physiologically important. The cAMP-dependent protein kinase is known to phosphorylate a number of substrates in even a single cell. Currently, very little is known about the requirements that must exist for one enzyme, the cAMP-dependent protein kinase, to "service" several substrates. Little is known, for example, regarding what dictates substrate preference following a cAMP signal. One possibility is that the different proteins are better or worse substrates for the protein kinase (in terms of K_m and V_{max}) so that they will be modulated in the correct sequence necessary to obtain the end physiological event. As noted above (Table 1), the protein substrates for the cAMP-dependent protein kinase frequently contain lysine as one of the residues in the basic subsite, even though with simple peptides a pair of arginines clearly results in the optimum kinetic parameters *in vitro*. This is seen with both Kemptide analogues (Table 1) and PKI peptides (Table 6). Furthermore, the spacing of this basic subsite from the phosphoryl acceptor serine in physiological protein substrates is also not always apparently optimal when compared to what is seen with the peptide substrates *in vitro* (Table 1). Perhaps all of these types of differences are related to what is required of physiological substrates, and there needs to be a preference for the order of substrate phosphorylation for the end physiological response. In studies of the catalytic mechanism of the cAMP-dependent protein kinase, therefore, it will be important to explore how several substrates bind at the catalytic site. There may be important dissimilarities between different protein substrates in their interactions with the protein kinase. These may reflect the different binding components that are seen with PKI peptides, and these differences may have important physiological ramifications.

D. THREE-DIMENSIONAL STRUCTURE OF PKI PEPTIDES

The early observation that the inhibitor protein was stable both to heat and to acid was thought to mean that in all likelihood it had no secondary or tertiary structure and that inhibitory potential was solely a consequence of a part or parts of the amino acid sequence. This is now known not to be true, at least for PKI peptides,[68] and thus it is unlikely to be true for the native inhibitor protein. Given this, it now seems most probable that either the inhibitor protein is stable to such denaturing conditions (high temperature or low pH), or it can readily regain its secondary and tertiary structure upon return to more physiological buffer conditions. Given its small size, the latter would not be surprising.

The first evidence that PKI peptides had structure was derived from circular dichroic (CD) investigations.[106] It was discerned that the far-UV CD spectra of PKI(5—22)amide differed quite noticeably from that of a peptide in an extended random coil, notably in possessing a negative ellipticity from 220 to 240 nm. Curve fitting was then applied (as reproduced in Figure 7), and a strong similarity was seen between the spectra of

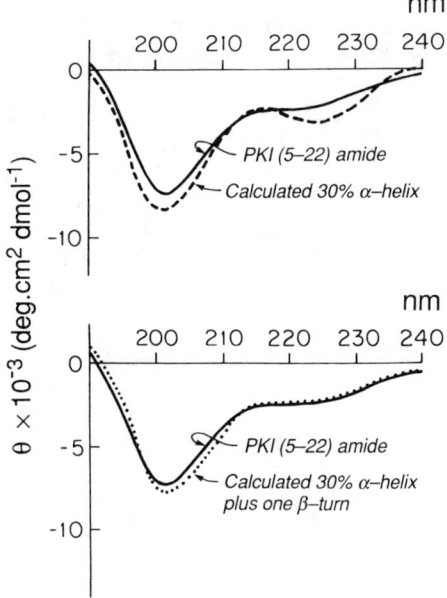

FIGURE 7. Circular dichroic spectra of PKI(5—22)amide. Upper: comparison of PKI(5—22)amide (——) with a calculated spectrum for 30% α-helix (- - -). Lower: comparison with a calculated spectrum containing 30% α-helix and one β-turn (····). (Data from Reference 106.)

PKI(5—22)amide and that of a model peptide with 30% α-helix. Such data were found to be compatible with Chou and Fasman analysis, which predicted a section of such structure in the NH_2-terminal portion of PKI(5—22)amide. The computed peptide CD spectrum with 30% α-helix, however, while clearly being quite a good fit, did not fully match that obtained experimentally with PKI(5—22)amide. Since Chou and Fasman analysis also suggested the likelihood that PKI(5—22)amide contains a β-bend, this parameter was added to the computed spectrum. With the summation of the spectra contributed by a Type III/I β-bend[106] and 30% α-helix with the remainder being extended coil, an excellent fit was observed with the experimental data (Figure 7). As discussed in Reference 106, certain reservations should always be borne in mind in the application of curve fitting to the interpretation of CD spectra. Nevertheless, the parallels between the CD data and Chou and Fasman analysis provided the first assessment of the structural components of PKI(5—22)amide. A compact structure for this peptide, as suggested by these CD studies, has since proved compatible with size and shape determinations deduced by X-ray solution scattering.[68] Viewed by this technique, PKI(5—22)amide exists in solution as a compact molecule with a maximum linear dimension of 25 Å and an average radius of gyration (R_g) of 9 Å. The X-ray scattering pair distance distribution function, P(r), was found to be quite symmetrical,[68] demonstrating the peptide to be a nearly globular molecule, and fitting well with its overall shape being that of a prolate ellipsoid of dimensions 25 × 17 Å.

The task of further refining the knowledge of the structure of PKI(5—22)amide has been approached by the application of NMR[68] and Fourier transform infrared (FTIR) spectroscopy. The use of the now extensive range of available peptide analogues, as discussed above, and the coupling of the structural data to information on inhibitory activity have provided an additional dimension for inquiry. So far, our best knowledge of the location of the α-helix in PKI(5—22)amide comes from NMR. Unfortunately, the size of

TABLE 8
Chemical Shifts of the Nonexchangeable Alanine and Isoleucine Methyl Protons in PKI(5—22)amide and Substituted Derivatives

		Native peptide	Monosubstituted peptides					
RCP[a]	Residue	(5—22)	Ala[6]	Ala[7]	Leu[8]	Ala[9]	Ala[10]	Ala[11]
0.89	Ile $CH_3\delta$	0.86[T]	0.86	0.87	0.86	0.86	0.87	0.86
		0.81[11]	0.81	0.85	0.83	0.81	0.83	abs[b]
0.95	Ile $CH_3\gamma$	0.93[T]	0.92	0.93	0.93	0.93	0.93	0.92
		0.84[11]	0.84	0.88	0.86	0.84	0.88	abs[b]
1.39	Ala CH_3	1.31[8]	1.31	1.34	abs[b]	1.31	1.35	1.27
		1.36	1.35	1.38	1.37	1.32	1.38	1.34
			1.37	1.42		1.37	1.40	1.37
		1.42	1.42	1.42	1.42	1.41	1.42	1.42

Note: Abbreviations for peptides are: peptide (5—22), PKI(5—22)amide; Ala[6], (Ala[6])PKI(5—22)amide; Ala[7], (Ala[7])PKI(5—22)amide; Leu[8], (Leu[8])PKI(5—22)amide; Ala[9], (Ala[9])PKI(5—22)amide; Ala[10], (Ala[10])PKI(5—22)amide; Ala[11], (Ala[11])PKI(5—22)amide. Superscripts to the resonance numbers denote residue number in the sequence, with T denoting the COOH terminal.

[a] Random coil position.
[b] Absent.

Data from Reference 68.

PKI(5—22)amide (~2000 mol wt) so far has prohibited the simple application of the more powerful nuclear Overhauser effect spectroscopy (NOESY) techniques to the solution of the structure, so more classical NMR approaches have been needed. For this peptide, two-dimensional NOESY did not reveal significant interresidue NOE signals above background, and even intraresidue NOEs, such as with tyrosine ring protons, were very weak. (With either a much smaller peptide or a larger protein this would not have been a difficulty; see discussion in Reference 68.) In the absence of what would otherwise have been the method of choice for making residue assignments and examining through-space interactions, use of the singly substituted peptides 29 to 34 proved to be a highly effective alternative. For example, depicted in Table 8 are the methyl proton resonances assigned from the NMR spectra for PKI(5—22)amide and six of the singly substituted NH$_2$-terminal analogues. (The inhibitory activities are given in Table 5.) The absence of specific proton resonances in those analogues where a particular residue has been replaced (i.e., Ala[8], 1.31 ppm or Ile[11], 0.84 ppm; Table 8) allows the assignment of resonances to the corresponding unsubstituted residues in the spectra of PKI(5—22)amide. The resonances presented in Table 8 also show interesting through-space interactions. For example, the elimination of tyrosine-7 caused the same degree of spectral shift on isoleucine-11 (0.84 to 0.88 ppm) as was observed with the elimination of phenylalanine-10. The latter is clearly because of the removal of the neighboring bulky hydrophobic group. However, that the elimination of tyrosine-7 had an effect of the same magnitude must mean that it is located very close to isoleucine-11. The distances between isoleucine and the phenylalanine and tyrosine rings must be quite similar even though tyrosine is located several residues along the peptide chain. Similarly, a shift in the isoleucine-11 methyl proton resonance is also seen when alanine-8 is replaced by the bulky hydrophobic residue leucine. The most likely cause for such ca. three to four residue interval interactions is that this stretch of peptide is in an α-helix. These and several related observations from NMR investigation[68] have provided good evidence that a primary element of the structure of PKI(5—22)amide is the existence of an α-helix in the NH$_2$-terminal third of the peptide. Such a hydrogen-bonded structure finds support from hydrogen exchange

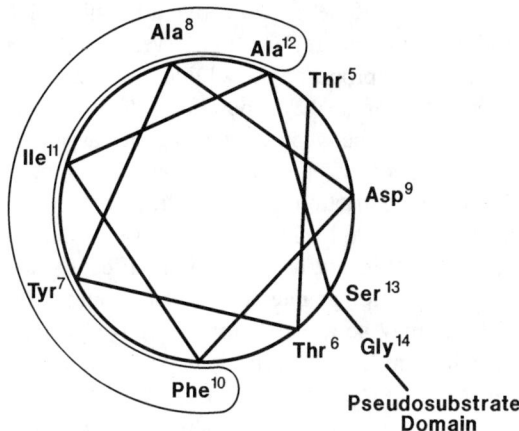

FIGURE 8. Helical wheel diagram of PKI(5—22)amide illustrating the amphiphilic nature of the NH$_2$-terminal α-helix. (From Reed, J., de Ropp, J. S., Trewhella, J., Glass, D. B., Liddle, W. K., Bradbury, E. M., Kinzel, V., and Walsh, D. A., *Biochem. J.*, Vol. 264, 1989. With permission.)

rates of the peptidyl amides, mutual shielding of areas in this part of the peptide by residues roughly three residues apart (as briefly discussed above), and a clustering of amide proton $^3J_{HN\alpha}$ values below 6 Hz for residues in this area of the peptide.[68] The helix seems to be located between threonine-5 or threonine-6 and isoleucine-11 or alanine-12, although the precise placement has not been established. This helix is seen to have a major hydrophobic face, in the middle of which (longitudinally) is phenylalanine-10 (Figure 8). Of interest, even though the substitution of phenylalanine-10 provoked a major decrease in inhibitory potency (Table 5), neither FTIR nor NMR data revealed any change in the secondary structure of the resultant peptide. This may indicate that the phenylalanine is recognized specifically by a hydrophobic pocket at the protein kinase catalytic site and that the hydrophobic face of the amphiphilic helix maximizes this interaction. In all of the substitution studies so far, such as illustrated in Tables 5 and 8, the type of substituent chosen (i.e., alanine or leucine) would not have been predicted to compromise the integrity of the α-helix, and by the NMR spectra no changes were observed. Other studies (now in progress) are needed where an inserted residue would be predicted to disrupt the helix. If the current structural conclusions are correct, a substitution that would disallow the helix would also have a major impact on inhibitory potency.

The NMR studies pursued to date[68] have also provided other insights. Most notably, the PKI(5—22)amide structure was found to be very stable and could only be converted into a disorganized conformation by very severe denaturation conditions. One other tentative possibility suggested by the NMR spectra, particularly from the substitution analogues and pH titration, is that aspartate-9 may be linked by a salt bridge to one of the arginine side chains.

The potential location of a β-bend, suggested by the CD spectra, has yet to be confirmed. The possible sites for a β-bend, as suggested by Chou and Fasman analysis, are either between glycine-17 and asparagine-20 or at one of the three possible combinations available in the region of alanine-12 to glycine-17. The substitution studies described above (Table 6) do not allow a distinction between these possibilities, since amino acid replacements in both of the locations that would have compromised the existence of the β-bend were found to diminish inhibitory potency.

From the various studies that are described above a possible picture of the structure of PKI(5—22)amide is beginning to emerge, although several of the specifics remain preliminary. Presented in Plate 1* is one proposed model that includes the localization of a stretch of α-helix from threonine-5 to alanine-12, a possible salt bridge between aspartate-9 and arginine-15, and a β-bend located between glycine-17 and asparagine-20 (one of the two possible regions suggested by Chou and Fasman analysis). From modeling studies it can be appreciated that even with such restrictions as these structures would impose there remains considerable flexibility in the peptide backbone, especially in the carboxy-terminal half of the peptide. (Such flexibility would not be reduced significantly with any of the other β-bend localizations.) There may be other interactions that would make for a more restricted structure, but such flexibility may be a reality for the native peptide in solution and could even contribute to the mechanism by which it interacts with the protein kinase. A question of importance, if this suggested structure were indeed correct, would be whether the structure remains intact when the peptide is bound in the active site. As discussed above, with the peptide substrate Kemptide, Granot et al.[44] and Bramson et al.[45] deduced that it does not contain a β-bend when bound at the catalytic site. By analogy, the location of a β-bend between glycine-17 and asparagine-20 would appear to be inconsistent with such a conclusion. Possibly, the presence or absence of a β-bend at such a location might distinguish between an active substrate and a potent inhibitor, but given that (Ser[21])PKI(14—22)amide is a very good substrate, such an alternative would not seem likely.

From all of the studies that are summarized above we are beginning to achieve a good understanding of how the peptide PKI(6—22)amide (and presumably also the native inhibitor protein) interacts with the catalytic site of the protein kinase. A current schematic is depicted in Figure 9. It appears that the secondary-structural elements of the peptide are critical for the high-affinity interaction. Presumably, they are essential for appropriately spacing and orienting those specific residues of the peptide that bind directly to key residue(s) within the catalytic site of the protein kinase. Residues of the peptide which appear to be recognized directly by the catalytic site include phenylalanine-10, arginine-18, arginine-19, and isoleucine-22. In addition, threonine-6, tyrosine-7, serine-13 and/or glycine-14, arginine-15, glycine-17, and asparagine-20 make major contributions to the structure and/or serve as additional recognition sites for the protein kinase. Threonine-6 and tyrosine-7 most likely play a structural role in the establishment of the α-helix (as suggested specifically from the analysis presented in Figure 6). The two glycine residues could be important for the structure, but it simply may be that a residue of greater bulk cannot be tolerated within the catalytic site. It appears likely that either serine-13 and glycine-14 or glycine-17 and asparagine-20 are components of the critical β-bend of the peptide. However, that being so, since a β-bend at both locations appears improbable, the nonglycine residue that is *not* in the β-bend (i.e., serine-13 or asparagine-20) is likely to be an additional residue recognized directly by the protein kinase.

As illustrated by Figure 5, PKI(5—24) in fact exhibits only 20% of the potency of the inhibitor protein; thus, the rest of the protein not yet examined (i.e., residues 31 to 75) also must be making some contribution to protein kinase binding. There are currently no data that address this question. The COOH-terminal two thirds of the inhibitor protein also makes an important contribution to its specificity. PKI peptides are able to inhibit the cGMP-dependent protein kinase[83] (Table 9), and although they are several orders of magnitude weaker inhibitors than for the cAMP-dependent enzyme, they nevertheless inhibit in the micromolar range. In distinct contrast, the inhibitor protein is totally ineffective in inhibiting the cGMP-dependent kinase, even when tested at 100 μM.[83] Of interest, PKI(14—24)amide

---

* Plate 1 follows Page 72.

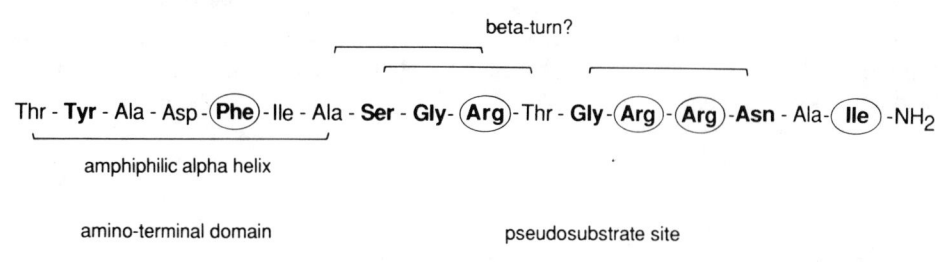

FIGURE 9. Schematic of the primary recognition residues and structure of PKI(6—22)amide.

TABLE 9
Comparison of the Inhibition of cAMP-Dependent
Protein Kinase and cGMP-Dependent Protein Kinase by
Inhibitor Protein and PKI Peptides

Peptide	$\dfrac{K_i \text{ cGMP-dependent PK}}{K_i \text{ cAMP-dependent PK}}$
PKI(14—24)amide	526
PKI(10—24)amide	739
PKI(7—22)amide	2,296
PKI(5—22)amide	55,500
Inhibitor protein	∞

Data from Reference 83.

was slightly more potent in inhibiting the cGMP-dependent kinase than PKI(5—22)amide, the reverse of what was seen with the cAMP-dependent enzyme. Thus, the high specificity of the inhibitor protein is provided both by a region in the active α-helix domain and (potentially more importantly) by some area or areas within residues 25 to 75.

V. THE USE OF INHIBITOR PROTEIN AS A TOOL TO STUDY cAMP ACTION

The inhibitor protein has been used extensively as an effective means to evaluate the role of cAMP in mediating cell regulation, both *in vitro* to classify protein kinases and cAMP-mediated protein phosphorylation and also by its introduction into intact cells. Its high specificity and high affinity make it very suitable for such purposes. Certain precautions are, however, necessary. Commercial preparations of the inhibitor protein that have been available are not highly purified, and impure preparations have been reported to contain several components that could readily compromise interpretations.[108-110] A pure preparation of PKI(5—22)amide is marketed under the name "Wiptide" (Peninsula Laboratories, Belmont, CA). This preparation should alleviate several of the problems that arise with partially purified protein, but given (as indicated above) that it does not exhibit the same apparent absolute specificity as the native inhibitor protein, more care is needed in selecting the concentrations utilized.

The three approaches that have been used so far to introduce the inhibitor protein into cells have been microinjection (see references below), liposome fusion,[20] and the introduction of a plasmid containing a synthetic gene encoding PKI(1—31).[21]* Each of these approaches has provided valuable information on the specific responses studied. These have included cAMP-regulated gene expression,[21] mitosis,[15] and oocyte maturation;[13,19,111] cardiac cell Ca^{2+} transport,[14] Na^+ efflux from barnacle muscle,[17,112-115] K^+ conductance in gill motor neurons of *Aplysia*,[16,116] and gap-junctional conductance of hepatocytes;[18] and corticotrophin-stimulated adrenocorticotrophic hormone (ACTH) release.[20,117] In each of these approaches the use of the inhibitor protein has allowed conclusions to be made about the role of the cAMP-dependent protein kinase in these various systems, and with the levels of inhibitor protein incorporated into the cells by these various techniques it has been possible to block the protein kinase for even up to several hours. Clearly the incorporation of the inhibitor protein peptide by means of a regulated plasmid allows for a more persistent and repetitive level of inhibitor to be attained, which has distinct advantages. Further work on the recognition factors inherent in PKI peptides might allow the synthesis of a peptide (or a compound based upon this structure) which could be taken up readily into cells and not subsequently inactivated. The strategy for this will be quite complex.

Currently, none of these experiments where the inhibitor protein was furnished exogenously has provided much information on the inhibitor protein *per se*, since it usually has not been readily feasible to equate the amount of inhibitor protein introduced exogenously with the amount present endogenously in the cell. At best, these experiments indicate that, when present, and if present in the appropriate cellular location, the inhibitor protein could modulate cAMP-dependent protein kinase activity; however, this is no more than would be expected from *in vitro* results.

In a second area of technique development the inhibitor protein has proven to be an effective tool to examine the cytochemical localization of the protein kinase catalytic subunit. This approach, pioneered by Fletcher and Byus,[118,119] is based upon using fluoresceinated inhibitor protein as a specific cytochemical probe to distinguish free catalytic subunit from that present as holoenzyme. This allows the examination of protein kinase dissociation as well as the cellular compartments in which or to which the catalytic subunit, once activated, becomes associated. As a companion to this technique, fluoresceinated catalytic subunit can be used to examine the localization of free regulatory subunit, again in distinction from regulatory subunit associated with holoenzyme.[120] These cytochemical probes for protein kinase localization have now been applied to several cell systems.[121-125]

VI. THE PHYSIOLOGICAL FUNCTION OF THE INHIBITOR PROTEIN

A. EXPERIMENTAL APPROACHES

Assay of the cellular content of the inhibitor protein has most frequently relied upon the measurement, in a denatured dialyzed cell extract, of an activity that blocks cAMP-dependent protein kinase catalysis.[126-130] Inherent in such an assay are several problems that can compromise an accurate determination. Because many components of a cell extract could spuriously inhibit the protein kinase, including nonspecific inhibitors[131] or possibly even fragments of the regulatory subunit of the protein kinase, it is doubtful that the inhibition caused by such extracts can be attributed to the inhibitor protein alone. An additional complication is the possible formation of nonphysiological protein kinase substrates as a consequence of denaturation.[132] At a minimum, the trypsin sensitivity of the inhibitory activity in the extract should be evaluated.[93,130] A better approach is to first remove the bulk of non-inhibitor protein through some type of separation methodology based upon the physicochemical characteristics of the inhibitor protein. Those employed have been either gel filtration[131] or DEAE chromatography.[93] Even this is not fully satisfactory since differences

---

* Day et al.[148] have recently described a plasmid that expresses the full inhibitor protein in mammalian cells.

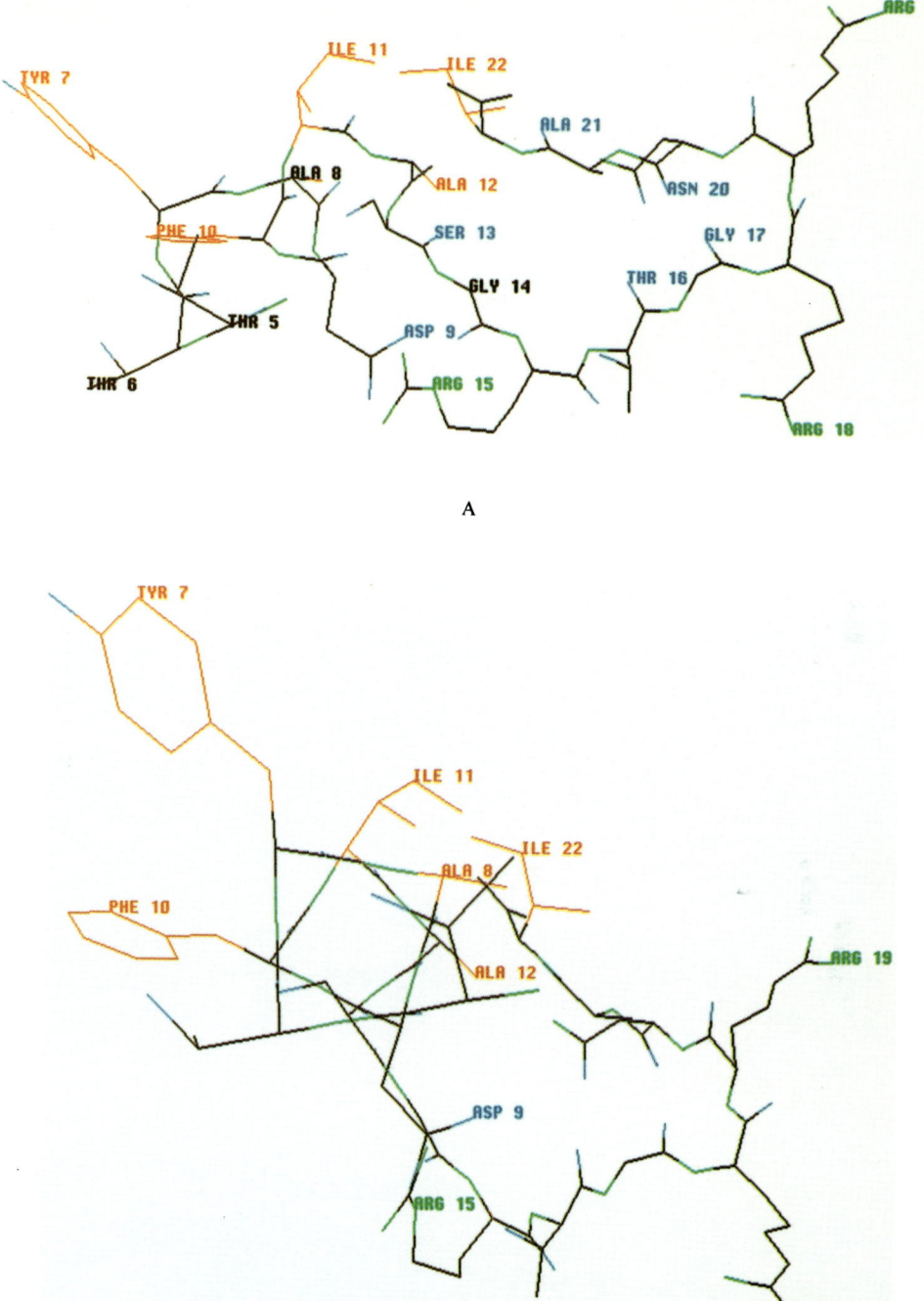

PLATE 1. Model of PKI (5—22) amide. This model was computed using the University of California-San Diego Molecular Modeling System with an Iris graphics terminal. Incorporated into the model is an α-helix from residues 5 to 12, a β-bend at residues 17 to 20, and a salt bridge between residues 9 and 15. The remainder of the molecule was conformed to provide a maximum linear dimension of 25 Å (as discerned by X-ray scattering) and shows the possibility of Leu-22 being located within the hydrophobic domain created by the α-helical NH_2-terminus. Panels A and B provide two views, with α-helix either nearly traverse (A) or end-on going into the page (B). Both views, in particular, illustrate the possible close relationship of Ile-22 with hydrophobic helical domain containing Phe-10.

in the inhibitor protein between tissues may exist, as clearly seems to be the case with the species as isolated from testis (see Section III.D above). Thus, it cannot simply be assumed that inhibitor protein from different tissues will necessarily always fractionate identically. In addition, yield through these partial purification steps should be monitored. In assaying the inhibitor protein it is also important to use a clearly defined protein kinase concentration. As a tightly bound inhibitor, the effect of inhibitor protein on activity is particularly sensitive to the concentrations of both the inhibitor protein and the protein kinase.[93] Furthermore, inactive protein kinase can also bind the inhibitor protein,[69] which can also lead to low estimates of inhibitor activity. Because of differences in assay conditions as well as in the purity and specific activity of the protein kinase utilized, assays of the relative amounts of the inhibitor protein performed by different laboratories have not so far been readily comparable. A radioimmunoassay for the inhibitor protein would be of benefit. Antibody to a 9.3-kDa species of testis inhibitor has been elicited,[77] but its cross-reactivity to other tissue species of inhibitor protein was not reported and, as discussed above, the best characterized form from testis (15-kDa molecular weight) appears distinct from that purified from skeletal muscle. So far, inhibitor protein from skeletal muscle has been refractory to extensive efforts to obtain a suitable antibody, and thus an RIA has not been possible.

Within the previously expressed caveats on the problems of the assay, some data on the cellular content of the inhibitor protein and its possible variation between tissues are available. The inhibitor protein has been found in a number of species, including rabbit, rat, beef, and chicken, and there is also a report of its presence in *Neurospora*.[133] In the rabbit[93] and chicken[134] the highest levels are present in brain, skeletal muscle, heart, and testis, with apparently much less in other tissues. As noted above, such assays may not have provided a truly accurate assessment, but in the study of rabbit tissues, where several physicochemical criteria of identity were applied,[93] the measurement (if in error) probably would have underestimated the amount. In rat heart and rabbit skeletal muscle the amount of inhibitor protein present has been determined to be sufficient to block about 20% of the total cellular content of cAMP-dependent protein kinase.[5] In subcellular distribution studies of the testis inhibitor it was shown to be primarily present in the cytosol, but also associated with other subcellular membrane fractions.[74] Cytochemical localization studies in dividing rat kangaroo PtK2 cells using anti-testis inhibitor antibody revealed inhibitor protein associated with cytoplasmic microtubules during interphase and with the spindle apparatus during mitosis.[15]

Again with the caveat of potential problems with the assay, changes in the amount of inhibitor protein in cells have been reported with several physiological perturbations. In Chinese hamster ovary cells stimulated by serum addition, inhibitor protein levels were noted to fluctuate during the cell cycle inversely with the level of protein kinase activity.[127] In chick kidney the amount of inhibitor protein was increased with vitamin D deficiency, also concomitant with a decrease in protein kinase activity.[126] On the other hand, in mouse kidney, vitamin D removal increased inhibitor protein levels, but was without effect on the protein kinase.[135] These effects of vitamin D were restricted to the kidney and not seen in other tissues.[134] Inhibitor protein levels in the pancreas have been reported to decrease in alloxan-induced diabetes.[136] One problem in the interpretation of these studies is that even though there appeared in some instances to be a reciprocal relationship between inhibitor protein and cAMP-dependent protein kinase activity, their activities relative to one another were not evaluated. In such tissues as kidney and pancreas the total amount of inhibitor protein appears to be too low to alter protein kinase activity significantly. Thus, the magnitude of the observed changes in inhibitor protein activity would likely make little contribution to modifying the activity of the endogenous cAMP-dependent protein kinase. In other studies with heart, brain, and testis, changes in the inhibitor protein have been seen which would seem more likely to affect protein kinase, as the level of inhibitor protein in these tissues is apparently sufficient to modulate protein kinase activity somewhat more markedly. It has been shown in normal rat heart that the level of inhibitor protein is sufficient to block about 20% of the endogenous cAMP-dependent protein kinase. This level has been reported to be depressed approximately twofold by either β-adrenergic agents,[131,137] exogenous dibutyryl cAMP addition,[137] or alloxan treatment[5] and with starvation was decreased about 30%, but then elevated to about double the initial normal value with subsequent refeeding.[5] Brain inhibitor protein, which is also present at levels sufficient to block about 20% of the total cAMP-dependent protein kinase, was likewise depressed ~50% by β-adrenergic stimulation when examined in hippocampus, brain stem, and pineal gland[1,138-140] and was decreased a similar amount by apomorphine, likely acting through the D_1-dopamine receptor. Another study of interest examined modulation of testis inhibitor protein (although as noted above it is likely different from the species present in muscle or brain). Rat testis inhibitor (concentration?) shows a marked increase during development, elevating several fold from birth to maturity.[130] Furthermore, it was reduced in the adult five- to sixfold by hypophysectomy, but was restored by follicle-stimulating hormone administration.[130] These are the most marked changes in inhibitor protein so far reported. A good correlation with reduction in cAMP-dependent protein kinase was reported with some of these changes, and it was calculated that as much as 41% of the protein kinase activity might be blocked in the cell as a consequence of the presence of inhibitor (determined at age 19 d). Using cycloheximide addition it was found that the $t_{1/2}$ of the testis inhibitor in otherwise normal animals was about 10 h.[74] As with testis, the inhibitor protein in rat adipose[128] and chick liver[65] has been reported to increase with ontogeny, whereas that in chick skeletal muscle[129] appears to remain fairly stable. All of these studies on possible fluctuations in inhibitor protein under various physiological manipulations should be regarded as preliminary. As noted above, the assay of inhibitor protein in crude cell extracts is fraught with possible inaccuracies. In most studies only an elementary cell extraction procedure was followed, and in none were follow-up studies undertaken to determine that the entity which was changing was indeed the inhibitor protein (in amount or activity).

B. POTENTIAL PHYSIOLOGICAL FUNCTIONS

The most outstanding question about the inhibitor protein is the following: what indeed is its physiological function and importance? The evidence that its role is to modulate the activity of the cAMP-dependent protein kinase remains only circumstantial and far from definitive. It is clearly of high potency and binds to the protein kinase at the catalytic site with several of the recognition features of a protein substrate. However, the remaining modes for its recognition by the protein kinase, which are equally if not more important for its high-affinity binding, are not known to be mirrored by other proteins that interact at or near the catalytic site, i.e., substrates or the regulatory subunit. Besides its high affinity, the remaining criterion for its role as a physiological modulator of the protein kinase is that in tissues such as brain, heart, skeletal muscle, and testis it is present in the cell at a level that would be effective. In other cells, however, of which the best studied so far is kidney, its levels appear to be such that it would make little or no significant contribution to altering the total activity of the protein kinase. With cAMP being such a broad-spectrum modulator, such a finding creates some doubt as to the inhibitor protein's role. This is especially so since it appears that for specific cellular processes, once the cAMP system has been turned on, the remaining controls are substrate specific and mediated by some regulation of the substrate (such as phosphorylation at an alternate site or the binding of an allosteric ligand) so that its efficacy as a substrate for the protein kinase is altered.

In the absence of specific knowledge of its function, several roles for the inhibitor protein can, however, be envisaged. These are presented schematically in Figure 10. The simplest possibility is that the inhibitor protein is acting to provide a threshold for the level of protein kinase which would be effective. Thus, in this mode, in a tissue such as heart where inhibitor

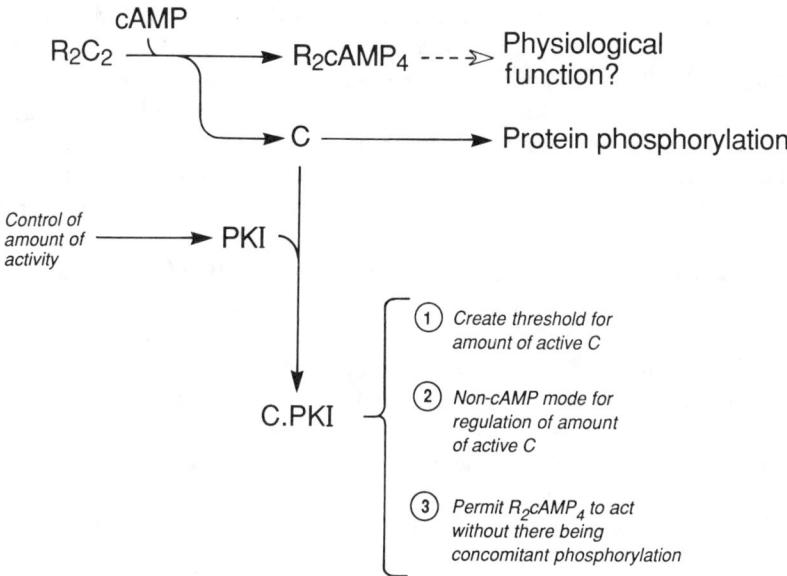

FIGURE 10. Potential physiological functions of the inhibitor protein.

protein levels could block ~20% of the cAMP-dependent protein kinase, it would require that cAMP be elevated to a higher level before any action of a hormone was manifested and that small fluctuations above basal levels would be without effect. Intuitively, it is not readily obvious why this would be advantageous since it would block any "fine tuning" of a system that a hormone would evoke, and the latter is the mode of regulation that is frequently observed. It might, however, be important to block the initiation of a cascade of reactions that basal levels of cAMP might otherwise cause. Creating a threshold would be an advantage if cAMP were to have actions other than via the protein kinase, but no evidence obtained to date would support this theory. An alternate possible function of the inhibitor protein would be to allow different concentrations of free regulatory subunit (i.e., R_2cAMP_4) and "active" catalytic subunit to be attained upon dissociation of the protein kinase. An important question that remains to be resolved is why the cAMP-dependent protein kinase has evolved so that its activation results in its separation into two components, i.e., catalytic and regulatory subunits. From a teleological standpoint this is not enzymologically necessary, as emphasized by the observation that this does not occur with the closely homologous cGMP-dependent protein kinase.[141,142] Furthermore, the mode of activation of the cGMP-dependent protein kinase, by the removal (without dissociation) of what is now termed the "prototope" region from the catalytic site,[143] is also the mechanism of activation of other protein kinases, such as protein kinase C and myosin light chain kinase. Such observations prompt the vexing question of whether the dissociation mechanism of the cAMP-dependent protein kinase also serves to produce free regulatory subunit, which could then have an independent function. This remains to be determined, although the specific binding of the regulatory subunit to other proteins supports such a speculation. If the regulatory subunit were to have an independent function, then the inhibitor protein could serve to allow for different degrees of expression of the action of cAMP. This would then give rise to a two-messenger system involving the separate actions of protein phosphorylation and whatever was (were) the function(s) of the free regulatory subunit. A third possibility for the physiological role of the inhibitor protein would be to allow an alternate means of regulating protein phosphorylation by the cAMP-dependent protein kinase, as a control auxiliary to that mediated by cAMP. If the amount of inhibitor protein and/or its ability to interact with the protein kinase catalytic subunit were regulated, then either would lead to an alteration of catalytic subunit activity. Under such circumstances the degree of cAMP-dependent protein kinase-catalyzed phosphorylation would reflect not only the hormonal signals regulating the concentration of cAMP, but also hormonal (or other) signals regulating the inhibitor protein. As reviewed above (see Section IV.A), some data have been obtained suggesting that the total amount of inhibitor protein may vary with hormonal status. Of the reports so far, the regulation of the testis inhibitor would have the greatest likelihood of producing significant changes in the degree of cAMP-mediated phosphorylation. Two other recent reports of possible inhibitor protein regulation are of interest. Van Patten et al.[144] found by in vitro experiments that epidermal growth factor (EGF) receptor catalyzed the phosphorylation of tyrosine-7 of the inhibitor protein, leading to a ninefold diminishment of inhibitory potency. This has yet to be shown to occur in vivo, but if it were it would indicate that the level of active catalytic subunit could be modulated by something other than a cAMP signal. Also of possible interest, it has been reported that PKI(11—30) binds to calmodulin in a Ca^{2+}-dependent manner, with a $K_d = 70$ nM.[78] Although this binding did not appear to diminish its interaction with the protein kinase and was not observed with the native inhibitor protein, it is possible that under some circumstance the inhibitor protein may be subject to some mode of calmodulin-dependent regulation. Both of these sets of experiments[78,144] point to the possibility that the inhibitor protein may be subject to additional controls that could modulate its function.

As reviewed above, the inhibitor protein has been central in many studies of the cAMP-dependent protein kinase, in elucidating its diverse roles within the cell and, more recently, as a probe of the topography of the catalytic site. The inhibitor protein also has the potential to be an important cellular mediator, but the full understanding of its physiological role awaits further investigation.

VII. APPENDIX

Purification of the Inhibitor Protein from Rabbit Skeletal Muscle*

Step 1 — Freshly excised rabbit skeletal muscle (4 kg) from eight 6- to 8-lb female New Zealand White Rabbits is ground and then homogenized using a Waring® Blendor (1 min, maximum speed) in 4 mM EDTA, pH 7.0 (2.5 l/kg). The homogenate is centrifuged for 40 min at 6200 × g, and the supernatant is decanted through glass wool.

Step 2 — The supernatant is adjusted to pH 7.0 by the addition of 6 N ammonium hydroxide, placed in a stainless steel bucket, and then stirred constantly while the temperature is raised to 90°C (approximate time = 50 min). The heated solution is immediately filtered through a double layer of cheese cloth (excess liquid is removed from the denatured protein by squeezing), and the filtrate is cooled by placing the container on ice. The denatured protein precipitate is homogenized in 2 l of 4 mM EDTA, pH 7.0. The supernatant of this homogenate is collected by centrifugation at 6200 × g for 30 min and then added to the cooled solution obtained from the initial filtration. The filtrate is then recentrifuged to remove residual precipitate and facilitate cooling.

Step 3 — To the supernatant, the temperature of which has been lowered to at least 12°C, 100% TCA is added rapidly and with vigorous stirring, to a final concentration of 5%. The suspension is stirred for 30 min, and then the precipitate is collected by centrifugation at 6200 × g for 30 min. The pellet is then resuspended in approximately 100 ml of distilled water and adjusted to pH 7.0 by the addition of 6 N ammonium hydroxide. At near neutral pH a dramatic clarification is observed, yielding a clear to barely cloudy solution. This solution is then dialyzed in Spectrapor 3 tubing against 2 × 10 l of 5 mM potassium phosphate buffer, pH 7.0 (with 1 mM EDTA) over a 16-h period. Steps 1 to 3 should be accomplished during the first day in approximately 6 h.

* Unless otherwise noted, all procedures are at 4°C.

Step 4 — Following dialysis, the pH of the solution should be ca. 7.0 and the conductivity <1 mMho. The pH of this solution is then decreased to 5.0 by the addition of 7% acetic acid; the solution is stirred for 30 min and then clarified by a 30-min centrifugation at 12,000 × g. The solution is applied to a 1.5 × 28 cm column of DE-52 (Whatman) equilibrated in 5 mM sodium acetate, pH 5.0. The column is washed with 250 ml of this same buffer and then the inhibitor protein eluted with a linear gradient of 5 to 350 mM sodium acetate, pH 5.0 (total volume = 1 l) at a flow rate of 70 ml/h. Typical elution profiles have been depicted previously.[67,69] The column separates the inhibitor protein into three peaks of activity, the major (and central) one (I-2) eluting at ca. 3.5 mMho. The fractions containing this peak are pooled, carefully neutralized with 1 N sodium hydroxide, and dialyzed in Spectrapor 3 tubing against 2 × 4 l of 30 mM potassium phosphate buffer, pH 6.8 (containing 0.1 mM EDTA and 15 mM 2-mercaptoethanol) over a 16-h period.

Step 5 — Preparation of the catalytic subunit-Sepharose® affinity column: All buffers and vessels used for this preparation are preflushed with N_2 and all steps are performed at 4°C. A 6-ml minicolumn of Sephadex® G-50 (in a 6-ml syringe casing) equilibrated in 50 mM potassium phosphate, pH 7.5, containing 500 mM NaCl and 5 mM EDTA (buffer A) is prepared as described by Fletcher et al.[120] To this column are applied 2.5 ml of freshly prepared pure bovine heart catalytic subunit of the cAMP-dependent protein kinase (8 mg/ml in 2.5 mM Tris chloride [pH 7.4], 0.1 mM EDTA, 7.5 mM 2-mercaptoethanol, 50 mM NaCl, and 50% glycerol; specific activity >2 U/mg), and the column is centrifuged according to the protocol described by Fletcher et al.[120] The eluted catalytic subunit is brought to 20% glycerol and immediately added to 1.0 g of CNBr-activated Sepharose® 4B (Pharmacia), which has been washed briefly in buffer A, in a tube of a size such that 85% of its capacity is occupied. The tube is mixed gently for 2 h at 4°C, and then 0.1 volume of 5 mM ethanolamine (pH 7.5) and 0.001 volume of 15 M 2-mercaptoethanol are added; mixing is continued for 2 h. The resin is collected by filtration on a sintered glass funnel and then washed three times with 10 ml of 30 mM potassium phosphate, pH 6.7, containing 0.1 mM EDTA and 15 mM 2-mercaptoethanol (buffer B) plus 1 M NaCl, never allowing the resin to go dry. It is then washed three times with buffer B containing 0.02% sodium azide. For storage, the resin is maintained in buffer B plus 0.02% sodium azide and 0.1% BSA. This column can be used repeatedly (for up to 2 years).

Step 6 — A column (1.3 × 3 cm) of catalytic subunit affinity resin is washed with 100 ml of buffer B, and then 100 ml of buffer B plus 1 mM ATP and 2 mM $MgCl_2$, by gravity flow at ~100 ml/h. The dialyzed preparation of inhibitor protein from Step 4 is brought to 1 mM ATP and 2 mM $MgCl_2$, and it is then applied to the affinity column using gravity flow (ca. 25 ml/h). The column is washed (again at the higher flow rate) with 10 ml of buffer B containing the same concentrations of ATP and $MgCl_2$, and it is then washed with a linear 50-ml 0 to 150 mM KCl gradient in buffer B plus ATP and $MgCl_2$. The inhibitor protein is eluted from the affinity resin (at the lower flow rate) with 40 ml of a buffer containing 30 mM potassium phosphate, pH 6.7, 5 mM EDTA, 15 mM 2-mercaptoethanol, 1 M KCl, and 0.3 M guanidinium hydrochloride. The fractions containing the inhibitor protein (i.e., the first 30 ml) are pooled, dialyzed in Spectrapor 3 tubing against 4 × 10 l of 1 mM MES (pH 6.8) over a 24-h period, and lyophilized. Inhibitor protein is obtained in a yield of ca. 400 μg from 4 kg of tissue, with an overall yield of ca. 10 to 20%. The final specific activity is 3 to 6 × 10^5 U/mg, representing approximately a 25,000-fold purification.

ACKNOWLEDGMENTS

I (D.A.W.) should like to acknowledge my many co-investigators who have contributed to our knowledge of the inhibitor protein. This work was supported by grant DK 21019 from the National Institutes of Health.

REFERENCES

1. **Szmigielski, A.**, Regulation of the activity of protein kinases by endogenous heat stable protein inhibitors, *Pol. J. Pharmacol. Pharm.*, 37, 273, 1985.
2. **Flockhart, D. A. and Corbin, J. D.**, Regulatory mechanisms in the control of protein kinases, *Crit. Rev. Biochem.*, 12, 133, 1982.
3. **Carlson, G. M., Bechtel, P. J., and Graves, D. J.**, Chemical and regulatory properties of phosphorylase kinase and cyclic AMP-dependent protein kinase, *Adv. Enzymol.*, 50, 41, 1979.
4. **Krebs, E. G. and Beavo, J. A.**, Phosphorylation-dephosphorylation of enzymes, *Annu. Rev. Biochem.*, 48, 923, 1979.
5. **Walsh, D. A. and Ashby, C. D.**, Protein kinases: aspects of their regulation and diversity, in *Recent Progress in Hormone Research*, Vol. 29, Greep, R. O., Academic Press, New York, 1973, 329.
6. **Posner, J. B., Hammermeister, K. E., Bratvold, G. E., and Krebs, E. G.**, The assay of adenosine-3',5'-phosphate in skeletal muscle, *Biochemistry*, 3, 1040, 1964.
7. **Gonzalez, G.**, Activation Characteristics of Phosphorylase Kinase, M.S. thesis, University of Washington, Seattle, 1962.
8. **Appleman, M. M., Birnbaumer, L., and Torres, H. N.**, Factors affecting the activity of muscle glycogen synthetase. III. The reaction with adenosine triphosphate, Mg^{2+}, and cyclic 3',5'-adenosine monophosphate, *Arch. Biochem. Biophys.*, 116, 39, 1966.
9. **Walsh, D. A., Perkins, J. P., Brostrom, C. O., Ho, E. S., and Krebs, E. G.**, Catalysis of the phosphorylase kinase activation reaction, *J. Biol. Chem.*, 246, p. 1968, 1971.
10. **Walsh, D. A., Ashby, C. D., Gonzalez, C., Calkins, D., Fischer, E. H., and Krebs, E. G.**, Purification and characterization of a protein inhibitor of adenosine 3',5'-monophosphate-dependent protein kinases, *J. Biol. Chem.*, 246, p. 1977, 1971.
11. **Walsh, D. A., Perkins, J. P., and Krebs, E. G.**, An adenosine 3',5'-monophosphate-dependent protein kinase from rabbit skeletal muscle, *J. Biol. Chem.*, 243, 3763, 1968.
12. **Traugh, J. A., Ashby, C. D., and Walsh, D. A.**, Criteria for the classification of protein kinases, *Methods Enzymol.*, 38, 290, 1973.
13. **Maller, J. L. and Krebs, E. G.**, Progesterone-stimulated meiotic cell division in *Xenopus* oocytes, *J. Biol. Chem.*, 252, 1712, 1977.
14. **Bkaily, G. and Sperelakis, N.**, Injection of protein kinase inhibitor into cultured heart cells blocks calcium slow channels, *Am. J. Physiol.*, 246, H630, 1984.
15. **Browne, C. L., Bird, M. L., and Bower, W.**, Effect of inhibition of the catalytic activity of cyclic AMP-dependent protein kinase on mitosis in PtK1 cells, *Cell Motility Cytoskeleton*, 7, 248, 1987.
16. **Adams, W. B. and Levitan, I. B.**, Intracellular injection of protein kinase inhibitor blocks the serotonin-induced increase in K^+ conductance in *Aplysia* neuron R15, *Proc. Natl. Acad. Sci. U.S.A.*, 79, 3877, 1982.
17. **Bittar, E. E. and Nwoga, J.**, Sensitivity to injected cholera toxin of the sodium efflux in single barnacle muscle fibers, *Comp. Biochem. Physiol.*, 78C, 89, 1984.
18. **Saez, J. C., Spray, D. C., Nairn, A. C., Hertzberg, E., Greengard, P., and Bennett, M. V. L.**, cAMP increases junctional conductance and stimulates phosphorylation of the 27-kDa principal gap junction polypeptide, *Proc. Natl. Acad. Sci. U.S.A.*, 83, 2473, 1986.
19. **Huchon, D., Ozon, R., Fischer, E. H., and Demaille, J. G.**, The pure inhibitor of cAMP-dependent protein kinase initiates *Xenopus laevis* meiotic maturation, *Mol. Cell. Endocrinol.*, 22, 211, 1981.
20. **Reisine, T., Rougon, G., and Barbet, J.**, Liposome delivery of cyclic AMP-dependent protein kinase inhibitor into intact cells: specific blockade of cyclic AMP-mediated adrenocorticotropin release from mouse anterior pituitary tumor cells, *J. Cell Biol.*, 102, 1630, 1986.
21. **Grove, J. R., Price, D. J., Goodman, H. M., and Avruch, J.**, Recombinant fragment of protein kinase inhibitor blocks cyclic AMP-dependent gene transcription, *Science*, 238, 530, 1987.
22. **Beebe, S. J. and Carlson, J. D.**, Cyclic nucleotide-dependent protein kinases, *Enzymes*, 17, 43, 1986.
23. **Taylor, S. S., Buechler, J. A., and Knighton, D. R.**, Cyclic AMP-dependent protein kinase: mechanism for ATP:protein phosphotransfer, in *Peptides and Protein Phosphorylation*, Kemp, B. E., Ed., CRC Press, Boca Raton, FL, 1990, chap. 1.
23a. **Zetterqvist, Ö., Ragnarsson, U., and Engström, L.**, Substrate specificity of cyclic AMP-dependent protein kinase, in *Peptides and Protein Phosphorylation*, Kemp, B. E., Ed., CRC Press, Boca Raton, FL, 1990, chap. 7.
24. **Hofmann, F., Beavo, J. A., Bechtel, P. J., and Krebs, E. G.**, Comparison of adenosine 3':5'-monophosphate-dependent protein kinases from rabbit skeletal and bovine heart muscle, *J. Biol. Chem.*, 250, 7795, 1975.
25. **Beavo, J. A., Bechtel, P. J., and Krebs, E. G.**, Mechanisms of control for cAMP-dependent protein kinase from skeletal muscle, *Adv. Cyclic Nucleotide Res.*, 5, 241, 1974.

26. **Beavo, J. A., Bechtel, P. J., and Krebs, E. G.**, Activation of protein kinase by physiological concentrations of cyclic AMP, *Proc. Natl. Acad. Sci. U.S.A.*, 71, 3580, 1974.
27. **Erlichman, J., Rosenfeld, R., and Rosen, O. M.**, Phosphorylation of a cyclic adenosine 3':5'-monophosphate-dependent protein kinase from bovine cardiac muscle, *J. Biol. Chem.*, 249, 5000, 1974.
28. **Whitehouse, S., Feramisco, J. R., Casnellie, J. E., Krebs, E. G., and Walsh, D. A.**, Studies on the kinetic mechanism of the catalytic subunit of the cAMP-dependent protein kinase, *J. Biol. Chem.*, 258, 3693, 1983.
29. **Hoppe, J., Marutzky, R., Freist, W., and Wagner, K. G.**, Mechanism of activation of the protein kinase I from rabbit skeletal muscle, *Eur. J. Biochem.*, 80, 369, 1977.
30. **Hoppe, J., Freist, W., Marutzky, R., and Shaltiel, S.**, Mapping the ATP-binding site in the catalytic subunit of adenosine-3':5'-monophosphate-dependent protein kinase, *Eur. J. Biochem.*, 90, 427, 1978.
31. **Uhler, M. D., Carmichael, D. F., Lee, D. C., Chrivia, J. C., Krebs, E. G., and McKnight, G. S.**, Isolation of cDNA clones coding for the catalytic subunit of mouse cAMP-dependent protein kinase, *Proc. Natl. Acad. Sci. U.S.A.*, 83, 1300, 1986.
32. **Uhler, M. D., Chrivia, J. C., and McKnight, G. S.**, Evidence for a second isoform of the catalytic subunit of cAMP-dependent protein kinase, *J. Biol. Chem.*, 261, 15360, 1986.
33. **Showers, M. O. and Maurer, R. A.**, A cloned bovine cDNA encodes an alternate form of the catalytic subunit of cAMP-dependent protein kinase, *J. Biol. Chem.*, 261, 16288, 1986.
34. **Chen, L.-J. and Walsh, D. A.**, Multiple forms of hepatic adenosine 3':5'-monophosphate dependent protein kinase, *Biochemistry*, 10, 3614, 1971.
35. **Bechtel, P. J., Beavo, J. A., and Krebs, E. G.**, Purification and characterization of catalytic subunit of skeletal muscle adenosine 3':5'-monophosphate-dependent protein kinase, *J. Biol. Chem.*, 252, 2691, 1977.
36. **Reed, J., Gagelmann, M., and Kinzel, V.**, Isolation and elucidation of some functional properties of the "mute" catalytic subunit of cAMP-dependent protein kinase, *Arch. Biochem. Biophys.*, 222, 276, 1983.
37. **Kinzel, V., Hotz, A., Konig, N., Gagelmann, M., Pyerin, W., Reed, J., Kubler, D., Hofmann, F., Obst, C., Gensheimer, H. D., Goldblatt, D., and Shaltiel, S.**, Chromatographic separation of two heterogeneous forms of the catalytic subunit of cyclic AMP-dependent protein kinase holoenzyme type I and II, *Arch. Biochem. Biophys.*, 253, 341, 1987.
38. **Van Patten, S. M., Hotz, A., Kinzel, V., and Walsh, D. A.**, The inhibitor protein of the cAMP-dependent protein kinase — catalytic subunit interaction: composition of multiple complexes, *Biochem. J.*, 256, 785, 1988.
39. **Gagelmann, M., Reed, J., Kubler, D., Pyerin, W., and Kinzel, V.**, Evidence of a "mute" catalytic subunit of cyclic AMP-dependent protein kinase from rat muscle and its mode of activation, *Proc. Natl. Acad. Sci. U.S.A.*, 77, 2492, 1980.
40. **Granot, J., Mildvan, A. S., and Kaiser, E. T.**, Studies of the mechanism of action and regulation of cAMP-dependent protein kinase, *Arch. Biochem. Biophys.*, 205, 1, 1980.
41. **Granot, J., Mildvan, A. S., Hiyama, K., Kondo, H., and Kaiser, E. T.**, Magnetic resonance studies of the effect of the regulatory subunit on metal and substrate binding to the catalytic subunit of bovine heart protein kinase, *J. Biol. Chem.*, 255, 4569, 1980.
42. **Cook, P. F., Neville, M. E., Vrana, K. E., Hartl, F. T., and Roskoski, R., Jr.**, Adenosine cyclic 3',5'-monophosphate dependent protein kinase, kinetic mechanism for the bovine skeletal muscle catalytic subunit, *Biochemistry*, 21, 5794, 1982.
43. **Thomas, N. E., Bramson, H. N., Miller, W. T., and Kaiser, T. E.**, Role of enzyme-peptide substrate backbone hydrogen bonding in determining protein kinase substrate specificities, *Biochemistry*, 26, 4461, 1987.
44. **Granot, J., Mildvan, A. S., Bramson, N., Thomas, N., and Kaiser, E. T.**, Nuclear magnetic resonance studies of the conformation and kinetics of the peptide-substrate at the active site of bovine heart protein kinase, *Biochemistry*, 20, 602, 1981.
45. **Bramson, H. N., Thomas, N. E., Miller, W. T., Fry, D. C., Mildvan, A. S., and Kaiser, E. T.**, Conformation of Leu–Arg–Arg–Ala–Ser–Leu–Gly bound in the active site of adenosine cyclic 3',5'-phosphate dependent protein kinase, *Biochemistry*, 26, 4466, 1987.
46. **Small, D., Chou, P. Y., and Fasman, G. D.**, Occurrence of phosphorylated residues in predicted β-turns: implications for β-turn participation in control mechanisms, *Biochem. Biophys. Res. Commun.*, 79, 341, 1977.
47. **Matsuo, M., Huang, C.-H., and Huang, L. C.**, Evidence for an essential arginine recognition site on adenosine 3':5'-cyclic monophosphate-dependent protein kinase of rabbit skeletal muscle, *Biochem. J.*, 173, 441, 1978.
48. **Reed, J., Kinzel, V., Kemp, B. E., Cheng, H.-C., and Walsh, D. A.**, Circular dichroic evidence for an ordered sequence of ligand/binding site interactions in the catalytic reaction of the cAMP-dependent protein kinase, *Biochemistry*, 24, 2967, 1985.
49. **Sowadski, J. M., Xuong, N., Anderson, D., and Taylor, S. S.**, Crystallization studies of cAMP-dependent protein kinase, *J. Mol. Biol.*, 182, 617, 1984.

50. **Kemp, B. E., Graves, D. J., Benjamini, E., and Krebs, E. G.,** Role of multiple basic residues in determining the substrate specificity of cyclic AMP-dependent protein kinase, *J. Biol. Chem.,* 252, 4888, 1977.
51. **Zetterqvist, O., Ragnarsson, U., Humble, E., Berglund, L., and Engstrom, L.,** The minimum substrate of cyclic AMP-stimulated protein kinase, as studied by synthetic peptides representing the phosphorylatable site of pyruvate kinase (type L) of rat liver, *Biochem. Biophys. Res. Commun.,* 70, 696, 1976.
52. **Feramisco, J. R., Glass, D. B., and Krebs, E. G.,** Optimal spatial requirements for the location of basic residues in peptide substrates for the cyclic AMP-dependent protein kinase, *J. Biol. Chem.,* 255, 4240, 1980.
53. **Glass, D. B. and Krebs, E. G.,** Protein phosphorylation catalyzed by cyclic AMP-dependent and cyclic GMP-dependent protein kinases, *Annu. Rev. Pharmacol. Toxicol.,* 20, 363, 1980.
54. **Glass, D. B. and May, J. M.,** *In vitro* phosphorylation of a synthetic collagen peptide by cyclic AMP-dependent protein kinase, *Collagen Rel. Res.,* 4, 63, 1984.
55. **Aitken, A., Bilham, T., and Cohen, P.,** Complete primary structure of protein phosphatase inhibitor-1 from rabbit skeletal muscle, *Eur. J. Biochem.,* 126, 235, 1982.
56. **Zetterqvist, O. and Ragnarsson, U.,** The structural requirements of substrates of cyclic AMP-dependent protein kinase, *FEBS Lett.,* 139, 287, 1982.
57. **Meggio, F., Chessa, G., Borin, G., Pinna, L. A., and Marchiori, F.,** Synthetic fragments of protamines as model substrates for rat liver cyclic AMP-dependent protein kinase, *Biochim. Biophys. Acta,* 662, 94, 1981.
58. **Glass, D. B., Cheng, H.-C., Mueller, L. M., Reed, J., and Walsh, D. A.,** Primary structural determinants essential for potent inhibition of cAMP-dependent protein kinase by inhibitory peptides corresponding to the active portion of the heat stable inhibitor protein, *J. Biol. Chem.,* 264, 8802, 1989.
59. **Demaille, J. G., Peters, K. A., and Fischer, E. H.,** Isolation and properties of the rabbit skeletal muscle protein inhibitor of adenosine 3',5'-monophosphate dependent protein kinases, *Biochemistry,* 16, 3080, 1977.
60. **Whitehouse, S. and Walsh, D. A.,** Purification of a physiological form of the inhibitor protein of the cAMP-dependent protein kinase, *J. Biol. Chem.,* 257, 6028, 1982.
61. **Demaille, J. G., Peters, K. A., Strandjord, T. P., and Fischer, E. H.,** Isolation and properties of the bovine brain protein inhibitor of adenosine 3':5'-monophosphate-dependent protein kinases, *FEBS Lett.,* 86, 113, 1978.
62. **Weber, H. and Rosen, O. M.,** Purification of a protein inhibitor of adenosine 3':5'-monophosphate-dependent protein kinase from bovine myocardium by a non-denaturing procedure, *J. Cyclic Nucleotide Res.,* 3, 415, 1977.
63. **Hashizume, K. and DeGroot, L. J.,** Cyclic nucleotide-dependent protein kinase modulators in thyroid tissue, *Endocrinology,* 105, 204, 1979.
64. **Szmigielski, A., Guidotti, A., and Costa, E.,** Endogenous protein kinase inhibitors. Purification, characterization, and distribution in different tissues, *J. Biol. Chem.,* 252, 3848, 1977.
65. **Kwast-Welfeld, J. and Kaniuga, Z.,** A protein kinase inhibitory modulator in liver of developing chick, *Int. J. Biochem.,* 9, 331, 1978.
66. **Van Patten, S. M., Fletcher, W. H., and Walsh, D. A.,** The inhibitor protein of cAMP-dependent protein kinase-catalytic subunit interaction. Parameters of complex formation, *J. Biol. Chem.,* 261, 5514, 1986.
67. **Whitehouse, S., McPherson, J. M., and Walsh, D. A.,** Characterization of multiple charge isomers of the inhibitor protein of the cyclic AMP-dependent protein kinase from bovine heart and rabbit skeletal muscle, *Arch. Biochem. Biophys.,* 203, 734, 1980.
68. **Reed, J., de Ropp, J. S., Trewhella, J., Glass, D. B., Liddle, W. K., Bradbury, E. M., Kinzel, V., and Walsh, D. A.,** Conformational analysis of PKI(5—22)amide, the active inhibitory fragment of the inhibitor protein of the protein kinase, *Biochem. J.,* Vol. 264, 1989.
69. **McPherson, J. M., Whitehouse, S., and Walsh, D. A.,** Possibility of shape conformers of the protein inhibitor of the cyclic adenosine monophosphate dependent protein kinase, *Biochemistry,* 18, 4835, 1979.
70. **Scott, J. D., Fischer, E. H., Demaille, J. G., and Krebs, E. G.,** Identification of an inhibitory region of the heat-stable protein inhibitor of the cAMP-dependent protein kinase, *Proc. Natl. Acad. Sci. U.S.A.,* 82, 4379, 1985.
71. **Scott, J. D., Fischer, E. H., Takio, K., Demaille, J. G., and Krebs, E. G.,** Amino acid sequence of the heat-stable inhibitor of the cAMP-dependent protein kinase from rabbit skeletal muscle, *Proc. Natl. Acad. Sci. U.S.A.,* 82, 5732, 1985.
72. **Ferraz, C., Demaille, J. G., and Fischer, E. H.,** The protein inhibitor of adenosine 3':5'-monophosphate-dependent protein kinases, *Biochimie,* 61, 645, 1979.
73. **Whitehouse, S. and Walsh, D. A.,** Mg·ATP-dependent interaction of the inhibitor protein of the cAMP-dependent protein kinase with the catalytic subunit, *J. Biol. Chem.,* 258, 3682, 1983.

74. **Beale, E. G., Dedman, J. R., and Means, A. R.**, Isolation and regulation of the protein kinase inhibitor and the calcium-dependent cyclic nucleotide phosphodiesterase regulator in the sertoli cell-enriched testis, *Endocrinology*, 101, 1621, 1977.
75. **Tash, J. S., Dedman, J. R., and Means, A. R.**, Protein kinase inhibitor in sertoli cell-enriched rat testis. Specific regulation by follicle-stimulating hormone, *J. Biol. Chem.*, 254, 1241, 1979.
76. **Beale, E. G., Dedman, J. R., and Means, A. R.**, Isolation and characterization of a protein from rat testis which inhibits cyclic AMP-dependent protein kinase and phosphodiesterase, *J. Biol. Chem.*, 252, 6322, 1977.
77. **Tash, J. S., Welsh, M. J., and Means, A. R.**, Protein inhibitor of cAMP-dependent protein kinase: production and characterization of antibodies and intracellular localization, *Cell*, 21, 57, 1980.
78. **Malencik, D. A., Scott, J. D., Fischer, E. H., Krebs, E. G., and Anderson, S. R.**, Association of calmodulin with peptide analogues of the inhibitory region of the heat-stable protein inhibitor of adenosine cyclic 3',5'-phosphate dependent protein kinase, *Biochemistry*, 25, 3502, 1986.
79. **Donnelly, T. E., Jr., Kuo, J. F., Reyes, P. L., Liu, Y.-P., and Greengard, P.**, Protein kinase modulator from lobster tail muscle. I. Stimulatory and inhibitory effects of the modulator on the phosphorylation of substrate proteins by guanosine 3',5'-monophosphate-dependent and adenosine 3',5'-monophosphate-dependent protein kinases, *J. Biol. Chem.*, 248, 190, 1973.
80. **Donnelly, T. E., Jr., Kuo, J. F., Miyamoto, E., and Greengard, P.**, Protein kinase modulator from lobster tail muscle. II. Effects of the modulator on holoenzyme and catalytic subunit of guanosine 3',5'-monophosphate-dependent and adenosine 3',5'-monophosphate-dependent protein kinases, *J. Biol. Chem.*, 248, 199, 1973.
81. **Kuo, W.-N. and Kuo, J. F.**, Isolation of stimulatory modulator of guanosine 3':5'-monophosphate-dependent protein kinase from mammalian heart devoid of inhibitory modulator of adenosine 3':5'-monophosphate-dependent protein kinase, *J. Biol. Chem.*, 251, 4283, 1976.
82. **Shoji, M., Brackett, N. L., Tse, J., Shapira, R., and Kuo, J. F.**, Molecular properties and mode of action of homogeneous preparation of stimulatory modulator of cyclic GMP-dependent protein kinase from the heart, *J. Biol. Chem.*, 253, 3427, 1978.
83. **Glass, D. B., Cheng, H.-C., Kemp, B. E., and Walsh, D. A.**, Differential and common recognition of the catalytic sites of the cGMP-dependent and cAMP-dependent protein kinases by inhibitory peptides derived from the heat-stable inhibitor protein, *J. Biol. Chem.*, 261, 12166, 1986.
84. **Szmigielski, A. and Guidotti, A.**, Action of harmaline and diazepam on the cerebellar content of cyclic GMP and on the activities of two endogenous inhibitors of protein kinase, *Neurochem. Res.*, 4, 189, 1979.
85. **Szmigielski, A., Szadowska, A., Szmigielska, H., and Starke, K.**, Changed sensitivity of α_2-adrenoreceptors mediating a decrease in protein kinase inhibitor activity in the brain of vasopressin-hypertensive rats, *Eur. J. Pharmacol.*, 122, 1, 1986.
86. **Szmigielski, A., Szadowska, A., Szmigielska, H., and Zalewska, J.**, The responsiveness of α_2-adrenergic receptors in hypothalamus of vasopressin-hypertensive rats, *Pol. J. Pharmacol. Pharm.*, 37, 773, 1985.
87. **Job, D., Pirollet, F., Cochet, C., and Chambaz, E. M.**, Interaction of a casein kinase (G-type) with a specific endogenous inhibitor, *FEBS Lett.*, 108, 508, 1979.
88. **Farron-Furstenthal, F.**, Two naturally occurring inhibitors of nuclear protein kinase, *J. Biol. Chem.*, 255, 4589, 1980.
89. **Morishita, Y., Sahai, A., Akogyeram, C., Hollis, V., Oka, T., and Criss, W.**, Identification and characterization of endogenous inhibitors of polyamine-responsive protein kinase activity, *J. Cyclic Nucleotide Res.*, 8, 173, 1982.
90. **Farron-Furstenthal, F.**, An inhibitor protein of nuclear protein kinases, *Nature (London)*, 280, 415, 1979.
91. **Pliego, J. F., Van-Arsdalen, K., and Kopf, G. S.**, Distribution of a seminal plasma-associated protein kinase inhibitor in normal, oligozoospermic, and vasectomized men, *Physiol. Reprod.*, 34, 885, 1986.
92. **McDonald, J. R. and Walsh, M. P.**, Ca^{2+}-binding proteins from bovine brain including a potent inhibitor of protein kinase C, *Biochem. J.*, 232, 559, 1985.
93. **Ashby, C. D. and Walsh, D. A.**, Characterization of the interaction of a protein inhibitor with adenosine 3',5'-monophosphate-dependent protein kinases. Interaction with the catalytic subunit of the protein kinase, *J. Biol. Chem.*, 247, 6637, 1972.
94. **Ashby, C. D. and Walsh, D. A.**, Characterization of the interaction of a protein inhibitor with adenosine 3',5'-monophosphate-dependent protein kinases, *J. Biol. Chem.*, 248, 1255, 1973.
95. **Witt, J. J. and Roskoski, R., Jr.**, Adenosine 3',5'-monophosphate-dependent protein kinase: interaction with guanidinium compounds, *Arch. Biochem. Biophys.*, 201, 36, 1980.
96. **Hofmann, F.**, Apparent constants for the interaction of regulatory and catalytic subunit of cAMP-dependent protein kinase I and II, *J. Biol. Chem.*, 255, 1559, 1980.
97. **Builder, S. E., Beavo, J. A., and Krebs, E. G.**, Mechanism of activation of protein kinase by cAMP, *J. Supramol. Struct.*, Suppl. 3, 58, 1979.

98. **Armstrong, R. N., Kondo, H., Granot, J., Kaiser, E. T., and Mildvan, A. S.**, Magnetic resonance and kinetic studies of the manganese(II) ion and substrate complexes of the catalytic subunit of adenosine 3′,5′-monophosphate dependent protein kinase from bovine heart, *Biochemistry*, 18, 1230, 1979.
99. **Moll, G. W. and Kaiser, E. T.**, Phosphorylation of histone catalyzed by a bovine brain protein kinase, *J. Biol. Chem.*, 251, 3993, 1976.
100. **Kochetkov, S. N., Bulargina, T. V., Sashchenko, L. P., and Severin, E. S.**, Studies on the mechanism of action of histone kinase dependent on cAMP. Evidence for involvement of histidine and lysine residues in the phosphotransferase reaction, *Eur. J. Biochem.*, 81, 111, 1977.
101. **Bolen, D. W., Stingelin, J., Bramson, H. N., and Kaiser, E. T.**, Stereochemical and kinetic studies of the action of the catalytic subunit of bovine cardiac muscle cAMP-dependent protein kinase using metal ion complexes of ATPβS, *Biochemistry*, 19, 1176, 1980.
102. **Demaille, J. G., Ferraz, C., and Fischer, E. H.**, The protein inhibitor of adenosine 3′,5′-monophosphate-dependent protein kinases. The NH_2-terminal portion of the peptide chain contains the inhibitory site, *Biochim. Biophys. Acta*, 586, 374, 1979.
103. **Cheng, H.-C., Van Patten, S. M., Smith, A. J., and Walsh, D. A.**, An active twenty-amino-acid-residue peptide derived from the inhibitor protein of the cyclic AMP-dependent protein kinase, *Biochem. J.*, 231, 655, 1985.
104. **Scott, J. D., Glaccum, M. B., Fischer, E. H., and Krebs, E. G.**, Primary-structure requirements for inhibition by the heat-stable inhibitor of the cAMP-dependent protein kinase, *Proc. Natl. Acad. Sci. U.S.A.*, 83, 1613, 1986.
105. **Cheng, H.-C., Kemp, B. E., Pearson, R. B., Smith, A. J., Misconi, L., Van Patten, S. M., and Walsh, D. A.**, A potent synthetic peptide inhibitor of the cAMP-dependent protein kinase, *J. Biol. Chem.*, 261, 989, 1986.
106. **Reed, J., Kinzel, V., Cheng, H.-C., and Walsh, D. A.**, Circular dichroic investigations of secondary structure in synthetic model peptide inhibitors of cAMP-dependent protein kinase, *Biochemistry*, 26, 7641, 1987.
107. **Pilkis, S. J., El-Maghrabi, M. R., Coven, B., Claus, T. H., Tager, H. S., Steiner, D. F., Keim, P. S., and Heinrikson, R. L.**, Phosphorylation of rat hepatic fructose-1,6-bisphosphatase and pyruvate kinase, *J. Biol. Chem.*, 255, 2770, 1980.
108. **Minocherhomjee, A.-V. and Roufogalis, B. D.**, Activation of erythrocyte Ca^{2+}-plus-Mg^{2+}-stimulated adenosine triphosphatase by protein kinase (cyclic AMP-dependent) inhibitor, *Biochem. J.*, 206, 517, 1982.
109. **Kanter, J. R. and Brunton, L. L.**, Cautions on the use of the heat stable inhibitor of protein kinase: studies with S49 lymphoma cells, *J. Cyclic Nucleotide Res.*, 7, 259, 1981.
110. **Rousseau, G. G. and De Visscher, M.**, Artifactual stimulation of cyclic AMP-dependent protein kinase activity by the heat-stable protein kinase "inhibitor", *Biochem. Biophys. Res. Commun.*, 72, 1423, 1976.
111. **Bornslaeger, E. A., Mattei, P., and Schultz, R. M.**, Involvement of cAMP-dependent protein kinase and protein phosphorylation in regulation of mouse oocyte maturation, *Dev. Biol.*, 114, 453, 1986.
112. **Bittar, E. E. and Chambers, G.**, The modulatory action of 5-hydroxytryptamine on sodium efflux: the barnacle muscle fibre as a model system, *Comp. Biochem. Physiol.*, 80C, 421, 1985.
113. **Nwoga, J. and Bittar, E. E.**, Increased sensitivity to injected 5′-guanylylimidodiphosphate of the sodium efflux in barnacle muscle fibres preexposed to aldosterone, *Comp. Biochem. Physiol.*, 74C, 177, 1983.
114. **Nwoga, J. and Bittar, E. E.**, Stimulation by proctolin of the ouabain-insensitive sodium efflux in single barnacle muscle fibers, *Comp. Biochem. Physiol.*, 81C, 345, 1985.
115. **Bittar, E. E., Chambers, G., and Fischer, E. H.**, The influence of injected cyclic AMP dependent protein kinase catalytic subunit on the sodium efflux in barnacle muscle fibres, *Am. J. Physiol.*, 333, 39, 1982.
116. **Castellucci, V. F., Nairn, A., Greengard, P., Schwartz, J. H., and Kandel, E. R.**, Inhibitor of adenosine 3′:5′-monophosphate-dependent protein kinase blocks presynaptic facilitation in Aplysia, *J. Neurosci.*, 2, 1673, 1982.
117. **Reisine, T., Rougon, G., Barbet, J., and Affolter, H.-U.**, Corticotropin-releasing factor-induced adrenocorticotropin hormone release and synthesis is blocked by incorporation of the inhibitor of cyclic AMP-dependent protein kinase into anterior pituitary tumor cells by liposomes, *Proc. Natl. Acad. Sci. U.S.A.*, 82, 8261, 1985.
118. **Fletcher, W. H. and Byus, C. V.**, Direct cytochemical localization of catalytic subunits dissociated from cAMP-dependent protein kinase in reuber H-35 hepatoma cells. I. Development and validation of fluoresceinated inhibitor, *J. Cell Biol.*, 93, 719, 1982.
119. **Byus, C. V. and Fletcher, W. H.**, Direct cytochemical localization of catalytic subunits dissociated from cAMP-dependent protein kinase in reuber H-35 hepatoma cells. II. Temporal and spatial kinetics, *J. Cell Biol.*, 93, 727, 1982.

120. **Fletcher, W. H., Van Patten, S. M., Cheng, H.-C., and Walsh, D. A.,** Cytochemical identification of the regulatory subunit of the cAMP-dependent protein kinase by use of fluorescently labeled catalytic subunit. Examination of protein kinase dissociation in hepatoma cells responding to 8-Br-cAMP stimulation, *J. Biol. Chem.*, 261, 5504, 1986.
121. **Murray, S. A. and Fletcher, W. H.,** Hormone-induced intercellular signal transfer dissociates cyclic AMP-dependent protein kinase, *J. Cell Biol.*, 98, 1710, 1984.
122. **Murray, S. A., Byus, C. V., and Fletcher, W. H.,** Intracellular kinetics of free catalytic subunit dissociated from adenosine 3',5'-monophosphate-dependent protein kinase in adrenocortical tumor cells (Y-1), *Endocrinology*, 116, 364, 1985.
123. **Fletcher, W. H. and Greenan, J. R. T.,** Receptor mediated action without receptor occupancy, *Endocrinology*, 116, 1660, 1985.
124. **Murray, S. A. and Fletcher, W. H.,** Cyclic AMP-dependent protein kinase mediated desensitization of adrenal tumor cells, *Mol. Cell. Endocrinol.*, 47, 153, 1986.
125. **Murray, S. A., Champ, C., and Lagenaur, C.,** Ultrastructural localization of cyclic adenosine 3',5'-monophosphate-dependent protein kinase after adrenocorticotropin stimulation in adrenal cortical tumor cells, *Endocrinology*, 120, 1921, 1987.
126. **Rudack-Garcia, D. and Henry, H. L.,** Effect of vitamin D status on cyclic AMP-dependent protein kinase activity and its heat stable inhibitor in chick kidney, *J. Biol. Chem.*, 256, 10781, 1981.
127. **Costa, M.,** Endogenous protein kinase inhibitor levels regulate changes in specific activity of protein kinase in quiescent cells stimulated to proliferate, *Biochem. Biophys. Res. Commun.*, 78, 1311, 1977.
128. **Skala, J. P., Drummond, G. I., and Hahn, P.,** A protein kinase inhibitor in brown adipose tissue of developing rats, *Biochem. J.*, 138, 195, 1974.
129. **Le Peuch, C. J., Ferraz, C., Walsh, M. P., Demaille, J. G., and Fischer, E. H.,** Calcium and cyclic nucleotide dependent regulatory mechanisms during development of chick embryo skeletal muscle, *Biochemistry*, 18, 5267, 1979.
130. **Tash, J. S., Dedman, J. R., and Means, A. R.,** Protein kinase inhibitor in sertoli cell-enriched rat testis, *J. Biol. Chem.*, 254, 1241, 1979.
131. **Szmigielski, A.,** The effect of beta-receptor stimulation on the activity of the inhibitor of cAMP-dependent protein kinase, *Acta Physiol. Pol.*, 32, 501, 1981.
132. **Bylund, D. B. and Krebs, E. G.,** Effect of denaturation on the susceptibility of proteins to enzymic phosphorylation, *J. Biol. Chem.*, 250, 6355, 1975.
133. **Judewicz, N. D., Glikin, G. C., and Torres, H. N.,** Protein kinase activities in *Neurospora crassa*, *Arch. Biochem. Biophys.*, 206, 87, 1981.
134. **Henry, H. L., Al-Abdaly, F. A., and Noland, T. A., Jr.,** Cyclic AMP dependent protein kinase and its endogenous inhibitor protein: tissue distribution and effect of vitamin D status in the chick, *Comp. Biochem. Physiol.*, 74B, 715, 1983.
135. **Tenenhouse, H. S. and Henry, H. L.,** Protein kinase activity and protein kinase inhibitor in mouse kidney: effect of the X-linked *Hyp* mutation and vitamin D status, *Endocrinology*, 117, 1719, 1985.
136. **Kuo, J. F.,** Changes in activities of modulators of cyclic AMP-dependent and cyclic GMP-dependent protein kinases in pancreas and adipose tissue from alloxan-induced diabetic rats, *Biochem. Biophys. Res. Commun.*, 65, 1214, 1975.
137. **Szmigielski, A.,** Modulation of the activity of endogenous protein kinase inhibitors in rat heart by the beta adrenergic receptor, *Arch. Int. Pharmacodyn.*, 249, 64, 1981.
138. **Zawilska, J. and Szmigielski, A.,** Isoprenaline-induced changes in activity of the endogenous inhibitor of cAMP-dependent protein kinases under conditions of α-adrenoreceptor supersensitivity, *J. Pharm. Pharmacol.*, 38, 239, 1986.
139. **Zawilska, J. and Szmigielski, A.,** Effect of repeated electroconvulsive shocks on isoprenaline-induced changes of the endogenous inhibitor of cAMP-dependent protein kinase in rat brain, *Eur. J. Pharmacol.*, 125, 273, 1986.
140. **Szmigielski, A., Zawilska, J., and Kondracki, K.,** Isoprenaline-induced changes in type I inhibitor activity as an index of beta-adrenergic receptor subsensitivity, *Pol. J. Pharmacol. Pharm.*, 36, 281, 1984.
141. **Lincoln, T. M., Flockhart, D. A., and Corbin, J. D.,** Studies on the structure and mechanism of activation of the guanosine 3':5'-monophosphate-dependent protein kinase, *J. Biol. Chem.*, 253, 6002, 1978.
142. **Lincoln, T. M., Dills, W. L., Jr., and Corbin, J. D.,** Purification and subunit composition of guanosine 3':5'-monophosphate-dependent protein kinase from bovine lung, *J. Biol. Chem.*, 252, 4269, 1977.
143. **House, C. and Kemp, B. E.,** Protein kinase C contains a pseudosubstrate prototope in its regulatory region, *Science*, 238, 1720, 1987.
144. **Van Patten, S. M., Heisermann, G. J., Cheng, H.-C., and Walsh, D. A.,** Tyrosine kinase catalyzed phosphorylation and inactivation of the inhibitor protein of the cAMP-dependent protein kinase, *J. Biol. Chem.*, 262, 3398, 1987.
145. **Hedrick, J. L. and Smith, A. J.,** Size and charge isomer separation and estimate of molecular weight of proteins by disc gel eletrophoresis, *Arch. Biochem. Biophys.*, 126, 155, 1968.

146. **Glass, D. B., Lundquist, L. J., Katz, B. M., and Walsh, D. A.,** Protein kinase inhibitor (6—22) amide peptide analogs with standard and non-standard substitutions for phenylalanine-10, *J. Biol. Chem.*, 264, 14579, 1989.
147. **Liddle, W. K., Trewhella, J., Glass, D. B., and Walsh, D. A.,** unpublished data.
148. **Day, R. N., Walder, J. A., and Maurer, R. A.,** A protein kinase inhibitor gene that reduces both basal and multihormone stimulated prolactin gene transcription, *J. Biol. Chem.*, 264, 431, 1989.

Chapter 3

THE *SRC* FAMILY OF PROTEIN-TYROSINE KINASES

Jonathan A. Cooper

TABLE OF CONTENTS

I. Introduction 86

II. Detection of *src*-Like Proto-Oncogenes 86

III. Evolution 88

IV. Structure of *src* Family Kinases 88
    A. Primary Sequences: Conserved and Variable Regions 89
    B. Higher Order Structure 90

V. Oncogenic Activation 92

VI. Oncogenicity Entails Derepression of Kinase Activity 93
    A. Mutant Forms of c-*src* 93
    B. Kinase Activation by Polyoma Virus 93

VII. Repression of Protein Kinase Activity by C-Terminal Phosphorylation 94

VIII. Physiological Regulation of Activity 95

IX. Mechanism of C-Terminal Phosphorylation 95

X. Activation via Phosphorylation of the Kinase Domain 96

XI. Additional Phosphorylation Sites 97
    A. Amino-Terminal Tyrosine Phosphorylations 97
    B. Serine Phosphorylations 98

XII. Altered Gene Products 99

XIII. Possible Roles for the Amino-Terminal Region 99
    A. The *src/fps/abl* Homology Region 100
    B. The N-Terminal Variable Region 100
    C. Subcellular Localization Signals and Mechanism 100

XIV. Substrates for *src* Family Kinases 101

XV. Possible Functions of *src* Family Kinases 103

Acknowledgments 104

References 104

I. INTRODUCTION

As the sequences of many protein-tyrosine kinases have been determined by analysis of cDNA clones, it has become clear that most protein-tyrosine kinases fall into one of two groups.[1] One group, typified by p60$^{c\text{-}src}$, the product of the cellular *src* gene, has members that are between 505 and 543 residues in size, show extensive sequence identity with other family members over a contiguous 460-residue stretch toward the carboxy terminus, and lack a pronounced hydrophobic region that could cross a membrane. The second group comprises the receptor kinases, which are much larger than p60$^{c\text{-}src}$ and are able to span the membrane by means of a transmembrane domain that separates an external, ligand-binding region from an internal kinase domain.[2] A few protein-tyrosine kinases — e.g., the products of the c-*abl* and c-*fps* (synonymous with c-*fes*) genes — do not fall into either of these tidy classifications.[1-3]

Most, probably all, of the *src* family genes have the potential to mutate into oncogenes — that is, dominant alleles that cause malignant transformation.[4] Indeed, the first genetic demonstration of an oncogene made use of Rous sarcoma virus (RSV), which carries an oncogenic form of c-*src* known as viral (v)-*src*.[5] The product of the v-*src* gene, p60$^{v\text{-}src}$, was the first oncoprotein shown to have protein-tyrosine kinase activity.[6-8] Oncogenic retroviruses carrying two other *src* family genes, v-*fgr* and v-*yes*, were soon recognized. Another *src* family gene (*lck*, previously named *lsk*^{T} or *tck*) is overexpressed in a cell line established from a mouse thymic lymphoma, and the unusually high expression may well have caused the original tumor.[9,10] Site-directed mutagenesis has been used to make variants of *lck* that are oncogenic for fibroblasts.[11,12] Oncogenic forms of *hck* have also been constructed.[13] Oncogenic versions of *fyn* have been isolated by selection for transformed cells in tissue culture.[14] Therefore, each of these *src* family genes is a proto-oncogene.

There have been several recent reviews on protein-tyrosine kinases,[2,3,15] growth factor receptors,[16,17] oncogenes,[4] and cell transformation and growth control.[18,19] In this chapter we will introduce the presently known (as of October 1988) members of the *src* family, describe in detail what is known about the regulation, substrates, and possible functions of the best characterized member (p60$^{c\text{-}src}$), and speculate on the significance of the conserved and divergent structural features of other family members.

II. DETECTION OF *SRC*-LIKE PROTO-ONCOGENES

Oncogenic forms of c-*src*, c-*yes*, and c-*fgr* were initially detected in the genomes of RSV (oncogene v-*src*), Yamaguchi 73 avian sarcoma virus (Y73-ASV, v-*yes*), and Gardner-Rasheed strain feline sarcoma virus (GR-FeSV, v-*fgr*), and their similarities were evident from the oncogene sequences.[20-23] The cellular proto-oncogenes were then isolated using the viral genes as probes. Close homologies also were revealed in the sequences of the proto-oncogenes[24-31] (Figure 1). Some other family members were detected by cross-hybridization at low stringency. Thus, for example, the c-*fgr* proto-oncogene was identified as a cellular gene that hybridized with v-*src* (hence, the preliminary name c-*src*-2).[26,32] The v-*yes* probe pulled out two related sequences, *fyn* (previously known as *syn* or *slk*) and *lyn*.[33-35] The *hck* gene was identified at low stringency with an *lck* probe[36] and independently with *src* probes.[37,38] A novel cross-hybridizing complementary DNA (cDNA) from a chicken library is known as c-*tkl*.[39] It may be a novel family member, although its sequence is very similar to that of *lck*. It seems likely that additional family members will be identified by cross-hybridization in the future.

The *lck* gene is atypical in that its protein product was identified before the gene was characterized. Casnellie et al.[40,41] observed that the murine thymic lymphoma, LSTRA, has unusually high levels of protein-tyrosine kinase activity. A 56-kDa protein is phosphor-

```
C-SRC    MSSSKSKP-KDP---SQRRESLEPPDSTHRYGGFPASQTPNKTAAPDTHRTPS-RESF-GTVATEPKL--
C-YES    MGCIKSKENKSPAIKYRPENTPEPVSTVSHYGAEPTTVSPCPSSSAKGTAVNFSSLSMTPRGSSSGVTPRGASSSFSVVPSSYPAGLTGVTIFVALYDYEARTTEDDLSFKKGERLQIVNNT-EGEMWWARSLATGEN
FYN      MGCVQCKD-KEATKLTEERDGSL---MGSSGYR--YGTDPTPQHYPSFGVTSIPNYNNFHAAGGQGLTVFGGVNSSSHTGLRTRGG-TGVTLFVALYDYEARTEDDLSFHKGEKFQILNSS-EGDWWEARSLTTGET
C-FGR    MGCVFCK-KL---------EPVSTAKEDAGLE-GDFRSYGAADHYGPDPT-KARPASSFABIPWYSNFSSQAIRRGFLDSGTIRGVSGIGVTLFIALYDYEARTEDDLTFTGEKPHILNNT-EGDWWEARSLSSGGT
HCK      MGCVC-----------------SSNKPEDDM-MENYDVCEHCHYPIVPLDSKISLPIRNGSEVRDPLVTYESSLPPASPL--QDNLVIALHSYEPSHDGDLGPEFEKQLRILEQS---GEMMKAQSLTTGQE
LYN      MGCMKSK---------------FLQVGGRTYSKETETSASPHCPVYVP-DFTSTIKPGNSEHSHTPGIREAQS--EDIIVVALYDYEATHHEDLSPQKGQDMVVLEE--SGEMWKARSLATRKE
                                         GDSLSDGVDLKTQPVRNTERTIYVRDPTSNKQQRPVPESQLLPGQRFQTKDPEEQGDIVVALYPYDGIHPDDLSFKKGERKVVLEE--EGEMWKAKSLLTKKE

          :   :::::: :++:  ++   +++  :::+ + +++++++     + +++++++++   ++  + +++:  :: :++:    +  +:+::   ::   :    +   :   [ +
C-SRC    GYIPSNYVAPSDSIQAEEWYFGKITRRESERLLLNPENPRGTFLVRESETTKGAYCLSVSDFDNAKGLNVKHYKIRKLDSGGFYITSRTQFSSLQQLVAYYSKHADGLCHRLTTVCPTSKPQTQGL---AKDAWEIPRES
C-YES    GYIPSNYVAPADSIQAEEWYFGKMGRKDAERRLLLNPGNPRGTFLVRESETTKGAYCLSVSDFDNAKGLNVKHYKIRKLDSGNVAHYKIRKLDNGSYYITTRAQFDTLQKLVRHYTEHADGLCHKLTTVCPTVKPQTQGL---AKDAWEIPRES
FYN      GYIPSNYVAPVDSIQAEEWYFGKLGRKDAERQLLSPGNPRGTFLVRESETTKGAYSLSIRDWDPMSKGDHVGHYKIRKLDNGSYYITTRAQFETLQQLVQHYSERAAGLCCRLVVPCHKGMPRLTDLSVKTKDWWE IPRES
C-FGR    GIPSNYVAPVDSIQAEEWYFGKIGRRDAERQLLSPGHRGTFLVRESETTKGAYSLSIRDWDDQTEGDHVAHYKIRKLDGSYYITRVQFNSVQELVQHYMEVNDGLCRLLIAPCTIMKQTLGL---AKDAWEISRSS
LCK      GFIPFNFVAKANSLEPEPWFFKNLSRKDAERQLLAPGNTRGSFLIRESESTAGSFLSVRDFDQNQGEVVKHYKIRNLDNGGFYISPRITFPGLHDLVRHYTNASDGLCTKLSRPCQTQKPQKP---WWEDEWEVPRET
HCK      GYIPSNYVARVDSLETEEWFFKGISRKDAERQLLAPGNMLGSFMIRDSETTKGSYSLSVRDYDPRQGDTVKHYKIRTLDNGGFYISPRSTFSTLQELVRHYKKGRDGLQLKLSVPCVSSKPQKP---WEXDAWEIPRES
LYN      GFIPSNYVAKLNTLETEEWFFKDITRDAERQLLAPGNSAGAFLIRESETLKGSFSLSVRDFDPVBGDVIKHYKIRSLDNGGYYISPRLITFCSDMIKHYKQADGLCRRLEKACISPKPQKP---MDKDAWEIPRES

         ++  ++ +++++   +    +++=+ +  +++    +    + + + ++++ ++    +++     ++++ + +++ +++     /+    ++++++ + ++++ +++ ++++
C-SRC    LRLEVKLGQGCFGEVWM*GTWNGTTRVAIKTLKPGTM*SPEAFLQEAQVM*KKLRHEKLVQLYAVVS--EEPIYIVTEYM*SKGSLLDFLKGEM*GKYLRLPQLVDMAAQIASGM*AYVERM*NYVHRDLRAANILVGENLVCKVADF
C-YES    LRLEVKLGQGCFGEVWM*GTWNGTTKVAIKTLKPGTM*FTEAFLQEAQIM*KKLRHDKLVPLAVVS--EEPIYIVTEFM*SKGSLLDFLKEGDGKYLKLPQLVDMAAQIADGM*AYIERM*NYIHRDLRAANILVCKADF
FYN      LQLIKRLGRGQFGEVWM*GTWNGNTKVAIKTLKPGTM*MFEAFLQEAQIM*KKLRHDKLVQLYAVVS--EEPIYIVTEYM*KGSLLDFLKNPEQGRALKLPWLVDMAAVAAGM*AYIERM*NYIHRDLRESANILVGNGLICKLADF
C-FGR    ITLERRLGTGCFGDVWLGTWNGSTKVAVKTLKPGTM*SPESFLEEAQVM*KLRRLGHDMGSLLDFLKPEQGDLRLPLVDMAAQVAEGM*AYVERM*ERYIHRDLRAANILVGERLACKLADF
LCK      IKL VERLGAQGFGEVWM*GYNGHVAVESLKQGSM*SPDAFLAEANLM*KQLQHFRLVRLYAVT--QEPIYIIEYM*ENGSLVDFLKTPSGIKLNVNKLLDMAAQIAEGM*AFIERM*NYHRDLRAANTLVSDTLSCKIADF
HCK      IKLEKKLGAGQFGEVYTKGHVAVKTKGSM*VFAFLAEANVM*TLQHKLVKLHVVTKE-PIYIITERM*AKGSLLDFLKSDEGSKQPLPKLIDFSAQIAGM*AFIEQRM*YIHRDLRAANILVSASLVCKLADF
LYN      IKLVRLGAGQFGEVWRSTKAVKTKPGTM*SVQAFLEEANLM*KTLQHDKLVRLYAVVREEPIYIITEYM*AKGSLLDFLKSDEGSKVLLPKLIDFSAQIAEGM*AYIERKNYIHRDLRAANVLVSESLM*CKLADF

         ++++ ***+++  ++++++ + +  +++++++++    +  +++        +++    + ++  ++  ++++   +++ + +]
C-SRC    GLARLIEDNEYTARQGAKFPIKM*TAPEAALYGRFTIKSDVWSFGILLTELTTKGRVPYPGMVNREVLDQVERGYRMPCPPECPESLHDLM*CQCW*RKDPEERPTFEYLQAFLEDYFTSTEPQYQPGENL.*
C-YES    GLARLIEDNEYTARQGAKFPIKM*TAPEAALYGRFTIKSDVWSFGILLTELVTKGRVPYPGM*VNREVLEQVERGYRM*PCQGCPESLHELM*NLCW*KKDPDERPTFEYIQSFLEDYFTATEPQYQPGENL.*
FYN      GLARLIEDNEYTARQGAKFPIKM*TAPEAALYGRFTIKSDVWSFGILLTEIVTKGRVPYPGM*NNREVLEQVERGYRM*PCPOCPISLHELM*IRCW*KKDPEERPTFEYLQSFLEDYFTATEPQYQPGENL.*
C-FGR    GLARLIS*DEYNPCQGSKFPIKM*TAPEAAALFGRFTIKSDVWSFGILLTEIITKGRIPYPGM*WRREVLEQVEQGYRM*PCPPGCPASLYEAM*QYTM*RLDPEERPTFEYLQSFLEDYFTSAEPQYQPGDQT.*
LCK      GLARLIEDNEYTAREGAKFPIKM*TAPEALM*YGFTIKSDVWSFGILLM*EIVTYGRIPYPGRTNADVM*TALSQGYRM*PRVEM*CPDELYDIM*KM*CWKEKAERPTFDYLQSVLDDFYTATEGQYQPQP*
HCK      GLARVIEDNEYTAREGAKFPIKM*TAEPALM*GFTFKSDVWSFGILLM*EIVTYGRIPYPGRTNADVHTALSQGYRM*PRVEM*CPDELYDIM*KM*CWKEKAERPTFDYLQSVLDDFYTATEGQYQQP*
LYN      GLARVIEDNEYTAREGAKFPIKM*TAPEALM*YGFIKSDVWSFGILLM*EIVTYGRIPYPGRTNADVHTALSQGYRM*PRVENCPDELYDIM*KM*CWKKAERPTFDYLQSVLDDFYTATEGQYQQP*
```

FIGURE 1. Predicted sequences of gene products of avian c-src,[24,120] human c-yes,[29] human fyn,[33,34] human c-fgr,[26-28] mouse lck,[9,10] human hck,[36,37] and human lyn,[35] aligned for maximum homology. Symbols: colon (:) indicates conserved residue in amino-terminal region that is also found in c-abl, but not in c-fps; plus (+) indicates conserved residue found also in both c-fps and c-abl; brackets ([ ]) indicate boundaries of kinase domain; equal (=) indicates lysine at ATP binding site; slash (/) indicates position of additional sequences in c-fms, c-kit, and the PDGF receptor; asterisk (*) indicates principal site of phosphorylation of active kinase; exclamation (!) indicates tyrosine phosphorylated in c-src and lck proteins.

TABLE 1
Chromosomal Locations and Expression Patterns of *src* Family Genes

Gene	Human chromosome	Tissues and cell lines where RNA or protein detected
c-src	20q13.3	Ubiquitous; highest in developing nervous system and blood platelets, increased in myeloid cell lines HL60 and U937 upon differentiation.
c-yes	18q21.3	Highest in brain and placenta; less in liver, kidney, and lung. Cell lines: high in A431 and KB (carcinomas), less in K562 (myelogenous leukemia), least in fibroblasts.
fyn	6q21	Highest in brain; less in placenta, lung, liver, and kidney. Cell lines: high in IM9 (B cells), less in fibroblasts, least in A431 and K562.
c-fgr	1p36.1	IM9 and other B cell lines.
lck	1p32-p35	Highest in T cells, detectable in B cells.
hck	20q11-q12	Highest in granulocytes, detectable in B cells. Expression in ML-1 and HL60 increases upon myeloid differentiation.
lyn	8q13-qter.	Highest in placenta; less in liver; least in brain, lung, and kidney.

Data from References 27, 33, 35—37, 146—151, 185, 186, 237—242.

ylated at tyrosine when LSTRA cell membranes are incubated with ATP *in vitro*.[42,43] The tryptic peptide phosphorylated in the 56-kDa protein (p56^{lck}) is identical to a tryptic phosphopeptide of known sequence from p60^{v-src}.[42,44] Therefore, oligonucleotides encoding the p60^{v-src} phosphorylation site sequence were synthesized and used to isolate p56^{lck} cDNA clones.[9,10] The *lck* gene has also been isolated by cross-hybridization.[45]

The protein products of most of the *src* family proto-oncogenes have been identified using specific antibodies derived from animals bearing oncogene-induced tumors or from animals immunized with synthetic peptide or bacterially synthesized antigens. The protein products of the *fyn*,[46,47] *hck*,[48] c-*src*,[49-51] c-*yes*,[52] c-*fgr*,[31] and *lck*[42,44] genes have the molecular sizes predicted from their cDNA sequences and, where tested, have the expected protein-tyrosine kinase activity.

III. EVOLUTION

Protein-tyrosine kinases recognized by and able to phosphorylate a certain polyclonal anti-p60^{v-src} antibody have been found in primitive eukaryotes,[53] but these kinases have not been characterized adequately to judge their relationship to p60^{c-src}. A closely related gene has been found in *Drosophila*.[54,55] The predicted *Drosophila src*-like protein is no more similar to vertebrate c-*src* than it is to other members of the vertebrate *src* family,[1] suggesting that the vertebrate family diverged after the evolutionary separation between arthropods and vertebrates. Indeed, since several members of the *src* family are restricted in their expression to certain specialized cell types, it seems likely that they evolved in concert with cellular specialization. In mammals, *lck* and c-*fgr* are found close together on one chromosome and *hck* and c-*src* close together on another, suggesting recent gene duplication events (Table 1). The positions of most exon/intron boundaries are conserved in the coding regions of the avian[23] and human[25,30] c-*src* genes, the human c-*fgr* gene,[31,32] and the mouse *lck* gene,[56] also suggesting a common origin.

IV. STRUCTURE OF *SRC* FAMILY KINASES

Almost our entire knowledge of *src* family kinase structure is based on their cDNA sequences and radiolabeling studies. None of the proteins has been obtained in sufficient quantity to permit the sorts of analyses that reveal details of three-dimensional structure.

Sequence comparisons and site-directed mutagenesis have been useful for identifying regions of the molecules involved in catalytic and regulatory functions, and portions of the sequence that are exposed in the native molecules have been identified by mapping protease cleavage sites and antigenic determinants.

A. PRIMARY SEQUENCES: CONSERVED AND VARIABLE REGIONS

The predicted amino acid sequences of seven *src* family members are shown in Figure 1. Working from N- to C-terminus, the following features are notable.*

All family members start with the dipeptide Met–Gly. Both p60$^{c\text{-}src}$ and p56^{lck} have Met-1 cleaved and replaced with a myristic acid group, in amide linkage to the α amino group of Gly-2.[44,57-59] The presence of Gly-2 seems to be critical for myristylation. Four other myristylated proteins, all outside the *src* family, have N-terminal glycines,[60-63] and substitution of Gly-2 in p60$^{v\text{-}src}$ prevents myristylation.[64,65] Other features in the ten N-terminal residues of p60$^{v\text{-}src}$ are also required for myristylation.[66-68] Even though the sequences of residues three to ten differ between family members, the invariance of Gly-2 means it is likely that all family members are myristylated.

The invariant Gly-2 is followed by a totally divergent region of 75 residues or so before sequence similarities are evident. The variable N-terminal region is encoded entirely by a single exon in the c-*src* gene,[24,30] so "exon shuffling" may have contributed to sequence divergence between family members. The significance of the divergent segment is presently uncertain. It may contain recognition sequences for interaction with other proteins — substrates, regulators, or binding proteins — and thereby give each family member unique properties. Comparison of human and chicken c-*src* sequences shows several substitutions and deletions in this region, with the greatest divergence between residues 21 and 29 and between 36 and 56 (chicken c-*src* numbering).[30] Perhaps these hypervariable regions are unimportant for whatever specialized functions of p60$^{c\text{-}src}$ that are the same in birds and mammals.

Following the unique region, the remaining 400 or so residues of all family members are virtually identical (Figure 2). Residues 80 to 260 are unimportant for catalysis, but may be important for some function common to all *src* family members. These residues are also conserved, albeit weakly, in a distantly related protein-tyrosine kinase, p150$^{c\text{-}abl}$. The greatest similarity with p150$^{c\text{-}abl}$ lies between positions 140 and 190, where p90$^{c\text{-}fps}$ is also homologous[3,69] (Figure 1). Some mutations of p60$^{v\text{-}src}$ that lead to thermolability or altered cell morphology map to this noncatalytic, conserved region.[70-73] A host range mutation of v-*src* maps to position 172,[74] so the *src*/*fps*/*abl* homology region may be involved in interactions with other cellular proteins. Possible roles of the N-terminal region will be discussed below.

Residues 260 to 516 encompass the catalytic domain. This is defined by (1) its homology with other protein kinases,[1,3] (2) the kinase activity of protein fragments containing this region,[75,76] and (3) the mapping of mutations that decrease kinase activity.[77-79] The kinase region contains a glycine-rich motif thought to form an ATP-binding pocket,[3] a lysine residue (Lys-295) that reacts with an ATP analogue,[80] and a tyrosine residue (Tyr-416) that can be phosphorylated.[81,82] This tyrosine lies in a region homologous to a portion of the cAMP-dependent protein kinase that can be cross-linked to a reactive peptide substrate,[83,84] placing it close to the peptide binding site.

Although it is clear that the sequence of the catalytic domain is constrained by the requirements for enzymatic activity, there are many residues in the catalytic region that are not conserved between all protein-tyrosine kinases (Figure 1).[1,3] For example, three of the receptor kinases contain a 76- to 107-residue insert between the ATP binding site and the presumed peptide binding site, not found in *src* family kinases.[85] Therefore, the high con-

---

* Unless otherwise stated, residue numbers apply to the chicken pp60$^{c\text{-}src}$ sequence[24] (corrected[120]).

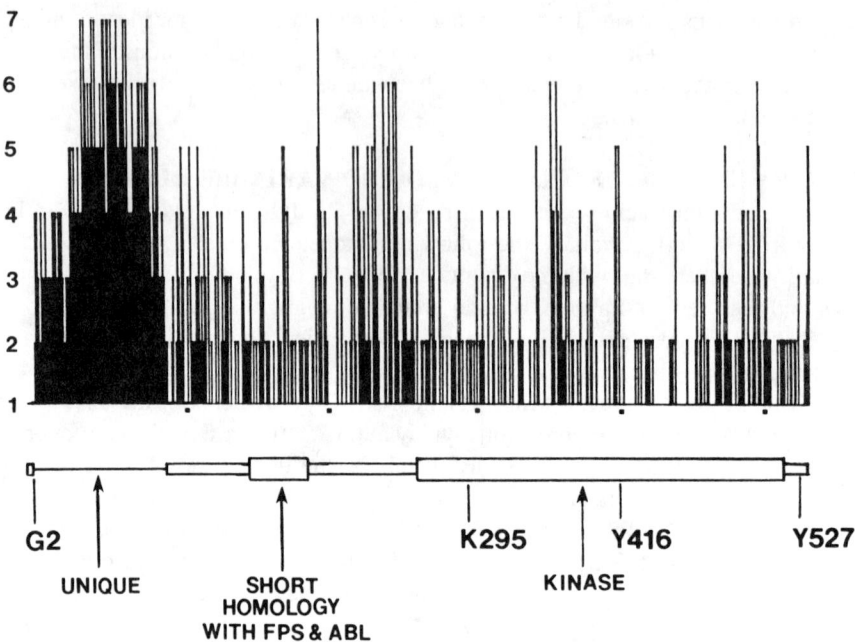

FIGURE 2. Divergence of sequences of *src* family kinases. The sequences were gapped and aligned as in Figure 1. Ordinate: residue number; abscissa: number of different amino acids found at a given position. A small value on the abscissa indicates high conservation; a large value (maximum seven) indicates divergence. A gap is counted as a residue. Various sequence features are indicated.

servation of the catalytic domain sequence between the *src*-like kinases suggests that it may play other roles besides catalysis and emphasizes the extreme divergence of the N-terminal unique regions.

The last residue that is essential for catalytic function is a conserved leucine (Leu-516 in p60$^{c\text{-}src}$).[86,87] However, the *src* family kinases have strong sequence homology almost to the C-terminus. This sequence contains a tyrosine (Tyr-527 in p60$^{c\text{-}src}$) that is phosphorylated in p60$^{c\text{-}src}$ and p56^{lck} (Figure 3).[11,88,89] The phosphorylation state of this tyrosine seems to play a major role in regulation of kinase activity (see Section II below), which may account for its invariance.

B. HIGHER ORDER STRUCTURE

Kinases p60$^{v\text{-}src}$ and p60$^{c\text{-}src}$ are both susceptible to membrane-associated proteases that cleave about 8 kDa from the N-terminus.[76] Further proteolysis of p60$^{v\text{-}src}$, p60$^{c\text{-}src}$, or the v-*yes* protein with chymotrypsin or trypsin releases a 29- to 30-kDa fragment containing protein kinase activity.[75,76] This fragment subsumes the kinase domain through the extreme C-terminus,[90] so the position of the protease-sensitive region must lie at about position 260. This is the start of the kinase domain defined by mutagenesis and is only 14 residues before the Gly–Xaa–Gly–Xaa–Xaa–Gly nucleotide-binding pocket.[3] This suggests that the N- and C-terminal halves of the protein fold separately from each other, linked by an exposed "bridge" at about position 260. The N-terminal half of p60$^{v\text{-}src}$ may inhibit the kinase activity of the C-terminal half,[75,76] suggesting important interactions between the two domains. A monoclonal antibody recognizes an epitope assembled from residues contained within both N- and C-terminal regions, suggesting that they lie in close proximity.[91]

```
                                        -           +
                      ... 3 2 1 0 9 8 7 6 5 4 3 2 1 0 1 2 3 4 5 6
                      _____

         c-src        ... A:F L E D Y:F T S T E P:Q Y Q P G E N L   (533)
         fyn          ...   S F L E D Y F T A T E P Q Y Q P G E N L (537)
         c-yes        ...   S F L E D Y F T A:T E P Q Y Q P G D N L (540)
         c-fgr        ...   S F L E D Y F:T S A E P Q Y Q P G D Q T (529)
         lck          ...   S V L D D F F T A T E G Q Y Q P Q P     (509)
         lyn          ...   S V L D D F Y T A T E G Q Y Q Q Q P     (512)
         hck          ...   S V L D D F Y T A T E S Q Y Q Q Q P     (505)

         CONSENSUS    ...   S:F L E D Y:F:T A:T E P:Q Y Q P G P N L
                            A:V * D   F:    S:A     G:   *   Q Q D * *
                                :           : :     :
                               RSV         S2 GR   Y73  S1
                                            : :
                                          FYN1-1 FYN1-2
```

FIGURE 3. Predicted sequences of C-termini of products of *src* family genes. Positions of the end of the kinase domain and the phosphorylated tyrosine are shown. Positions of sequence divergence or truncation of viral oncogenes are also indicated. S2, S-ASV; GR, GR-FeSV; S1, S1-ASV; FYN1-1 and FYN1-2, transforming variants of human *fyn* selected by passage of a retroviral vector through cell cultures.[14] Colon (:) indicates position of divergence of viral oncogene; single asterisk (*) indicates end of kinase domain; double asterisk (:) indicates phosphorylated tyrosine.

Recent data suggest that interactions between N- and C-terminal halves may be complex. Removal of the N-terminal half of p60^{c-src} by proteolysis stimulates phosphorylation of some substrates, but inhibits phosphorylation of others.[90] Thus, the N-terminal region may confer substrate specificity upon the catalytic domain.

Proteolytic digestion with thermolysin or pronase E can cleave about 1.5 kDa from the extreme C-terminus of p60^{c-src},[90] suggesting that this region, C-terminal to the kinase domain (bounded by Leu-516), forms a separate "tail". These studies suggest a model for p60^{c-src} in which the kinase domain is compactly folded and is linked to the N-terminal region and the tail by protease-sensitive regions. Since the fates of the N-terminal region and the tail have not been followed during proteolysis, it is possible that they are readily digested.

SDS-denatured p60^{c-src} is cleaved by *Staphylococcus aureus* strain V8 protease at positions 305 and 310.[92,93] Cleavage of p60^{c-src} and p56^{lck} occurs in similar positions.[42,49-51,94] Secondary cleavages of p60^{src}s occur about 14 and 18 kDa from the N-terminus.[93] It is not evident why some regions are more susceptible to proteolysis than others in the denatured protein.

The most immunogenic regions of p60^{v-src} lie in the N-terminal half.[95,96] Immunization of mice with p60^{v-src} allows production of monoclonal antibodies to residues 28 to 38 and 92 to 128,[95] suggesting that these regions are exposed to solvent. The region containing residues 28 to 38 differs between chicken and human p60^{c-src},[30] and antibodies to this region discriminate between chicken and human p60^{c-src}.[95] Synthetic peptides modeled on the p60^{v-src} sequence also have been used as immunogens, and some of the antipeptide sera recognize the native protein, allowing mapping of accessible regions. Antibodies to peptides derived from the N- and C-termini of p60^{v-src} recognize the complete protein.[97,98] Several other hydrophilic regions are also accessible — e.g., residues 155 to 160, 315 to 321, 409 to 420 (including the phosphorylation site at Tyr-416), 458 to 463, and 498 to 512 (antibodies to this region inhibit kinase activity)[99-101] — but some are not (e.g., residues 103 to 108, 326 to 333, and 419 to 427).[99] The C-termini of p60^{c-src} and p56^{lck} are also accessible to antipeptide antibodies.[11,94,102]

Mapping of temperature-sensitive mutations yields information regarding primary sequence elements involved in initiating or maintaining an active conformation. Certain mutant RSVs cause a temperature-sensitive transformed phenotype. Since phenotypic changes occur in the presence of protein synthesis inhibitors, these mutations must create thermolabile $p60^{v-src}$s. Some temperature-sensitive mutations map to the N-terminal region,[72,73] although the majority map to the kinase domain,[74,103,104] suggesting that both halves of the molecule need to be folded correctly for transformation to occur.

V. ONCOGENIC ACTIVATION

The *src* family genes appear to lack transforming activity, even when overexpressed,[11,12,14,105-109] but are prone to oncogenic mutations. Evidence will be presented in Section VI showing that many oncogenic mutations appear to stimulate kinase activity. The normal state of the kinase thus may be considered to be repressed and the oncogenic form derepressed. Furthermore, positions of oncogenic mutations delineate parts of the molecule implicated in negative regulation. The availability of mutations that affect kinase activity has been vital for understanding mechanisms of enzymatic regulation.

The c-*src* gene has been acquired by three different retroviruses (RSV and avian sarcoma viruses [ASVs] S1 and S2), c-*fgr* by GR-FeSV, and c-*yes* by Y73-ASV. Comparing the sequence of each oncogene with its benign parent shows that a variety of changes have occurred. The v-*fgr* and v-*yes* proteins have greatly altered N-termini, owing to the presence of sequences encoded by viral genes.[20,21] Most of the oncogenes, with the exception of the v-*src* found in S2-ASV,[110,111] also contain scattered point mutations. The only feature common to all five oncoproteins is the substitution of several residues at the C-termini of the cognate proto-oncoproteins (Figure 3). Also, transforming variants of *fyn* isolated by passage of a retrovirus vector through tissue culture cells have altered C-termini.[14] In each case the tyrosine homologous to Tyr-527 of $p60^{c-src}$ has been replaced, together with variable numbers of flanking residues.

In vitro recombination experiments show that, for $p60^{c-src}$, substitution of the C-terminus with the RSV $p60^{v-src}$ C-terminus is sufficient for transformation.[105,106,108] Significantly, the oncoprotein of S2-ASV is identical with $p60^{c-src}$, but for the replacement of the C-terminus, and the sequence substituted differs from that found in the oncoproteins of RSV or S1-ASV.[111] Therefore, replacement of the normal tail with either of two different sequences is oncogenic. Site-directed mutagenesis has shown that simply docking the "tail" of $p60^{c-src}$ is, in fact, sufficient for transformation.[112,113] A special role for Tyr-527 has been suggested by the use of site-directed mutagenesis to change this residue to phenylalanine. This single change is sufficient for transformation.[112-116] The corresponding mutation changes the benign *lck* gene into an oncogene able to transform mouse fibroblasts.[11,12]

Even though the normal C-terminus of $p60^{c-src}$ is necessary for its benign nature, placing this terminus on the end of $p60^{v-src}$ does not reduce its oncogenicity for chick cells.[108] Therefore, $p60^{v-src}$ contains mutations outside the tail region that are sufficient for transformation. (Whereas the C-terminal mutants are oncogenic in both chicken and mouse fibroblasts, the severity of the oncogenic effects of N-terminal mutants depends on the cell type and expression level.[117,118]) Introduction of each $p60^{v-src}$ mutation singly into $p60^{c-src}$ and analysis of spontaneously transforming variants of c-*src* have allowed the identification of specific single amino acid changes in the kinase domain and the N-terminal region that are oncogenic. Kinase domain mutations Thr-338 to Ile, Glu-378 to Gly, and Ile-441 to Phe are all oncogenic.[119,120] Simultaneously changing three residues in the N-terminal region in $p60^{c-src}$ (Gly-63, Arg-95, and Thr-96) to their cognates in $p60^{v-src}$ (Asp, Trp, and Ile, respectively) is sufficient for transformation of chick cells.[119] Further analysis suggests that Trp-95 suffices for transformation.[121] It also appears that certain deletions in the N-terminal region may be oncogenic.[122]

Because S2-ASV v-*src* and the oncogenic *fyn* viruses differ from their proto-oncogenes only in their C-terminal coding regions,[14,111] it is probable that C-terminal alterations generally occurred first in the acquisition of *src* family proto-oncogenes by viruses and are alone sufficient to allow proliferation of infected cells. The other mutations in the oncogenes may allow more complete transformation and thereby give further growth advantage to transformed cells.

VI. ONCOGENICITY ENTAILS DEREPRESSION OF KINASE ACTIVITY

A. MUTANT FORMS OF C-*SRC*

Early *in vitro* experiments detected no differences in kinase activity between p60$^{v\text{-}src}$ and p60$^{c\text{-}src}$. Because RSV-transformed chicken cells have more p60$^{v\text{-}src}$ than p60$^{c\text{-}src}$, expression level alone could account for the five- to tenfold increase in tyrosine phosphorylation in the transformed cells.[123] It is now clear, however, that (1) even low levels of p60$^{v\text{-}src}$ are sufficient for transformation;[124] (2) overexpression of p60$^{c\text{-}src}$ to tenfold the normal level neither transforms fibroblasts nor leads to a large increase in phosphotyrosine in cell proteins;[125,126] (3) in immunoprecipitates formed with monoclonal antibodies that recognize epitopes toward its N-terminus, p60$^{v\text{-}src}$ has 10 to 20 times the specific activity of p60$^{c\text{-}src}$.[126,127] Thus, the newer data are consistent with the hypothesis that v-*src* encodes a more active kinase than c-*src*, increasing phosphorylation of cell proteins at tyrosine and thereby causing transformation. The correlation between oncogenicity and kinase specific activity extends to recombinant c-*src*/v-*src* genes, specific amino acid substitutions, and spontaneous "activating" mutations in c-*src*.[126] As a corollary, mutations in v-*src* that reduce its kinase activity render a less aggressively transformed, or normal, phenotype. Moreover, the kinase activity *in vitro* generally correlates with the phosphotyrosine content of cell proteins *in vivo*. It should be noted, however, that the mutations also may alter the substrate specificity of the kinase — perhaps mutations that increase the kinase activity also render the kinase less selective in the proteins it phosphorylates *in vivo*. This may explain why certain mutant v-*src* genes fail to transform even though they encode active kinases.[64,72,73] The substrate specificity question is discussed in more detail below.

B. KINASE ACTIVATION BY POLYOMA VIRUS

Polyoma virus is a DNA tumor virus that encodes three related transforming proteins — the large, medium, and small tumor (T) antigens. All three cooperate for complete transformation of primary fibroblasts, but medium T (mT) is sufficient to transform some established fibroblast cell lines. Immunoprecipitates of mT contain protein-tyrosine kinases able to phosphorylate mT.[128] Gradient sedimentation and immunoprecipitation experiments show that mT itself is not a kinase, but is phosphorylated by associated *src* family kinases.[102,129,130] mT that lacks bound kinase has a low sedimentation coefficient and does not become phosphorylated *in vitro*. A small population of mT has a high sedimentation coefficient and has associated kinase activity.[102,129,130] p60$^{c\text{-}src}$,[130] p62$^{c\text{-}yes}$,[131] and p59$^{c\text{-}fyn}$[46,47] are found complexed with mT, probably each with different mT molecules. Binding of mT to p60$^{c\text{-}src}$ requires sequences both in the tail[112,132] and in the body[133] of the protein. Not all p60$^{c\text{-}src}$ becomes complexed with mT,[130] but the associated population has greatly increased kinase activity.[134] The magnitude of increase is difficult to estimate because of the presence of other activated *src* family kinases. mT is as effective as p60$^{v\text{-}src}$ as a transforming agent and is rather more potent than p60$^{c\text{-}src}$ with single amino acid mutations. This may be due to stimulation of several kinase activities.

Many mutations in mT affect transformation. In general, mutations that lead to a failure of mT to associate with or to activate p60$^{c\text{-}src}$ (and presumably p62$^{c\text{-}yes}$ and p59$^{c\text{-}fyn}$) are not transforming.[130,134] Furthermore, if the level of p60$^{c\text{-}src}$ in mT-transformed cells is reduced by expression of an antisense c-*src* clone, the amount of activated p60$^{c\text{-}src}$ is reduced and the cells become less transformed.[135] There are, however, mT mutants that fail to transform despite association with, and stimulation of, p60$^{c\text{-}src}$ (for example).[136] These mutants suggest that increased p60$^{c\text{-}src}$ specific activity alone is insufficient for transformation. However, the status of association of these mutant mT proteins with p62$^{c\text{-}yes}$ is unknown. It is also possible that these mT mutations have dominant, *cis*-acting, negative influences. For example, the mutations may prevent localization of the activated p60$^{c\text{-}src}$-mT complex to the correct parts of the membrane.

VII. REPRESSION OF PROTEIN KINASE ACTIVITY BY C-TERMINAL PHOSPHORYLATION

Since increased activity is necessary for transformation by p60$^{c\text{-}src}$, what do oncogenic mutations suggest about enzymatic regulation? First, the normal C-terminus is "special", in that only the natural C-terminus can repress the protein kinase and transforming activity. Second, mutations in the kinase domain and N-terminus also can derepress activity.

Phosphopeptide analysis shows that p60$^{c\text{-}src}$ from nontransformed fibroblasts normally is phosphorylated extensively at Tyr-527.[88,89] In contrast, the activated p60$^{c\text{-}src}$ complexed with mT lacks phosphate at Tyr-527.[137] Oncogenic mutations in the tail replace Tyr-527, blocking phosphorylation and increasing protein kinase specific activity. Studies of p60$^{c\text{-}src}$s activated by mutations in the body of the protein (including v-*src*/c-*src* chimeras and spontaneous transforming mutant c-*src*s) have shown that some of these p60^{src}s also lack phosphate at Tyr-527.[126] Mutations in the body of p60$^{c\text{-}src}$ or association with mT could reduce the stoichiometry of phosphorylation at Tyr-527 by increasing the rate of dephosphorylation or by reducing the rate of phosphorylation.

These observations suggest that phosphorylation of Tyr-527 in p60$^{c\text{-}src}$ represses its kinase activity *in vivo* and *in vitro*. Biochemical experiments confirm that the phosphorylation state of Tyr-527 regulates kinase activity *in vitro*. Selective dephosphorylation of Tyr-527 can occur spontaneously under some conditions of cell lysis and immunoprecipitation, and the resultant p60$^{c\text{-}src}$ is derepressed.[138] Treatment of isolated p60$^{c\text{-}src}$ with phosphatase also relieves inhibition.[94] Avian p60$^{c\text{-}src}$ synthesized in yeast has low phosphorylation stoichiometry at Tyr-527 and is derepressed.[139,140]

The phosphorylated C-terminus of p60$^{c\text{-}src}$ may be thought of as an allosteric effector of kinase activity. Analysis of the kinetics of substrate phosphorylation by inhibited p60$^{c\text{-}src}$ reveals changes in Michaelis constants and maximum rates[94] consistent with inhibition at a late step in the proposed reaction mechanism.[141] One possibility is that the phosphorylated C-terminus interacts with the active site, perhaps by behaving as a product analogue. An analogy may be drawn with the cAMP-dependent protein kinase, where the regulatory subunit binds to the active site of the catalytic subunit in the absence of cAMP.[84] Another model is provided by myosin light chain kinase (MLCK), which is activated by calmodulin. The C-terminus of MLCK normally acts as an inhibitor (perhaps by acting as a substrate analogue), and inhibition is relieved (i.e., the kinase is activated) when calmodulin binds to the C-terminus.[142] Similarly, binding of an antipeptide antibody to the phosphorylated C-terminus of p60$^{c\text{-}src}$ relieves repression,[94] perhaps by competing with the active site for binding the C-terminus. The true situation may be more complicated, however. A proteolytic fragment of p60$^{c\text{-}src}$ containing the kinase domain and tail appears not to be activated by the antipeptide serum that activates intact p60$^{c\text{-}src}$, suggesting that the N-terminal half of p60$^{c\text{-}src}$ is involved in inhibition of the kinase domain by the phosphorylated C-terminus.[90] The N-terminal region may be required for the catalytic domain to assume a conformation that can be inhibited by the phosphorylated C-terminus. Some oncogenic mutations in the

N-terminal region thus could derepress the kinase domain, even though the C-terminus is phosphorylated normally.

Current evidence suggests that other *src* family kinases are also regulated by C-terminal phosphorylation. The tryptic peptide p56[lck] is phosphorylated at Tyr-505,[11] the residue homologous to Tyr-527 of p60[c-src] (Figure 3), and replacement of this residue by phenylalanine activates phosphorylation of cellular proteins at tyrosine.[11,12] The corresponding mutation in *hck* is also oncogenic.[13] In addition to the C-terminus, other regions of the kinase molecule may serve as binding sites for effectors and for regulatory posttranslational modifications. Certain mutations implicate the N-terminal region in such modulation. This will be discussed in more detail in Section XIII.

VIII. PHYSIOLOGICAL REGULATION OF ACTIVITY

Given the paramount role of the phosphorylated tail, p60[c-src] and other *src* family kinases could be regulated in the cell by proteins that interact with the tail, including, perhaps, phosphatases and kinases that alter the phosphorylation state of the C-terminal tyrosine and binding proteins (like mT) that could modify the level of phosphorylation or interfere with interactions between the phosphorylated tail and the kinase domain. So far, however, little is known about cellular factors that regulate activity. Phosphorylation of Tyr-527 is normally almost complete[88] and is only known to be reduced in mT-transformed cells.[137] Two conditions — exposure of fibroblasts to platelet-derived growth factor (PDGF)[143,144] and mitosis[145] — appear to stimulate p60[c-src] kinase activity reversibly, but in neither case does the phosphorylation status of Tyr-527 appear to change.[144,145] The specific activity of p60[c-src] also may vary according to the cell type from which it is obtained. p60[c-src] appears to have higher specific kinase activity when obtained from cells of neuronal origin,[146-151] colon carcinoma cell lines,[152,153] and tumor-derived Syrian hamster embryo cell lines[154] when compared with p60[c-src] from glial cell lines, normal colon cell lines, and fibroblasts, respectively. In each of these cases the p60[c-src] appears to be phosphorylated at Tyr-527, although the stoichiometry is not known.

There is one instance where a nonmutant *src* family kinase is underphosphorylated at its C-terminus and is derepressed. This is the p56[lck] present in the lymphoma cell line LSTRA. This p56[lck] appears to be a mixture of two populations of similar abundance: one population is phosphorylated at Tyr-505, but the other is not.[11] In contrast, p56[lck] from normal lymphocytes is only phosphorylated at Tyr-505.[155] The presence of a significant population of p56[lck] molecules lacking phosphate at Tyr-505 is consistent with the high level of protein-tyrosine kinase activity in LSTRA cells.[40,41]

IX. MECHANISM OF C-TERMINAL PHOSPHORYLATION

As yet, there are few clues regarding which cellular kinases phosphorylate the tail tyrosine. One simple possibility is that Tyr-527 can be phosphorylated by p60[c-src] itself. The major autophosphorylation site of p60[c-src] is Tyr-416 (see Section X), but limited phosphorylation of Tyr-527 occurs when immunoprecipitated p60[c-src] is incubated with ATP, provided that Tyr-527 is not phosphorylated already.[94,139] This suggests that Tyr-527 is a second "autophosphorylation" site.

Further evidence that Tyr-527 phosphorylation can be catalyzed by p60[c-src] is provided by experiments using yeast cells. Both wild-type and kinase-inactive forms (mutated at Lys-295) of chicken p60[c-src] have been expressed in baker's yeast.[156,157] In yeast only the wild-type p60[c-src] is phosphorylated at Tyr-527,[156,157] even though both proteins are phosphorylated at Tyr-527 when expressed in fibroblasts.[156] The lack of phosphorylation of the kinase-inactive p60[c-src] suggests that yeast protein-tyrosine kinases cannot phosphorylate p60[c-src] and that the phosphorylation of wild-type p60$^{c\text{-}src}$ in yeast is self-catalyzed. The yeast cell thus provides an environment free of other protein-tyrosine kinases in which to study the mechanism of Tyr-527 phosphorylation. The kinase-inactive mutant p60$^{c\text{-}src}$ has been expressed together with kinase-active p60$^{c\text{-}src}$ in which the phosphorylation site at Tyr-527 has been mutated to Phe.[157] In these cells the kinase-inactive p60$^{c\text{-}src}$ is phosphorylated at Tyr-527 as extensively as wild-type p60$^{c\text{-}src}$ expressed alone. This shows that kinase-inactive p60$^{c\text{-}src}$ can serve as a substrate for kinase-active p60$^{c\text{-}src}$ and suggests that the phosphorylation of wild-type p60$^{c\text{-}src}$ at Tyr-527 is also an intermolecular reaction, in which one p60$^{c\text{-}src}$ molecule phosphorylates another.[157] It is not known whether intramolecular autophosphorylation is also possible.

The yeast experiments suggest that Tyr-527 of p60$^{c\text{-}src}$ is accessible for intermolecular phosphorylation catalyzed by p60$^{c\text{-}src}$ and, perhaps, other protein-tyrosine kinases. In fibroblasts, kinase-inactive p60$^{c\text{-}src}$ is phosphorylated extensively at Tyr-527.[156] This could be catalyzed by the endogenous p60$^{c\text{-}src}$, perhaps together with one or more of the other protein-tyrosine kinases present. If the main enzyme transferring phosphate to Tyr-527 is p60$^{c\text{-}src}$, a simple model for p60$^{c\text{-}src}$ autoregulation can be proposed. In this model, p60$^{c\text{-}src}$ could be activated by dephosphorylation of Tyr-527 and could then rephosphorylate at Tyr-527 and return to the repressed state. It is important for a responsive regulatory system to contain a negative feedback loop, and "autophosphorylation" at Tyr-527 could fulfill this role for p60$^{c\text{-}src}$.

Are other kinases able to phosphorylate p60$^{c\text{-}src}$ at Tyr-527? The rates of phosphorylation of Tyr-527 catalyzed by p60$^{c\text{-}src}$ in an immunoprecipitate or in yeast cells appear to be insufficient to account for the extensive phosphorylation of this residue in fibroblasts. The extent of phosphorylation is determined by the rates of phosphate hydrolysis as well as esterification, however, and the large size of the phosphate pool in cells makes it difficult to determine these rates. Thus, the extent of Tyr-527 phosphorylation in fibroblasts could be increased by a variety of mechanisms. In addition to the possibility that kinases besides p60$^{c\text{-}src}$ may phosphorylate Tyr-527, cellular components could protect phosphotyrosine-527 from phosphatases or could augment p60$^{c\text{-}src}$-catalyzed phosphorylation at Tyr-527. The existence of cellular factors that promote C-terminal phosphorylation is suggested by observations on p56^{lck}. This kinase is expressed at extremely high levels in the LSTRA cell line and is underphosphorylated at its C-terminus (Tyr-505).[11] This may result from some other change in LSTRA cells, e.g., increased phosphatase activity, but is explained parsimoniously by proposing that the positive factors (kinases or binding proteins) that maintain maximal Tyr-505 phosphorylation in normal T cell lines are inadequate to cope with the large quantity of p56^{lck} in LSTRA cells.

The recent finding that p60$^{c\text{-}src}$ can be phosphorylated at Tyr-527 even when it is removed from cell membranes[158] lends hope to the possibility that agents regulating C-terminal phosphorylation may be detectable by biochemical experiments in cell-free systems. Both dephosphorylation[138] and rephosphorylation[158] of Tyr-527 have been detected in crude cell lysates. It also may be possible to define the sequences at the C-terminus that are permissive for phosphorylation of Tyr-527 using site-directed mutagenesis[132,159] and, hence, learn the sequence specificity of the kinases involved.

X. ACTIVATION VIA PHOSPHORYLATION OF THE KINASE DOMAIN

All activated *src* family kinases are phosphorylated *in vivo* at a tyrosine in the catalytic domain (Tyr-416 in p60^{src}, Tyr-394 in p56^{lck}).[42,81,82] This includes the viral transforming proteins, chimeras with their N-terminal three-quarters coded by v-*src*, p60$^{c\text{-}src}$s with altered C-termini, p60$^{c\text{-}src}$ complexed with mT, and over-expressed p56^{lck} from LSTRA cells.[11,112,114,126,137] The same tyrosine is also the primary site of phosphorylation when immunoprecipitated or chromatographically purified src family kinases are incubated with ATP *in vitro*, suggesting that the reaction is autocatalytic.[42,81,82] However, studies of the phosphorylation state of p60[c-src] mutants expressed in yeast cells, and of the phosphorylation reactions of these yeast-synthesized molecules following immunoprecipitation, suggest that phosphorylation of Tyr-416 is not necessarily intramolecular. Instead, one derepressed p60[c-src] molecule phosphorylates another.[157] The affinity of the interaction must be very high since the reaction is independent of protein concentration over a large range,[157] perhaps down to 10^{-9} M.[160] It is not known whether Tyr-416 of a derepressed src family kinase can be phosphorylated only by another molecule of the same provenance, by other active kinases of the src family, or by any other protein-tyrosine kinase that happens to be active in the cell. Endogenous p60[c-src] is not phosphorylated at Tyr-416 in fibroblasts also expressing p60[v-src],[126] so the repressed conformation of p60[c-src] normally may prohibit access by derepressed p60[c-src]s to Tyr-416.

Phosphorylation of Tyr-416 has a curious feature. The phosphate on the corresponding tyrosine of a purified active fragment of the v-*abl* kinase (a protein-tyrosine kinase outside the src family) can be either passed to water or exchanged with ATP.[161] Thus, phosphate turnover at Tyr-416 potentially could be catalyzed by kinases as well as by phosphatases. This may explain why p60[v-src] is phosphorylated to only about 30% stoichiometry in the cell,[162] even though all p60[v-src] molecules are presumably in the derepressed conformation.

Site-directed mutagenesis shows that Tyr-416 is important; replacement of Tyr-416 in p60[v-src] by Phe,[163] or its deletion,[164] reduces kinase activity *in vitro*[165] and ameliorates some aspects of transformation.[166] The same mutation in p60[c-src] derepressed by a Phe-527 mutation reduces kinase activity by a factor of two and attenuates oncogenicity.[114,115] The partial activity of Phe-416 mutants suggests that any role which Tyr-416 may play in catalysis or substrate binding is not obligatory. The results do not distinguish whether the Tyr-416 hydroxyl group per se or its phosphorylation is necessary for full activity.

Direct evidence that Tyr-416 phosphorylation augments activity has recently been obtained.[165] Incubation of fibroblasts with *ortho*-vanadate, a phosphatase inhibitor, causes p60[c-src] to become extensively phosphorylated at Tyr-416, in addition to the usual phosphorylation at Tyr-527. This double-phosphorylated p60[c-src] has three times the *in vitro* kinase activity of normal p60[c-src]. Therefore, it appears that src family kinases have three distinct activity states: (1) phosphorylated at the C-terminus and repressed (5 to 10% activity *in vitro*), (2) dephosphorylated and derepressed (100% activity), and (3) phosphorylated in the kinase domain and fully activated (200 to 300% activity) (Figure 4). These three states are interconvertible by intermolecular phosphorylation catalyzed by another molecule of the same kinase. How the choice between phosphorylating the tail vs. the kinase domain is made is not clear, but presumably the conformation of the substrate is important.

XI. ADDITIONAL PHOSPHORYLATION SITES

A. AMINO-TERMINAL TYROSINE PHOSPHORYLATIONS

Other tyrosines besides Tyr-416 may become phosphorylated in derepressed forms of p60[c-src]. Tyrosines in the N-terminal 34 kDa are phosphorylated *in vitro* when p60[v-src] is incubated with high ATP concentrations.[160,167-169] The hyperphosphorylated p60[v-src] has tenfold increased kinase activity (over its usual derepressed level)[167] and has retarded migration upon SDS polyacrylamide gel electrophoresis. Similar slowly migrating forms of p60[src] have been observed with (1) p60[v-src] and p60[c-src] from cells exposed to *ortho*-vanadate, a phosphatase inhibitor;[170,171] (2) p60[c-src] complexed with mT and incubated with ATP *in vitro*;[172,173] (3) p60[c-src] kinase reactions stimulated with anti-C-terminal peptide antibody;[94] and (4) p60[c-src] from cells treated with PDGF.[143,144] In some of the above cases the mobility change

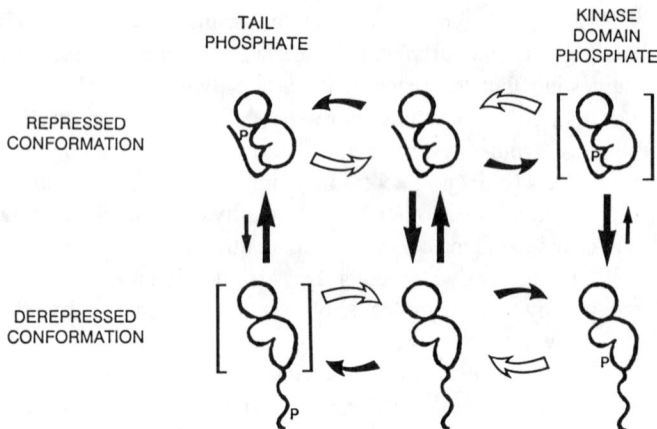

FIGURE 4. Model for regulation of *src* family kinases by stabilization of different activity states by phosphorylation in the tail and in the kinase domain. Straight arrows, conformational equilibria; curved closed arrows, reactions catalyzed by kinases; curved open arrows, reactions catalyzed by phosphatases; square brackets enclose unstable species. According to this model, dephosphorylated kinase molecules exist in free equilibrium between repressed and derepressed conformations. The equilibrium constant is not known, but is shown here as close to unity. Phosphorylation of the tail tyrosine stabilizes the repressed conformation. Phosphorylation in the kinase domain stabilizes the derepressed conformation. It is not known which of the two conformations is able to be phosphorylated in the tail, nor which is able to be phosphorylated in the kinase domain. Likewise, the accessibility of the tyrosines to phosphatases may depend upon the conformation. The proportion of molecules in either state is regulated by the activity of the specific kinases and phosphatases that operate on the tail and kinase domain tyrosines and by the relative accessibility of these tyrosines in the two conformations.

may be due to serine, not tyrosine, phosphorylation.[144,171,180] Phosphorylations have been localized to the N-terminal 18 kDa,[172-174] but whether the same tyrosine residue (or residues) is phosphorylated in each case is not clear. The only tyrosines in the first 130 residues of p60^{v-src} lie at positions 90 and 92, so these are candidates for phosphorylation. Tyr-92 is conserved between different *src* family members (Figure 1). In no case is it known whether the N-terminally phosphorylated species is activated relative to the remainder of the population: because N-terminal phosphorylation typically accompanies activation of p60, it is generally assumed to be a cause or consequence of increased activity in a subpopulation. However, in some of the aforementioned conditions, evidence for increased activity is weak. Sugimoto et al.[160] have shown that activation by preincubation with ATP does not require phosphate transfer and may merely be due to stabilization of an otherwise labile enzyme, and Collett and Belzer[175] have reported that both the normal mobility and hyperphosphorylated species of p60^{v-src} from *ortho*-vanadate-treated cells are equally active.

B. SERINE PHOSPHORYLATIONS

Both p60^{c-src} and p60^{v-src} are phosphorylated at several serine residues, all in the amino-terminal region. Ser-17 is a major site of phosphorylation (about 60% stoichiometry) in both proteins.[93,162,176] Ser-17 lies in the canonical phosphorylation sequence for the cAMP-dependent protein kinase (Arg–Arg–Xaa–Ser), and both *in vitro* and *in vivo* experiments show that this kinase can phosphorylate Ser-17.[93,177] Phosphorylation of Ser-17 has little effect on the kinase activity of p60^{v-src}, and this residue can be deleted or substituted with a nonphosphorylatable residue without affect.[164,178]

Ser-12 and Ser-48 in $p60^{v-src}$ and chicken $p60^{c-src}$ are usually not phosphorylated. Phosphorylation is greatly increased, to 50 to 100% stoichiometry, when cells are exposed to agents that stimulate protein kinase C (PKC).[179-182] (Human $p60^{c-src}$ contains several mutations close to Ser-48, and this residue seems to be unphosphorylated when PKC is active in mammalian cells.[180]) Phosphorylation by PKC slows electrophoresis on SDS polyacrylamide gels. Purified PKC can phosphorylate immunoprecipitated $p60^{src}$s at Ser-12 and Ser-48, as well as a synthetic peptide containing residues 2 to 20 at Ser-12.[180] PKC phosphorylation of the epidermal growth factor (EGF) receptor inhibits its EGF-stimulated activity,[3] but the effects of PKC phosphorylation on $p60^{c-src}$ or $p60^{v-src}$ are not known.[180-182]

$p60^{c-src}$ is also subject to phosphorylation at a novel serine and threonine residue under conditions of PDGF stimulation[144] and mitosis,[145] respectively. These residues have not been located in the sequence.

$p56^{lck}$ also becomes modified in cells exposed to activators of PKC, such as phorbol diesters,[183,184] and the modified protein is dramatically retarded in mobility during SDS gel electrophoresis. The phosphorylation sites lie in the amino-terminal region, but have not been mapped precisely.[155]

XII. ALTERED GENE PRODUCTS

High amounts of $p60^{c-src}$ kinase activity are found in neurons,[146-149] some neuroblastomas,[150] partially differentiated neuroendocrine cells derived from certain embryonal carcinomas,[151] some colon carcinoma cell lines,[152,153] and blood platelets.[185] Comparisons of $p60^{c-src}$ activities and quantities have suggested that $p60^{c-src}$ specific activity is elevated in these cell types. In each case, *in vitro* phosphorylation labels a form of $p60^{c-src}$ with reduced mobility on SDS polyacrylamide gels;[150,151,174,186-188] $p60^{c-src}$ of normal mobility is also found, although whether individual cells contain only one or both types of $p60^{c-src}$ is not known.

The altered gel mobility of neuronal $p60^{c-src}$ could result from anomalous sites of *in vitro* phosphorylation or from normal *in vitro* phosphorylation of a preexisting population of slowly migrating $p60^{c-src}$. The latter explanation seems to be correct because (1) biosynthetic labeling of neuronal cultures reveals two species of $p60^{c-src}$;[188] (2) even though neuronal $p60^{c-src}$ contains a novel serine phosphorylation,[151,186,188] its removal with phosphatase does not restore normal gel mobility;[187,189] (3) *in vitro* translation of neuronal mRNA generates both forms of $p60^{c-src}$, suggesting two mRNAs;[187,189] and (4) a c-*src* cDNA from brain mRNA predicts a sequence with six additional amino acids following the first coding exon (position 114).[190,191] This mRNA may code for an altered $p60^{c-src}$ that is subject to a novel serine phosphorylation.

There is some question whether neuronal $p60^{c-src}$ is truly more active than normal $p60^{c-src}$ *in vitro*. It is possible that the original estimates of specific kinase activity, which in most cases utilized metabolic labeling to estimate the quantity of $p60^{c-src}$ present, may have underestimated the quantity of $p60^{c-src}$ in neuronal cells owing to its low turnover rate. Using immunoblotting instead, several authors now agree that the specific kinase activity of neuronal $p60^{c-src}$ is not much greater than that of normal $p60^{c-src}$.[189,192] Neuronal $p60^{c-src}$ lacks phosphate at Tyr-416,[188] like other inactive forms of $p60^{c-src}$. What remains to be explained is why neuronal $p60^{c-src}$ is a better substrate for tyrosine phosphorylation when immunoprecipitates containing both the neuronal and normal forms of $p60^{c-src}$ are incubated with ATP *in vitro*. Most importantly, the possible functions of neuronal $p60^{c-src}$ remain a mystery.

XIII. POSSIBLE ROLES FOR THE AMINO-TERMINAL REGION

While covalent and noncovalent interactions at the C-terminus are cardinal determinants of $p60^{c-src}$ kinase activity, interactions elsewhere could also be important. In the cell, activity and substrate specificity may be modulated in ways that are not presently amenable to analysis *in vitro*. In particular, compartmentalization of protein kinases, substrates, and protein phosphatases could be important determinants of substrate phosphorylation *in vivo*. Deletion and substitution mutants of p60^{v-src} suggest that the N-terminal region contains localization signals and perhaps also has binding sites for other effectors.

A. THE *SRC/FPS/ABL* HOMOLOGY REGION

Recently, attention has been focused on the portion of the N-terminus that shares homology with *fps* and *abl* by the discovery of further sequence homologies between this region and proteins that are not kinases. The *src/fps/abl* homology is also shared with a member of the phospholipase C family (phospholipase C-II),[193] with the oncoprotein p47$^{gag-crk}$,[194] with spectrin,[195] and with GAP, the protein that stimulates the GTPase activity of p21^{c-ras}.[196] These proteins seem to have little in common with *src* family kinases.

Oncoprotein p47$^{gag-crk}$ contains 232 residues of *crk*-coded sequence of which nearly one half is related to two 50-residue blocks found in *src* family kinases (residues 86 to 139 and 148 to 205 of p60^{c-src}).[194] It has no kinase domain, yet it transforms chicken fibroblasts causing increased cell protein phosphorylation at tyrosine. Immunoprecipitates of p47$^{gag-crk}$ contain protein-tyrosine kinase activity.[194] These observations lead to the hypothesis that p47$^{gag-crk}$ interacts directly or indirectly with one or more protein-tyrosine kinases and regulates their activity. The transforming protein could be a direct activator of an *src*-like kinase, perhaps via homotypic interactions with the amino-terminal region, or may compete for a repressor of an *src*-like kinase that normally binds to the amino-terminal region.

The biological significance of the sequence similarities with phospholipase C-II, spectrin, and GAP is unclear. Other phospholipase Cs do not contain the region of homology, suggesting it is not involved in catalysis. One hypothesis is that the sequence similarity confers interaction with a common cell protein or proteins. The nonerythroid spectrins and active forms of p60^{c-src} are similarly localized to the submembranous cytoskeleton (see Section XIII.C); the whereabouts of GAP and phospholipase C-II are not known. Deletion of residues 169 to 264[200] or of residue 172 alone in p60^{v-src} compromises membrane localization,[74] so perhaps the homologous region is involved in cytoskeletal interactions.

B. THE N-TERMINAL VARIABLE REGION

N-terminal to the *src/fps/abl* region the *src* family kinases show the greatest variation. This variable region could be the raison d'etre for multiple *src*-like proteins. Although commonalities are presently more evident than idiosyncrasies, there are preliminary indications that different *src*-like proteins interact distinctively with other cell proteins. For example, whereas p60^{c-src}, p62^{c-yes}, and p59^{c-fyn} bind to mT, p56^{lck} appears not to, and p56^{lck} interacts strongly with the cytoskeleton, but p60^{c-src} does not.[133] Site-directed mutagenesis or the construction of hybrids between different family members may allow the regions involved in these interactions to be mapped to the protein sequence.

C. SUBCELLULAR LOCALIZATION SIGNALS AND MECHANISM

Where investigated, most *src* family kinases have been found associated with cellular membranes;[197] p60^{v-src} is found in the plasma membrane, primarily in regions in close contact with the substratum.[197,198] A significant population of p60^{v-src} is also found in perinuclear membranes.[199] Additionally, p60^{c-src}, p56^{lck}, and the v-*fgr* and v-*yes* proteins are also membrane bound.[41,197]

p60^{v-src} is synthesized on polyribosomes that are not associated with membranes[76] and is myristylated during or soon after synthesis.[201] Following a delay of about 5 min, newly synthesized p60^{v-src} associates with membranes.[76] Mutants that lack sequences required for myristylation remain free in the cytoplasm.[64,65] This suggests that myristylation is necessary (but not sufficient[200,201]) for localization to the membrane. Limited proteolysis studies have shown that regions within the N-terminal 8 kDa of p60[v-src] are bound to the membrane;[76] this would include the N-terminal myristoyl group.

The mechanism by which myristylated proteins are transported to membranes is not known. p60[v-src] and the v-*yes* protein associate with a 90-kDa heat-shock protein and a 50-kDa protein shortly after synthesis.[202] These proteins were once thought to ferry the protein to the membrane. However, several pieces of evidence suggest that the main role of the complex is to tag badly folded proteins. The proportion of thermolabile p60[v-src] mutant molecules found complexed with the 90- and 50-kDa proteins is greater than the proportion of wild-type p60[v-src] molecules so complexed.[202] However, p60[c-src] appears not to bind to these proteins. The 90-kDa protein may bind to the altered C-terminus of p60[v-src],[98] perhaps because it fails to integrate properly into the protein structure.

Since the conservation of Gly-2 suggests that all *src* family kinases are myristylated, all *src* family kinases may also be membrane bound. We can speculate that the substrates that each kinase phosphorylates as part of its normal function are also membrane proteins.

Membrane localization seems to be important for transformation by p60[v-src]. N-terminal mutations that block myristylation and membrane localization fail to transform, even though many tyrosine phosphorylation events proceed as they do in cells transformed by wild-type p60[v-src].[64,65] Perhaps one or more substrates whose phosphorylation is critical for transformation lie in the membrane. Since myristylation of p60[v-src] appears to be cotranslational, and there is no significant unmyristylated population,[201] it seems unlikely that modulation of the extent of myristylation will be a physiological mechanism for relocating *src* family kinases.

Because p60[c-src] can be extracted with mild detergents, it appears to be bound primarily to the phospholipids in the membrane.[203,204] In contrast, p60[v-src] is not extracted with detergents, but remains associated with the cortical cytoskeleton.[203-205] The activated p60[c-src] that is complexed with mT is also detergent insoluble.[206] Studies with mutant forms of p60[c-src] and p60[v-src] show that cytoskeletal localization requires derepressed kinase activity as well as intact membrane localization signals. Kinase-negative mutants of p60[v-src] are soluble.[203] Long N-terminal deletions in p60[v-src] reduce, but do not abolish, cytoskeletal association.[203] This is consistent with a model in which the derepressed conformation of p60[c-src] exposes a region in the N-terminus for binding to the cytoskeleton. Curiously, wild-type, repressed p56[lck] appears to be cytoskeletal when expressed in fibroblasts,[133] suggesting that the relocalization of p60[c-src] upon derepression may not be a general feature of *src* family kinases.

XIV. SUBSTRATES FOR *SRC* FAMILY KINASES

What are the substrates through which *src* family kinases relay their signal? As our knowledge of physiological regulators of *src* kinase activity improves, we may be able to identify target proteins by their specific phosphorylation on tyrosine under conditions when a *src* family kinase is active. Historically, the best approach has been to compare the phosphoproteins of cells in which these kinases are constitutively active with the phosphoproteins of control cells. The proteins phosphorylated under conditions of acute stimulation could be quite different, however.

Fibroblasts transformed by v-*src*, v-*yes*, and v-*fgr* are characterized by a five- to tenfold increase in overall protein phosphorylation at tyrosine.[123] Many different proteins undergo increased tyrosine phosphorylation.[207] Particular phosphotyrosine-containing proteins include structural proteins (including vinculin, talin,[208,209] the fibronectin receptor,[210] p81 [ezrin], and p36 [calpactin I, lipocortin II]), metabolic enzymes (enolase, lactate dehydrogenase, and phosphoglycerate mutase), calmodulin,[211] and a number of proteins for which only size is known.[207] Their increased phosphotyrosine content suggests that they are substrates for viral protein-tyrosine kinases, but it is also possible that they are phosphorylated by another protein-tyrosine kinase activated indirectly. Many of these proteins are also phosphorylated on tyrosine in cells transformed by protein-tyrosine kinases outside the immediate *src* family.[207]

One way to sort out which substrates may be involved in transformation is to study effects of mutant v-*src* genes. Some mutants stimulate high levels of tyrosine phosphorylation *in vivo*, yet are not transforming. Particularly interesting are mutations that block membrane localization.[64,65] These mutants do not transform despite extensive phosphorylation of many of the usual $p60^{v\text{-}src}$ substrates, suggesting that phosphorylation of these substrates is not sufficient for transformation.[212] Recently, a subset of proteins that are phosphorylated in cells expressing wild-type, but not mutant, $p60^{v\text{-}src}$ have been identified, and these are candidates for membrane-localized substrates that may be important for transformation.[213-215]

Given the complexity of tyrosine phosphorylation in v-*onc*-transformed fibroblasts, we may ask whether the pattern is any simpler when the cellular kinases are active. This has been investigated in fibroblasts transformed by mT (i.e., containing activated $p60^{c\text{-}src}$, $p59^{c\text{-}fyn}$, and $p62^{c\text{-}yes}$) and in cells overexpressing various *src* family kinases.

Polyoma-transformed or polyoma-infected cells do not exhibit gross increases in phosphotyrosine content. However, if the phosphatase inhibitor *ortho*-vanadate is added to the culture media, protein-tyrosine phosphorylation is increased to a greater degree in cells expressing mT than in control cells.[216] Many of the same proteins accumulate phosphotyrosine when *ortho*-vanadate is added to the two cell types; the differences are quantitative, not qualitative. This is reasonable, as the control cells probably contain a small number of active *src* kinase molecules, and mT-expressing cells contain more. The particular proteins phosphorylated (e.g., p36) are similar to those phosphorylated in cells expressing $p60^{v\text{-}src}$.[216] Unfortunately, it is not possible to be sure whether they are phosphorylated directly by mT-activated $p60^{c\text{-}src}$. Also, *ortho*-vanadate could affect the substrate specificity of the kinases.

In LSTRA cells, overexpression of $p56^{lck}$ allows some $p56^{lck}$ molecules to become active, witnessed by the phosphorylation of these molecules at Tyr-394. There is also an increase in phosphotyrosine content in total proteins.[40] Which proteins are phosphorylated has not been described in detail.

Overexpression of $p60^{c\text{-}src}$ does not increase overall phosphotyrosine content detectably in most cell lines studied, presumably because the level of $p60^{c\text{-}src}$ is inadequate to overwhelm the negative regulatory factors. However, one line, which was selected as a focus of denser cells from a monolayer of cells transfected with a c-*src* cDNA expression clone,[217] expresses $p60^{c\text{-}src}$ at about 15 times the normal level and exhibits a roughly threefold increase in overall protein-tyrosine phosphorylation (one half to one third the increase caused by $p60^{v\text{-}src}$).[127] These cells are only partially transformed, and phosphorylation of the usual $p60^{v\text{-}src}$ substrates, p36 and enolase, is scarcely detectable.[127] This suggests that most of the increase in protein-tyrosine phosphorylation is due to phosphorylation of quantitatively minor, undetected substrates. The low levels of phosphorylation of enolase and p36 may imply that $p60^{c\text{-}src}$ and $p60^{v\text{-}src}$ have different specificities. It is known, however, that some v-*src* mutants significantly increase total phosphotyrosine content without noticeable phosphorylation of p36 and enolase.[218] Other mutants (and wild-type virus) further elevate total tyrosine phosphorylation, and in these cells tyrosine phosphorylation of p36 and enolase is detected easily.[218] Therefore, the substrates phosphorylated when *src* kinases are active at a low level may be the same for $p60^{c\text{-}src}$ and $p60^{v\text{-}src}$.

There are recent indications that alternative approaches, such as immunoblotting with antibodies to phosphotyrosine, may reveal additional substrates. This particular technique has allowed the detection of an 81-kDa phosphotyrosine-containing protein in cells transformed by mT, without the use of *ortho*-vanadate.[219,220] This protein could well be a natural substrate of derepressed $p60^{c\text{-}src}$.

XV. POSSIBLE FUNCTIONS OF *SRC* FAMILY KINASES

Despite considerable progress in understanding regulation, mapping mutants, and cataloging substrates of *src* family kinases, as yet no member of the family has a firmly ascribable biological function. Clues to the functions of *src* kinases may lie in the properties of the cells in which they are expressed naturally, the factors which modulate their activity in the cell, and in the phenotypes of cells in which they are overexpressed. Different family members could have different functions, determined by the specific sequences in their N-terminal regions. The physiological effects of chronically derepressed *src* family kinases may not be the same as the consequences of transient activation. Confusingly, v-*src* transforms fibroblasts, acting as a complete mitogen,[221] yet it drives PC12 pheochromocytoma cells to differentiate,[222] and mT, presumably acting via *src* family kinases, causes differentiation of ML1 myeloid cells and embryonal carcinomas.[223,224] Possibly, derepression of each *src* kinase may have multiple sequelae, depending upon which substrates are present in the cell type of interest at the time of derepression.

Many *src* family kinases are found naturally in differentiated cell types of limited proliferative potential (Table 1). While c-*src* expression is detectable in almost every cell type examined, the level of expression is especially high in blood platelets[185] and neurons,[146] both of which are postmitotic. Expression of c-*src* increases during terminal differentiation of monocytes[225] and neurons.[186,188] Both *lck* and *hck* are expressed in specific hematopoietic lineages, and *hck* expression increases during terminal differentiation of macrophages.[37] In nonmitotic cell types, *src* family kinases may control differentiated functions.

Under what conditions are *src* family kinases active *in vivo*? Exposure of fibroblasts to PDGF stimulates $p60^{c-src}$.[143,144] It might seem redundant for one protein-tyrosine kinase to stimulate another: the effect may be signal amplification, or spread of the signal over the entire cell surface, as the activated receptors are gathered into coated pits, ferried into the cell interior, and degraded. The recent observation that $p60^{c-src}$ is activated, and subject to a novel phosphorylation, at mitosis[145] suggests a function at a step in the cell cycle different from that stimulated by PDGF or $p60^{v-src}$ in fibroblasts.

It is interesting that one effect of c-*src* overexpression is to increase the saturation density of fibroblasts, perhaps indicating a role in the response to growth factors.[217] Indeed, C3H10T$\frac{1}{2}$ mouse fibroblasts that overexpress $p60^{c-src}$ show increased DNA synthesis in response to EGF.[226] The increased saturation density also could be a consequence of reduced gap-junctional communication between cells,[227,228] since control cells at low density and overexpressors at high density form an equal number of junctions with their neighbors. In Madin-Darby canine kidney (MDCK) epithelial cells, $p60^{c-src}$ overexpression does not affect gap junctions, but alters overall cell architecture.[229]

Exposing T lymphocytes to a mitogen mixture (lectin plus phorbol diester) results in the disappearance of *lck* mRNA and the extensive modification of $p56^{lck}$.[183,230] If this response is a means of turning $p56^{lck}$ kinase activity off (c.f., increased degradation of activated EGF receptors[2]), perhaps mitogen treatment of T lymphocytes transiently stimulates $p56^{lck}$ kinase activity. It is interesting that several T cell growth factors appear to stimulate protein phosphorylation at tyrosine.[231,232] A response of T cells to mitogens, such as interleukin 2 secretion, could be mediated by $p56^{lck}$.[183] Excitingly, $p56^{lck}$ recently was found to complex with the T lymphocyte cell surface proteins CD4 and CD8.[233,234] Interactions between T cells and their targets could cause changes in CD4 or CD8 that could alter the localization or activity of $p56^{lck}$.

The membrane localization of *src*-like kinases suggests that their effectors and their substrates must lie close to (or in), or be freely able to approach, the membrane. Many of the changes in a malignantly transformed fibroblast are membrane changes — increased membrane ruffling, loss of adhesion to the extracellular matrix, release of actin fibers from their point of attachment to the cortical cytoskeleton, increased metabolite and ion transport, increased phosphoinositide degradation and resynthesis — many of which occur rapidly after p60$^{v\text{-}src}$ is expressed and do not require the cell nucleus. The cell types in which c-*src*, *lck*, and *hck* are expressed at the highest level (neurons and platelets, T cells, and monocytes, respectively) are specialized for (among other things) the coordinated exocytosis of export vesicles in response to specific stimuli. Perhaps the activity of *src* family kinases is required for membrane fusion events and cytoskeletal changes in these cells.

The several kinases in the *src* family may differ in substrate specificity or regulation, or both. If all are regulated by C-terminal tyrosine phosphorylation, as p60$^{c\text{-}src}$ and p56^{lck} appear to be, then a common kinase and phosphatase could regulate the entire family. If this were the case, expression of one member at a high enough level to saturate the regulatory system should activate all family members in that cell. Other controls could be specific to each family member. It seems that there must be specificity either in substrates or in regulation; otherwise, some family members would be redundant.

When thinking of functions for *src* family proto-oncogenes it may be important not to overrate the significance of the finding that their chronic activity in fibroblasts causes malignant transformation. Whereas fibroblasts respond to many external agents by increased proliferation and other changes reminiscent of malignancy, some other cell types respond with differentiation and decreased proliferation. For example, EGF can "transform" fibroblasts reversibly (in cooperation with transforming growth factor [TGF] β),[235] but induces characteristic morphological changes in PC12 pheochromocytoma cells.[236] Similarly, p60$^{v\text{-}src}$ transforms fibroblasts, but induces PC12 cells to differentiate.[222] Therefore, the "enigma" — that some *src* family kinases are expressed at the highest level in specialized, postmitotic cells, yet their mutant derivatives transform fibroblasts — may be reconcilable if it is found that different substrates are expressed in the different cell types or that these substrates are the same but their phosphorylation regulates different phenomena.

ACKNOWLEDGMENTS

I thank numerous investigators for communicating unpublished material; my associates, especially Alasdair MacAuley and Kathy Gould, for their criticisms of the manuscript; and Ann Lootens for secretarial assistance. Supported by U.S. Public Health Service grant CA-41072.

REFERENCES

1. **Hanks, S. K., Quinn, A. M., and Hunter, T.,** The protein kinase family: conserved features and deduced phylogeny of the catalytic domains, *Science*, 241, 42, 1988.
2. **Hunter, T. and Cooper, J. A.,** Protein-tyrosine kinases, *Annu. Rev. Biochem.*, 54, 897, 1985.
3. **Hunter, T. and Cooper, J. A.,** Viral oncogenes and tyrosine phosphorylation, in *The Enzymes,* Vol. 17, Boyer, P. D. and Krebs, E. G., Eds., Academic Press, Orlando, FL, 1986, 191.
4. **Bishop, J. M.,** The molecular genetics of cancer, *Science*, 235, 305, 1987.
5. **Martin, G. S.,** Rous sarcoma virus: a function required for the maintenance of the transformed state, *Nature (London)*, 227, 1021, 1970.
6. **Levinson, A. D., Oppermann, H., Levintow, L., Varmus, H. E., and Bishop, J. M.,** Evidence that the transforming gene of avian sarcoma virus encodes a protein kinase associated with a phosphoprotein, *Cell*, 15, 561, 1978.
7. **Collett, M. S. and Erikson, R. L.,** Protein kinase activity associated with avian sarcoma virus *src* gene product, *Proc. Natl. Acad. Sci. U.S.A.*, 75, 2021, 1978.
8. **Hunter, T. and Sefton, B. M.,** Transforming gene product of Rous sarcoma virus phosphorylates tyrosine, *Proc. Natl. Acad. Sci. U.S.A.*, 77, 1311, 1980.

9. **Marth, J. D., Peet, R., Krebs, E. G., and Perlmutter, R. M.**, A lymphocyte-specific protein-tyrosine kinase gene is rearranged and overexpressed in the murine T cell lymphoma LSTRA, *Cell*, 43, 393, 1985.
10. **Voronova, A. F. and Sefton, B. M.**, Expression of a new tyrosine protein kinase is stimulated by retrovirus promoter insertion, *Nature (London)*, 319, 682, 1986.
11. **Marth, J. D., Cooper, J. A., King, C. S., Ziegler, S. F., Tinker, D. A., Overell, R. W., Krebs, E. G., and Perlmutter, R. M.**, Neoplastic transformation induced by an activated lymphocyte-specific protein-tyrosine kinase (pp56^{lck}), *Mol. Cell. Biol.*, 8, 540, 1988.
12. **Amrein, K. E. and Sefton, B. M.**, Mutation of a site of tyrosine phosphorylation in the lymphocyte-specific tyrosine protein kinase, p56^{lck}, reveals its oncogenic potential in fibroblasts, *Proc. Natl. Acad. Sci. U.S.A.*, 85, 4247, 1988.
13. **Ziegler, S. F., Levin, S. D., and Perlmutter, R. M.**, Transformation of fibroblasts by an activated form of p59^{hck}, *Mol. Cell. Biol.*, 9, 2724, 1989.
14. **Kawakami, T., Kawakami, Y., Aaronson, S. A., and Robbins, K. C.**, Acquisition of transforming properties by *FYN*, a normal *SRC*-related human gene, *Proc. Natl. Acad. Sci. U.S.A.*, 85, 3870, 1988.
15. **Sefton, B. M.**, The viral tyrosine protein kinases, *Curr. Top. Microbiol. Immunol.*, 123, 40, 1986.
16. **Martin, G. S.**, ErbB and the EGF-receptor, *Cancer Surv.*, 5, 199, 1986.
17. **Herschman, H. R.**, The EGF receptor, in *Control of Animal Cell Proliferation*, Vol. 1, Boyton, A. L. and Leffert, H. L., Eds., Academic Press, New York, 1985, 169.
18. **Foulkes, J. G. and Rosner, M. R.**, Tyrosine-specific protein kinases as mediators of growth potential, in *Molecular Mechanisms of Transmembrane Signalling*, Cohen, P. and Houslay, M. D., Eds., Elsevier, New York, 1985, 217.
19. **Sefton, B. M.**, The role of tyrosine protein kinases in the action of growth factors, in *Control of Animal Cell Proliferation*, Vol. 1, Boynton, A. L. and Leffert, H. L., Eds., Academic Press, Orlando, FL, 1985, 315.
20. **Naharro, G., Robbins, K. C., and Reddy, E. P.**, Gene product of v-*fgr onc*: hybrid protein containing a portion of actin and a tyrosine-specific protein kinase, *Science*, 223, 63, 1984.
21. **Kitamura, N., Kitamura, A., Toyoshima, K., Hirayama, Y., and Yoshida, M.**, Avian sarcoma virus Y73 genome sequence and structural similarity of its transforming gene product to that of Rous sarcoma virus, *Nature (London)*, 297, 205, 1982.
22. **Schwartz, D. E., Tizard, R., and Gilbert, W.**, Nucleotide sequence of Rous sarcoma virus, *Cell*, 32, 853, 1983.
23. **Takeya, T. and Hanafusa, H.**, DNA sequence of the viral and cellular *src* gene of chickens. Comparison of the *src* genes of two strains of avian sarcoma virus and of the cellular homolog, *J. Virol.*, 44, 12, 1982.
24. **Takeya, T. and Hanafusa, H.**, Structure and sequence of the cellular gene homologous to the RSV *src* gene and the mechanism for generating the transforming virus, *Cell*, 32, 881, 1983.
25. **Anderson, S. K., Gibbs, C. P., Tanaka, A., Kung, H., and Fujita, D. J.**, Human cellular *src* gene: nucleotide sequence and derived amino acid sequence of the region coding for the carboxy-terminal two-thirds of pp60$^{c\text{-}src}$, *Mol. Cell. Biol.*, 5, 1122, 1985.
26. **Parker, R. C., Mardon, G., Lebo, R. V., Varmus, H. E., and Bishop, J. M.**, Isolation of duplicated human c-*src* genes located on chromosomes 1 and 20, *Mol. Cell. Biol.*, 5, 831, 1985.
27. **Nishizawa, M., Semba, K., Yoshida, M. C., Yamamoto, T., Sasaki, M., and Toyoshima, K.**, Structure, expression, and chromosomal location of the human c-*fgr* gene, *Mol. Cell. Biol.*, 6, 511, 1986.
28. **Inoue, K., Ikawa, S., Semba, K., Sukegawa, J., Yamamoto, T., and Toyoshima, K.**, Isolation and sequencing of cDNA clones homologous to the v-*fgr* oncogene from a human B lymphocyte cell line, IM-9, *Oncogene*, 1, 301, 1987.
29. **Sukegawa, J., Semba, K., Yamanashi, Y., Nishizawa, M., Miyajima, N., Yamamoto, T., and Toyoshima, K.**, Characterization of cDNA clones for the human c-*yes* gene, *Mol. Cell. Biol.*, 7, 41, 1987.
30. **Tanaka, A., Gibbs, C. P., Arthur, R. R., Anderson, S. K., Kung, H., and Fujita, D. J.**, DNA sequence encoding the NH$_2$-terminal region of the human c-*src* protein: implications of sequence divergence among *src*-type kinase oncogenes, *Mol. Cell. Biol.*, 7, 1978, 1987.
31. **Katakime, S., Notario, V., Rao, C. D., Miki, T., Cheah, M. S. C., Tronick, S. R., and Robbins, K. C.**, Primary structure of the human *fgr* proto-oncogene product p55$^{c\text{-}fgr}$, *Mol. Cell. Biol.*, 8, 259, 1988.
32. **Tronick, S. R., Popescu, N. C., Cheah, M. S. C., Swan, D. C., Amsbaugh, S. C., Lengel, C. R., DiPaolo, J. A., and Robbins, K. C.**, Isolation and chromosomal localization of the human *fgr* protoon-cogene, a distinct member of the tyrosine kinase gene family, *Proc. Natl. Acad. Sci. U.S.A.*, 82, 6595, 1985.
33. **Semba, K., Nishizawa, M., Miyajima, N., Yoshida, M. C., Sukegawa, J., Yamanashi, Y., Sasaki, M., Yamamoto, T., and Toyoshima, K.**, *yes*-related protooncogene, *syn*, belongs to the protein-tyrosine kinase family, *Proc. Natl. Acad. Sci. U.S.A.*, 83, 5459, 1986.
34. **Kawakami, T., Pennington, C. Y., and Robbins, K. C.**, Isolation and oncogenic potential of a novel human *src*-like gene, *Mol. Cell. Biol.*, 6, 4195, 1986.

35. Yamanashi, Y., Fukushige, S., Semba, K., Sukegawa, J., Miyajima, N., Matsubara, K., Yamamoto, T., and Toyoshima, K., The *yes*-related cellular gene *lyn* encodes a possible tyrosine kinase similar to p56^{lck}, *Mol. Cell. Biol.*, 7, 237, 1987.
36. Ziegler, S. F., Marth, J. D., Lewis, D. B., and Perlmutter, R. M., A novel protein tyrosine kinase gene (*hck*) preferentially expressed in cell of hematopoietic origin, *Mol. Cell. Biol.*, 7, 2276, 1987.
37. Quintrell, N., Lebo, R., Varmus, H., Bishop, J. M., Pettenati, M. J., Le Beau, M. M., Diaz, M. O., and Rowley, J. D., Identification of a human gene (*hck*) that encodes a protein-tyrosine kinase and is expressed in hemopoietic cells, *Mol. Cell. Biol.*, 7, 2267, 1987.
38. Holtzman, D. A., Cook, W. D., and Dunn, A. R., Isolation and sequence of a cDNA corresponding to a *src*-related gene expressed in murine hemopoietic cells, *Proc. Natl. Acad. Sci. U.S.A.*, 84, 8325, 1987.
39. Strebhardt, K., Mullins, J. I., Bruck, C., and Rubsamen-Waigmann, H., Additional member of the protein-tyrosine kinase family: the *src*- and *lck*-related protooncogene c-*tkl*, *Proc. Natl. Acad. Sci. U.S.A.*, 84, 8778, 1988.
40. Casnellie, J. E., Harrison, M. L., Hellstrom, K. E., and Krebs, E. G., A lymphoma cell line expressing elevated levels of tyrosine protein kinase activity, *J. Biol. Chem.*, 258, 10738, 1983.
41. Casnellie, J. E., Harrison, M. L., Pike, L. J., Hellstrom, K. E., and Krebs, E. G., Phosphorylation of synthetic peptides by a tyrosine protein kinase from the particulate fraction of a lymphoma cell line, *Proc. Natl. Acad. Sci. U.S.A.*, 78, 282, 1982.
42. Casnellie, J. E., Harrison, M. L., Hellstrom, K. E., and Krebs, E. G., A lymphoma protein with an *in vitro* site of tyrosine phosphorylation homologous to that in pp60^{src}, *J. Biol. Chem.*, 257, 13877, 1982.
43. Gacon, G., Gisselbrecht, S., Piau, J. P., Boissel, J. P., Tolle, J., and Fischer, S., High level of tyrosine protein kinase in a murine lymphoma cell line induced by Moloney leukemia virus, *EMBO J.*, 1, 1579, 1982.
44. Voronova, A. F., Buss, J. E., Patschinsky, T., Hunter, T., and Sefton, B. M., Characterization of the protein apparently responsible for the elevated tyrosine protein kinase activity in LSTRA cells, *Mol. Cell. Biol.*, 4, 2705, 1984.
45. Koga, Y., Caccia, N., Toyonaga, B., Spolski, R., Yanagi, Y., Yoshiaki, Y., and Mak, T. W., A human T cell-specific cDNA clone (YT16) encodes a protein with extensive homology to a family of protein-tyrosine kinases, *Eur. J. Immunol.*, 16, 1643, 1986.
46. Kypta, R. M., Hemming, A., and Courtneidge, S., Identification and characterization of p59^{fyn} in normal and polyoma virus transformed cells, *EMBO J.*, 7, 3837, 1988.
47. Cheng, S. H., Harvey, R., Espino, P. C., Semba, K., Yamamoto, T., Toyoshima, K., and Smith, A. E., Peptide antibodies to the human c-*fyn* gene product demonstrate pp59^{c-fyn} is capable of complex formation with the middle-T antigen of polyoma virus, *EMBO J.*, 7, 3845, 1988.
48. Ziegler, S. F., Wilson, C. B., and Perlmutter, R. M., Augmented expression of a myeloid-specific protein-tyrosine kinase gene (*hck*) after macrophage activation, *J. Exp. Med.*, 168, 1801, 1988.
49. Collett, M. S., Brugge, J. S., and Erikson, R. L., Characterization of a normal avian cell protein related to the avian sarcoma virus transforming gene product, *Cell*, 15, 1363, 1978.
50. Rohrschneider, L. R., Eisenman, R. N., and Leitch, C. R., Identification of a Rous sarcoma virus transformation related protein in normal avian and mammalian cells, *Proc. Natl. Acad. Sci. U.S.A.*, 76, 4479, 1979.
51. Oppermann, H., Levinson, A. D., Levintow, L., Varmus, H. E., and Bishop, J. M., Uninfected vertebrate cells contain a protein that is closely related to the product of the avian sarcoma virus transforming gene (*src*), *Proc. Natl. Acad. Sci. U.S.A.*, 76, 1804, 1979.
52. Sudol, M. and Hanafusa, H., Cellular proteins homologous to the viral *yes* gene product, *Mol. Cell. Biol.*, 6, 2839, 1986.
53. Barnekow, A. and Muller, W. A., An *src*-related tyrosine kinase activity in the hydroid, *Hydractinia, Differentiation*, 33, 29, 1986.
54. Shilo, B., Proto-oncogenes in *Drosophilia melanogaster, Trends Genet.*, 3, 69, 1987.
55. Simon, M. A., Dress, B., Kornberg, T., and Bishop, J. M., The nucleotide sequence and the tissue-specific expression of *Drosophilia* c-*src*, *Cell*, 42, 831, 1985.
56. Garvin, A., Pawar, S., Marth, J. D., and Perlmutter, R. M., Structure of the murine *lck* gene and its rearrangement in a murine lymphoma cell line, *Mol. Cell. Biol.*, 8, 3058, 1988.
57. Shulz, A. M., Henderson, L. E., Oroszlan, S., Garber, E. A., and Hanafusa, H., Amino terminal myristylation of the protein kinase p60^{src}, a retroviral transforming protein, *Science*, 227, 427, 1985.
58. Buss, J. E. and Sefton, B. M., Myristic acid, a rare fatty acid, is the lipid attached to the transforming protein of Rous sarcoma virus and its cellular homolog, *J. Virol.*, 53, 7, 1985.
59. Marchildon, G. A., Casnellie, J. E., Walsh, K. A., and Krebs, E. G., Covalently bound myristate in a lymphoma tyrosine protein kinase, *Proc. Natl. Acad. Sci. U.S.A.*, 81, 7679, 1984.
60. Carr, S. A., Biemann, K., Shoji, S., Parmlee, D. C., and Titani, K., *n*-Tetradecanoyl is the NH$_2$-terminal blocking group of the catalytic subunit of cyclic AMP-dependent protein kinase from bovine cardiac muscle, *Proc. Natl. Acad. Sci. U.S.A.*, 79, 6128, 1982.

61. **Aitken, A., Cohen, P., Santikarn, S., Williams, D. H., Calder, A. G., Smith, A., and Klee, C. B.**, Identification of the NH$_2$-terminal blocking group of calcineurin B as myristic acid, *FEBS Lett.*, 150, 314, 1982.
62. **Henderson, L. E., Krutsch, H. C., and Oroszlan, S.**, Myristyl amino-terminal acylation of murine retroviral proteins: an unusual protein modification, *Proc. Natl. Acad. Sci. U.S.A.*, 77, 1311, 1983.
63. **Ozols, J., Carr, S. A., and Strittmatter, P.**, Identification of NH$_2$-terminal blocking group of NADH-cytochrome b$_5$ reductase as myristic acid and the complete amino acid sequence of the membrane binding domain, *J. Biol. Chem.*, 259, 13349, 1984.
64. **Buss, J. E., Kamps, M. P., Gould, K., and Sefton, B. M.**, The absence of myristic acid decreases membrane binding of p60^{src} but does not affect tyrosine protein kinase activity, *J. Virol.*, 58, 468, 1986.
65. **Cross, F. R., Garber, E. A., Pellman, D., and Hanafusa, H.**, A short sequence in the p60^{src} N terminus is required for p60^{src} myristylation and membrane association and for cell transformation, *Mol. Cell. Biol.*, 4, 1834, 1984.
66. **Pellman, D., Garber, E. A., Cross, F. R., and Hanafusa, H.**, An N-terminal peptide from p60^{src} can direct myristylation and plasma membrane localization when fused to heterologous proteins, *Nature (London)*, 314, 374, 1985.
67. **Pellman, D., Garber, E. A., Cross, F. R., and Hanafusa, H.**, Fine structural mapping of a critical NH$_2$-terminal region of p60^{src}, *Proc. Natl. Acad. Sci. U.S.A.*, 82, 1623, 1985.
68. **Kaplan, J. M., Mardon, G., Bishop, J. M., and Varmus, H. E.**, The first seven amino acids encoded by the v-*src* oncogene act as a myristylation signal: lysine 7 is a critical determinant, *Mol. Cell. Biol.*, 8, 2435, 1988.
69. **Sadowski, I., Stone, J. C., and Pawson, T.**, A noncatalytic domain conserved among cytoplasmic protein-tyrosine kinases modifies the kinase function and transforming activity of Fujinami sarcoma virus P130$^{gag-fps}$, *Mol. Cell. Biol.*, 6, 4396, 1986.
70. **Fincham, V. J., Chiswell, D. J., and Wyke, J. A.**, Mapping of nonconditional and conditional mutants in the *src* gene of Prague strain Rous sarcoma virus, *Virology*, 116, 72, 1982.
71. **Kitamura, N. and Yoshida, M.**, Small deletion in *src* of Rous sarcoma virus modifying transformation phenotypes: identification of 207-nucleotide deletion and its smaller product with protein kinase activity, *J. Virol.*, 46, 985, 1983.
72. **Bryant, D. and Parsons, J. T.**, Site-directed mutagenesis of the *src* gene of Rous sarcoma virus: construction and characterization of a deletion mutant temperature sensitive for transformation, *J. Virol.*, 44, 683, 1982.
73. **Stoker, A. W., Enrietto, P. J., and Wyke, J. A.**, Functional domains of the pp60^{v-src} protein as revealed by analysis of temperature-sensitive Rous sarcoma virus mutants, *Mol. Cell. Biol.*, 4, 1508, 1984.
74. **Wyke, J. A. and Stoker, A. W.**, Genetic analysis of the form and function of the viral *src* oncogene product, *Biochim. Biophys. Acta*, 907, 47, 1987.
75. **Brugge, J. S. and Darrow, D.**, Analysis of the catalytic domain of phosphotransferase activity of two avian sarcoma virus-transforming proteins, *J. Biol. Chem.*, 259, 4550, 1984.
76. **Levinson, A. D., Courtneidge, S. A., and Bishop, J. M.**, Structural and functional domains of the Rous sarcoma virus transforming protein (pp60^{src}), *Proc. Natl. Acad. Sci. U.S.A.*, 78, 1624, 1981.
77. **Bryant, D. L. and Parsons, J. T.**, Amino acid alterations within a highly conserved region of the Rous sarcoma virus *src* gene product pp60^{src} inactivate tyrosine protein kinase activity, *Mol. Cell. Biol.*, 4, 862, 1984.
78. **Snyder, M. A., Bishop, J. M., McGrath, J. P., and Levinson, A. D.**, A mutation at the ATP-binding site of pp60^{v-src} abolishes kinase activity, transformation, and tumorigenicity, *Mol. Cell. Biol.*, 5, 1772, 1985.
79. **Kamps, M. P. and Sefton, B. M.**, Neither arginine nor histidine can carry out the function of lysine-295 in the ATP-binding site of p60^{src}, *Mol. Cell. Biol.*, 6, 751, 1986.
80. **Kamps, M. P., Taylor, S. S., and Sefton, B. M.**, Direct evidence that oncogenic tyrosine kinases and cyclic AMP-dependent protein kinase have homologous ATP-binding sites, *Nature (London)*, 310, 589, 1984.
81. **Patschinsky, T., Hunter, T., Esch, F. S., Cooper, J. A., and Sefton, B. M.**, Analysis of the sequence of amino acids surrounding sites of tyrosine phosphorylation, *Proc. Natl. Acad. Sci. U.S.A.*, 79, 973, 1982.
82. **Smart, J. E., Opperman, H., Czernilofsky, A. P., Purchio, A. F., Erikson, R. L., and Bishop, J. M.**, Characterization of sites for tyrosine phosphorylation in the transforming protein of Rous sarcoma virus (pp60^{v-src}) and its normal cellular homologue (pp60^{c-src}), *Proc. Natl. Acad. Sci. U.S.A.*, 78, 6013, 1981.
83. **Bramson, H. N., Thomas, N., Matsueda, R., Nelson, N. C., Taylor, S. S., and Kaiser, E. T.**, Modification of the catalytic subunit of bovine heart cAMP-dependent protein kinase with affinity labels related to peptide substrates, *J. Biol. Chem.*, 257, 10575, 1982.

84. Beebe, S. J. and Corbin, J. D., Cyclic nucleotide-dependent protein kinases, in *The Enzymes*, Vol. 17, Boyer, P. D. and Krebs, E. G., Eds., Academic Press, Orlando, FL, 1986, 44.
85. Yarden, Y., Escobedo, J. A., Kuang, W. J., Yang-Feng, T. L., Daniel, T. P., Tremble, P. M., Chen, E. Y., Ando, M. E., Harkins, R. N., Francke, U., Fried, V. A., Ullrich, A., and Williams, L. T., Structure of the receptor for platelet-derived growth factor helps define a family of closely related growth factor receptors, *Nature (London)*, 323, 226, 1986.
86. Yaciuk, P. and Shalloway, D., Features of the pp60^{v-src} carboxyl terminus that are required for transformation, *Mol. Cell. Biol.*, 6, 2807, 1986.
87. Wilkerson, V. W., Bryant, D. L., and Parsons, J. T., Rous sarcoma virus variants that encode *src* proteins with an altered carboxy terminus are defective for cellular transformation, *J. Virol.*, 55, 314, 1985.
88. Cooper, J. A., Gould, K. L., Cartwright, C. A., and Hunter, T., Tyr^{527} is phosphorylated in pp60^{c-src}: implications for regulation, *Science*, 231, 1431, 1986.
89. Laudano, A. P. and Buchanan, J. M., Phosphorylation of tyrosine in the carboxyl-terminal tryptic peptide of pp60^{c-src}, *Proc. Natl. Acad. Sci. U.S.A.*, 83, 892, 1986.
90. MacAuley, A. and Cooper, J. A., Structural differences between repressed and derepressed forms of p60^{c-src}, *Mol. Cell. Biol.*, 9, 2648, 1989.
91. McCarley, D. J., Parsons, J. T., Benjamin, D. C., and Parsons, S. J., Inhibition of the tyrosine kinase activity of v-*src*, v-*fgr* and v-*yes* gene products by a monoclonal antibody which binds both amino and carboxy peptide fragments of pp60^{v-src}, *J. Virol.*, 61, 1927, 1987.
92. Beemon, K., personal communication.
93. Collett, M. S., Erikson, E., and Erikson, R. L., Structural analysis of the avian sarcoma virus transforming protein: sites of phosphorylation, *J. Virol.*, 29, 770, 1979.
94. Cooper, J. A. and King, C. S., Dephosphorylation or antibody binding to the carboxy terminus stimulates pp60^{c-src}, *Mol. Cell. Biol.*, 6, 4467, 1986.
95. Parsons, S. J., McCarley, D. J., Raymond, V. W., and Parsons, J. T., Localization of conserved and nonconserved epitopes within the Rous sarcoma virus-encoded *src* protein, *J. Virol.*, 59, 755, 1986.
96. Resh, M. D. and Erikson, R. L., Development and characterization of antisera specific for amino- and carboxy-terminal regions of pp60^{src}, *J. Virol.*, 55, 242, 1985.
97. Walter, G., personal communication.
98. Sefton, B. M. and Walter, G., Antiserum specific for the carboxy terminus of the transforming protein of Rous sarcoma virus, *J. Virol.*, 44, 467, 1982.
99. Tamura, T., Bauer, H., Birr, C., and Pipkorn, R., Antibodies against synthetic peptides as a tool for functional analysis of the transforming protein pp60^{src}, *Cell*, 34, 587, 1983.
100. Gentry, L. E., Rohrschneider, L. R., Casnellie, J. E., and Krebs, E. G., Antibodies to a defined region of pp60^{src} neutralize the tyrosine-specific kinase activity, *J. Biol. Chem.*, 258, 11219, 1983.
101. Hunter, T., Synthetic peptide substrates for a tyrosine protein kinase, *J. Biol. Chem.*, 257, 4843, 1982.
102. Courtneidge, S. A. and Smith, A. E., The complex of polyoma virus middle-T antigen and pp60^{c-src}, *EMBO J.*, 3, 585, 1984.
103. Mayer, B. J., Jove, R., Krane, J. F., Poirier, F., Calothy, G., and Hanafusa, H., Genetic lesions involved in temperature-sensitivity of the *src* gene of four Rous sarcoma virus mutants, *J. Virol.*, 60, 858, 1986.
104. Welham, M. J. and Wyke, J. A., A single point mutation has pleiotropic effects on pp60^{v-src} function, *J. Virol.*, 62, 1898, 1988.
105. Tanaka, A. and Fujita, D. J., Expression of a molecularly cloned human c-*src* oncogene by using a replication-competent retroviral vector, *Mol. Cell. Biol.*, 6, 3900, 1986.
106. Shalloway, D., Coussens, P. M., and Yaciuk, P., Overexpression of the c-src protein does not induce transformation of NIH 3T3 cells, *Proc. Natl. Acad. Sci. U.S.A.*, 81, 7071, 1984.
107. Parker, R. C., Varmus, H. E., and Bishop, J. M., Expression of v-src and chicken c-src in rat cells demonstrates qualitative differences between pp60^{v-src} and pp60^{c-src}, *Cell*, 37, 131, 1984.
108. Iba, H., Takeya, T., Cross, F. R., Hanafusa, T., and Hanafusa, H., Rous sarcoma virus variants that carry the cellular *src* gene instead of the viral *src* gene cannot transform chicken embryo fibroblasts, *Proc. Natl. Acad. Sci. U.S.A.*, 81, 4424, 1984.
109. Piwnica-Worms, H., Kaplan, D. R., Whitman, M., and Roberts, T. M., Retrovirus shuttle vector for study of kinase activities of pp60^{c-src} synthesized *in vitro* and over-produced *in vivo*, *Mol. Cell. Biol.*, 6, 2033, 1986.
110. Ikawa, S., Hagino-Yamagishi, K., Kawai, S., Yamamoto, T., and Toyoshima, K., Activation of the cellular *src* gene by transducing retrovirus, *Mol. Cell. Biol.*, 6, 2420, 1986.
111. Shuntaro, I., Yamamoto, T., and Toyoshima, K., Modification of carboxyl-terminal region is the cause of activation of the *src* gene in avian sarcoma virus, *Jpn. J. Cancer Res.*, 77, 611, 1986.
112. Cartwright, C. A., Eckhart, W., Simon, S., and Kaplan, P. L., Cell transformation by pp60^{c-src} mutated in the carboxy-terminal regulatory domain, *Cell*, 49, 83, 1987.

113. **Reynolds, A. B., Vila, J., Lansing, T. J., Potts, W. M., Weber, M. J., and Parsons, J. T.**, Activation of the oncogenic potential of the avian cellular *src* protein by specific structural alteration of the carboxyterminus, *EMBO J.*, 6, 2359, 1987.
114. **Kmiecik, T. E. and Shalloway, D.**, Activation and suppression of pp60$^{c\text{-}src}$ transforming ability by mutation of its primary sites of tyrosine phosphorylation, *Cell*, 49, 65, 1987.
115. **Piwnica-Worms, H., Saunders, K. B., Roberts, T. M., Smith, A. E., and Cheng, S. H.**, Tyrosine phosphorylation regulates the biochemical and biological properties of pp60$^{c\text{-}src}$, *Cell*, 49, 75, 1987.
116. **Kato, J., Hirota, Y., Nakamura, N., Nakamura, N., and Takeya, T.**, Structural features of the carboxy terminus of p60$^{c\text{-}src}$ that are required for the regulation of its intrinsic kinase activity, *Jpn. J. Cancer Res.*, 78, 1354, 1987.
117. **Shalloway, D., Coussens, P. M., and Yaciuk, P.**, c-*src* and *src* overexpression in mouse cells, *Cancer Cells*, 2, 9, 1984.
118. **Reddy, S., Yaciuk, P., Kmiecik, T. E., Coussens, P. M., and Shalloway, D.**, v-*src* mutations outside the carboxyl-coding region are not sufficient to fully activate transformation by pp60$^{c\text{-}src}$ in NIH 3T3 cells, *Mol. Cell. Biol.*, 8, 704, 1988.
119. **Kato, J., Takeya, T., Grandori, C., Iba, H., Levy, J. B., and Hanafusa, H.**, Amino acid substitutions sufficient to convert the nontransforming p60$^{c\text{-}src}$ protein to a transforming protein, *Mol. Cell. Biol.*, 6, 4155, 1986.
120. **Levy, J. B., Iba, H., and Hanafusa, H.**, Activation of the transforming potential of p60$^{c\text{-}src}$ by a single amino acid change, *Proc. Natl. Acad. Sci. U.S.A.*, 83, 4228, 1986.
121. **Potts, W. M., Reynolds, A. B., Lansing, T. J., and Parsons, J. T.**, Activation of pp60$^{c\text{-}src}$ transforming potential by mutations altering the structure of an amino-terminal domain containing residues 90—95, *Oncogene Res.*, 3, 343, 1988.
122. **Nemeth, S. P., Fox, L. G., DeMarco, M., and Brugge, J. S.**, Deletions within the amino-terminal half of the c-*src* gene product that alter the functional activity of the protein, *Mol. Cell. Biol.*, 9, 1109, 1989.
123. **Sefton, B. M., Hunter, T., Beemon, K., and Eckhart, W.**, Evidence that the phosphorylation of tyrosine is essential for cellular transformation by Rous sarcoma virus, *Cell*, 20, 807, 1980.
124. **Jakobovits, E. B., Majors, J. E., and Varmus, H. E.**, Hormonal regulation of the Rous sarcoma virus *src* gene via a heterologous promoter defines a threshold dose for cellular transformation, *Cell*, 38, 757, 1984.
125. **Cooper, J. A., Hunter, T., and Shalloway, D.**, Protein-tyrosine kinase activity of pp60$^{c\text{-}src}$ is restricted in intact cells, in *Cancer Cells*, Vol. 3, Feramisco, J., Ozanne, B., and Stiles, C., Eds., Cold Spring Harbor Laboratory, Cold Spring Harbor, New York, 1985, 321.
126. **Iba, H., Cross, F. R., Garber, E. A., and Hanafusa, H.**, Low level of cellular protein phosphorylation by nontransforming overproduced pp60$^{c\text{-}src}$, *Mol. Cell. Biol.*, 5, 1058, 1985.
127. **Coussens, P. M., Cooper, J. A., Hunter, T., and Shalloway, D.**, Restriction of the *in vitro* and *in vivo* tyrosine protein kinase activities of pp60$^{c\text{-}src}$ relative to pp60$^{v\text{-}src}$, *Mol. Cell. Biol.*, 5, 2753, 1985.
128. **Eckhart, W., Hutchinson, M. A., and Hunter, T.**, An activity phosphorylating tyrosine in polyoma T antigen immunoprecipitates, *Cell*, 18, 925, 1979.
129. **Grussenmeyer, T., Scheidtmann, K. H., Hutchinson, M. A., Eckhart, W., and Walter, G.**, Complexes of polyoma virus medium T antigen and cellular proteins, *Proc. Natl. Acad. Sci. U.S.A.*, 82, 7952, 1985.
130. **Courtneidge, S. A. and Smith, A. E.**, Polyoma virus transforming protein associates with the product of the c-*src* cellular gene, *Nature (London)*, 303, 435, 1983.
131. **Kornbluth, S., Sudol, M., and Hanafusa, H.**, Association of the polyomavirus middle-T antigen with c-*yes* protein, *Nature (London)*, 325, 171, 1987.
132. **Cheng, S. H., Piwnica-Worms, H., Harvey, R. W., Roberts, T. M., and Smith, A. E.**, The carboxy-terminus of pp60$^{c\text{-}src}$ is a regulatory domain and is involved in complex formation with the middle-T antigen of polyoma virus, *Mol. Cell. Biol.*, 8, 1736, 1988.
133. **Louie, R. R., King, C. S., MacAuley, A., Marth, J. D., Perlmutter, R. M., Eckhart, W., and Cooper, J. A.**, p56^{lck} protein-tyrosine kinase is cytoskeletal and does not bind to polyoma virus middle T antigen, *J. Virol.*, 62, 4673, 1988.
134. **Bolen, J. B., Thiele, C. J., Israel, M. A., Yonemoto, W., Lipsich, L. A., and Brugge, J. S.**, Enhancement of cellular *src* gene product associated tyrosyl kinase activity following polyoma virus infection and transformation, *Cell*, 38, 767, 1984.
135. **Amini, S., DeSeau, V., Reddy, S., Shalloway, D., and Bolen, J. B.**, Regulation of pp60$^{c\text{-}src}$ synthesis by inducible mRNA complementary to c-*src* mRNA in polyoma virus transformed rat cells, *Mol. Cell. Biol.*, 6, 2305, 1986.
136. **Markland, W., Cheng, S. H., Oostra, B. A., and Smith, A. E.**, *In vitro* mutagenesis of the putative membrane-binding domain of polyomavirus middle-T antigen, *J. Virol.*, 59, 82, 1986.
137. **Cartwright, C. A., Kaplan, P. L., Cooper, J. A., Hunter, T., and Eckhart, W.**, Altered sites of tyrosine phosphorylation in pp60$^{c\text{-}src}$ associated with polyoma middle tumor antigen, *Mol. Cell. Biol.*, 6, 1562, 1986.

138. **Courtneidge, S. A.,** Activation of the pp60$^{c\text{-}src}$ kinase by middle T antigen binding or by dephosphorylation, *EMBO J.,* 4, 1471, 1985.
139. **Cooper, J. A. and Runge, K.,** Avian pp60$^{c\text{-}src}$ is more active when expressed in yeast than in vertebrate fibroblasts, *Oncogene Res.,* 1, 297, 1987.
140. **Kornbluth, S., Jove, R., and Hanafusa, H.,** Characterization of avian and viral p60^{src} proteins expressed in yeast, *Proc. Natl. Acad. Sci. U.S.A.,* 84, 4455, 1987.
141. **Wong, T. W. and Goldberg, A. R.,** Kinetics and mechanism of angiotensin phosphorylation by the transforming gene product of Rous sarcoma virus, *J. Biol. Chem.,* 259, 3127, 1984.
142. **Stull, J. T., Nunnally, M. H., and Michnoff, C. H.,** Calmodulin-dependent protein kinases, in *The Enzymes,* Vol. 17, Boyer, P. D. and Krebs, E. G., Eds., Academic Press, Orlando, FL, 1986, 114.
143. **Ralston, R. and Bishop, J. M.,** The product of the protooncogene c-*src* is modified during the cellular response to platelet-derived growth factor, *Proc. Natl. Acad. Sci. U.S.A.,* 82, 7845, 1985.
144. **Gould, K. L. and Hunter, T.,** Platelet-derived growth factor induces multisite phosphorylation of pp60$^{c\text{-}src}$ and increases its protein-tyrosine kinase activity, *Mol. Cell. Biol.,* 8, 345, 1988.
145. **Chakalaparampil, I. and Shalloway, D.,** Altered phosphorylation and activation of pp60$^{c\text{-}src}$ during fibroblast mitosis, *Cell,* 52, 801, 1988.
146. **Cotton, P. C. and Brugge, J. S.,** Neural tissues express high levels of the cellular *src* gene product pp60$^{c\text{-}src}$, *Mol. Cell. Biol.,* 3, 1157, 1983.
147. **Fults, D. W., Towle, A. C., Lauder, J. M., and Maness, P. F.,** pp60$^{c\text{-}src}$ in the developing cerebellum, *Mol. Cell. Biol.,* 5, 27, 1985.
148. **Maness, P. F., Sorge, L. K., and Fults, D. W.,** An early developmental phase of pp60$^{c\text{-}src}$ expression in the neural ectoderm, *Dev. Biol.,* 117, 83, 1986.
149. **Sorge, L. K., Levy, B. T., and Maness, P. F.,** pp60$^{c\text{-}src}$ is developmentally regulated in the neural retina, *Cell,* 36, 249, 1984.
150. **Rosen, N., Bolen, J. B., Schwartz, A. M., Cohen, P., DeSeau, V., and Israel, M. A.,** Analysis of pp60$^{c\text{-}src}$ protein kinase activity in human tumor cell lines and tissues, *J. Biol. Chem.,* 261, 13754, 1986.
151. **Lynch, S. A., Brugge, J. S., and Levine, J. M.,** Induction of altered c-*src* product during neural differentiation of embryonal carcinoma cells, *Science,* 234, 873, 1986.
152. **Bolen, J. B., Veillette, A., Schwartz, A. M., DeSeau, V., and Rosen, N.,** Activation of pp60$^{c\text{-}src}$ protein kinase activity in human colon carcinoma, *Proc. Natl. Acad. Sci. U.S.A.,* 84, 2251, 1987.
153. **Bolen, J. B., Veillette, A., Schwartz, A. M., DeSeau, V., and Rosen, N.,** Analysis of pp60$^{c\text{-}src}$ in human colon carcinoma and normal human colon mucosal cells, *Oncogene Res.,* 1, 149, 1987.
154. **Kanner, S. B., Parsons, S. J., Parsons, J. T., and Gilmer, T. M.,** Activation of pp60$^{c\text{-}src}$ tyrosine kinase specific activity in tumor-derived Syrian hamster embryo cells, *Oncogene Res.,* 2, 327, 1988.
155. **Veillette, A., Horak, I. D., and Bolen, J. B.,** Post-translational alterations of the tyrosine kinase p56^{lck} in response to activators of protein kinase C, *Oncogene Res.,* 2, 385, 1988.
156. **Jove, R., Kornbluth, S., and Hanafusa, H.,** Enzymatically inactive p60$^{c\text{-}src}$ mutant with altered ATP-binding site is fully phosphorylated in its carboxy-terminal region, *Cell,* 50, 937, 1987.
157. **Cooper, J. A. and MacAuley, A.,** Potential positive and negative autoregulation of p60$^{c\text{-}src}$ by intermolecular autophosphorylation, *Proc. Natl. Acad. Sci. U.S.A.,* 85, 4232, 1988.
158. **Schuh, S. M. and Brugge, J. S.,** Investigation of factors that influence phosphorylation of pp60$^{c\text{-}src}$ on tyrosine 527, *Mol. Cell. Biol.,* 8, 2465, 1988.
159. **MacAuley, A. and Cooper, J.,** The carboxy-terminal sequence of p56^{lck} can regulate p60$^{c\text{-}src}$, *Mol. Cell. Biol.,* 8, 560, 1988.
160. **Sugimoto, Y., Erikson, E., Graziani, Y., and Erikson, R. L.,** Inter- and intramolecular interactions of highly purified Rous sarcoma virus-transforming protein, pp60$^{v\text{-}src}$, *J. Biol. Chem.,* 260, 13838, 1985.
161. **Foulkes, J. G., Chow, M., Gorka, C., Frackelton, A. R., and Baltimore, D.,** Purification and characterization of a protein-tyrosine kinase encoded by Abelson murine leukemia virus, *J. Biol. Chem.,* 260, 8070, 1985.
162. **Sefton, B. M., Patschinsky, T., Berdot, C., Hunter, T., and Elliot, T.,** Phosphorylation and metabolism of the transforming protein of Rous sarcoma virus, *J. Virol.,* 41, 813, 1982.
163. **Snyder, M. A., Bishop, J. M., Colby, W. W., and Levinson, A. D.,** Phosphorylation of tyrosine-416 is not required for the transforming properties and kinase activity of pp60$^{v\text{-}src}$, *Cell,* 32, 891, 1983.
164. **Cross, F. R. and Hanafusa, H.,** Local mutagenesis of Rous sarcoma virus: the major sites of tyrosine and serine phosphorylation of p60^{src} are dispensable for transformation, *Cell,* 34, 597, 1983.
165. **Kmiecik, T. E., Johnson, P. J., and Shalloway, D.,** Regulation by the autophosphorylation site in overexpressed pp60$^{c\text{-}src}$, *Mol. Cell. Biol.,* 8, 4541, 1988.
166. **Snyder, M. A. and Bishop, J. M.,** A mutation at the major phosphotyrosine in pp60$^{v\text{-}src}$ alters oncogenic potential, *J. Virol.,* 136, 375, 1984.
167. **Collett, M. S., Belzer, S. K., and Purchio, A. F.,** Structurally and functionally modified forms of pp60$^{v\text{-}src}$ in Rous sarcoma virus-transformed cell lysates, *Mol. Cell. Biol.,* 4, 1213, 1984.
168. **Resh, M. D. and Erikson, R. L.,** Characterization of pp60^{src} phosphorylation *in vitro* in Rous sarcoma virus-transformed cell membranes, *Mol. Cell. Biol.,* 5, 916, 1985.

169. **Purchio, A. F., Wells, S. K., and Collett, M. S.**, Increase in the phosphotransferase specific activity of purified Rous sarcoma virus pp60$^{v\text{-}src}$ protein after incubation with ATP plus Mg^{2+}, *Mol. Cell. Biol.*, 3, 1589, 1983.
170. **Brown, D. J. and Gordon, J. A.**, The stimulation of pp60$^{v\text{-}src}$ kinase activity by vandate in intact cells accompanies a new phosphorylation state of the enzyme, *J. Biol. Chem.*, 259, 9580, 1984.
171. **Ryder, J. W. and Gordon, J. A.**, In vivo effect of sodium orthovanadate on pp60$^{c\text{-}src}$ kinase, *Mol. Cell. Biol.*, 7, 1139, 1987.
172. **Cartwright, C. A., Hutchinson, M., and Eckhart, W.**, Structural and functional modification of pp60$^{c\text{-}src}$ associated with polyoma middle tumor antigen from infected or transformed cells, *Mol. Cell. Biol.*, 5, 2647, 1985.
173. **Yonemoto, W., Jarvis-Morar, M., Brugge, J. S., Bolen, J. B., and Israel, M. A.**, Tyrosine phosphorylation within the amino-terminal domain of pp60$^{c\text{-}src}$ molecules associated with polyoma virus middle-sized tumor antigens, *Proc. Natl. Acad. Sci. U.S.A.*, 82, 4568, 1985.
174. **Bolen, J. B., Rosen, N., and Israel, M. A.**, Increased pp60$^{c\text{-}src}$ tyrosyl kinase activity in human neuroblastomas is associated with amino-terminal tyrosine phosphorylation of the *src* gene product, *Proc. Natl. Acad. Sci. U.S.A.*, 82, 7275, 1985.
175. **Collett, M. S. and Belzer, S. K.**, Forms of pp60$^{v\text{-}src}$ isolated from Rous sarcoma virus-transformed cells, *J. Virol.*, 61, 1593, 1987.
176. **Patschinsky, T., Hunter, T., and Sefton, B. M.**, Phosphorylation of the transforming protein of Rous sarcoma virus: direct demonstration of the phosphorylation of serine 17 and identification of an additional site of tyrosine phosphorylation in p60$^{v\text{-}src}$ of Prague Rous sarcoma virus, *J. Virol.*, 59, 73, 1986.
177. **Roth, C. W., Richert, N. D., Pastan, I., and Gottesman, M. M.**, Cyclic AMP treatment of Rous sarcoma virus-transformed Chinese hamster ovary cells increases phosphorylation of pp60^{src} and increases pp60^{src} kinase activity, *J. Biol. Chem.*, 258, 10768, 1983.
178. **Hirota, Y., Kato, J., and Takeya, T.**, Substitution of Ser-17 of pp60$^{c\text{-}src}$: biological and biochemical characterization in chicken embryo fibroblasts, *Mol. Cell. Biol.*, 8, 1826, 1988.
179. **Tamura, T., Friis, R. R., and Bauer, H.**, pp60$^{c\text{-}src}$ is a substrate for phosphorylation when cells are stimulated to enter cycle, *FEBS Lett.*, 177, 151, 1984.
180. **Gould, K. L., Woodgett, J. R., Cooper, J. A., Buss, J. E., Shalloway, D., and Hunter, T.**, Protein kinase C phosphorylates pp60$^{c\text{-}src}$ at a novel site, *Cell*, 42, 849, 1985.
181. **Gentry, L. E., Chaffin, K. E., Shoyab, M., and Purchio, A. F.**, Novel serine phosphorylation of pp60$^{c\text{-}src}$ in intact cells after tumor promoter treatment, *Mol. Cell. Biol.*, 6, 735, 1986.
182. **Purchio, A. F., Shoyab, M., and Gentry, L. E.**, Site-specific increased phosphorylation of pp60$^{v\text{-}src}$ after treatment of RSV-transformed cells with a tumor promoter, *Science*, 229, 1393, 1985.
183. **Marth, J. D., Lewis, D. B., Wilson, C. B., Gearn, M. E., Krebs, E. G., and Perlmutter, R. M.**, Regulation of pp56^{lck} during T cell activation: functional implications for the *src*-like protein tyrosine kinases, *EMBO J.*, 6, 2727, 1987.
184. **Casnellie, J. E. and Lamberts, R. J.**, Tumor promoters cause changes in the state of phosphorylation and apparent molecular weight of a tyrosine protein kinase in T lymphocytes, *J. Biol. Chem.*, 261, 4921, 1986.
185. **Golden, A., Nemeth, S. P., and Brugge, J. S.**, Blood platelets express high levels of the pp60$^{c\text{-}src}$ specific tyrosine kinase activity, *Proc. Natl. Acad. Sci. U.S.A.*, 83, 852, 1986.
186. **Brugge, J. S., Cotton, P. C., Queral, A. E., Barrett, J. N., Nonner, D., and Keane, R. W.**, Neurones express high levels of a structurally modified activated form of pp60$^{c\text{-}src}$, *Nature (London)*, 316, 554, 1985.
187. **Brugge, J., Cotton, P., Lustig, A., Yonemoto, W., Lipsich, L., Coussens, P., Barrett, J. N., Nonner, D., and Keane, R. W.**, Characterization of the altered form of the c-*src* gene product in neuronal cells, *Genes Dev.*, 1, 287, 1987.
188. **Cartwright, C. A., Simantov, R., Kaplan, P. L., Hunter, T., and Eckhart, W.**, Alterations in pp60$^{c\text{-}src}$ accompany differentiation of neurons from rat embryo striatum, *Mol. Cell. Biol.*, 7, 1830, 1987.
189. **O'Shaughnessy, J., DeSeau, V., Amini, S., Rosen, N., and Bolen, J. B.**, Analysis of the c-*src* gene product structure, abundance, and protein kinase activity in human neuroblastoma and glioblastoma cells, *Oncogene Res.*, 2, 1, 1987.
190. **Martinez, R., Mathey-Prevot, B., Bernards, A., and Baltimore, D.**, cDNA cloning of the neuronal form of pp60$^{c\text{-}src}$, *Science*, 237, 411, 1987.
191. **Levy, J. B., Dorai, T., Wang, L.-H., and Brugge, J. S.**, The structurally distinct form of pp60$^{c\text{-}src}$ detected in neuronal cells is encoded by a unique c-*src* mRNA, *Mol. Cell. Biol.*, 7, 4142, 1987.
192. **Yang, X. and Walter, G.**, Specific kinase activity and phosphorylation state of pp60$^{c\text{-}src}$ from neuroblastomas and fibroblasts, *Oncogene*, 3, 237, 1988.
193. **Stahl, M. L., Ferenz, C. R., Kelleher, K. L., Kriz, R. W., and Knopf, J. L.**, Sequence similarity of phospholipase C with the non-catalytic region of *src*, *Nature (London)*, 332, 269, 1988.
194. **Mayer, B. J., Hamaguchi, M., and Hanafusa, H.**, A novel viral oncogene with structural similarity to phospholipase C, *Nature (London)*, 32, 272, 1988.

195. **Lehto, V.-P., Wasenius, V.-M., Salven, P., and Saraste, M.,** Transforming and membrane proteins, *Nature (London)*, 334, 388, 1988.
196. **Vogel, U. S., Dixon, R. A. F., Schaber, M. D., Diehl, R. E., Marshall, M. S., Scolnick, E. M., Sigal, I. S., and Gibbs, J. B.,** Cloning of bovine GAP and its interaction with oncogenic *ras* p21, *Nature (London)*, 335, 90, 1988.
197. **Rohrschneider, L. R. and Gentry, L. E.,** Subcellular locations of retroviral transforming proteins define multiple mechanisms of transformation, *Adv. Viral Oncol.*, 4, 269, 1984.
198. **Krueger, J. G., Garber, E. A., and Goldberg, A. R.,** Subcellular localization of pp60^{src} in RSV-transformed cells, *Curr. Top. Microbiol. Immunol.*, 107, 52, 1983.
199. **Resh, M. D. and Erikson, R. L.,** Highly specific antibody to Rous sarcoma virus *src* gene product recognizes a novel population of pp60^{v-src} and pp60^{c-src} molecules, *J. Cell Biol.*, 100, 409, 1985.
200. **Garber, E. A., Cross, F. R., and Hanafusa, H.,** Processing of p60^{v-src} to its myristylated membrane-bound form, *Mol. Cell. Biol.*, 5, 2781, 1985.
201. **Buss, J. E., Kamps, M. P., and Sefton, B. M.,** Myristic acid is attached to the transforming protein of Rous sarcoma virus during or immediately after synthesis and is present in both soluble and membrane-bound forms of the protein, *Mol. Cell. Biol.*, 4, 2697, 1984.
202. **Brugge, J. S.,** Interaction of the Rous sarcoma virus protein pp60^{src} with the cellular proteins pp50 and pp90, *Curr. Top. Microbiol. Immunol.*, 123, 1, 1986.
203. **Hamaguchi, M. and Hanafusa, H.,** Association of p60^{c-src} with Triton X-100-resistant cellular structure correlates with morphological transformation, *Proc. Natl. Acad. Sci. U.S.A.*, 84, 2312, 1987.
204. **Loeb, D. M., Woolford, J., and Beemon, K.,** pp60^{c-src} has less affinity for the detergent-insoluble cellular matrix than do pp60^{v-src} and other viral protein-tyrosine kinases, *J. Virol.*, 61, 2420, 1987.
205. **Burr, J. G., Dreyfuss, G., Penman, S., and Buchanan, J. M.,** Association of the *src* gene product of Rous sarcoma virus with cytoskeletal structures of chicken embryo fibroblasts, *Proc. Natl. Acad. Sci. U.S.A.*, 77, 3484, 1980.
206. **Schaffhausen, B. S., Dorai, H., Arakere, G., and Benjamin, T. L.,** Polyoma virus middle T antigen: relationship to cell membranes and apparent lack of ATP-binding activity, *Mol. Cell. Biol.*, 2, 1187, 1982.
207. **Cooper, J. A. and Hunter, T.,** Regulation of cell growth and transformation by tyrosine-specific protein kinases, the search for important cellular substrate proteins, *Curr. Top. Microbiol. Immunol.*, 107, 125, 1983.
208. **DeClue, J. E. and Martin, G. S.,** Phosphorylation of talin at tyrosine in Rous sarcoma virus-transformed cells, *Mol. Cell. Biol.*, 7, 371, 1987.
209. **Pasquale, E. B., Maher, P. A., and Singer, S. J.,** Talin is phosphorylated on tyrosine in chicken embryo fibroblasts transformed by Rous sarcoma virus, *Proc. Natl. Acad. Sci. U.S.A.*, 83, 5507, 1986.
210. **Hirst, R., Horwitz, A., Buck, C., and Rohrschneider, L.,** Phosphorylation of the fibronectin receptor complex in cells transformed by oncogenes that encode tyrosine kinases, *Proc. Natl. Acad. Sci. U.S.A.*, 83, 6470, 1986.
211. **Fukami, Y., Nakamura, T., Nakayama, A., and Kanehisa, T.,** Phosphorylation of tyrosine residues of calmodulin in Rous sarcoma virus-transformed cells, *Proc. Natl. Acad. Sci. U.S.A.*, 83, 4190, 1986.
212. **Kamps, M. P., Buss, J. E., and Sefton, B. M.,** Rous sarcoma virus transforming protein lacking myristic acid phosphorylates known polypeptide substrates without inducing transformation, *Cell*, 45, 105, 1986.
213. **Linder, M. E. and Burr, J. G.,** Nonmyristolyated p60^{v-src} fails to phosphorylate proteins of 115—120 kDa in chicken embryo fibroblasts, *Proc. Natl. Acad. Sci. U.S.A.*, 85, 2608, 1988.
214. **Kamps, M. P. and Sefton, B. M.,** Identification of multiple novel polypeptide substrates of the v-*src*, v-*yes*, v-*fps*, v-*ros*, and v-*erb*B oncogenic protein tyrosine kinases utilizing antisera against phosphotyrosine, *Oncogene*, 2, 305, 1988.
215. **Hamaguchi, M., Grandori, C., and Hanafusa, H.,** Phosphorylation of cellular proteins in Rous sarcoma virus infected cells: analysis by use of anti-phosphotyrosine antibodies, *Mol. Cell. Biol.*, 8, 3035, 1988.
216. **Yonemoto, W., Filson, A. J., Queral-Lustig, A. E., Wang, J. Y., and Brugge, J. S.,** Detection of phosphotyrosine-containing proteins in polyomavirus middle tumor antigen-transformed cells after treatment with a phosphotyrosine phosphatase inhibitor, *Mol. Cell. Biol.*, 7, 905, 1987.
217. **Johnson, P. J., Coussens, P. M., Danko, A. V., and Shalloway, D.,** Overexpressed c-*src* can induce focus formation without complete transformation of NIH/3T3 cells, *Mol. Cell. Biol.*, 5, 1073, 1985.
218. **Cooper, J. A., Nakamura, K. D., Hunter, T., and Weber, M. J.,** Phosphotyrosine-containing proteins and the expression of transformation parameters in cells infected with partial transformation mutants of Rous sarcoma virus, *J. Virol.*, 46, 15, 1983.
219. **Courtneidge, S. A. and Heber, A.,** An 81 kd protein complexed with middle T antigen and pp60^{c-src}: a possible phosphatidylinositol kinase, *Cell*, 50, 1031, 1988.
220. **Kaplan, D. R., Whitman, M., Schaffhausen, B., Pallas, D. C., White, M., Cantley, L., and Roberts, T. M.,** Common elements in growth factor stimulation and oncogenic transformation: 85 kd phosphoprotein and phosphatidylinositol kinase activity, *Cell*, 50, 1021, 1987.

221. **Durkin, J. P. and Whitfield, J. F.,** The mitogenic activity of pp60$^{v\text{-}src}$, the oncogenic protein product of the *src* gene of avian sarcoma virus, is independent of external serum growth factors, *Biochem. Biophys. Res. Commun.,* 123, 411, 1984.
222. **Alema, S., Casalbore, P., Agostini, E., and Tato, F.,** Differentiation of PC12 phaeochromocytoma cells induced by v-*src* oncogene, *Nature (London),* 316, 557, 1985.
223. **Gee, C. E., Griffin, J., Sastre, L., Miller, L. J., Springer, T. A., Piwnica-Worms, H., and Roberts, T. M.,** Differentiation of myeloid cells is accompanied by increased levels of pp60$^{c\text{-}src}$ protein and kinase activity, *Proc. Natl. Acad. Sci. U.S.A.,* 83, 5131, 1986.
224. **Boulter, C. A. and Wagner, E. F.,** The effects of v-*src* expression on the differentiation of embryonal carcinoma cells, *Oncogene,* 2, 207, 1988.
225. **Barnekow, A. and Gessler, M.,** Activation of the pp60$^{c\text{-}src}$ kinase during differentiation of monomyelocytic cells *in vitro*, *EMBO J.,* 5, 701, 1986.
226. **Luttrell, D. K., Luttrell, L. M., and Parsons, S. J.,** Augmented mitogenic responsiveness to epidermal growth factor in murine fibroblasts that over-express pp60$^{c\text{-}src}$, *Mol. Cell. Biol.,* 8, 497, 1988.
227. **Azarnia, R., Reddy, S., Kmiecik, T. E., Shalloway, D., and Loewenstein, W. R.,** The cellular *src* gene product regulates junctional cell-to-cell communication, *Science,* 239, 398, 1988.
228. **Chang, C.-C., Trosko, J. E., Kung, H.-J., Bombick, D., and Matsumura, F.,** Potential role of the *src* gene product in inhibition of gap-junctional communication in NIH/3T3 cells, *Proc. Natl. Acad. Sci. U.S.A.,* 82, 5360, 1985.
229. **Warren, S. L., Handel, L. M., and Nelson, W. J.,** Elevated expression of pp60$^{c\text{-}src}$ alters a selective morphogenetic property of epithelial cells *in vitro* without a mitogenic effect, *Mol. Cell. Biol.,* 8, 632, 1988.
230. **Veillette, A., Horak, I. D., Horak, E. M., Bookman, M. A., and Bolen, J. B.,** Alterations of the lymphocyte-specific protein tyrosine kinase (p56^{lck}) during T-cell activation, *Mol. Cell. Biol.,* 8, 4353, 1988.
231. **Saltzman, E. M., Thom, R. R., and Casnellie, J. E.,** Activation of tyrosine phosphorylation is an early event in the stimulation of T lymphocytes by interleukin-2, *J. Biol. Chem.,* 263, 656, 1988.
232. **Morla, A. O., Schreurs, J., Miyajima, A., and Wang, J. Y. J.,** Hematopoietic growth factors activate the tyrosine phosphorylation of distinct sets of proteins in interleukin-3-dependent cell lines, *Mol. Cell. Biol.,* 8, 2214, 1988.
233. **Rudd, C. E., Trevillyan, J. M., Dasgupta, J. D., Wong, L. L., and Schlossman, S. F.,** The CD4 receptor is complexed in detergent lysates to a protein-tyrosine kinase (pp58) from human T lymphocytes, *Proc. Natl. Acad. Sci. U.S.A.,* 85, 5190, 1988.
234. **Veillette, A., Bookman, M. A., Horak, E. M., and Bolen, J. B.,** The CD4 and CD8 T cell surface antigens are associated with the internal membrane protein-tyrosine kinase p56^{lck}, *Cell,* 55, 301, 1988.
235. **Roberts, A. B., Anzano, M. A., Lamb, L. C., Smith, J. M., Frolik, C. A., Marquardt, H., Todaro, G. J., and Sporn, M. B.,** Isolation from murine sarcoma cells of novel transforming growth factors potentiated by EGF, *Nature (London),* 295, 417, 1982.
236. **Connolly, J. L., Green, S. A., and Greene, L. A.,** Comparison of rapid changes in surface morphology and coated pit formation of PC12 cells in response to nerve growth factor, epidermal growth factor and dibutyryl cyclic AMP, *J. Cell Biol.,* 98, 457, 1979.
237. **Semba, K., Yamanashi, Y., Nishizawa, M., Sukegawa, J., Yoshida, M., Sasaki, M., Yamamoto, T., and Toyoshima, K.,** Location of the c-*yes* gene on the human chromosome and its expression in various tissues, *Science,* 227, 1038, 1985.
238. **Yoshida, M. C., Satoh, H., Sasaki, M., Semba, K., and Yamamoto, T.,** Regional location of a novel *yes*-related proto-oncogene, *syn*, on human chromosome 6 at band q21, *Jpn. J. Cancer Res.,* 77, 1059, 1986.
239. **Yoshida, M. C., Sasaki, M., Mise, K., Semba, K., Nishizawa, M., Yamamoto, T., and Toyoshima, K.,** Regional mapping of the human proto-oncogene c-*yes*-1 to chromosome 18 at band q21.3, *Jpn. J. Cancer Res.,* 76, 559, 1985.
240. **Marth, J. D., Disteche, C., Pravtcheva, D., Ruddle, F., Krebs, E. G., and Perlmutter, R. M.,** Localization of a lymphocyte-specific protein tyrosine kinase gene (*lck*) at a site of frequent chromosomal abnormalities in human lymphomas, *Proc. Natl. Acad. Sci. U.S.A.,* 83, 7400, 1986.
241. **Cheah, M. S., Ley, T. J., Tronick, S. R., and Robbins, K. C.,** *fgr* proto-oncogene mRNA induced in B lymphocytes by Epstein-Barr virus infection, *Nature (London),* 319, 238, 1986.
242. **Muller, R. and Verma, I. M.,** Expression of cellular oncogenes, *Curr. Top. Microbiol. Immunol.,* 112, 74, 1984.

Chapter 4

MYOSIN LIGHT CHAIN KINASES

Bruce E. Kemp and James T. Stull

TABLE OF CONTENTS

I.	Introduction		116
	A.	Contractile Elements	116
		1. Skeletal Muscle	116
		2. Smooth Muscle	116
	B.	Functions of Myosin Phosphorylation	117
		1. Smooth Muscle	117
		2. Skeletal Muscle	120
II.	Myosin Light Chain Kinases		121
	A.	Physiochemical Properties	121
	B.	Catalysis	122
III.	Calmodulin Activation and Pseudosubstrate Control		125
IV.	Summary		127
References			129

I. INTRODUCTION

Ca^{2+} is a second messenger that activates many Ca^{2+}-dependent cellular processes. Ca^{2+} has been recognized as an important component in muscle contraction and cell motility because it regulates a distinct class of protein kinases, myosin light chain kinases.[1-3] Activation of myosin light chain kinase by Ca^{2+}/calmodulin results in phosphorylation of a regulatory light chain subunit of myosin. Myosin light chain phosphorylation leads to potentiation of contraction in striated muscles, but initiates contraction in smooth muscles. Before reviewing the catalytic properties of myosin light chain kinases, a general discussion of contractile proteins and the role of myosin phosphorylation will be presented so that the reader will have some appreciation of the biological importance of this class of kinases. The main topic, however, will deal with their respective biochemical properties as elucidated with peptides.

A. CONTRACTILE ELEMENTS

Striated muscles, including cardiac and skeletal muscle tissues, have served as useful models for studying the mechanism by which Ca^{2+} regulates contractile activity via actin-myosin interactions.[4] Myosin and actin also play important roles in nonmuscle contractile functions.[5,6] Thin filaments consist of 42-kDa actin monomers which are polymerized into a double helix. Thin filaments also contain a rod-shaped protein, tropomyosin, which spans seven actin monomers. Tropomyosin is polymerized end to end and forms part of the double helix of the thin filament. In striated muscle cells another thin filament protein complex, troponin, is present for each tropomyosin and for every seven actin monomers.

Thick filaments primarily contain myosin. Myosin is a hexamer consisting of two heavy chains (200 kDa each) and two pairs of light chain subunits. The masses of the pairs of light chains vary from 16 to 28 kDa and are dependent upon the type of muscle tissue.[7] One class of light chains, referred to as regulatory light chains, have similar biochemical properties: masses range from 18 to 20 kDa, they bind divalent cations, and vertebrate light chains are phosphorylated by Ca^{2+}/calmodulin-dependent myosin light chain kinases. Thus, they have also been referred to as phosphorylatable light chains. The carboxy terminus of the heavy chains forms a coiled-coil, α-helical, rod-like region which comprises the backbone of thick filaments. The amino-terminal portions of the two heavy chains form globular head regions which project from the surface of the thick filament. Each globular head has an actin-binding domain, an ATP hydrolysis site, and one each of two types of light chains.

1. Skeletal Muscle

Ca^{2+} activation of contraction in striated muscles is via the thin-filament-linked regulatory complex, troponin-tropomyosin.[4] When the cytoplasmic Ca^{2+} concentration increases as a result of a stimulus, Ca^{2+} binds to troponin, which leads to myosin binding to actin, hydrolysis of ATP, and force or shortening developed by the muscles.[8]

2. Smooth Muscle

Troponin is not present in smooth muscle, and the regulation of actin-myosin interactions by Ca^{2+} is more complex.[1,6,9] Different biochemical mechanisms have been proposed and may be differentiated into two general classes involving thick and thin filament regulatory processes, respectively. One thick filament regulatory scheme involves Ca^{2+} binding to calmodulin and activation of myosin light chain kinase (Figure 1). The resultant phosphorylation of the 20-kDa light chain subunit of smooth muscle myosin plays a prominent role in the contractile process.

In addition to thick-filament Ca^{2+} regulatory systems, it has been proposed that Ca^{2+} may also regulate actin-myosin interactions in smooth muscle via thin-filament proteins.

MYOSIN PHOSPHORYLATION

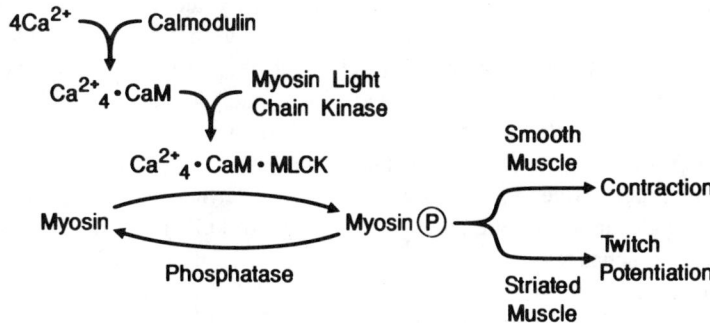

FIGURE 1. General scheme for myosin phosphorylation in skeletal and smooth muscles. Ca^{2+} binds to calmodulin which then binds to inactive myosin light chain kinase. The activated myosin light chain kinase phosphorylates a regulatory light chain subunit on myosin. The reverse reaction is catalyzed by protein phosphatases. Phosphorylation of myosin in smooth muscle initiates contraction. However, in striated muscles, phosphorylation leads to potentiation of contractile force. MLCK = myosin light chain kinase; CaM = calmodulin.

More recent investigations have focused on a potential regulatory system involving caldesmon.[9] Caldesmon binds to actin filaments and inhibits the actin activation of myosin ATPase activity. This inhibitory effect is reversed by calmodulin. Thus, biochemical evidence along with other physiological studies indicate that there may be more than one mechanism for Ca^{2+} regulation of smooth-muscle contraction.

B. FUNCTIONS OF MYOSIN PHOSPHORYLATION

The general properties of myosin light chain kinase activation and myosin light chain phosphorylation are similar in skeletal and smooth cells (Figure 1). However, the effect of light chain phosphorylation on the biochemical properties of myosin and the physiological properties of contraction are different in skeletal and smooth muscles.

1. Smooth Muscle

The kinetic mechanism by which myosin is phosphorylated has been investigated in smooth muscle cells. Electric field stimulation of bovine trachealis allows smooth muscle cells to be activated rapidly and synchronously via endogenous neurotransmitter (acetylcholine) released by intrinsic nerves.[10] Myosin light chain was phosphorylated with a pseudo-first-order rate constant of 1.1 per second in an apparently random, noncooperative mechanism.[11] A technique for measuring nonphosphorylated, monophosphorylated, and diphosphorylated forms of myosin was developed to compare with measurements on the extent of light chain phosphorylation.[12] Measurements of the relative amount of diphosphorylated myosin vs. extent of light chain phosphorylation directly demonstrated that myosin phosphorylation followed a random process. These results on myosin light chain phosphorylation in smooth muscle cells are consistent with recent biochemical studies on purified myosin.[13-15]

The kinetic mechanisms by which myosin light chain phosphorylation stimulates MgATPase activity of myosin have been the subject of many investigations. It has been suggested that nonphosphorylated light chains of smooth muscle myosin inhibit the actin-activated MgATPase activity and that the inhibition is removed by phosphorylation. Investigations on the effect of avian gizzard myosin binding to actin have revealed the following findings: (1) in the absence of MgATP there is only a fourfold difference in the binding constants of phosphorylated and nonphosphorylated gizzard heavy meromyosin (HMM) to actin; and (2) at high actin concentrations the majority of both phosphorylated and nonphosphorylated HMM is bound to actin, yet the actin-activated myosin MgATPase activity of nonphosphorylated heavy meromyosin is much less than that of phosphorylated HMM.[16] Detailed kinetic studies have attempted to determine the step in the actomyosin ATPase cycle which is influenced by phosphorylation. These analyses have excluded a number of possibilities and have resulted in the suggestion that the most likely step to be regulated by phosphorylation is the release of inorganic phosphate from the myosin heads after hydrolysis of ATP.[17] This product release is associated with a large increase in the strength of actin binding and is understood to be related to the "work stroke" in the intact muscle.[8]

Conformational changes in smooth muscle myosin have also been documented as a result of the phosphorylation of the regulatory light chains.[6] Some investigators have proposed that smooth muscle contractile mechanisms involve ATP-induced changes in the morphology of myosin filaments which are altered reversibly by phosphorylation and dephosphorylation of the myosin light chains. The conformation of myosin as well as the filamentous state of myosin are under the influence of a number of factors. Based upon sedimentation properties, smooth muscle myosin molecules in solution may exist in two conformations: 6S and 10S forms. Electron microscopic studies have demonstrated that the 6S form has a characteristic elongated configuration, while the 10S structure resembles a hairpin in which the tail portion has folded back between the heads.[18-20] The two conformational forms, 10S and 6S, may represent different functional states of the myosin molecule, with the 6S and 10S myosin forms representing active and inactive species, respectively.[21-24] In the smooth muscle cell, myosin is present predominantly in a filamentous state whether levels of light chain phosphorylation are low or high.[25] Thus, changes in the conformation of the myosin, such as those resulting from reorientation of the heads in relation to the rod portion of myosin, may induce changes in the MgATPase activity. Phosphorylation may not affect a 10S to 6S conformational transition, per se, in living cells, but may activate myosin by making the heads more mobile and extended from a constrained conformation.[26,27] This conformational change would increase the ability of myosin heads to interact with actin in a manner that increases MgATPase activity.

Studies in a variety of functionally skinned and intact smooth muscle preparations have demonstrated the importance of the myosin phosphorylation system in smooth muscle contraction. Results from numerous experiments with skinned fibers are consistent with the scheme that the Ca^{2+} dependence of force is conferred by calmodulin activation of myosin light chain kinase and the resulting phosphorylation of myosin. The addition of Ca^{2+} to permeabilized preparations results in net increases in the phosphate content of the regulatory light chain.[6,28] In such preparations the phosphorylation of myosin is associated with contraction. While the range of values of light chain phosphorylation varies among preparations, a positive correlation is seen between the degree of phosphorylation of the myosin light chains and tension development under conditions where Ca^{2+} concentrations are varied.[29-31] The Ca^{2+} sensitivity of contractile force can be increased by the addition of exogenous calmodulin and decreased by the addition of compounds, such as the phenothiazine derivatives, that bind to calmodulin and inhibit activation of calmodulin-dependent enzymes.[28,32] Exposure of skinned smooth muscle fibers to ATP-γS, an ATP analogue used as a substrate by the myosin light chain kinase, results in thiophosphorylation of the light chains. Thiophosphorylated light chains are poor substrates for protein phosphatase, and this treatment results in the irreversible activation of tension even in the absence of Ca^{2+}.[28,30,33] Furthermore, a Ca^{2+}/calmodulin-independent myosin light chain kinase produces a Ca^{2+}-independent contraction.[34] These and other types of observations in permeable smooth muscle fibers are consistent with a primary role for Ca^{2+}-dependent myosin phosphorylation in activation of smooth muscle contraction.

As expected for a Ca^{2+}-dependent process, low levels of light chain phosphorylation are generally found in unstimulated intact smooth muscle tissues. Typical resting values range between 0.02 and 0.10 mol phosphate per mole light chain for muscles that do not exhibit spontaneous contractile activity or active tone.[10,25,35-39] Stimulation of intact smooth muscles by various pharmacological and physiological interventions is associated with increases in the extent of myosin phosphorylation to levels above basal values. Although the maximum amount of phosphate incorporated into myosin light chain differs among smooth muscle preparations, changes in myosin phosphorylation precede changes in isometric force.[1] The addition of pharmacological agents which interfere with the Ca^{2+}-dependent activation of myosin light chain kinase results in decreases in the amount of phosphate incorporated into the light chain and proportional decreases in isometric force development.[1] Moreover, smooth muscle relaxation is associated with light chain dephosphorylation. The rate of dephosphorylation is as fast or faster than the rate of relaxation.

Studies with smooth muscle have demonstrated a dependency of force generation on myosin phosphorylation, thus confirming an important role of this process in contraction.[1,6,28] However, there is not always a simple quantitative relationship between developed force and extent of myosin phosphorylation. A general scheme of events has emerged on the processes leading to smooth muscle contraction. In a resting smooth muscle cell the cytoplasmic Ca^{2+} concentration is about 0.18 μM, a concentration that results in little light chain phosphorylation.[1] The extent of light chain phosphorylation under resting conditions is generally 10% or less. The small extent of light chain phosphorylation means that some of the myosin cross bridges may be bound to actin, which could account for resistance to stretch observed in resting muscle.[40] Following excitation, there is an increase in cytoplasmic free Ca^{2+} concentration. It is apparent that events leading to increases in cytoplasmic Ca^{2+} concentrations and activation of myosin light chain kinase may take up to 500 ms.[11] Following activation of myosin light chain kinase, myosin light chain is rapidly phosphorylated, with a coincident increase in muscle stiffness (attachment of myosin cross bridge to actin). The maximal velocity of shortening also increases. When a maximal increase in cytoplasmic Ca^{2+} concentration has been reached, there is a subsequent decrease to some intermediate level that is greater than resting levels and is maintained with continuous excitation.[41] With the decrease in cytoplasmic Ca^{2+} concentration there is also a reduction in the extent of myosin light chain phosphorylation and a reduction in the maximal shortening velocity. Nevertheless, force developed by such an activated smooth muscle will be maintained during a sustained period of stimulation. This sustained phase of contraction is thus characterized by relatively low levels of free Ca^{2+} and myosin light chain phosphorylation, low maximal shortening velocities (hence, low rates of cross-bridge cycling), with maintenance of force. Murphy and colleagues[42-44] proposed that force was maintained due to a different state of the myosin cross bridge ("latch bridge") whereby it remained attached longer and cycled more slowly. The "latch bridge" appears to be dependent upon dephosphorylation of phosphorylated myosin light chain.[1,45]

There are conditions in which myosin phosphorylation and, presumably, cytoplasmic Ca^{2+} concentrations do not decline to low levels following stimulation. For example, Silver and Stull[46] showed that agonist stimulation of smooth muscle resulted in a transient increase in Ca^{2+}-dependent myosin and phosphorylase phosphorylation. However, stimulation with 60 mM KCl led to only a small decline in light chain phosphorylation after 2 h of maintained force. The steady state maintenance of light chain phosphorylation during contraction is presumably related to specific agonist effects on Ca^{2+} turnover and to maintenance of cytoplasmic Ca^{2+} concentrations at different levels. Because of the dynamic properties of myosin phosphorylation and force in relation to cytoplasmic Ca^{2+} concentrations, it is apparent that there is no one specific relationship between developed force and extent of myosin light chain phosphorylation. However, there does appear to be a single relationship between cytoplasmic Ca^{2+} concentration and myosin light chain phosphorylation.[47-49]

2. Skeletal Muscle

A role for myosin phosphorylation in skeletal muscle contraction has been difficult to elucidate. It is clear that myosin phosphorylation does not play a dominant or obligatory role in contraction of this muscle as it does in smooth muscle contraction. However, recent biochemical and physiological experiments indicate that myosin light chain phosphorylation does have a regulatory function in skeletal muscle contraction.

Phosphorylation of myosin light chain increased MgATPase activity by decreasing the K_{app} of actin for myosin.[50,51] There was no effect on the V_{max} value. These observations were made with both myosin and HMM. Phosphorylated myosin also dissociates from actin at higher pyrophosphate concentrations than does nonphosphorylated myosin.[52] Thus, phosphorylated myosin has an apparent higher affinity for actin. Phosphorylation of light chain results in conformational changes of myosin and release of cross bridges from the myosin filament surface.[53,54] The effect of phosphorylation may be related to the altered charge balance of the myosin head due to the introduction of negative charges from the phosphate moiety. These studies demonstrate an effect of myosin light chain phosphorylation on the biochemical properties of myosin. The effects are significantly smaller than the effects of phosphorylation on smooth muscle myosin.[6]

The effect of myosin light chain phosphorylation on isometric force has been examined in permeable rabbit skeletal muscle fibers at 0.6 and 10 μM Ca^{2+}.[55] At the lower Ca^{2+} concentration, phosphorylation of myosin light chain resulted in a twofold increase in isometric force. Addition of a protein phosphatase dephosphorylated light chain and reversed the contraction response. At the higher Ca^{2+} concentration, light chain phosphorylation had no effect on force. Myosin phosphorylation increased force at <50% of maximal activation by Ca^{2+}, with a leftward shift of the Ca^{2+} concentration-force relationship.[56] This shift was due to a decrease in the slope of the Ca^{2+}-force relationship. Myosin light chain phosphorylation does not appear to alter the kinetics of the cross-bridge cycle as assessed by the force-velocity relationship, which is different from the effects seen in smooth muscle.[55,57] Thus, phosphorylation of myosin is not required for contraction in skeletal muscle fibers, but appears to affect the response to Ca^{2+} activation.

In intact mammalian skeletal muscles there are several important physiological properties of the myosin phosphorylation system: an increase in the phosphate content of light chain is dependent upon the frequency of muscle stimulation; the maximal rate of light chain phosphorylation exceeds the maximal rate of light chain dephosphorylation by 10 to 60 fold; the maximal rates of phosphorylation and dephosphorylation are three to four orders of magnitude slower than the rates of contraction and relaxation.[58-60] Based upon these properties, a quantitative model has been proposed for explaining the contraction frequency dependence of myosin light chain phosphorylation.[60,61] When previously quiescent muscle is stimulated in a very rapid and repetitive fashion (200 Hz) to produce sustained contraction such as tetany, myosin light chain kinase is fully activated due to a rapid and sustained increase in cytoplasmic Ca^{2+} concentration. However, a rapid increase and decrease in cytoplasmic Ca^{2+} concentration as a result of a single twitch contraction lasting only 20 ms results in the rapid activation of a small portion (<1%) of myosin light chain kinase. This activation is followed by a relatively slow rate of inactivation (1 s^{-1}) after the decrease in cytoplasmic Ca^{2+} concentration and relaxation. The critical feature of this model is related to the slow rate of inactivation of myosin light chain kinase (seconds) relative to the rapid rate of decrease in cytoplasmic Ca^{2+} concentration and relaxation (milliseconds). The fraction of myosin light chain kinase in the active form will thus be dependent upon the relative magnitudes of the activation and inactivation processes. An increase in the frequency of contraction at physiological rates (0.5 to 30 Hz) increases the extent of fractional activation because a greater number of Ca^{2+} transients results in a greater incremental increase in the active form of the enzyme for a given period of time. In addition, the time between Ca^{2+}

transients, during which myosin light chain kinase inactivation occurs, is decreased. With a continuous repetitive stimulation, therefore, the fractional activation of myosin light chain kinase increases until an equilibrium is established between the activation and inactivation processes.

It is clear that myosin light chain phosphorylation is not required for actin activation of myosin MgATPase activity or for contraction of permeable striated muscle fibers. Therefore, myosin light chain phosphorylation *in vivo* would not be expected to be necessary for contraction. Myosin light chain phosphorylation in intact skeletal muscle fibers has no significant effect on maximal force development produced during tetany or the maximal shortening velocity.[37,58,59,62,63]

The only correlation of myosin light chain phosphorylation to a physiological property of striated muscles is to the contraction-induced potentiation of isometric twitch tension in fast-twitch fibers.[58-60,64-66] Potentiation of isometric twitch tension occurs either following a high-frequency stimulation producing tetanus (post-tetanic) or upon repetitive slow-frequency stimulation of contractile twitches (staircase). A correlation between the extent of potentiation of isometric twitch tension following a tetanus or during the staircase response has been found under a variety of physiological conditions.[58-60,64-66] This correlation has been found in muscles from rat, mouse, rabbit, and human.[61] In relation to recent data obtained with permeable fibers, Ca^{2+} activation of thin filaments during a twitch would have to be less than maximal for myosin phosphorylation to be a mechanism for potentiation of isometric twitch tension. Recent modeling of Ca^{2+} movements in skeletal muscle upon activation suggests that the maximal amount of Ca^{2+} binding to troponin during an isometric twitch is probably not saturating the troponin Ca^{2+} binding sites.[67] The Ca^{2+} binding-time integral during a twitch shows <50% saturation during the transient increase and decrease in cytoplasmic Ca^{2+} concentration. Thus, phosphorylation of myosin could result in potentiation of isometric tension during twitches because <50% saturation of troponin is achieved during much of the Ca^{2+} transient time. In the case of a tetanus, phosphorylation would not result in potentiation because Ca^{2+} binding sites on troponin are >50% saturated.

In summary, myosin is phosphorylated under physiological conditions of repetitive, low-frequency stimulation in skeletal muscles. Phosphorylation under these conditions can be accounted for by the activation-inactivation properties of the kinase in addition to the relative and total amounts of myosin light chain kinase and myosin light chain phosphatase activities, respectively. In fast-twitch skeletal muscle, myosin phosphorylation increases actin-activated myosin ATPase activity, increases tension at levels of Ca^{2+} that produce partial activation in permeable fibers, and correlates to potentiation of isometric twitch tension following repetitive stimuli.

II. MYOSIN LIGHT CHAIN KINASES

Myosin light chain kinases are protein serine and threonine kinases that catalyze the Ca^{2+}- and calmodulin-dependent phosphorylation of the regulatory light chains of myosin.[68] These enzymes have been described in a number of vertebrate muscles, as well as in vertebrate nonmuscle tissues.[2]

A. PHYSIOCHEMICAL PROPERTIES

Avian gizzard myosin light chain kinase is an asymmetrical molecule that exists as a 130-kDa monomer.[69] The mammalian smooth muscle myosin light chain kinases appear to be slightly larger in size as compared to the purified avian kinases.[2,70] The structural domains of chicken gizzard myosin light chain kinase have been characterized by limited and sequential digestions with different proteolytic enzymes.[71] Within the primary structure of the molecule they found no overlap between the catalytic and calmodulin-binding domains, and a linear model was proposed, with the catalytic site in the center of the enzyme and the calmodulin domain toward the carboxy-terminal end. A synthetic peptide analogue of myosin light chain kinase from chicken gizzard was shown to have a high affinity for calmodulin and to contain a phosphorylation site for cyclic AMP-dependent protein kinase.[72] The peptide had structural properties similar to the previously reported calmodulin-binding peptide from rabbit skeletal muscle myosin light chain kinase.[73] The partial sequence of avian smooth muscle myosin light chain kinase has been derived recently from cDNA where a catalytic domain was identified by homology to another protein kinase.[74]

Myosin light chain kinases from mammalian striated muscles have similar but also unique biochemical properties compared to the avian smooth muscle enzyme. The complete amino acid sequence of rabbit skeletal muscle myosin light chain kinase has been determined recently.[75] The calculated molecular weight of 65,040 is significantly smaller than values determined by sodium dodecyl sulfate-polyacrylamide gel electrophoresis (SDS-PAGE), which are about 90,000. Determination of the native molecular weight of the rabbit skeletal muscle kinase by centrifugation has resulted in values in the range of about 70,000.[2] Thus, the rabbit skeletal muscle myosin light chain kinase also appears to be a monomeric enzyme.

It is noteworthy that the masses of the avian gizzard smooth and rabbit skeletal muscle kinases are so different. Western blot analysis demonstrated that the mass of the kinase in rabbit skeletal muscle *in situ* was identical to that of the purified enzyme, with no evidence of proteolysis.[76] It was also found that the relative masses of skeletal muscle myosin light chain kinases varied from 68 kDa (human) to 150 kDa (chicken) and were dependent upon the animal species. The relative mass of myosin light chain kinase was constant within an animal species regardless of skeletal muscle fiber type. The chicken skeletal muscle myosin light chain kinase appears to be significantly larger than the enzyme from gizzard smooth muscle.[77] Affinity-purified antibodies to rabbit skeletal muscle myosin light chain kinase cross-react with the chicken skeletal muscle kinase, but not with the avian gizzard or mammalian smooth muscle enzymes.[70,77]

Rabbit skeletal muscle myosin light chain kinase can also be cleaved proteolytically to separate specific domains. After partial digestion with trypsin a globular, 36-kDa, catalytically active fragment was obtained.[78] This fragment required calmodulin for activity. Another 33-kDa fragment was catalytically inactive and highly asymmetrical. These two proteolytic products were referred to as head and tail fragments, respectively. Based upon these and other analyses, these investigators proposed a model with an end-to-end arrangement of the head and the tail fragments and an overall length of 380 Å.

Blumenthal and co-workers[73] purified and sequenced a 27-residue fragment of rabbit skeletal muscle myosin light chain kinase that bound calmodulin in a Ca^{2+}-dependent manner with a high affinity. The peptide sequence showed that it contained a single tryptophan residue with no acidic or propyl residues and had a high probability of α-helix formation. These data corroborate the hypothesis that skeletal muscle myosin light chain kinase can be divided into at least two functional domains involving catalytic activity and calmodulin binding with no overlap within the primary structure. The calmodulin-binding domain represents the carboxy terminus of the enzyme.[79,80]

In summary, myosin light chain kinases are asymmetrical, monomeric enzymes that vary considerably in mass depending upon the type of muscle and the animal species. Two classes of myosin light chain kinases represented by smooth and skeletal muscles, respectively, are defined by unique physical and antigenic properties. A relatively large portion of the kinase may be removed by limited proteolysis without loss of calmodulin-dependent enzyme activity. The calmodulin-binding domain is distinct from the catalytic site.

B. CATALYSIS

A unique feature of the protein kinase reaction catalyzed by myosin light chain kinases is the substrate specificity. They phosphorylate other protein substrates at rates <3% relative to the rates of light chain phosphorylation.[2]

The catalytic properties of smooth muscle myosin light chain kinase have been examined with myosin and free myosin light chain. The gizzard smooth muscle myosin light chain kinase randomly phosphorylates gizzard myosin with apparent K_m and V_{max} values of approximately 15 μM and 15 μmol phosphate incorporated per minute per milligram, respectively.[14,81] These values are similar to the kinetic constants for isolated gizzard myosin light chain.

Skeletal muscle myosin is also phosphorylated *in vitro* by a random mechanism.[50] Kinetic constants for phosphorylation of rabbit skeletal muscle myosin by rabbit skeletal muscle myosin light chain kinase are 19 μM for K_m and 47 μmol phosphate incorporated per minute per milligram for V_{max} values, respectively. The properties of phosphorylation for free light chain and light chain bound to myosin are similar. It has been proposed from detailed studies that the skeletal muscle kinase has a rapid-equilibrium bi-bi kinetic mechanism, and it forms a dead-end complex with ADP and nonphosphorylated myosin light chain.[82] Although a detailed kinetic analysis has not been performed with smooth muscle myosin light chain kinase, there are certain similarities regarding product inhibition and reversibility of the phosphorylation reaction.[82,83]

Myosin light chain kinase from gizzard smooth muscle demonstrates specificity for myosin light chains from smooth muscle, in contrast to phosphorylation of light chains from striated muscles. There is a marked decrease in V_{max} values and an increase in K_m values for striated muscle light chain substrates compared to the gizzard light chain with gizzard myosin light chain kinase.[77] Rabbit skeletal muscle myosin light chain kinase phosphorylates skeletal and smooth myosin light chains with similar K_m and V_{max} values.[77] Thus, in contrast to the smooth muscle kinase, skeletal muscle myosin light chain kinase shows no marked specificity in phosphorylating myosin light chains from different tissues. The differences in specificity of the skeletal and smooth muscle enzymes are also apparent from studies with model synthetic peptide substrates.

Recently, an additional phosphorylation site in gizzard myosin light chain has been described for myosin light chain kinase.[84,85] The 20-kDa light chain of myosin can be phosphorylated at both Ser-19 and Thr-18, with Ser-19 phosphorylated much faster than the threonine residue. Phosphorylation of the light chain at both sites is associated with alterations in the enzymatic properties of myosin; i.e., the incorporation of phosphate into threonine causes a greater increase in the actin-activated MgATPase activity of myosin than that seen with phosphorylation of Ser-19 alone. The physiological role for phosphorylation of Thr-18 by myosin light chain kinase is not clear. The amount of Thr-18 phosphorylation is substantially less than Ser-19 phosphorylation in contracting smooth muscle.[86]

Isolated smooth muscle myosin light chain may be phosphorylated by a number of protein kinases in addition to myosin light chain kinase. These kinases include, for example, protein kinase C, protease-activated kinase I, casein kinase II, cyclic AMP-dependent protein kinase, phosphorylase kinase, epidermal growth factor receptor kinase, and a multifunctional calmodulin-dependent protein kinase.[87-92] However, most of these kinases do not catalyze phosphorylation when the light chain is bound to the heavy chain of myosin.[91] During recent years, attention has focused on the effects of phosphorylation by the Ca^{2+}-activated, phospholipid-dependent protein kinase C. Since this kinase is widely distributed among mammalian tissues and catalyzes the phosphorylation of various protein substrates, investigators have examined a potential role for protein kinase C in the regulation of smooth muscle contraction by myosin light chain phosphorylation.[93]

Endo et al.[94] first described the phosphorylation of the isolated light chains from gizzard myosin and intact myosin by protein kinase C. Nishikawa et al.[87] subsequently reported that a threonine residue in the light chain was the site of phosphorylation by protein kinase C.

Investigations of the functional aspects of phosphorylation by protein kinase C revealed that the phosphorylation of HMM by protein kinase C subsequent to phosphorylation by myosin light chain kinase resulted in a decrease in the actin-activated myosin MgATPase activity.[95] These studies also demonstrated that phosphorylation of HMM by protein kinase C caused a decrease in the rate of HMM phosphorylation by myosin light chain kinase. A study evaluating conformational changes in myosin structure induced by phosphorylation mediated by myosin light chain kinase vs. protein kinase C indicated that the phosphorylation of myosin by protein kinase C inhibited the conformational change induced by myosin light chain kinase phosphorylation.[96] The sites phosphorylated in myosin light chains by protein kinase C are distinct from the two sites phosphorylated by myosin light chain kinase.[97,98] Ser-1, Ser-2, and Thr-9 are phosphorylated by protein kinase C, whereas Thr-18 and Ser-19 are phosphorylated by myosin light chain kinase. Based on these biochemical observations a modulatory role has been proposed for protein kinase C in smooth muscle contraction at the level of the contractile proteins. However, recent studies demonstrate that during the initial transient and sustained phases of smooth muscle contraction, respectively, protein kinase C does not phosphorylate myosin light chain.[86]

Studies with synthetic peptides and chemically modified myosin light chains have revealed the broad specificity requirements of the smooth muscle myosin light chain kinase. Initially it was found that the 23-residue peptide

$$S^1SKRAKAKTTK^{11}KRPQR^{16}ATS^{19}(PO_4)NVFS^{23}$$

corresponding to the amino-terminal region of the gizzard myosin light chains was phosphorylated stoichiometrically on Ser-19, the same residue phosphorylated in the parent smooth muscle protein.[99,100] Using a series of truncated and substituted synthetic peptides it was found that the region between residues 11 and 22 contained critical specificity determinants. The four basic residues Lys-11, Lys-12, Arg-13, and Arg-16 were essential, having a strong effect on the K_m of peptide phosphorylation at Ser-19. The V_{max} for peptide phosphorylation was influenced by Arg-16 and the hydrophobic residues Val-21 and Phe-23.[101] Since the apparent binding of the peptide substrates to the smooth muscle myosin light chain kinase depended on the basic residues and not on the serine phosphate acceptor site, it was possible to exploit this feature and prepare basic peptide inhibitors of the enzyme.[101] The myosin light chain kinase activity was strongly influenced by the spatial arrangements of the basic residue specificity determinants in synthetic peptide substrates. Shifting Arg-16 from position 16 to position 15 caused a complete switch in the site of phosphorylation, from Ser-19 to Thr-18.[102] The spatial relationship between Arg-16 and the three adjacent basic residues Lys-11, Lys-12, Arg-13 also influenced whether phosphorylation occurred preferentially on Ser-19 or Thr-18. The apparent K_m for Thr-18 phosphorylation was similar to the value for Ser-19 phosphorylation.

The myosin light chain kinase specificity requirements are more complex and depend on determinants extending over a larger linear sequence around the phosphorylation site than is the case for other protein kinases such as the cyclic AMP-dependent protein kinase and the calmodulin-dependent protein kinase II.[103,104] The complexity of the specificity requirements of the myosin light chain kinase seen with peptide substrates is consistent with the restricted specificity observed for this enzyme for protein substrates.

It seems likely that the specificity requirements revealed from the studies with synthetic peptides reflect those operating for the native myosin light chains. Chemical modification of either lysine or arginine residues in the myosin light chains inhibited their phosphorylation, with the K_m being the kinetic parameter influenced most strongly, as expected from the synthetic peptide analogue studies.[105] The myosin light chain kinase exhibited absolute specificity for L-serine and did not phosphorylate D-serine-containing synthetic peptides at measurable rates, comparable to other serine and threonine protein kinases.[106]

TABLE 1
Comparison of Kinetics of Peptide Phosphorylation by Chicken Skeletal and
Smooth Muscle Myosin Light Chain Kinases

		Enzyme			
		Skeletal		Smooth	
Myosin light chain	Sequence	K_m (μM)	V_{max} (units)	K_m (μM)	V_{max} (units)
Gizzard MLC(11—23)	KKRPQRATSNVFS	12	31.0	20	2.5
	KKRAARATSNVFS	9*	44.0*	8	1.4
	KKRAAEATSNVFS	592	1.9	490	0.1
Gizzard MLC	Native protein	10	69.0	9	36.0
Skeletal MLC(5—17)	KRRAAEGSSMVFS	550	0.5	ND	ND
Skeletal MLC	Native protein	8	42.0	96	4

Note: Data from Nunnally et al.,[77] Kemp and Pearson,[102] and Michnoff et al.[107] The units of enzyme activity are $\mu mol\ min^{-1}\ mg^{-1}$, with data for the rabbit skeletal muscle myosin light chain kinase indicated by an asterisk (*), not determined indicated by ND, and myosin light chain indicated by MLC.

The smooth muscle and skeletal muscle myosin light chain kinases exhibited differences in specificity when compared using a variety of synthetic peptides. The smooth muscle myosin light chain peptide MLC(11—23) is phosphorylated by either enzyme (Table 1). The V_{max} with the skeletal enzyme (31 $\mu mol\ min^{-1}\ mg^{-1}$) is 45% of that obtained with the native protein, whereas the V_{max} with the smooth muscle enzyme was only 7% (2.5 $\mu mol\ min^{-1}\ mg^{-1}$ compared to 36 $\mu mol\ min^{-1}\ mg^{-1}$). The apparent K_m values with this peptide as substrate are similar to those with the native myosin light chains, in the range 9 to 20 μM. Surprisingly, the skeletal muscle peptide is a poor substrate for both enzymes, with a high K_m and a low V_{max} (Table 1). The reason for this is Glu-10 in the skeletal muscle myosin light chain peptide, which corresponds to Arg-16 in the smooth muscle sequence. Substitution of Arg-16 in the smooth muscle sequence causes correspondingly unfavorable kinetics of peptide phosphorylation comparable to those of the skeletal muscle peptide (Table 1). However, these considerations do not explain why the native skeletal muscle myosin light chains with Glu-10 are readily phosphorylated by the skeletal muscle myosin light chain kinase, but not the smooth muscle counterpart. It has been found that additional basic residues toward the amino terminus from the cluster of three basic amino acids will lower the K_m value.[107] These basic amino acids are present in the native skeletal muscle myosin light chain kinase.

Further differences in the properties of the smooth muscle and skeletal muscle enzymes were observed with a series of synthetic peptide analogues. Substitution in the smooth muscle peptide of any of the three basic residues Lys-11, Lys-12, or Arg-13 with alanine dramatically increased (tenfold) the K_m of the smooth muscle enzyme (Table 2). In contrast, substitution of Lys-11 and Lys-12 had correspondingly less of an effect on the skeletal muscle enzyme. This difference permits the smooth muscle myosin light chain analogue [Ala[11]]MLC(11—23) to act as a relatively specific substrate for the skeletal muscle myosin light chain kinase.[107] The skeletal muscle myosin light chain kinases from either the rabbit or chicken displayed essentially identical specificity toward synthetic peptides, indicating a lack of species differences.

III. CALMODULIN ACTIVATION AND PSEUDOSUBSTRATE CONTROL

Myosin light chain kinase was the first protein kinase found to be regulated by cal-

TABLE 2
Comparison of the Specificity of Skeletal and Smooth Muscle Myosin Light Chain Kinases Using Synthetic Peptides

Sequence	Skeletal MLCK		Smooth MLCK	
	K_m (μM)	V_{max} (units)	K_m (μM)	V_{max} (units)
KKRPQRATSNVFS	12	31	20	2.5
KKRAARATSNVFS	9	44	8	1.4
AKRAARATSNVFS	10	31	463	2.1
KAKAARATSNVFS	52	36	445	1.6
KKAAARATSNVFS	107	41	951	2.7

Note: Data from Kemp and Pearson[102] and Michnoff et al.[107] The units of enzyme activity are μmol min^{-1} mg^{-1}. The peptide sequences shown correspond to analogues of the smooth muscle myosin light chain MLC(11—23) with alanine substitutions as indicated.

Protein	Sequence
MLC(1-23)	S S K R A K A K T T K K R P Q R A T S^{19}(P) N V F A
SkMLCK(328-360)	S Q R L L K K Y L M K R R W K K N F I A V S A A N R F K K
SmMLCK(484-512)	S K D R M K K Y M A R R K W Q K T G H A V R A I G R L S S

FIGURE 2. Comparison of the myosin light chain phosphorylation site sequence and the calmodulin-binding region. Conserved residues are shown in bold. MLCK = chicken gizzard myosin light chain kinase (Sk, skeletal muscle; Sm, smooth muscle); the substrate MLC(1—23) = the chicken gizzard myosin light chain. Ser-19 is the site of phosphorylation in MLC.

modulin.[108,109] As indicated earlier, Blumenthal et al.[73] found that calmodulin binding to skeletal muscle myosin light chain kinase depended on a peptide segment of 26 residues present on the carboxy-terminal end of the enzyme. Subsequently, a structurally related calmodulin-binding region was found in the smooth muscle enzyme.[72,74,100,110] This led to the concept that a sequence within the calmodulin-binding region might be responsible for maintaining the enzyme in the inactive form by mimicking the substrate, i.e., by acting as a pseudosubstrate.[110] Studies with synthetic peptides indicated that the corresponding synthetic peptides were in fact potent substrate antagonists. Although the calmodulin-binding and pseudosubstrate sequences overlap, they are not identical (Figure 2). Maximal calmodulin binding activity depends on residues carboxy-terminal to the region showing homology with the substrate phosphorylation site.

More direct support for the concept of pseudosubstrate regulation has been obtained recently by studying the structures of proteolyzed forms of the myosin light chain kinase. This enzyme, like many other protein kinases, can be activated by limited proteolysis to generate calmodulin-independent forms (Figure 3).[111-114] The calmodulin-independent 61-kDa tryptic fragment of the smooth muscle myosin light chain kinase is inhibited by synthetic peptides corresponding to the pseudosubstrate sequence associated with the calmodulin-binding region.[113] Similar findings have now been made for the skeletal muscle myosin light chain kinase.[113]

During trypsin activation of the smooth muscle myosin light chain kinase to the calmodulin-independent form, an intermediate inactive form is generated that does not bind calmodulin (Figure 3). The 64-kDa inactive form of the enzyme can be converted to a

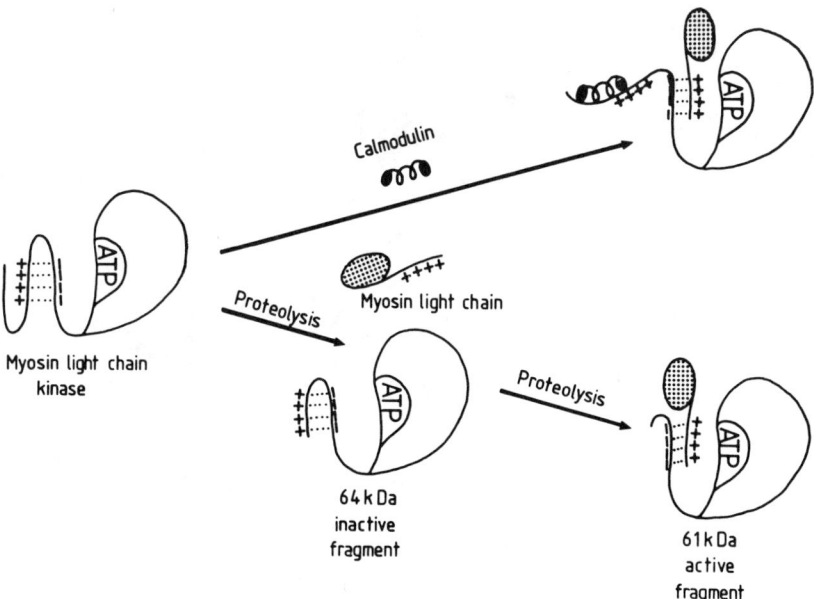

FIGURE 3. Activation of myosin light chain by calmodulin or proteolysis. Activation of myosin light chain kinase by binding calmodulin exposes the active site allowing access to the substrate. Activation by proteolysis proceeds via an intermediate inactive form of the enzyme that does not bind calmodulin.

calmodulin-independent active 61-kDa form with further tryptic cleavage at the carboxy-terminal end.[115] This observation indicates that the calmodulin-binding domain could be destroyed initially by trypsin without disrupting the inhibitory region. Sequencing studies of the 64-kDa inactive fragment revealed that Arg-505 was the carboxy-terminal residue. Cleavage of the myosin light chain kinase at Arg-505 to generate the 64-kDa fragment occurs within the calmodulin-binding region and therefore explains why this fragment no longer binds calmodulin. The 64-kDa fragment, however, retains the pseudosubstrate sequence which ends at Val-504 and may therefore account for the lack of catalytic activity of this fragment (Figure 4). Recent sequencing studies have provided evidence that the carboxy terminus of the 61-kDa active fragment of the myosin light chain kinase is Lys-476.[119] This cleavage site is amino-terminal to Ser-484, the commencement of the pseudosubstrate region, by only seven residues and provides compelling evidence that the pseudosubstrate region is responsible for the autoregulation of the enzyme.

The concept of a pseudosubstrate inhibitory region associated with the calmodulin-binding region has recently been extended to other calmodulin-dependent protein kinases and may represent a general mechanism of control for these enzymes.[116-118]

IV. SUMMARY

Synthetic peptides have played a major role in understanding the structure-function relationships of the myosin light chain kinases as well their substrate specificity. The study of these enzymes has been particularly significant both in terms of understanding their physiological function and because they have provided important precedents for understanding the regulation of other enzymes. The myosin light chain kinases were the first protein kinases found to be regulated by calmodulin. They also provided the first example of the structure of a calmodulin binding domain being specified by a short segment of the enzyme's amino acid sequence. More recently, work on the myosin light chain kinase has led to the

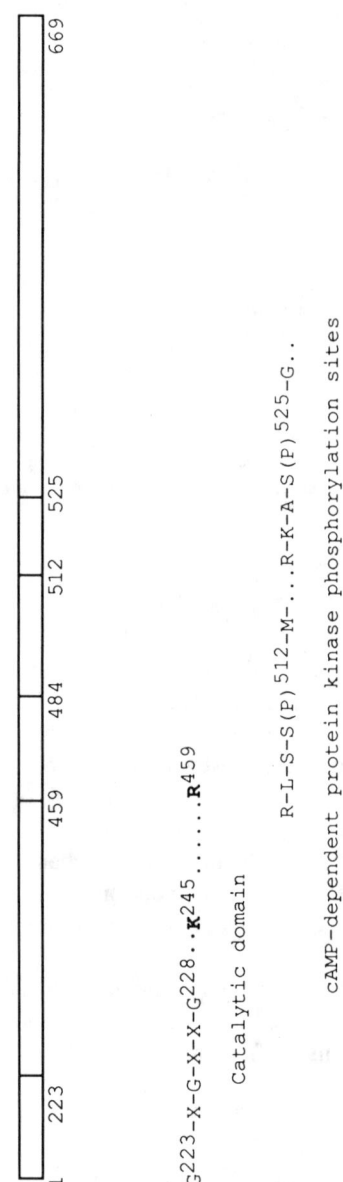

FIGURE 4. Summary of the locations of the catalytic and regulatory features of the myosin light chain kinase. The sequence of 1 to 669 corresponding to the partial cDNA derived sequence of myosin light chain kinase is shown.[74] Residues conserved in the catalytic domain of all protein kinases, including the glycine-rich ATP-binding motif, the essential Lys-245, and Arg-459, are shown. The regulatory features, including the pseudosubstrate, the calmodulin-binding regions, and the phosphorylation sites, are located on the carboxy-terminal side of the catalytic domain.

concept of an autoregulatory pseudosubstrate sequence overlapping the calmodulin binding domain that has provided a general hypothesis for understanding how other calmodulin-dependent protein kinases are regulated.

REFERENCES

1. **Kamm, K. E. and Stull, J. T.**, The function of myosin and myosin light chain kinase phosphorylation in smooth muscle, *Annu. Rev. Pharmacol. Toxicol.*, 25, 593, 1985.
2. **Stull, J. T., Nunnally, M. H., and Michnoff, C. H.**, Calmodulin-dependent protein kinases, in *The Enzymes*, Vol. 17, Boyer, P. D. and Krebs, E. G., Eds., Academic Press, New York, 1986, 113.
3. **Stull, J. T.**, Myosin light chain kinases and caldesmon: biochemical properties and roles in skeletal and smooth muscle contractions, in *Molecular Aspects of Cellular Regulation: Calmodulin*, Vol. 5, Cohen, P. and Klee, C., Eds., Elsevier, Amsterdam, 1988, 91.
4. **Leavis, P. C. and Gergely, J.**, Thin filament proteins and thin filament-linked regulation of vertebrate muscle contraction, *Crit. Rev. Biochem.*, 16, 235, 1984.
5. **Korn, E. D. and Hammer, J. A., III,** Myosins of nonmuscle cells, *Annu. Rev. Biophys. Chem.*, 17, 23, 1988.
6. **Sellers, J. R. and Adelstein, R. S.**, Regulation of contractile activity, in *The Enzymes*, Vol. 18, Boyer, P. D. and Krebs, E. G., Eds., Academic Press, Orlando, FL, 1987, 382.
7. **Synghedauw, B.**, Developmental and functional adaptation of contractile proteins in cardiac and skeletal muscles, *Physiol. Rev.*, 66, 710, 1986.
8. **Eisenberg, E. and Hill, T. L.**, Muscle contraction and free energy transduction in biological systems, *Science*, 227, 999, 1985.
9. **Marston, S. B. and Smith, C. W. J.**, The thin filaments of smooth muscles, *J. Muscle Res. Cell Motility*, 6, 669, 1985.
10. **Kamm, K. E. and Stull, J. T.**, Myosin phosphorylation, force, and maximal shortening velocity in neurally stimulated tracheal smooth muscle, *Am. J. Physiol.*, 249, C238, 1985.
11. **Kamm, K. E. and Stull, J. T.**, Activation of smooth muscle contraction: relation between myosin phosphorylation and stiffness, *Science*, 232, 80, 1986.
12. **Persechini, A., Kamm, K. E., and Stull, J. T.**, Different phosphorylated forms of myosin in contracting tracheal smooth muscle, *J. Biol. Chem.*, 261, 6293, 1986.
13. **Trybus, K. M. and Lowey, S.**, Mechanism of smooth muscle myosin phosphorylation, *J. Biol. Chem.*, 260, 15988, 1985.
14. **Sobieszek, A.**, Phosphorylation reaction of vertebrate smooth muscle myosin: an enzyme kinetic analysis, *Biochemistry*, 24, 1266, 1985.
15. **Wagner, P. D., Vu, N.-D., and George, J. N.**, Random phosphorylation of the two heads of thymus myosin and the independent stimulation of their actin-activated ATPase, *J. Biol. Chem.*, 260, 8084, 1985.
16. **Greene, L. E. and Sellers, J. R.**, Effect of phosphorylation on the binding of smooth muscle heavy meromyosin X ADP to actin, *J. Biol. Chem.*, 262, 4177, 1987.
17. **Sellers, J. R.**, Mechanism of the phosphorylation-dependent regulation of smooth muscle heavy meromyosin, *J. Biol. Chem.*, 260, 15815, 1985.
18. **Craig, R., Smith, R., and Kendrick-Jones, J.**, Light-chain phosphorylation controls the conformation of vertebrate non-muscle and smooth muscle myosin molecules, *Nature (London)*, 302, 436, 1983.
19. **Onishi, H. and Wakabayashi, T.**, Electron microscopic studies on myosin molecules from chicken gizzard muscle. III. Myosin dimers, *J. Biochem. (Tokyo)*, 95, 903, 1984.
20. **Trybus, K. M. and Lowey, S.**, Conformational states of smooth muscle myosin, *J. Biol. Chem.*, 259, 8564, 1984.
21. **Ikebe, M., Hinkins, S., and Hartshorne, D. J.**, Correlation of intrinsic fluorescence and conformation of smooth muscle myosin, *J. Biol. Chem.*, 258, 14770, 1983.
22. **Ikebe, M., Barsotti, R. J., Hinkins, S., and Hartshorne, D. J.**, Effects of magnesium chloride on smooth muscle actomyosin adenosine-5'-triphosphatase activity, myosin conformation, and tension development in glycerinated smooth muscle fibers, *Biochemistry*, 23, 5062, 1984.
23. **Ikebe, M. and Hartshorne, D. J.**, Conformation-dependent proteolysis of smooth-muscle myosin, *J. Biol. Chem.*, 259, 11639, 1984.
24. **Ikebe, M. and Hartshorne, D. J.**, Phosphorylation of smooth muscle myosin at two distinct sites by myosin light chain kinase, *J. Biol. Chem.*, 260, 10027, 1985.

25. Somlyo, A. V., Butler, T. M., Bond, M., and Somlyo, A. P., Myosin filaments have non-phosphorylated light chains in relaxed smooth muscle, *Nature (London)*, 294, 567, 1981.
26. Sommerville, L. E. and Hartshorne, D. J., Intracellular calcium and smooth muscle contraction, *Cell Calcium*, 7, 353, 1986.
27. Hartshorne, D. J., Biochemistry of the contractile process in smooth muscle, in *Physiology of the Gastrointestinal Tract*, 2nd ed., Johnson, L. R., Ed., Raven Press, New York, 1987, 423.
28. Kerrick, W. G. L. and Hoar, P. E., Regulation of contraction in skinned smooth muscle cells by Ca^{2+} and protein phosphorylation, in *Advances in Protein Phosphatases*, Vol. 2, Merlevede, W. and DiSalvo, J., Eds., Leuven University Press, Leuven, Belgium, 1985, 133.
29. Hoar, P. E., Kerrick, W. G. L., and Cassidy, P. S., Chicken gizzard: relation between calcium-activated phosphorylation and contraction, *Science*, 204, 503, 1979.
30. Haeberle, J. R., Hathaway, D. R., and DePaoli-Roach, A. A., Dephosphorylation of myosin by the catalytic subunit of a type-2 phosphatase produces relaxation of chemically skinned uterine smooth muscle, *J. Biol. Chem.*, 260, 9965, 1985.
31. Hellstrand, P. and Arner, A., Myosin light chain phosphorylation and the cross-bridge cycle at low substrate concentration in chemically skinned guinea pig *Taenia coli*, *Pflügers Arch.*, 405, 323, 1985.
32. Meisheri, K. D., Rüegg, J. C., and Paul, R. J., Studies on skinned fiber preparations, in *Calcium and Contractility*, Grover, A. K. and Daniel, E. E., Eds., Humana Press, Clifton, NJ, 1985, 191.
33. Kerrick, W. G. L., Hoar, P. E., Cassidy, P. S., Bolles, L., and Malencik, D. A., Calcium-regulatory mechanisms: functional classification using skinned fibers, *J. Gen. Physiol.*, 77, 177, 1981.
34. Walsh, M. P., Bridenbaugh, R., Hartshorne, D. J., and Kerrick, W. G. L., Phosphorylation-dependent activated tension in skinned gizzard muscle fibers in the absence of Ca^{2+}, *J. Biol. Chem.*, 257, 5987, 1982.
35. Silver, P. J. and Stull, J. T., Regulation of myosin light chain and phosphorylase phosphorylation in tracheal smooth muscle, *J. Biol. Chem.*, 257, 6145, 1982.
36. Aksoy, M. O., Mras, S., Kamm, K. E., and Murphy, R. A., Ca^{2+}, cAMP, and changes in myosin phosphorylation during contraction of smooth muscle, *Am. J. Physiol.*, 245, C255, 1983.
37. Butler, T. M., Siegman, M. J., and Mooers, S. U., Chemical energy usage during shortening and work production in mammalian smooth muscle, *Am. J. Physiol.*, 244, C234, 1983.
38. Gerthoffer, W. T. and Murphy, R. A., Myosin phosphorylation and regulation of cross-bridge cycle in tracheal smooth muscle, *Am. J. Physiol.*, 244, C182, 1983.
39. Haeberle, J. R., Hoyt, J. W., and Hathaway, D. R., Regulation of isometric force and isotonic shortening velocity by phosphorylation of the 20,000 dalton myosin light chain of rat uterine smooth muscle, *Pflügers Arch.*, 403, 215, 1985.
40. Siegman, M. J., Butler, T. M., Mooers, S. U., and Davies, R. E., Crossbridge attachment, resistance to stretch, and viscoelasticity in resting mammalian smooth muscle, *Science*, 191, 383, 1976.
41. Rembold, C. M. and Murphy, R. A., Myoplasmic calcium, myosin phosphorylation, and regulation of the crossbridge cycle in swine arterial smooth muscle, *Circ. Res.*, 58, 803, 1986.
42. Dillon, P. F., Aksoy, M. O., Driska, S. P., and Murphy, R. A., Myosin phosphorylation and the crossbridge cycle in arterial smooth muscle, *Science*, 211, 495, 1981.
43. Driska, S. P., Aksoy, M. O., and Murphy, R. A., Myosin light chain phosphorylation associated with contraction in arterial smooth muscle, *Am. J. Physiol.*, 240, C222, 1981.
44. Aksoy, M. O., Murphy, R. A., and Kamm, K. E., Role of Ca^{2+} and myosin light chain phosphorylation in regulation of smooth muscle, *Am. J. Physiol.*, 242, C109, 1982.
45. Hai, C.-M. and Murphy, R. A., Cross-bridge phosphorylation and regulation of latch state in smooth muscle, *Am. J. Physiol.*, 254, C99, 1988.
46. Silver, P. J. and Stull, J. T., Phosphorylation of myosin light chain and phosphorylase in tracheal smooth muscle in response to KCl and carbachol, *Mol. Pharmacol.*, 25, 267, 1984.
47. Rembold, C. M. and Murphy, R. A., Myoplasmic $[Ca^{2+}]$ determines myosin phosphorylation in agonist-stimulated swine arterial smooth muscle, *Circ. Res.*, 63, 593, 1988.
48. Taylor, D. A. and Stull, J. T., Calcium dependence of myosin light chain phosphorylation in smooth muscle cells, *J. Biol. Chem.*, 263, 14456, 1988.
49. Taylor, D. A., Bowman, B. F., and Stull, J. T., Cytoplasmic Ca^{2+} is a primary determinant for myosin phosphorylation in smooth muscle cells, *J. Biol. Chem.*, 264, 6207, 1989.
50. Persechini, A. and Stull, J. T., Phosphorylation kinetics of skeletal muscle myosin and the effect of phosphorylation on actomyosin adenosinetriphosphatase activity, *Biochemistry*, 23, 4144, 1984.
51. Pemrick, S. M., The phosphorylated L_2 light chain of skeletal myosin is a modifier of the actomyosin ATPase, *J. Biol. Chem.*, 255, 8836, 1980.
52. Michnicka, M., Kasman, K., and Kakol, I., The binding of actin to phosphorylated and dephosphorylated myosin, *Biochim. Biophys. Acta*, 704, 470, 1982.
53. Ritz-Gold, C. J., Cooke, R., Blumenthal, D. K., and Stull, J. T., Light chain phosphorylation alters the conformation of skeletal muscle myosin, *Biochem. Biophys. Res. Commun.*, 93, 209, 1980.

54. **Mrakovcic-Zenic, A. and Reisler, E.,** Light-chain phosphorylation and cross-bridge conformation in myosin from vertebrate skeletal muscle, *Biochemistry,* 22, 525, 1983.
55. **Persechini, A., Stull, J. T., and Cooke, R.,** The effect of myosin phosphorylation on the contractile properties of skinned rabbit skeletal muscle fibers, *J. Biol. Chem.,* 260, 7951, 1985.
56. **Sweeney, H. L. and Stull, J. T.,** Phosphorylation of myosin in permeabilized mammalian cardiac and skeletal muscle cells, *Am. J. Physiol.,* 250, C657, 1986.
57. **Sweeney, H. L. and Kushmerick, M. J.,** Myosin phosphorylation in permeabilized rabbit psoas fibers, *Am. J. Physiol.,* 249, C362, 1985.
58. **Manning, D. R. and Stull, J. T.,** Myosin light chain phosphorylation and phosphorylase A activity in rat extensor digitorum longus muscle, *Biochem. Biophys. Res. Commun.,* 90, 164, 1979.
59. **Manning, D. R. and Stull, J. T.,** Myosin light chain phosphorylation-dephosphorylation in mammalian skeletal muscle, *Am. J. Physiol.,* 242, C234, 1982.
60. **Moore, R. L. and Stull, J. T.,** Myosin light chain phosphorylation in fast and slow skeletal muscles *in situ, Am. J. Physiol.,* 247, C462, 1984.
61. **Stull, J. T., Nunnally, M. H., Moore, R. L., and Blumenthal, D. K.,** Myosin light chain kinases and myosin phosphorylation in skeletal muscle, *Adv. Enzyme Regul.,* 23, 123, 1985.
62. **Barsotti, R. J. and Butler, T. M.,** Chemical energy usage and myosin light chain phosphorylation in mammalian skeletal muscle, *J. Muscle Res. Cell Motility,* 5, 45, 1984.
63. **Crow, M. T. and Kushmerick, M. J.,** Myosin light chain phosphorylation is associated with a decrease in the energy cost for contraction in fast twitch mouse muscle, *J. Biol. Chem.,* 257, 2121, 1982.
64. **Klug, G. A., Botterman, B. R., and Stull, J. T.,** The effect of low frequency stimulation on myosin light chain phosphorylation in skeletal muscle, *J. Biol. Chem.,* 257, 4688, 1982.
65. **Houston, M. E., Green, H. J., and Stull, J. T.,** Myosin light chain phosphorylation and isometric twitch potentiation in intact human muscle, *Pflügers Arch.,* 403, 348, 1985.
66. **Moore, R. L., Houston, M. E., Iwamoto, G. A., and Stull, J. T.,** Phosphorylation of rabbit skeletal muscle myosin *in situ, J. Cell. Physiol.,* 125, 301, 1985.
67. **Cannell, M. B. and Allen, D. G.,** Model of calcium movements during activation in the sarcomere of frog skeletal muscle, *Biophys. J.,* 45, 913, 1984.
68. **Edelman, A. M., Blumenthal, D. K., and Krebs, E. G.,** Protein serine/threonine kinases, *Annu. Rev. Biochem.,* 56, 567, 1987.
69. **Adelstein, R. S. and Klee, C. B.,** Purification and characterization of smooth muscle myosin light chain kinase, *J. Biol. Chem.,* 256, 7501, 1981.
70. **Kamm, K. E., Leachman, S. A., Michnoff, C. H., Nunnally, M. H., Persechini, A., Richardson, A. L., and Stull, J. T.,** Myosin light chain kinases and kinetics of myosin phosphorylation in smooth muscle cells, in *Proc. IUPS Conf.: Regulation and Contraction of Smooth Muscle,* Siegman, M. J., Somlyo, A. P., and Stephens, N. L., Eds., Alan R. Liss, New York, 1987, 183.
71. **Foyt, H. L., Guerriero, V., Jr., and Means, A. R.,** Functional domains of chicken gizzard myosin light chain kinase, *J. Biol. Chem.,* 260, 7765, 1985.
72. **Lukas, T. J., Burgess, W. H., Prendergast, F. G., Lau, W., and Watterson, D. M.,** Calmodulin binding domains: characterization of a phosphorylation and calmodulin binding site from myosin light chain kinase, *Biochemistry,* 25, 1458, 1986.
73. **Blumenthal, D. K., Takio, K., Edelman, A. M., Charbonneau, H., Titani, K., Walsh, K. A., and Krebs, E. G.,** Identification of the calmodulin-binding domain of skeletal muscle myosin light chain kinase, *Proc. Natl. Acad. Sci. U.S.A.,* 82, 3187, 1985.
74. **Guerriero, V., Jr., Russo, M. A., Olson, N. J., Putkey, J. A., and Means, A. R.,** Domain organization of chicken gizzard myosin light chain kinase deduced from a cloned cDNA, *Biochemistry,* 25, 8372, 1986.
75. **Takio, K., Blumenthal, D. K., Walsh, K. A., Titani, K., and Krebs, E. G.,** Amino acid sequence of rabbit skeletal muscle myosin light chain kinase, *Biochemistry,* 25, 8049, 1986.
76. **Nunnally, M. H. and Stull, J. T.,** Mammalian skeletal muscle myosin light chain kinases: a comparison by antiserum cross-reactivity, *J. Biol. Chem.,* 259, 1776, 1984.
77. **Nunnally, M. H., Rybicki, S. B., and Stull, J. T.,** Characterization of chicken skeletal muscle myosin light chain kinase: evidence for muscle-specific isozymes, *J. Biol. Chem.,* 260, 1020, 1985.
78. **Mayr, G. W. and Heilmeyer, L. M. G.,** Shape and substructure of skeletal muscle myosin light chain kinase, *Biochemistry,* 22, 4316, 1983.
79. **Edelman, A. M., Takio, K., Blumenthal, D. K., Hansen, R. S., Walsh, K. A., Titani, K., and Krebs, E. G.,** Characterization of the calmodulin-binding and catalytic domains in skeletal muscle myosin light chain kinase, *J. Biol. Chem.,* 260, 11275, 1985.
80. **Takio, K., Blumenthal, D. K., Edelman, A. M., Walsh, K. A., Krebs, E. G., and Titani, K.,** Amino acid sequence of an active fragment of rabbit skeletal muscle myosin light chain kinase, *Biochemistry,* 24, 6028, 1985.
81. **Persechini, A. and Hartshorne, D. J.,** Ordered phosphorylation of the two 20,000 molecular weight light chains of smooth muscle myosin, *Biochemistry,* 22, 470, 1983.

82. Geuss, U., Mayr, G. W., and Heilmeyer, L. M. G., Jr., Steady-state kinetics of skeletal muscle myosin light chain kinase indicate a strong down regulation by products, *Eur. J. Biochem.*, 153, 327, 1985.
83. Ikebe, M. and Hartshorne, D. J., Reverse reaction of smooth muscle myosin light chain kinase, *J. Biol. Chem.*, 261, 8249, 1986.
84. Ikebe, M. and Hartshorne, D. J., Effects of Ca^{2+} on the conformation and enzymatic activity of smooth muscle myosin, *J. Biol. Chem.*, 260, 13146, 1985.
85. Ikebe, M., Hartshorne, D. J., and Elzinga, M., Identification, phosphorylation, and dephosphorylation of a second site for myosin light chain kinase on the 20,000-Dalton light chain of smooth muscle myosin, *J. Biol. Chem.*, 261, 36, 1986.
86. Colburn, J. C., Michnoff, C. H., Hsu, L.-C., Slaughter, C. A., Kamm, K. E., and Stull, J. T., Sites phosphorylated in myosin light chain in contracting smooth muscle, *J. Biol. Chem.*, 263, 19166, 1989.
87. Nishikawa, M., Hidaka, H., and Adelstein, R. S., Phosphorylation of smooth muscle heavy meromyosin by calcium-activated, phospholipid-dependent protein kinase: the effect on actin-activated MgATPase activity, *J. Biol. Chem.*, 258, 14069, 1983.
88. Tuazon, P. T. and Traugh, J. A., Activation of actin-activated ATPase in smooth muscle by phosphorylation of myosin light chain with protease-activated kinase I, *J. Biol. Chem.*, 259, 541, 1984.
89. Tashiro, Y., Matsumura, S., Murakami, N., and Kumon, A., The phosphorylation site for casein kinase II on 20,000-Da light chain of gizzard myosin, *Arch. Biochem. Biophys.*, 233, 540, 1984.
90. Walsh, M. P., Hinkins, S., and Hartshorne, D. J., Phosphorylation of smooth muscle actin by the catalytic subunit of the cAMP-dependent protein kinase, *Biochem. Biophys. Res. Commun.*, 102, 149, 1981.
91. Singh, T. J., Akatsuka, A., and Huang, K.-P., Phosphorylation of smooth muscle myosin light chain by five different kinaes, *FEBS Lett.*, 159, 217, 1983.
92. Gallis, B., Edelman, A. M., Casnellie, J. E., and Krebs, E. G., Epidermal growth factor stimulates tyrosine phosphorylation of the myosin regulatory light chain from smooth muscle, *J. Biol. Chem.*, 258, 13089, 1983.
93. Kikkawa, U. and Nishizuka, Y., Protein kinase C, in *The Enzymes*, Vol. 17, Boyer, P. D. and Krebs, E. G., Eds., Academic Press, Orlando, FL, 1986, 167.
94. Endo, T., Naka, M., and Hidaka, H., Ca^{2+}-phospholipid dependent phosphorylation of smooth muscle myosin, *Biochem. Biophys. Res. Commun.*, 105, 942, 1982.
95. Nishikawa, M., Sellers, J. R., Adelstein, R. S., and Hidaka, H., Protein kinase C modulates *in vitro* phosphorylation of the smooth muscle heavy meromyosin by myosin light chain kinase, *J. Biol. Chem.*, 259, 8808, 1984.
96. Umekawa, H., Naka, M., Inagaki, M., Onishi, H., Wakabayashi, T., and Hidaka, H., Conformational studies of myosin phosphorylation by protein kinase C, *J. Biol. Chem.*, 260, 9833, 1985.
97. Bengur, A. R., Robinson, E. A., Apella, E., and Sellers, J. R., Sequence of the sites phosphorylated by protein kinase C in the smooth muscle myosin light chain, *J. Biol. Chem.*, 262, 7613, 1987.
98. Ikebe, M., Elzinga, M., and Hartshorne, D. J., Identification of the two phosphorylation sites for protein kinase C on the 20,000 Dalton light chain of smooth muscle myosin, *Biophys. J.*, 51, 361a, 1987.
99. Kemp, B. E., Pearson, R. B., and House, C., Role of basic residues in the phosphorylation of synthetic peptides by myosin light chain kinase, *Proc. Natl. Acad. Sci. U.S.A.*, 80, 7471, 1983.
100. Pearson, R. B., Jakes, R., John, M., Kendrick-Jones, J., and Kemp, B. E., Phosphorylation site sequence of smooth muscle myosin light chain (Mr = 20,000), *FEBS Lett.*, 168, 108, 1984.
101. Pearson, R. B., Misconi, L. Y., and Kemp, B. E., Smooth muscle myosin kinase requires residues on the COOH-terminal side of the phosphorylation site: peptide inhibitors, *J. Biol. Chem.*, 261, 251, 1986.
102. Kemp, B. E. and Pearson, R. B., Spatial requirements for location of basic residues in peptide substrates for smooth muscle myosin light chain kinase, *J. Biol. Chem.*, 260, 3355, 1985.
103. Pearson, R. B., Forrest, S., Davis, M., Martin, T. J., and Kemp, B. E., Comparison of the substrate specificity of the myosin light chain kinase and the cyclic AMP-dependent protein kinase, *Biochim. Biophys. Acta*, 786, 261, 1984.
104. Pearson, R. B., Woodgett, J. R., Cohen, P., and Kemp, B. E., Substrate specificity of a multifunctional calmodulin-dependent protein kinase, *J. Biol. Chem.*, 260, 14471, 1985.
105. Pearson, R. B. and Kemp, B. E., Chemical modification of lysine and arginine residues in the myosin light chain inhibits phosphorylation, *Biochim. Biophys. Acta*, 870, 312, 1986.
106. Pearson, R. B., Floyd, D. M., Hunt, J. T., Lee, V. G., and Kemp, B. E., Hydroxyaminoacid specificity of smooth muscle myosin light chain kinase, *Arch. Biochem. Biophys.*, 260, 37, 1987.
107. Michnoff, C. H., Kemp, B. E., and Stull, J. T., Phosphorylation of synthetic peptides by skeletal muscle myosin light chain kinases, *J. Biol. Chem.*, 261, 8320, 1986.
108. Dabrowska, R., Sherry, J. M. F., Aromatoria, D. K., and Hartshorne, D. J., Modulator protein as a component of the myosin light chain kinase from chicken gizzard, *Biochemistry*, 17, 253, 1978.
109. Yazawa, M., Kuwayama, H., and Yagi, K., Modulator protein as a Ca^{2+}-dependent activator of rabbit skeletal myosin light chain kinase, *J. Biochem.*, 84, 1253, 1978.

110. **Kemp, B. E., Pearson, R. B., Guerriero, V., Jr., Bagchi, I. C., and Means, A. R.**, The calmodulin binding domain of chicken smooth muscle myosin light chain kinase contains a pseudosubstrate sequence, *J. Biol. Chem.*, 262, 2542, 1987.
111. **Walsh, M. P., Dabrowska, R., Hinkins, S., and Hartshorne, D. J.**, Calcium independent myosin light chain kinase of smooth muscle: preparation by limited chymotryptic digestion of the Ca^{2+}-dependent enzyme, purification and characterization, *Biochemistry*, 21, 2019, 1982.
112. **Tanaka, T., Naka, M., and Hidaka, H.**, Activation of myosin light chain kinase by trypsin, *Biochem. Biophys. Res. Commun.*, 92, 313, 1980.
113. **Ikebe, M., Stepinska, M., Kemp, B. E., Means, A. R., and Hartshorne, D. J.**, Proteolysis of smooth muscle myosin light chain kinase: formation of inactive and calmodulin-independent fragments, *J. Biol. Chem.*, 260, 13828, 1987.
114. **Kennelly, P. J., Edelman, A. M., Blumenthal, D. K., and Krebs, E. G.**, Rabbit skeletal muscle myosin light chain kinase, the calmodulin binding domain as a potential active site-directed inhibitory domain, *J. Biol. Chem.*, 262, 11958, 1987.
115. **Pearson, R. B., Wettenhall, R. E. H., Means, A. R., Hartshorne, D. J., and Kemp, B. E.**, Autoregulation by enzyme pseudosubstrate prototopes: myosin light chain kinase, *Science*, 241, 970, 1988.
116. **Payne, M. E., Fong, Y. L., Ono, T., Colbran, R. J., Kemp, B. E., Soderling, T. R., and Means, A. R.**, Calcium/calmodulin-dependent protein kinase. II. Characterization of distinct calmodulin binding and inhibitory domains, *J. Biol. Chem.*, 263, 7190, 1988.
117. **Kelly, P. T., Weinberger, R. P., and Waxham, M. N.**, Active site-directed inhibition of Ca^{2+}/calmodulin-dependent protein kinase type II by a bifunctional calmodulin-binding peptide, *Proc. Natl. Acad. Sci. U.S.A.*, 85, 4991, 1988.
118. **Kemp, B. E., Pearson, R. B., House, C., Robinson, P. J., and Means, A. R.**, Regulation of protein kinases by pseudosubstrate prototopes, *Cell. Signalling*, 1, 303, 1989.
119. **Olson, N. J., Pearson, R. B., Needleman, D., Hurwitz, M. Y., Kemp, B. E., and Means, A. R.**, unpublished results.

Chapter 5

USE OF SYNTHETIC PEPTIDES TO STUDY SUBSTRATE SPECIFICITY OF PROTEIN PHOSPHATASES

Donald K. Blumenthal

TABLE OF CONTENTS

I. Introduction 136

II. General Considerations 136

III. Specific Examples 137

IV. Summary and Perspectives 141

References 141

I. INTRODUCTION

Synthetic peptides are widely used as models to study various aspects of protein structure and function. In the broad area of protein phosphorylation, synthetic peptides have been employed extensively as tools in studying various biochemical properties of protein kinases, such as determinants of substrate specificity and mechanisms of allosteric regulation. Comparable sorts of studies with protein phosphatases have been done to a limited extent, but until recently there has been a great deal of confusion regarding the identification and classification of different phosphatase activities. This confusion has made it difficult to compare different preparations of phosphatase from different laboratories and has therefore hampered efforts to exploit the various advantages of using synthetic peptides to study these enzymes.

Two schemes for classifying protein phosphatases have now been proposed independently that appear to clear up some of the earlier confusion and that should facilitate comparison of results between different laboratories.[1,2] The details of these two schemes will not be addressed here, except to say that the two taxonomies differ primarily with respect to nomenclature and that both systems classify the different protein phosphatase activities on the basis of their substrate specificities and their susceptibility to regulatory molecules. Thus, synthetic phosphopeptides should be ideally suited for characterizing the various phosphatase activities classified using either scheme, since synthetic peptides represent well-defined, readily available substrates.

The purpose of this chapter will be to consider the various advantages of using synthetic substrate peptides to study protein phosphatases, to review several specific examples of studies that have employed synthetic peptides, and to suggest future directions and applications of synthetic peptides in the study of protein phosphatase structure, function, and regulation. The reader who is interested in the general problem of protein kinase and protein phosphatase specificity is also referred to two recent reviews on the subject.[3,4]

II. GENERAL CONSIDERATIONS

The use of synthetic peptide substrates for work with protein phosphatases has a number of advantages compared to using protein substrates. Perhaps the most important overall advantage to using synthetic peptides as substrates is that they are chemically well-defined entities that can be obtained in highly purified form and in relatively large quantity. Many of the commonly used phosphoprotein substrates are rather poorly defined in terms of their phosphorylation sites (e.g., histones and casein) or are difficult to obtain in sufficient quantity to use on a routine basis as phosphatase substrates. Many protein substrates can also present various problems in handling, particularly with respect to solubility at high concentrations, and often these problems are minimized by using synthetic oligopeptides.

Another important advantage associated with using chemically synthesized peptides is that the sequence and length of the substrate peptide can be varied. The ability to systematically alter the structure around the site of phosphorylation is essential for studies of substrate specificity. The ability to manipulate the sequence of a peptide substrate might also allow one to develop an enzyme-specific substrate,[5,6] as well as to incorporate reporter groups at specific positions in the peptide. Synthetic substrate peptides with reporter groups that undergo phosphorylation-dependent absorbance and fluorescence changes have been developed for use in spectrophotometric assays of protein kinase activity by incorporating appropriate residues (tyrosine and tryptophan) near the phosphorylation site in several different peptide sequences.[7-9] One of these peptides[9] has also been used to assay protein phosphatase activity spectrophotometrically. Chromogenic and fluorogenic substrate peptides not only permit continuous observation of phosphatase activity, but also provide a means to avoid the major inconvenience associated with having to prepare fresh ^{32}P-labeled phosphopeptide substrate periodically because of the short half-life of ^{32}P.

Although ^{32}P-labeled substrates are widely used in assays of protein phosphatase activity because of the high sensitivity they afford, the preparation and characterization of ^{32}P-labeled substrates, both phosphoprotein and phosphopeptide, require a considerable expenditure of time and effort. Since labeling is performed enzymatically, the appropriate protein kinases must first be purified. Following incorporation of ^{32}P into the substrate, the phosphoprotein or phosphopeptide must be separated from any ^{32}P$_i$- and/or ^{32}P-labeled ATP to minimize background in the phosphatase assay. It is typically much quicker and easier to purify most phosphopeptides following phosphorylation than it is to purify most phosphoproteins because peptides generally are better able to withstand moderately harsh treatments. Thus, phosphopeptides are normally purified using procedures such as Dowex ion-exchange chromatography in 1 N acetic acid and concentrated by lyophilization, whereas phosphoproteins are usually purified by dialysis or electrodialysis. The procedures for removal of unincorporated ^{32}P from phosphopeptides might take several hours, while complete dialysis of phosphoproteins may require several days.

One of the major technical problems associated with the use of synthetic peptides is that a given peptide may be a poor substrate for any available protein kinase, making it difficult or impossible to phosphorylate that peptide enzymatically. Although this also can be a problem with some protein substrates, it is a serious consideration when preparing a series of variant synthetic peptides for use in substrate-specificity studies. Several laboratories have recently reported different procedures for the incorporation of phosphoamino acids during synthesis by using protected phosphoserine and phosphothreonine derivatives (see also Chapter 12 of this volume).[10-12] Although these procedures have not been worked out fully, this general methodology would allow the incorporation of phosphoamino acid derivatives into specific positions during synthesis and, thus, avoid the necessity of enzymatic phosphorylation. Even though it probably would not be practical to prepare ^{32}P-labeled phosphopeptides by this procedure because of the difficulties in handling the large quantities of radioactivity involved, it may be possible to use this strategy in conjunction with one of the spectrophotometric assays described above. In addition, this methodology would be useful in the preparation of phosphopeptides with multiple phosphate-accepting residues where it is important that only selected sites be phosphorylated.

III. SPECIFIC EXAMPLES

The first use of phosphopeptides to study the substrate specificity of protein phosphatases was reported by Graves et al. in 1960.[13] In this study, phosphopeptides obtained by proteolytic digestion of phosphorylase a were used to characterize a preparation of phosphorylase phosphatase from rabbit skeletal muscle. Several critical (and noncritical) determinants of specificity were identified using proteolytic fragments of different length and by chemically modifying specific residues in the phosphopeptides. It is worth noting that most studies of substrate specificity using synthetic peptides have begun by using protease-generated phosphopeptides in a manner pioneered by this study.

Synthetic peptides were first used to study the substrate specificity of protein phosphatases in 1977, when Titanji et al.[14] used phosphopeptides based on the cAMP-dependent protein kinase phosphorylation site in liver pyruvate kinase to study the specificity of a phosphatase preparation purified from rat liver. A subsequent report from this laboratory also included data obtained using synthetic peptides based on the phosphorylation site in phosphorylase a.[15] The conclusion derived from these studies was that only four residues, one to the amino-terminal side and two to the carboxy-terminal side, were required for rapid dephosphorylation by the phosphatase preparation. Although this work demonstrated some of the advantages of using synthetic peptides to study protein phosphatase specificity, this enzyme preparation was not well characterized by current standards. For instance, the preparation of this enzyme involved ethanol treatment, which is known to release the catalytic subunits of several different types of phosphatase. Thus, the conclusions drawn must be regarded as somewhat tentative because of the possibility that the phosphatase preparation used was actually a mixture of several different phosphatase activities (see References 3 and 4).

Synthetic peptides based on the structure of the phosphorylation site of pyruvate kinase have been the most extensively used phosphopeptides for studying protein phosphatases. This is largely because of their wide use in studying cyclic nucleotide-dependent protein kinases. For instance, Wright et al.[9] demonstrated the utility of the fluorescent substrate peptide Leu–Arg–Arg–Trp–Ser–Leu–Gly for monitoring both cAMP-dependent protein kinase activity and the activity of a preparation of bovine cardiac protein phosphatase. The same preparation of phosphatase was used in a subsequent study from the same laboratory, in which the synthetic phosphopeptide substrates Leu–Arg–Arg–Ala–Ser(P)–Val–Ala–Gln–Leu and Leu–Arg–Arg–Ala–Ser(P)–Leu–Gly were used to study the individual and combined effects of divalent cations, ATP, and fluoride on enzymatic activity.[16] The phosphatase preparation utilized in these two studies was obtained using a procedure involving ethanol treatment and therefore may suffer from the same criticism as the study described above. Enzymatic analysis using protein phosphatase inhibitor-2 indicated the preparation contained approximately 88% protein phosphatase-2 and 12% protein phosphatase-1 activity.[16] Immunochemical analysis using a monoclonal antibody has also indicated that this cardiac preparation is predominantly, if not entirely, composed of protein phosphatase-2A.[17]

Using a series of synthetic phosphopeptides that were variants of the pyruvate kinase and phosphatase inhibitor-1 phosphorylation site sequences, Donella-Deanna et al.,[5] and subsequently Agostinis et al.,[6] compared the substrate specificities of a number of different phosphatase preparations. It was found that several different forms of rabbit skeletal muscle polycation-stimulated (PCS) phosphorylase phosphatase (also termed protein phosphatase-2A)[6] and a partially purified rat liver phosphatase preparation termed phosphatase-T[5] (which may represent a form of PCS phosphatase; see Pinna et al.[3]) preferentially dephosphorylated Arg–Arg–Ala–Thr(P)–Val–Ala compared with Arg–Arg–Ala–Ser(P)–Val–Ala and five other phosphoseryl and phosphothreonyl peptides. Interestingly, the holoenzyme form, PCS_H, showed the greatest preference for the phosphothreonyl peptide (approximately 50-fold difference in rates between the two indicated peptides), while the catalytic subunit, PCS_C, showed the least (8-fold).[6] This indicates the importance of the intact enzyme in determining substrate specificity. In contrast to the PCS phosphatase activities, a preparation of catalytic subunit of rabbit skeletal muscle MgATP-dependent protein phosphatase (also referred to as protein phosphatase-1) was unable to efficiently dephosphorylate any of the seven phosphopeptides used, as compared to its ability to dephosphorylate phosphorylase *a*.[6] A similar finding has been reported by Fischer and co-workers[18,19] using this enzyme (discussed later in this section), although it is not clear whether the catalytic subunits used in the two studies are identical or not. Partially purified rat liver phosphatase-1 (which presumably represents an enzymatic form similar to that purified from skeletal muscle) apparently dephosphorylated the two phosphothreonyl and phosphoseryl peptides shown above at comparable rates; its ability to dephosphorylate phosphorylase *a*, however, was not reported.[5] The substrate preferences of calf intestine alkaline phosphatase and potato acid phosphatase, two "nonspecific" phosphatases that can catalyze the dephosphorylation of low molecular weight compounds as well as peptides and proteins, were also tested using the two phosphopeptides described above, and both enzymes were found to dephosphorylate the phosphoseryl peptide preferentially.

A series of synthetic peptide analogues of the sequence around phosphorylation sites 1 and 2 in pyruvate dehydrogenase have recently been employed by Mullinax et al.[20] to study the substrate specificity of pyruvate dehydrogenase phosphatase. This well-characterized mitochondrial protein phosphatase had previously been shown to dephosphorylate a phosphotetradecapeptide obtained by tryptic digestion of phosphorylated pyruvate dehydrogenase.[21] Phosphate was only incorporated into site 1 of the synthetic tetradecapeptide by pyruvate dehydrogenase kinase, yielding the phosphopeptide Tyr–His–Gly–His–Ser(P)–Met–Ser–Asp–Pro–Gly–Val–Ser–Tyr–Arg. The V_{max} value of the peptide (2.3 μmol/min/mg) was 10-fold greater and the K_m value (0.28 mM) was 40-fold poorer than that for the intact protein. Methylation of the aspartate residue improved the K_m by about fourfold, but lowered the V_{max} by about the same amount. Modification of the N-terminus of the peptide by N-acetylation or by the addition of a lysyl residue had little effect on the substrate's kinetic parameters. Thermolytic cleavage of the glycyl-valyl bond was used to remove the four C-terminal residues, and this also had little effect on the competency of the peptide. Acid cleavage at the aspartyl-prolyl bond yielded an phosphooctapeptide whose K_m value was about fourfold poorer than the parent phosphotetradecapeptide, but which retained a similar V_{max} value. Thus, site 1 phosphopeptides as short as eight residues can still serve as reasonably good substrates for the pyruvate dehydrogenase phosphatase. Additional studies are required to determine the absolute minimal length phosphopeptide that will serve as a substrate, the effect of altering specific residues in the sequence, and the specificity of the enzyme toward peptides phosphorylated at site 2 and at both sites 1 and 2.

Probably the most substrate-specific protein phosphatase is the calmodulin-dependent protein phosphatase, also known as calcineurin or protein phosphatase-2B.[1,22] The substrate specificity of this phosphatase has been studied using synthetic phosphopeptides based on one of its few known substrates, the type II regulatory subunit (R^{II}) of the cAMP-dependent protein kinase.[23-25] The R^{II} subunit is phosphorylated at Ser-95 by the catalytic subunit of the cAMP-dependent protein kinase, and this site is therefore termed the autophosphorylation site. Phosphopeptides spanning residues 81 to 99 of R^{II} (Asp–Leu–Asp–Val–Pro–Ile–Pro–Gly–Arg–Phe–Asp–Arg–Arg–Val–Ser(P)–Val–Cys–Ala–Glu) that were generated by proteolysis with *Staphylococcus aureus* V8 protease were found to be dephosphorylated as rapidly as the intact protein, whereas tryptic and chymotryptic phosphopeptides (residues 93 to 400 and 91 to 400, respectively) were dephosphorylated considerably slower.[23,25]

In order to determine the minimum length phosphopeptide that could serve as an optimal substrate for the calmodulin-dependent protein phosphatase, synthetic phosphopeptides corresponding to residues 81 to 99, 85 to 99, 90 to 99, and 91 to 99 of R^{II} were prepared and their abilities to serve as substrates tested.[23,25] Only the nonadecapeptide (81—99) had K_m (26 μM) and V_{max} (1.7 μmol/min/mg) values comparable to those of the intact protein (K_m = 20 μM, V_{max} = 2 μmol/min/mg). The phosphopentadecapeptide (85—99) had K_m (130 μM) and V_{max} (0.24 μmol/min/mg) values that were considerably worse than the nonadecapeptide, indicating that the four amino-terminal residues (81 to 84) are important determinants of substrate specificity. The data are best explained if it is hypothesized that the phosphatase requires an amphipathic β-sheet structure at the amino terminus of this peptide (i.e., residues 81 to 90). Although at least one of the other known substrates of the calmodulin-dependent protein phosphatase, protein phosphatase inhibitor-1, also appears to contain some β-structure in the vicinity of its phosphorylation site,[26] it is not clear whether this is true for all of this phosphatase's substrates. For example, the α-subunit of phosphorylase kinase is an excellent substrate for the phosphatase, yet its phosphorylation site is located at Ser-5 of the protein. Thus, it has virtually no structure to the amino-terminal side of the phosphorylation site, nor is there any indication of β-structure in the first 30 residues of the protein. However, some sort of higher order structure appears to be important for recognition of this protein substrate, since a phosphopeptide based on the first 19 residues of the sequence of the α-subunit is not detectably dephosphorylated by the calmodulin-dependent phosphatase.[25] Thus, if a β-structure is involved in α-subunit recognition by the phosphatase, this structure is probably formed by a noncontiguous segment of the polypeptide chain.

An unusual feature of the calmodulin-dependent protein phosphatase is its ability to dephosphorylate p-nitrophenyl phosphate, free phosphotyrosine,[27] and several phosphotyrosyl-containing polypeptides.[24,28,29] Specificity studies using a variety of defined, unrelated oligopeptide and polypeptide substrates indicated that a major substrate specificity determinant of the phosphatase appeared to be the phosphotyrosyl moiety.[24] Of particular interest was the use of the synthetic R^{II} peptide (81—99), where Ser-95 was replaced by a tyrosyl residue. This peptide was phosphorylated by the protein-tyrosine kinase activity present in A431 cell membranes in order to form the phosphotyrosyl analogue of R^{II}-(81—99). Surprisingly, this phosphotyrosyl peptide substrate was not a significantly better substrate than any of the other phosphotyrosyl substrates studied, indicating that the determinants that were important for recognition of phosphoseryl-R^{II}-(81—99) were not necessarily determinants for phosphotyrosyl phosphatase activity. Substrate competition experiments indicated that the phosphotyrosyl and phosphoseryl/phosphothreonyl phosphatase activities shared a common catalytic site. Thus, the limited quantity of information obtained to date indicates that the calmodulin-dependent phosphatase exhibits a very narrow substrate specificity with phosphothreonyl/seryl substrates, but a very broad specificity with phosphotyrosyl substrates.

Studies by Fischer and co-workers[18,19] using synthetic phosphopeptides corresponding to the phosphorylation sites in R^{II} (residues 91 to 99, 90 to 99, 85 to 99, and 81 to 99), phosphorylase a (residues 5 to 18), and the α-subunit (residues 1 to 19) and β-subunit (Arg–Thr–Lys–Arg–Ser–Gly–Ser(P)–Val–Tyr–Glu–Pro–Leu) of phosphorylase kinase indicate that the various conformers of phosphorylase phosphatase (protein phosphatase-1, MgATP-dependent protein phosphatase) exhibit quite different substrate specificities. The inactive holoenzyme form of the enzyme, E_h, can be activated in three ways. Treatment of E_h with trypsin in the presence on Mn^{2+} yields the form E_a^{Tr-Mn}, which has a broad substrate specificity and is able to dephosphorylate synthetic phosphopeptides with the same efficiency as the native proteins upon which they were based. This form, which has been activated in a nonphysiological manner, has lost its regulatory subunit and has suffered a slight proteolysis of the catalytic subunit. The form E_a^{Mn} is the free catalytic subunit which has not been proteolyzed, but which requires Mn^{2+} for activity. This form also dephosphorylates phosphoproteins and phosphopeptide substrates with comparable efficiency. Form E_h^{Fa} is prepared by phosphorylating E_h with the enzyme F_A (also termed glycogen synthase kinase-3). Activation by F_A is presumed to be the physiological mode of activation for the phosphatase. This F_A-activated holoenzyme form of the phosphatase rapidly dephosphorylates R^{II}, phosphorylase a, and phosphorylase kinase, but does not dephosphorylate their corresponding synthetic phosphopeptides. The same enzymatic behavior is exhibited by E_a^{Fa}, the free catalytic subunit form of E_h^{Fa}. Thus, the regulatory subunit of the enzyme does not appear to play a role in the observed specificities of the F_A-activated forms of the enzyme. Moreover, the specificity of the F_A-activated enzyme appears to involve higher order determinants as well as primary structural features. The higher order determinants important for this phosphatase appear to be different than those for the calmodulin-dependent protein phosphatase.

Synthetic phosphopeptide substrates also have proven useful in developing assays to distinguish between protein phosphotyrosine phosphatases and "nonspecific" alkaline and acid phosphatases.[30,31] Synthetic phosphopeptides based on the phosphorylation site sequence of the *src* gene product (referred to as RR-src: Arg–Arg–Leu–Ile–Glu–Asp–Ala–Glu–Tyr(P)–Ala–Ala–Arg–Gly) and angiotensin II (Asp–Arg–Val–Tyr(P)–Val–His–Pro–Phe) were shown to be poor substrates for cytosolic phosphotyrosine phosphatases purified from rabbit kidney, but good substrates for calf intestine alkaline phosphatase.[30] Similarly, phosphotyrosine-angiotensin II, phosphotyrosine-poly(Glu,Tyr)4:1, and another peptide related to the phosphorylation site of *src*, Asp–Ala–Glu–Tyr(P)–Ala–Ala–Arg–Arg–Arg–Gly, were found to be dephosphorylated by yeast repressible acid phosphatase with kinetic constants comparable to free phosphotyrosine.[31] The phosphoseryl peptide Arg–Arg–Ala–Ser(P)–Val–Ala was also rapidly dephosphorylated by this "nonspecific" yeast enzyme, but several similar phosphoseryl and phosphothreonyl peptides and proteins were not.[31]

IV. SUMMARY AND PERSPECTIVES

Synthetic peptides are ideally suited for characterizing the biochemical properties of protein phosphatases, as well as for developing enzyme-specific assays of phosphatase activity. However, because of past confusion regarding the identification of phosphatases purified in different laboratories, synthetic phosphopeptides have not been widely employed. With the advent of new classification schemes, much of this confusion has been eliminated and synthetic phosphopeptides undoubtedly will be used to a much greater extent to study and compare the substrate specificities of different phosphatase preparations. These efforts will be greatly facilitated by current developments in techniques to incorporate phosphoamino acid derivatives during the synthesis of peptides. An important aspect of understanding the substrate specificity of an enzyme is the determination of the three-dimensional structure of the substrate (and enzyme) in the enzyme-substrate complex. Bramson and co-workers[32,33] have recently demonstrated the feasibility of determining the solution structure of protein kinase synthetic peptide substrates using a combined approach involving nuclear magnetic resonance (NMR) spectroscopy and synthetic peptides containing N-methylated peptide bonds. Similar approaches should be applicable to determining the three-dimensional structures of synthetic phosphatase substrates. X-ray crystallographic studies of phosphopeptide-phosphatase complexes will be another means of approaching this important problem.

In addition to being used as probes in studies of substrate specificity, synthetic peptides should have applications in other studies of protein phosphatase structure and function. For instance, Hiraga et al.[34] recently used synthetic peptides based on the phosphorylation site sequence of the G-subunit of the glycogen-bound form of protein phosphatase-1 as antigens to raise monoclonal antibodies. These antibodies were used to confirm that the G-subunit protein is indeed a subunit of this form of the enzyme and that the G-subunit and its proteolytic fragments interact with both glycogen and the catalytic subunit of the enzyme. As additional sequence data from protein phosphatases become available, synthetic peptides will become increasingly useful in similar immunochemical studies designed to elucidate the relationships of phosphatase subunit structure and function. Synthetic peptides also should be valuable in developing enzyme-specific peptide inhibitors based on the sequences of known polypeptide inhibitors such as protein phosphatase inhibitor-1 and -2, the cAMP-dependent protein kinase inhibitor (which inhibits protein phosphatase-1 with micromolar affinity),[18,19] and substrate analogue inhibitors. A library of such inhibitors should prove invaluable in trying to sort out the physiological function of each phosphoprotein phosphatase in a given tissue.

V. REFERENCES

1. **Ingebritsen, T. S. and Cohen, P.,** The protein phosphatases involved in cellular regulation. I. Classification and substrate specificities, *Eur. J. Biochem.*, 132, 255, 1983.
2. **Merlevede, W.,** Protein phosphates and protein phosphatases. Landmarks in an eventful century, *Adv. Protein Phosphatases*, 1, 1, 1985.
3. **Pinna, L. A., Agostinis, P., and Ferrari, S.,** Selectivity of protein kinases and protein phosphatases: a comparative analysis, *Adv. Protein Phosphatases*, 3, 327, 1986.
4. **Sparks, J. W. and Brautigan, D. L.,** Molecular basis for substrate specificity of protein kinases and phosphatases, *Int. J. Biochem.*, 18, 497, 1986.

5. **Donella-Deanna, A., Marchiori, F., Meggio, F., and Pinna, L. A.**, Dephosphorylation of synthetic phosphopeptides by protein phosphatase-T, a phosphothreonyl protein phosphatase, *J. Biol. Chem.*, 257, 8565, 1982.
6. **Agostinis, P., Goris, J., Waelkens, E., Pinna, L. A., Marchiori, F., and Merlevede, W.**, Dephosphorylation of phosphoproteins and synthetic phosphopeptides: study of the specificity of the polycation-stimulated and MgATP-dependent phosphorylase phosphatases, *J. Biol. Chem.*, 262, 1060, 1987.
7. **Malencik, D. A. and Anderson, S. R.**, Characterization of a fluorescent substrate for the adenosine 3',5'-cyclic monophosphate-dependent protein kinase, *Anal. Biochem.*, 132, 34, 1983.
8. **Bramson, H. N., Thomas, N., DeGrado, W. F., and Kaiser, E. T.**, Development of a convenient spectrophotometric assay for peptide phosphorylation catalyzed by adenosine 3',5'-monophosphate dependent protein kinase, *J. Am. Chem. Soc.*, 102, 7156, 1980.
9. **Wright, D. E., Noiman, E. S., Chock, P. B., and Chau, V.**, Fluorometric assay for adenosine 3',5'-cyclic monophosphate-dependent protein kinase and phosphoprotein phosphatase activities, *Proc. Natl. Acad. Sci. U.S.A.*, 78, 6048, 1981.
10. **Grehn, L., Fransson, B., and Ragnarsson, U.**, Synthesis of substrates of cyclic AMP-dependent protein kinase and use of their protected precursors for the convenient preparation of phosphoserine peptides, *J. Chem. Soc. Perkin Trans. 1*, p. 529, 1987.
11. **Arendt, A. and Hargrave, P. A.**, Solid phase synthesis of phosphopeptides from the carboxyl terminus of rhodopsin, in *Peptides: Structure and Function*, Deber, C. M., Hruby, V. J., and Kopple, K. D., Eds., Pierce Chemical Co., Rockford, Il, 1985, 237.
12. **Perich, J. W., Valerio, R. M., and Johns, R. B.**, Solid-phase synthesis of an *o*-phosphoseryl-containing peptide using phenyl phosphotriester protection, *Tetrahedron Lett.*, 27, 1377, 1986.
13. **Graves, D. J., Fischer, E. H., and Krebs, E. G.**, Specificity studies on muscle phosphorylase phosphatase, *J. Biol. Chem.*, 235, 805, 1960.
14. **Titanji, V. P. K., Zetterqvist, O., and Ragnarsson, U.**, Activity of rat-liver phosphoprotein phosphatase on phosphopeptides formed in the cyclic AMP-dependent protein kinase reaction, *FEBS Lett.*, 78, 86, 1977.
15. **Titanji, V. P. K., Ragnarsson, U., Humble, E., and Zetterqvist, O.**, Phosphopeptide substrates of a phosphoprotein phosphatase from rat liver, *J. Biol. Chem.*, 255, 11339, 1980.
16. **Shacter-Noiman, E. and Chock, P. B.**, Properties of a M_r = 38,000 phosphoprotein phosphatase: modulation by divalent cations, ATP, and fluoride, *J. Biol. Chem.*, 258, 4214, 1983.
17. **Mumby, M. C., Green, D. D., and Russell, K. L.**, Structural characterization of cardiac protein phosphatase with a monoclonal antibody: evidence that the M_r = 38,000 phosphatase is the catalytic subunit of the native enzyme(s), *J. Biol. Chem.*, 260, 13763, 1985.
18. **Ballou, L. M., Villa-Moruzzi, E., McNall, S. J., Scott, J. D., Blumenthal, D. K., Krebs, E. G., and Fischer, E. H.**, Structure, properties and regulation of phosphorylase phosphatase, *Adv. Protein Phosphatases*, 1, 21, 1985.
19. **McNall, S. J. and Fischer, E. H.**, Phosphorylase phosphatase: comparison of active forms using peptide substrates, *J. Biol. Chem.*, 263, 1893, 1988.
20. **Mullinax, T. R., Stepp, L. R., Brown, J. R., and Reed, L. J.**, Synthetic peptide substrates for mammalian pyruvate dehydrogenase kinase and pyruvate dehydrogenase phosphatase, *Arch. Biochem. Biophys.*, 243, 655, 1985.
21. **Davis, P. F., Pettit, F. H., and Reed, L. J.**, Peptides derived from pyruvate dehydrogenase as substrates for pyruvate dehydrogenase kinase and phosphatase, *Biochem. Biophys. Res. Commun.*, 75, 541, 1977.
22. **Stewart, A. A., Ingebritsen, T. S., and Cohen, P.**, The protein phosphatases involved in cellular regulation. V. Purification and properties of a Ca^{++}/calmodulin-dependent protein phosphatase (2B) from skeletal muscle, *Eur. J. Biochem.*, 132, 289, 1983.
23. **Blumenthal, D. K., Takio, K., Hansen, R. S., and Krebs, E. G.**, Dephosphorylation of cAMP-dependent protein kinase regulatory subunit (type II) by calmodulin-dependent protein phosphatase: determinants of substrate specificity, *J. Biol. Chem.*, 261, 8140, 1986.
24. **Chan, C. P., Gallis, B., Blumenthal, D. K., Pallen, C. J., Wang, J. H., and Krebs, E. G.**, Characterization of the phosphotyrosyl protein phosphatase activity of calmodulin-dependent protein phosphatase, *J. Biol. Chem.*, 261, 9890, 1986.
25. **Blumenthal, D. K., Chan, C. P., Takio, K., Gallis, B., Hansen, R. S., and Krebs, E. G.**, Substrate specificity of calmodulin-dependent protein phosphatases, *Adv. Protein Phosphatases*, 1, 163, 1985.
26. **Cohen, P., Nimmo, G. A., Shenolikar, S., and Foulkes, G. A.**, The role of inhibitor-1 in the cyclic AMP mediated control of glycogen metabolism in skeletal muscle, *FEBS Symp.*, 54, 161, 1978.
27. **Pallen, C. J. and Wang, J. H.**, Calmodulin-stimulated dephosphorylation of *p*-nitrophenyl phosphate and free phosphotyrosine by calcineurin, *J. Biol. Chem.*, 258, 8550, 1983.
28. **Chernoff, J., Sells, M. A., and Li, H.-C.**, Characterization of phosphotyrosyl-protein phosphatase activity associated with calcineurin, *Biochem. Biophys. Res. Commun.*, 121, 141, 1984.

29. **Pallen, C. J., Valentine, K. A., Wang, J. H., and Hollenberg, M. D.**, Calcineurin-mediated dephosphorylation of the human placental membrane receptor for epidermal growth factor urogastrone, *Biochemistry,* 24, 4727, 1985.
30. **Sparks, J. W. and Brautigan, D. L.**, Specificity of protein phosphotyrosine phosphatases: comparison with mammalian alkaline phosphatase using polypeptide substrates, *J. Biol. Chem.,* 260, 2042, 1985.
31. **Donella-Deana, A., Lopandic, K., Barbaric, S., and Pinna, L. A.**, Distinct specificities of repressible acid phosphatase from yeast toward phosphotyrosyl and phosphotyrosyl phosphopeptides, *Biochem. Biophys. Res. Commun.,* 139, 1202, 1986.
32. **Bramson, H. N., Thomas, N. E., Miller, W. T., Fry, D. C., Mildvan, A. S., and Kaiser, E. T.**, Conformation of Leu–Arg–Arg–Ala–Ser–Leu–Gly bound in the active site of adenosine cyclic 3′,5′-phosphate dependent protein kinase, *Biochemistry,* 26, 4466, 1987.
33. **Thomas, N. E., Bramson, H. N., Miller, W. T., and Kaiser, E. T.**, Role of enzyme-peptide substrate backbone hydrogen-bonding in determining protein kinase substrate specificities, *Biochemistry,* 26, 4461, 1987.
34. **Hiraga, A., Kemp, B. E., and Cohen, P.**, Further studies on the structure of the glycogen-bound form of protein phosphatase-1 from rabbit skeletal muscle, *Eur. J. Biochem.,* 163, 253, 1987.

Chapter 6

TYPE-2 CASEIN KINASES: GENERAL PROPERTIES AND SUBSTRATE SPECIFICITY

Lorenzo A. Pinna, Flavio Meggio, and Fernando Marchiori

TABLE OF CONTENTS

I. History and Classification of "Casein Kinases" 146

II. Oligomeric Structure of Type-2 Casein Kinases 147

III. Regulation of Type-2 Casein Kinases 149

IV. Endogenous Substrates and Physiological Implications 150

V. Specificity Determinants 152
    A. Primary Structure of the Phosphorylation Sites 152
    B. Studies with Synthetic Peptides 154
    C. Comparison with Casein Kinase-1, Mammary Gland Casein Kinase, and Other Protein Kinases 158

Addendum 161

Acknowledgments 163

References 163

I. HISTORY AND CLASSIFICATION OF "CASEIN KINASES"

The ubiquitous enzymes which nowadays are operationally termed "casein kinases" were the first protein kinases to be detected, as early as in 1954.[1] Their discovery was accelerated by the use of casein and phosvitin as model substrates, based on the implicit supposition that these molecules — which were the only well-known phosphoproteins at that time — might be endowed with intrinsic features enabling them to undergo enzymatic phosphorylation in the presence of ATP. Although such a concept was later proved to be devoid of general validity, these proteins nevertheless played a crucial role in the first detection of ATP:protein phosphotransferase activity in different organisms and tissues.[1,2]

For some years, such an activity could be unambiguously referred to as "protein (phospho)kinase", as the only other protein kinase known at that time — namely, phosphorylase b kinase[3] — had been shown to be inactive on casein and phosvitin while the enzymes phosphorylating casein and phosvitin were unable to use phosphorylase b as a substrate.[2] Approximately one decade later, however, the discovery of cAMP-dependent protein kinase[4] and of other kinases stimulated by second messengers and the exciting physiological implications of such new enzymes diverted the general attention from the casein/phosvitin phosphorylating kinases and *de facto* defrauded them of the name protein kinase, thenceforth more commonly intended to be synonymous with cAMP-dependent protein kinase. Consequently, a new name was desired to indicate the "casein/phosvitin-specific" protein kinases. Neither a nomenclature based on physiological effectors nor a name derived from one of their targets was easily applicable to this class of "independent" and multisubstrate protein kinases. It was felt preferable to adopt a conventional term coined after their preference for casein and phosvitin as artificial substrates, the name "casein kinase" soon prevailing over "phosvitin kinase". It should always be borne in mind, however, that such multisubstrate and ubiquitous "casein kinases" affecting casein *in vitro* are distinct enzymes from the physiological casein kinase(s) responsible for the *in vivo* phosphorylation of casein within the mammary gland.

Although a thorough characterization of the different subsets of casein kinases was achieved only in the recent years, some of their distinctive features were anticipated by pioneer studies in the late 1960s and early 1970s. Thus, a brain "phosvitin kinase" also using GTP as phosphate donor was described in 1964,[5] and the existence of two distinct classes of casein kinases was proposed in a 1969 report showing that the properties of a microsomal, ATP-specific enzyme were substantially different from those of the "phosvitin kinase" predominant in the cytosolic fraction, also able to use GTP as phosphate donor.[6] Likewise, the preference for either seryl or threonyl residues of casein and the acidic nature of endogenous targets of casein kinases were described several years before they were definitely recognized as criteria for the classification of these enzymes.[7,8] Finally, the observation that casein kinases exhibit an outstanding capacity to adsorb on phosphocellulose provided a simple and effective method for readily separating these enzymes from most other protein kinases, which are active on histones and protamines.[9] Phosphocellulose subsequently was exploited in nearly all the purification procedures for casein kinases. As summarized in Table 1, such an unusual chromatographic behavior together with insensitivity to any known second messenger and preference for casein and phosvitin over histones as artificial substrate can be taken as typical features of ubiquitous enzymes referred to as "casein kinases". Based on these criteria they also differentiate from the mammary gland casein kinase(s) involved in the biosynthesis of casein which are inactive on phosvitin and do not bind to phosphocellulose so tightly.

The ability to utilize GTP as phosphate donor, the distinct site specificity, and the diverse chromatographic behavior (especially on DEAE-cellulose and gel filtration) afforded the main criteria for grouping casein kinases into two different subclasses.[10,11] The most con-

TABLE 1
Distinctive Properties of Ubiquitous "Casein Kinases"

Insensitivity to cAMP, cGMP, Ca^{2+}, diacylglycerols, and phospholipids

High activity *in vitro* toward casein and phosvitin

Negligible activity toward histones and protamines

High affinity for phosphocellulose

TABLE 2
Classification and Distinctive Properties of Animal Casein Kinases

	CK-1	CK-2
Elution from DEAE-cellulose	First	Second
Elution from gel filtration	Second	First
M_r (kDa)	28—38	120—180
Subunit composition	Monomer	α (α') (36—42 kDa)
		β (~25 kDa)
Phosphate donor	ATP	ATP and GTP
Residues affected in whole casein	Ser	Mainly Thr
Residue affected in β-casein A^2	Ser-22	Thr-41
Response to 1 μM heparin	Unaffected	Inhibited
Response to polyamines (spermine)	Unaffected	Stimulated

ventional among these criteria — the elution order from DEAE-cellulose — is the one almost universally adopted for the classification of casein kinases. Thus, casein kinases eluted *first* on DEAE-cellulose (albeit *second* on gel filtration), using only ATP as phosphate donor and affecting seryl residues of casein (in particular, Ser-22 in β-casein), are ascribed to the first type (and henceforth will be termed casein kinases-1, or "CK-1s"), whereas the casein kinases more retarded on DEAE-cellulose (but *less* retarded on gel filtration), using GTP as well as ATP as phosphate donor and affecting both threonyl and seryl residues of whole casein (in particular, Thr-41 of β-casein), are ascribed to the second type ("CK-2s"), as summarized in Table 2. Two more functional terminologies based on nucleotide specificity (casein kinases A and G) and on the nature of affected casein residues (casein kinases S and TS) were adopted by other groups,[10,12] but the numeral one nowadays is universally accepted. It should be added, however, that the S/TS nomenclature may suggest the incorrect concept of a different *residue* specificity for the two classes of enzymes, whereas the preference for either seryl or threonyl residues by CK-1s and CK-2s, respectively, is only evident using casein as substrate and, albeit operationally useful, it reflects distinct *site* (rather than *residue*) specificity. It should also be recalled that casein kinases of nuclear origin sharing the main properties of CK-1s and CK-2s are generally indicated as NI and NII,[21] respectively, the term "casein kinase" sometimes being replaced by the more general "protein kinase".

Besides the above distinctive features, additional properties of CK-2 enzymes have been outlined that are not shared by CK-1, like inhibition by submicromolar concentrations of heparin[11,13] and of polyglutamic and polyaspartic acids[14,15] and stimulation by polyamines (especially spermine)[16,18] and by polybasic peptides.[19,20]

II. OLIGOMERIC STRUCTURE OF TYPE-2 CASEIN KINASES

A recurrent feature of CK-2 is a relatively large molecular weight as compared to that of CK-1, a property which is often exploited for separating these two types of enzymes by gel filtration.

Whenever CK-2s of animal origin have been purified to homogeneity and analyzed by SDS-polyacrylamide gel electrophoresis (SDS-PAGE), they invariably have exhibited an oligomeric structure. In most instances they were shown to be heterotetramers of around 130,000 M_r composed of α- (and often α'-) subunits of ca. 40 kDa M_r and β-subunits of about 25 kDa.[11,21-24] Their proposed structures are either $\alpha_2\beta_2$ or $\alpha\alpha'\beta_2$. Similar peptide mappings and immunological responses[11,25] support the view that α and α' are closely related. Furthermore, spontaneous and trypsin-induced conversion of α into slightly smaller polypeptides comigrating with α'[11,25] suggested that α' might derive from α by limited proteolysis. In contrast, calf thymus α and α' are both detectable immunologically under conditions preventing proteolysis,[25] suggesting that they are physiologically relevant individual entities (see also below).

Several lines of evidence indicate that the α-subunits are the catalytic ones[24,26,27] while the β-subunit is devoid of catalytic activity; the latter undergoes fast autophosphorylation through an intramolecular mechanism,[24] and its function is not yet clear. In the presence of polycationic peptides, which also activate CK-2, autophosphorylation of the β-subunit is almost entirely prevented, while a considerable autophosphorylation of the α-subunit(s) takes place.[20]

Such a heterotetrameric structure is a common feature of all typical cytosolic CK-2s of animal origin described so far. In the case of nuclear enzymes, however, besides the canonical $\alpha_2\beta_2$ (or $\alpha\alpha'\beta_2$) CK-2, enzymes with different structures also have been described. Thus, the NII from pig liver, resembling in every other respect an orthodox type-2 casein kinase, revealed a dimeric structure consisting of two identical 95-kDa subunits.[28] More recently, Delpech et al.[29] have described the isolation of a dimeric rat liver CK-NII composed of two 72-kDa subunits by a procedure employing high concentrations of protease inhibitors.[29] They suggest that the heterotetrameric forms of NII described by others in the same material could derive from the homodimeric form through proteolysis, since the latter one, once incubated in the absence of protease inhibitors, gives rise to a 42-kDa fragment still retaining the MgATP binding capacity and a phosphorylatable 27-kDa fragment incapable of binding ATP.[29] While the precise understanding of the possible relationship between the two forms of nuclear NII kinases requires further study, the concept that *cytosolic* heterotetrameric CK-2s described in many reports are not generated by artifact during the isolation procedure is supported by the finding that cytosolic CK-2 from pig liver is an individual heterotetrameric enzyme quite distinct (even for antigenic properties) from its dimeric nuclear counterpart.[30] Consequently, a typical heterotetrameric CK-2 could be isolated from rat liver cytosol even in the presence of the high antiprotease concentrations employed by Delpech et al. for obtaining the dimeric NII from rat liver nuclei.[30a] A notable feature of this preparation, however, was the spontaneous autophosphorylation of an evident α' band which is fainter, if not absent, in the usual preparations. Interestingly, the slightly larger α-subunit is not affected by such an atypical autophosphorylation. This would sharply differentiate it from α', in agreement with the concept that they are distinct entities,[12] and might rather suggest an intriguing relationship between the α'- and β-subunits. In light of this result, anyway, the correspondence between α' and proteolytic derivative(s) of α displaying similar electrophoretic mobilities[11,24] should be reconsidered carefully.

While the subunit composition as well as the distinctive properties of animal CK-2 appear to be highly conserved, even in organisms as phylogenetically remote as mammals and insects,[25] the structure of type-2 casein kinases in non-animal cells could be significantly different. In the case of yeast[31] and molds,[32] typical CK-2s (as judged from their substrate and nucleotide specificities and response to polyanionic inhibitors) have been purified to near homogeneity. They turned out to be oligomeric enzymes of molecular weight similar

TABLE 3
Compounds which Are Known to Affect CK-2 Activity

Effect	Compound	Effective conc. (μM)[a]	Ref.
Inhibition, competitive with respect to protein substrate	Heparin	<0.01	11
	Inositol hexasulfate	40	38
	Pyridoxal phosphate	400	38
	Polyglutamate	2.0[b]	14
	Polyaspartate	0.6[b]	15
	2,3-DPG[c]	1500	41
	Polyphosphate	10	—[d]
Inhibition, competitive with respect to ATP	Quercetin	0.85	12
	DRB[e]	8	39
Stimulation	Polyamines	500	40
	Basic polypeptides	60[b]	19, 24

[a] Expressed by concentrations promoting half-maximal effects.
[b] Variable with the length of the polymer.
[c] 2,3-Bisphosphoglycerate can also act as a stimulator.[41]
[d] Unpublished data.
[e] 5,2-Dichloro-1-D-ribofuranosylbenzimidazole.

to that of animal CK-2, lacking, however, any component reminiscent of the noncatalytic 25-kDa β-subunit of their animal counterparts. Rather, a 39-kDa noncatalytic component, phosphorylatable like animal β-subunit, is associated with the catalytic 37-kDa subunits in *Saccharomyces cerevisiae* CK-2 preparations.[31] Recent immunological evidence obtained with highly purified yeast CK-2 supports the view that such a polypeptide actually represents an unusually large β-subunit.[33] As pointed out before, it cannot be ruled out that the animal CK-2 also might include α'-subunits with similar properties.

A comparative study in plants is often hindered by the difficulty of accommodating their casein kinases into the scheme devised for such enzymes of animal origin. Assuming, however, substrate and site specificity, heparin sensitivity, and GTP utilization as distinctive properties of type-2 casein kinases, one should conclude that CK-2-like enzymes of plants display quite different structures than their animal counterparts. PK_{85}, a nuclear NII casein kinase isolated from tobacco cells,[34] is reported to be a monomer,[35] while CK-II from soybean cotyledons is described as a 120-kDa protein composed of 52-, 37-, and 35-kDa subunits, the 35-kDa form undergoing autophosphorylation.[36] A GTP-utilizing, heparin-inhibited casein kinase from wheat germ appears to be a dimer easily dissociable into 38-kDa subunits.[37] In no instance, to our knowledge, was the 25-kDa autophosphorylatable β-subunit found in most CK-2s from animal tissues described in preparations of CK-2-like enzymes from plants.

III. REGULATION OF TYPE-2 CASEIN KINASES

Independence of any known second messenger is one of the criteria for the definition of casein kinases (see Table 1). Moreover, despite several other compounds being known to affect the activity of CK-2 *in vitro* (see Table 3), in no instance could the involvement of any of them in the physiological modulation of this class of protein kinase be proved unambiguously. One might actually surmise that CK-2 lacks any absolute requirement for specific activators, considering that most of the "uncontrolled" protein radiolabeling detectable in crude extracts incubated with ^{32}P-ATP in the absence of any specific activators is often prevented by heparin and by unlabeled GTP, as if endogenous CK-2 alone were entirely responsible for most of them. In such experiments, however, the concentrations of

CK-2 and of its substrates and effectors, if any exist, are deeply divergent from the intracellular ones. On this matter, it has recently been shown that at physiological ionic strength CK-2 can give rise to filamentous forms whose stability is critically dependent on both enzyme and Mg^{2+} concentrations.[42] Assuming enzyme activity is inhibited by filament formation, as suggested,[42] one could argue that CK-2 which is latent under intracellular conditions could be activated in crude homogenates even by simple dilution.

On the other hand, it should be recalled that the effectiveness of polycationic stimulators of CK-2 is largely dependent on the nature of the protein substrates.[43,44] Thus, the phosphorylation of p90, an endogenous target of rat liver CK-2 recently identified as a heat shock protein,[45] becomes greatly predominant over that of other protein substrates only if polycationic peptides are added;[46] likewise, histones or other basic proteins are absolutely required in order to ensure a fast and stoichiometric phosphorylation of calmodulin and clathrin β-chain, whose radiolabeling by CK-2 is otherwise hardly detectable.[47,48] Apparently, moreover, under intracellular conditions casein kinases are subjected to functional compartmentation through association with multimolecular systems.[46] It is conceivable, therefore, that *in vivo* the indiscriminate protein phosphorylation attributable *in vitro* to CK-2 is prevented by intracellular conditions, while the activity toward selected targets could be triggered by tissue- or even compartment-specific effectors, like polyamines, histones, and related compounds which have been shown to decrease the K_m of CK-2 for certain substrates.[31]

Likewise, modulation of CK-2 could occur through inhibition-deinhibition: macromolecular compounds of either polysaccharidic[49] or peptidic nature[50] capable of inhibiting CK-2 *in vitro* have been isolated from several tissues, and it is expectable that they are structurally related to the two well-known classes of artificial inhibitors exemplified by heparin and polyglutamate (Table 3). Endogenous compounds of this kind are undoubtedly responsible for the underestimation of casein kinases in many crude dialyzed extracts as compared with more purified preparations, but their actual involvement under *in vivo* conditions remains unproven. Other inhibitors listed in Table 3 could play a physiological role under special conditions and in particular organisms and tissues: thus, the relatively high I_{50} of 2,3-bisphosphoglycerate suggests that it may regulate CK-2 activity in erythroid cells,[41] while polyphosphates could play a physiological role in yeast[51] as well as in nuclei of animal cells,[52] where they reach concentrations that are inhibitory of CK-2 activity. In plants even inhibition by quercetin and related flavonoids might prove physiologically relevant, whereas in animals it could explain some of the pharmacological effects of these drugs.[12]

Finally, the possibility should also be mentioned that CK-2 could be controlled through reversible covalent modifications. The microheterogeneity of this enzyme[53] would be consistent with such a hypothesis, and the recent finding that rat liver CK-2 is stimulated significantly by preincubation with protein phosphatases[53a] would suggest a negative control of this enzyme by phosphorylation.

In short, while a single intracellular mediator that uses CK-2 for delivering its message either does not exist or is undiscovered, a multitude of candidates are available that could contribute to a flexible modulation of this pleiotropic enzyme.

IV. ENDOGENOUS SUBSTRATES AND PHYSIOLOGICAL IMPLICATIONS

Until the end of the 1970s the only known phosphorylatable substrates of CK-2 were foreign proteins like casein fractions and phosvitin, obviously unrelated to this enzyme. Actually, reports concerning the phosphorylation of translational components, erythrocyte membrane proteins, and glycogen synthase by cAMP-independent protein kinases also active on phosvitin and/or casein and reported in some cases to use GTP also and to bind tightly

TABLE 4
Possible Physiological Substrates Phosphorylated by CK-2

Protein[a]	Ref.[b]
Translation initiation factors (e.g., eIF-2)	58
HMG[c] protein 17	59
RNA polymerases I and II	60
Glycogen synthase	61, 62
Troponin-T	63, 64
Calsequestrin	62
Regulatory subunit of cAMP-dependent PK-II (R^{II})	65, 66
T-substrate (wheat germ)	67
HMG protein 14	68
Acetyl CoA carboxylase	69
Myosin light chain	70
Ornithine decarboxylase	71
C-proteins (hnRNP particles)	72
Myosin heavy chain (brain)	73
Androgen receptor[d]	74
Protein phosphatase inhibitor-2	85
DNA topoisomerase II	75
DNA topoisomerase I	76
Spermine binding protein	77
Spectrin	78
Glycophorin	78
Clathrin beta light chain	48
Nucleolar protein B23	79
Nucleolar protein C23	80
Fibrinogen	81
Low-density lipoprotein receptor	82
90-kDa heat shock protein	77
Calmodulin	47

[a] Listed chronologically.
[b] Due to the great number of reports, it is possible that some proteins and/or pertinent references have been omitted. In this case we apologize to the authors.
[c] HMG = high-mobility group.
[d] Androgen receptor has been reported to undergo phosphorylation only by the nuclear form of CK-2 (NII).[74]

to phosphocellulose had been published already in contexts other than the study of casein kinases.[54-57] Not always, however, was the identification of such enzymes with typical casein kinases immediately perceived. This helped make the nomenclature of casein kinases more confusing, since terms like troponin-T kinase and glycogen synthase kinase were sometimes adopted to indicate enzymes which turned out to correspond to CK-2.

In 1979, anyway, the identification of CK-2 as one of the independent enzymes responsible for the phosphorylation of initiation factors, notably of eIF-2, was definitely established,[58] and soon HMG protein 17, RNA polymerase II, glycogen synthase, and troponin-T were unambiguously recognized as substrates of CK-2.[59-64] The list of proteins that are phosphorylated by CK-2 grew rapidly in the following years, and it now comprises not fewer than 28 names (Table 4), including key enzymes of several metabolic pathways, regulatory subunits of other protein kinases and of protein phosphatases, translational components and nuclear polypeptides involved in gene expression and protein synthesis, structural proteins, and heat shock proteins.

Only for a few proteins could the actual phosphorylation *in vivo* by CK-2 be documented unambiguously. In several other instances, however, the physiological relevance of such a phosphorylation has been supported more or less strongly by coincidental observations, including kinetic constants compatible with intracellular conditions, copurification of protein substrates with CK-2, overlapping of *in vitro* and *in vivo* ^{32}P-peptide mappings, enhancement of CK-2-dependent phosphorylation after previous dephosphorylation with protein phosphatases, and recurrence of the typical structural determinants of CK-2 at the sites phosphorylated *in vivo*.

Except for RNA polymerases and DNA topoisomerases I and II, whose activities were reported to be stimulated,[60,75,76] and of ornithine decarboxylase, which seems to be inhibited under certain circumstances,[83] the phosphorylation of the other catalytic proteins by CK-2 appears to be "silent" in that it does not induce any appreciable change in activity. It has been pointed out, however, that the phosphorylation of glycogen synthase by CK-2, albeit ineffective per se, potentiates the subsequent phosphorylation by glycogen synthase kinase-3 which in turn dramatically affects the activity of the enzyme.[84] Likewise, a synergistic effect between these two protein kinases has also been observed on protein phosphatase inhibitor-2 (modulator protein)[85] and the regulatory subunit of type II cAMP-dependent protein kinase.[66] It is also conceivable, therefore, that in other instances the "silent" phosphorylation of protein substrates by CK-2 may represent a prerequisite for subsequent modifications leading to evident changes of properties. On this matter it should be recalled that although the glycogen synthase site phosphorylated by CK-2 is somewhat resistant to dephosphorylation,[86] suggesting that it has a structural role, other substrates of CK-2 are readily dephosphorylated, especially by protein phosphatases-2A (polycation stimulated),[87,88] supporting the view that CK-2 can be involved in the reversible induction of short-term effects as well.

In summary, the multiplicity and variety of its substrates, many of obvious importance in the regulation of gene expression and metabolism, together with its ubiquity among eukaryotic organisms, support the view that CK-2 plays a crucial role in the overall integration of cellular metabolism.

V. SPECIFICITY DETERMINANTS

A. PRIMARY STRUCTURE OF THE PHOSPHORYLATION SITES

As long as its physiological targets remained unknown, the specificity of CK-2 was necessarily explored with only the aid of foreign protein substrates. Casein fractions — α_{s1}, α_{s2}, β, κ, and sometimes their genetic variants — proved especially helpful as model substrates, as their primary structures had already been elucidated, thus facilitating the identification of the residues phosphorylated by CK-2 *in vitro*. Among the components of whole bovine casein, α_{s2}-casein and β-casein are greatly preferred as targets over α_{s1}- and κ-casein. Both are affected almost exclusively at threonyl residues identified as Thr-130 of α_{s2}-casein[89] and Thr-41 of β-casein.[90,91]

As shown in Table 5, both of these residues are adjacent to the N-terminal side of three consecutive acidic residues. Thr-49 of α_{s1}-casein, which is N-terminal to just two acidic residues, is also slowly phosphorylated, at a rate, however, that is about one tenth that of β-casein Thr-41.[10] The preference of CK-2 for threonyl over seryl residues of casein fractions is abolished by previous partial dephosphorylation of the protein substrate: such a treatment both increases the phosphorylation efficiency and promotes overwhelming radiolabeling of serines over threonines.[92] The residues affected were localized within the typical Ser(P)–Ser(P)–Ser(P)–Glu–Glu sequences recurrent in most casein fractions; in the case of the 17—19 seryl cluster of β-casein it was also shown that only the first serine undergoes fast rephosphorylation, the second one being slightly radiolabeled while Ser-19 was not affected at all.[93]

TABLE 5
Phosphorylation Sites for CK-2 In Artificial Protein Substrates

Protein	Site[a]
α_{s2}-Casein	130 -Leu-Ser(P)-<u>Thr</u>-Ser(P)-Glu-Glu-Asn-
β-Casein	41 -Gln-Gln-<u>Thr</u>-Glu-Asp-Glu-Leu-
Dephosphorylated α_{s1}-casein	66 -Ile-<u>Ser-Ser-Ser</u>-Glu-Glu-Ser-
Dephosphorylated β-casein	17 -Leu-<u>Ser</u>-Ser-Ser-Glu-Glu-Ile-
Soybean BBI antiprotease	65 -Pro-<u>Ser</u>-Glu-Asp-Asp-Lys-Glu-
Soybean C-II antiprotease	4 -Asp-His-<u>Ser-Ser-Ser</u>-Asp-Asp-Glu-Ser-
Hen phosvitin	21 -<u>Thr</u>-Ala-<u>Thr</u>-Ser(P)-Ser(P)-Ser(P)-Ser(P)-Ser(P)-Ser(P)-

[a] Ser(P) denotes phosphoseryl residues already phosphorylated in the protein substrate before incubation with CK-2. Residues affected by CK-2 are underlined. Dotted underlining indicates that individual residues affected are not identified. Data from References 89 to 95, or unpublished (phosvitin).

From these data it was inferred that serine, whenever available in the suitable sequence, is preferred over threonine and that the structural requirements of CK-2 must be similar (but not identical) to those of the casein kinase(s) that phosphorylate casein within the mammary gland. In particular, the concept that a cluster of acidic residues on the C-terminal side are critically required for site recognition by CK-2 was confirmed using another class of artificial substrates, namely, the low molecular weight soybean antiproteases that include acidic stretches adjacent to seryl residues. The phosphorylation sites for CK-2 in Bowmann-Birk and CII antiproteases,[94,95] also reported in Table 5, actually conform to the pattern of the phosphorylation sites identified in casein fractions to have multiple acidic residues on the C-terminal side of the phosphorylated serines.

As soon as the first "physiological" sites phosphorylated by CK-2 were identified in troponin-T, glycogen synthase, and the regulatory subunit of type II cAMP-dependent protein kinase (Table 6), the prediction based on artificial substrates proved true: invariably, in fact, four or more acidic residues are located on the C-terminal side of the phosphorylated amino acids. The same is true of several other phosphorylation sites for CK-2 subsequently identified in myosin light chain, HMG protein 14, protein phosphatase inhibitor-2, calmodulin, and ornithine decarboxylase (Table 6). Consequently, the finding that two nucleolar phosphoproteins (C23 and B23) included phosphoseryl residues followed by huge clusters of many acidic residues[79,80] was held as an indication that CK-2 is responsible for their phosphorylation. Likewise, it is reasonable to surmise that the phosphorylation sites for CK-2 in spermine binding protein, whose primary structure has recently been deduced from that of its cRNA,[101] are located in the extremely acidic C-terminal domain, probably corresponding to Ser-262 and/or Ser-281, which are both embedded between extended acidic stretches.

The elucidation of the primary structure of phosvitin,[102] a good artificial substrate of CK-2, disclosed the absence of any (Glu/Asp)-n sequences downstream from phosphorylatable residues. Phosphorylation by CK-2, therefore, would be expectable only assuming that the Ser(P)-n clusters typical of phosvitin may act as specificity determinants instead of carboxylic residues. Such a conclusion has been corroborated recently by the identification of two threonyl residues which are phosphorylated by CK-2 in the N-terminal CNBr fragment of phosvitin and which are sitting in the sequence Thr–Ala–Thr–Ser(P)–Ser(P)–Ser(P)–Ser(P)–Ser(P)–Ser(P).[102a]

TABLE 6
Phosphorylation Sites Identified in Putative Physiological Substrates of CK-2

Protein	Amino acid sequence[a]	Ref.
Troponin-T	AcSer-Asp-Glu-Glu-Val-Glu-	63
Glycogen synthase	-Gln-Ser-Glu-Asp-Glu-Glu-Glu-Glu-	96
R_{II}	-Asp-Ser-Glu-Ser-Glu-Asp-Glu-Glu-Glu-	65, 66
Myosin light chain	-Phe-Thr-Asp-Glu-Glu-Val-Asp-Glu-	97
Nucleolar protein B23	-Glu-Ser-Glu-Asp-Glu-Asp-Glu-Glu-Asp-	79
Nucleolar protein C23	-Ala-Ser-Glu-Asp-Glu-Asp-Glu-Glu-Asp-	80
HMG[b] protein 14	-Ala-Ser-Asp-Glu-Ala-Glu-Glu-	98
Inhibitor-2 of protein phosphatase	-Glu-Ser-Ser-Gly-Glu-Glu-Asp-Ser-Asp-	99
Calmodulin	-Asp-Thr-Asp-Ser-Glu-Glu-Glu-Ile-	47
Ornithine decarboxylase	-Gly-Ser-Asp-Asp-Glu-Asp-Glu-Ser-	100

[a] The residues that are phosphorylated by CK-2 are indicated by bold type.
[b] HMG = high-mobility group.

B. STUDIES WITH SYNTHETIC PEPTIDES

While inspection of the primary structure around the residues phosphorylated by CK-2 strongly suggested that a cluster of C-terminal acidic groups must represent a necessary condition for site recognition, studies with synthetic peptides proved that this is a sufficient condition as well and provided detailed information about the minimum structural requirements of CK-2.

Two synthetic hexapeptides were first prepared which are reminiscent of the sites phosphorylated by CK-2 in glycogen synthase and troponin-T — namely, SEEEEE* and acetyl-SEEEVE. Unlike the "basic" peptides affected by either cAMP-dependent protein kinase or protein kinase-C (or both), which are not substrates at all, both of these acidic peptides were phosphorylated readily by CK-2.[15,103] Parallel studies have shown that the same is true of the threonyl peptide RREEETEEE,[104] sharing with the two hexapeptides the presence of a C-terminal acidic block.

Next, more than 20 derivatives of the glycogen synthase and troponin-T analogue peptides were prepared, differing from each other in several respects, including the nature of the phosphorylatable residue, the length and position of the acidic cluster, its distance from the target residue, the substitution of individual glutamyl residues with neutral ones, and

* The amino acid sequences of synthetic peptides are denoted by one-letter abbreviations: A, alanine; D, aspartic acid; E, glutamic acid; K, lysine; L, leucine; M, methionine; P, proline; R, arginine; S, serine; T, threonine; V, valine; Y, tyrosine.

TABLE 7
Acidic Peptides as Substrates and Inhibitors of CK-2

Peptide[a]		V_{max}[b]	K_m (mM)	K_i (mM)
I.	Ser-**Glu-Glu-Glu-Glu-Glu**	100	0.27	0.25
II.	Thr-**Glu-Glu-Glu-Glu-Glu**	10	6.25	1.05
III.	Tyr-**Glu-Glu-Glu-Glu-Glu**	<1	—	0.70
IV.	Ser-**Glu-Glu-Glu-Glu**-Ala	35	4.52	2.30
V.	Ser-**Glu-Glu-Glu**-Ala-Ala	27	5.90	5.71
VI.	Ser-**Glu-Glu**-Ala-Ala-Ala	<1	—	25.00
VII.	Ser-Ala-Ala-**Glu-Glu-Glu**	62	6.60	1.55
VIII.	Ser-**Glu-Glu-Glu**-Ala-**Glu**	100	1.52	1.25
IX.	Ser-**Glu-Glu-Glu**-Val-**Glu**	100	3.23	5.25
X.	AcSer-**Glu-Glu-Glu**-Val-**Glu**	100	3.20	n.d.[c]
XI.	Ser-**Glu-Glu**-Ala-**Glu-Glu**	4	1.71	0.90
XII.	Ser-**Glu**-Ala-**Glu-Glu-Glu**	75	1.95	0.90
XIII.	Ser-Ala-**Glu-Glu-Glu-Glu**	89	5.55	1.10
XIV.	Ser-Ala-**Glu-Glu-Glu-Glu-Glu**	74	1.33	0.93
XV.	Ser-Ala-Ala-**Glu-Glu-Glu-Glu-Glu**	116	1.06	0.70
XVI.	**Glu-Glu-Glu-Glu-Glu**-Ser	<1	—	3.10
XVII.	**Glu-Glu-Glu-Glu-Glu**-Tyr	<1	—	4.70
XVIII.	AcSer-**Glu-Glu-Glu-Glu-Glu**	79	0.64	n.d.[c]
IXX.	Ala-Ser-**Glu-Glu-Glu-Glu-Glu**	117	1.74	0.95
XX.	**Glu**-Ser-**Glu-Glu-Glu-Glu-Glu**	90	0.14	0.51
XXI.	Arg-Ser-**Glu-Glu-Glu**-Val-**Glu**	<1	—	3.98
XXII.	Ala-**Glu-Glu-Glu-Glu-Glu**	—	—	8.19
XXIII.	**Glu-Glu-Glu-Glu-Glu**	—	—	2.35
	(Glu)$_{10}$	—	—	0.14
	(Glu)$_{130}$	—	—	0.0019
	(Asp)$_{90}$	—	—	0.0006

[a] Acidic residues are indicated by bold type.
[b] Values are expressed relative to that of peptide I (970 nmol P/min/mg CK-2). V_{max} < 1 means that the phosphorylation rate determined at 6 mM peptide concentration was <1% of that of peptide I under the same conditions.
[c] n.d. = not determined.

the addition of either a neutral, a basic, or an acidic residue on the N-terminal side. All of them were tested as substrates for CK-2 from either rat liver or other sources, and the kinetic constants were calculated whenever possible. Nonphosphorylatable derivatives and peptides with negligible phosphorylation rates were assayed as competitive inhibitors, and their K_i values were calculated.

The results of this partially published study[105] are summarized in Table 7. They allow one to draw conclusions about the following points:

1. *Nature of the phosphorylatable residue.* By comparing peptides I, II, and III it is evident that serine is greatly preferred over threonine while tyrosine is not affected at all. Substitution of threonine for serine causes both an increase in K_m and a fall in V_{max}, whereas the unsuitability of tyrosine is almost entirely due to its lack of catalytic competency, as the K_i value of the tyrosyl derivative is not much higher than that of the seryl one.
2. *Number of critical acidic residues.* When the number of C-terminal glutamic acids is progressively reduced from five to two by replacing them with alanines, the phos-

phorylation efficiency decreases as well, essentially through an increment of K_m and/or K_i values (compare peptides I, IV, V, VI, and VII). In particular, while SEEEAA is still a phosphorylatable substrate, albeit quite a poor one, no phosphorylation of SEEAAA could be detected at up to 6 mM concentrations and, accordingly, its inhibitory power is also negligible. This would indicate that three C-terminal acidic residues represent the minimum requirement of CK-2. It is still possible, however, that just two acidic residues, at positions other than +1 and +2, might be compatible with phosphorylation as well (see point 3). The synthesis of the peptide SAEEAA is in progress to check this point.

3. *Relevance of individual acidic residues.* This has been studied using a set of hexapeptides having the same amino acid composition (Ser, Glu-4, Ala), yet varying in the position of Ala relative to the N-terminal Ser (peptides IV, VIII, XI, XII, and XIII of Table 7). Clearly, a gap in the C-terminal cluster is more or less harmful depending on its distance from serine: positions +1, +2, and +4 can be occupied by single alanines without any dramatic fall in phosphorylation efficiency; position +5 is more critical, and position +3 is so crucial that the derivative SEEAEE is a much poorer substrate than analogues with three instead of four acidic residues. Such effects are invariably accounted for by differences in V_{max}, the K_m values being very similar for all the tetraglutamyl hexapeptides.

4. *Position of the acidic cluster.* The C-terminal location of the acidic determinants is compulsory, as clearly evidenced by comparing the excellent substrate SEEEEE (I) with the inversely sequenced peptide EEEEES (XVI), which is not a substrate at all. On the other hand, it is not required that the critical acidic residues should be adjacent to the C-terminal side of the target serine, as outlined by comparing, for example, peptides IV and XIII as well as peptides V and VII. Moreover, the good phosphorylation efficiency of the heptapeptide SAEEEEE (XIV) and octapeptide SAAEEEEE (XV) compared with the hexapeptides SAEEEE (XIII) and SAAEEE (VII), respectively, would indicate that the residues at positions +6 and +7 are still interacting with the substrate binding site of CK-2, as they can influence the phosphorylation efficiency.

5. *Influence of N-terminal residues.* The phosphorylation of acetyl-Ser-1 in troponin-T as well as that of synthetic peptides with a single N-terminal serine indicates unambiguously that the specificity determinants of CK-2 must be located on the C-terminal side of the target reside. It is expectable, however, that the nature of residues located on the opposite side also could influence the phosphorylation efficiency. This point was addressed with derivatives of SEEEEE bearing an additional residue bound to the amino group of serine. Apparently, elongation with a neutral residue (Ala) does not improve but actually decreases the phosphorylation efficiency through a remarkable increment of K_m (compare I with IXX). If, however, this N-terminal Ala is replaced with Glu, phosphorylation efficiency is increased even over the control thanks to a drop in K_m to the lowest value obtained with all the peptides of Table 7 (compare IXX and XX). It is likely that the optimizing effect of N-terminal acidic residues is even more evident if their number increases. This could explain the low K_m of the peptide RRREEETEEE, having only three acidic residues on its C-terminal side.[104] Conversely, an adjacent N-terminal Arg is deleterious, giving rise to derivatives that are hardly substrates of CK-2 (compare IX and X with XXI).

To sum up the experiments with model peptide substrates and outline the minimum structural requirement as well as several concurrent factors determining the site specificity of CK-2, the *sine qua non* condition for phosphorylation consists of an acidic cluster (not necessarily continuous) located on the C-terminal side of the target residue, starting from position +1 to +3; optimizing factors are the extension of such a cluster, the occupancy by acidic residues of especially critical positions (namely, the +3 one), and the presence of additional acidic amino acids also on the N-terminal side. Conversely, phosphorylation efficiency is hampered by the replacement of threonine for serine (tyrosine not being a target at all) and by the presence of basic residue(s) adjacent to the C-terminal side of the phosphorylatable amino acid. The effect of basic residues on the C-terminal side is still unexplored; the synthesis of SEEEKE is in progress to check this point.

Among these factors, the overall number of acidic residues is critical for determining the affinity for the substrate binding site by lowering the K_m (or K_i) value, whereas the individual acidic groups at positions +3 and, to a lesser extent, +5 are required for incrementing V_{max}. Both binding efficiency and, even more, phosphorylation rate are influenced by the nature of the phosphorylatable residue.

Although direct evidence is not yet available, the powerful competitive inhibition of CK-2 by polyaspartate[15] and the fast phosphorylation of phosvitin (see Section V.A and Table 5) support the view that either aspartic acid or phosphoserine can replace glutamic acid as an acidic determinant of CK-2-catalyzed phosphorylation.

All the phosphorylation sites identified to date in putative physiological substrates of CK-2 (Table 6) fulfill the structural requirements outlined with synthetic peptide substrates: most of them actually include serines followed by stretches of up to seven acidic residues. In only one case (calmodulin Ser-81) is the number of C-terminal acidic residues not more than three; such an unfavorable situation, however, is compensated for by the presence of two additional acidic residues on the N-terminal side, an optimizing feature that is not so frequent in the other phosphorylation sites. The only site lacking the critical acidic residue at position +3 is that of HMG protein 14, whose phosphorylation rate, however, is just one tenth that of casein.[98] This provides indirect support to the concept that an acidic residue in such a position is truly essential for a fast phosphorylation. On the other hand, the preference for serine over threonine, outlined by comparing peptides SEEEEE and TEEEEE, is corroborated by the observations that ^{32}P-Ser becomes by far predominant over ^{32}P-Thr as soon as the suitable seryl residues of casein are made available by previous dephosphorylation[92] and that, regarding Thr-79 and Ser-81 occupying the same phosphorylation site of calmodulin (Table 6), the latter is phosphorylated much more rapidly than the former[47] despite its less favorable structural determinants (three rather than four C-terminal acidic residues).

The substantial agreement between the structure of phosphorylation sites in physiological substrates and that of short phosphorylatable peptides supports the view that most of the information for target recognition by CK-2 lies in the amino acid sequences close to the phosphorylated residues, especially on the C-terminal side of them. Accordingly, the V_{max} values with the best peptide substrates are quite comparable to those of the corresponding phosphorylatable proteins.[15] The K_m values of peptides, however, are significantly higher, suggesting that peptide substrates may still lack certain structural features that increase the affinity of intact proteins for CK-2 (compare the lowest K_m of peptides — 140 μM, with ESEEEEE — with the K_ms of glycogen synthase, troponin-T, and calmodulin — 11, 34, and 12 μM, respectively[15,47]). This could be due to either conformational factors or, more simply, additional local features not yet reproduced in any of the model peptides. In order to scrutinize this point it will be necessary to elongate the peptide substrates on both sides, but in particular on the N-terminal one, which has been neglected up to now due to the feeling that the most critical determinants are downstream from the target residue.

A more puzzling discrepancy outlined by our studies concerns the high phosphorylation efficiency of casein fractions at sites that, based on model peptides, should be poorly affected by CK-2. This is especially true of the recurrent sequence Ser(P)–Ser(P)–Ser(P)–Glu–Glu, whose first serine is readily rephosphorylated by CK-2 after dephosphorylation, despite the

TABLE 8
Possible Role of Multiple Serines as β-Turn Inducers in Potentiating Phosphorylation by CK-2

Peptide[a]	p_t[b]	Predicted as β-turn	Phosphorylation efficiency (V_{max}/K_m)
E–S–L–[S–S–S–E]–E	1.33	YES	2000[d]
[S–A–A–E]–E–E	0.19	NO	9[e]

[a] Amino acids are indicated by their one-letter symbols. Brackets include the tetrapeptides displaying the highest p_t values.
[b] Calculated according to Chou and Fasman.[106]
[c] Prediction based on a cutoff value of $p_t = 0.75$.[106]
[d] Unpublished results in collaboration with J. W. Perich and R. B. Johns.
[e] Calculated from Table 7.

presence of just two acidic residues on its C-terminal side. Even assuming, in fact, that two acidic residues, one of which lies at the +3 position, are sufficient for ensuring phosphorylation, the competency of such a site should not exceed that of the peptide Ser–Ala–Ala–Glu–Glu–Glu, whose K_m, conversely, is two orders of magnitude higher than the values calculated for casein and casein fractions (20 to 50 μM). Considering, on the other hand, that caseins are relatively unfolded molecules, conformational effects seem quite unlikely. In agreement with this prediction, the synthetic peptide Glu–Ser–Leu–Ser–Ser–Ser–Glu–Glu, reproducing the site of dephosphorylated β-casein, is also an excellent substrate for CK-2.[105a] It is likely, therefore, that the cluster of three consecutive serines is responsible for the high suitability of these sites. Interestingly, the soybean antiprotease CII, an outstanding model substrate for CK-2,[95] is phosphorylated at a very similar site composed of three serines followed by three acidic residues (see Table 5). Such a favorable effect of adjacent serines could be due to their direct participation in the binding process, but also could be explained by the ability of serines to promote the formation of β-bends. Actually, serine is a powerful β-turn inducer according to the empirical predictive model of Chou and Fasman.[106] Consequently, the probability of β-turns is very high at sites like SSSEE, while it becomes negligible if the second and third serines are replaced by alanines (see Table 8).

Very recently the primary structure of the β-subunit of bovine lung CK-2 has been elucidated,[106a] revealing no existence in it of any phosphorylatable residue located on the N-terminal side of acidic sequences, like those determining the phosphorylation of CK-2 substrates. This finding, while accounting for the failure of CK-2 to phosphorylate β-subunits through an *inter*molecular mechanism,[24] would indicate that *intra*molecular autophosphorylation of CK-2 at its β-subunits doesn't need the acidic determinants required for the recognition of external targets. Actually, CK-2 autophosphorylation was also found to be entirely insensitive to competitive inhibition by a large excess of phosphorylatable protein substrate,[24] and the phosphate incorporated by such a process is inaccessible to protein phosphatases.[53a] It is conceivable, therefore, that catalytic site(s) other than those responsible for the phosphorylation of external substrates might be involved in the β-subunit phosphorylation process.

C. COMPARISON WITH CASEIN KINASE-1, MAMMARY GLAND CASEIN KINASE, AND OTHER PROTEIN KINASES

Experiments with both protein and peptide substrates unambiguously show that CK-2 is strictly specific for seryl and (to a lesser extent) threonyl residues, while it does not at all affect tyrosyl residues, even if they are disposed within the most suitable consensus sequences (e.g., in YEEEEE). This is sufficient for distinguishing CK-2 from all tyrosine-

TABLE 9
Grouping of Protein Kinases According to the Nature of Residue(s) Affected and Local Determinants[109]

Residue(s) phosphorylated	Local determinants	Protein kinase
Ser/Thr	Basic residues	cAMP-dependent PK
		cGMP-dependent PK
		Phosphorylase kinase?
		Myosin light chain kinase
		Ca/calmodulin-dependent multi-PK
		Protein kinase C
		Protease-activated PKs
		HSV- and PRV-induced PKs
		dsRNA-activated PK?
Ser/Thr	Acidic residues	Casein kinase-2
		Casein kinase-1
		Mammary gland casein kinase
		PK(s) phosphorylating pepsin, ACTH, etc.?
Tyr	Acidic residues?	Oncogene-coded Tyr-PKs
		Receptor Tyr-PKs
		"Other" Tyr-PKs

Note: PK = protein kinase.

protein kinases (reviewed in Reference 107), which in turn are entirely ineffective on serines and threonines. It also should be noted that although acidic residues are often included into the phosphorylation sites of tyrosine-protein kinases as well, they are mostly located on the N- rather than C-terminal side of the target tyrosine. Accordingly, four tyrosine-protein kinases recently isolated from spleen (partially described in Reference 108) proved inactive toward the peptide YEEEEE (i.e., the tyrosyl derivative of one of the best substrates of CK-2, SEEEEE), while all of them readily phosphorylated the inversely sequenced peptide EEEEEY, whose seryl derivative is in turn totally unaffected by CK-2[108a] (see also Table 7).

On the other hand, the acidic nature of the structural determinants of CK-2 sharply differentiates this enzyme from the numerous "basic-residue-requiring" protein kinases (Table 9), including (among others) cAMP- and cGMP-dependent protein kinases, phosphorylase kinase, myosin light chain kinase and other Ca^{2+}/calmodulin-dependent protein kinases, protein kinase C, the protease-activated H4 kinase, and virus-induced protein (protamine) kinases. As reviewed in Reference 109, all of these enzymes, although exhibiting variable structural requirements, share a preference for sites having basic residues on either the N- or C-terminal side or, eventually, on both sides of the target amino acid. Such sites are all obviously unsuitable for CK-2, whose efficiency depends on acidic residues and is actually compromised by the presence of basic amino acids (see Table 7). Conversely, acidic residues proved harmful for the activity of basic-residue-requiring protein kinases.[110,111] Consequently, it has been shown that the peptide substrates that are readily phosphorylated by cAMP-dependent protein kinase are entirely unaffected by CK-2, and vice versa.[103]

The protein kinases whose specificities are likely to be closer to that of CK-2 are CK-1 and the mammary gland enzyme(s) responsible for the *in vivo* phosphorylation of casein ("physiological" casein kinase[s]). Not only do both of these classes of kinases share with CK-2 the preference for casein, but they also are supposed to recognize acidic determinants

TABLE 10
Phosphorylation Sites for CK-1 in Casein Fractions and Structure of Related Peptides that Proved Inactive as Substrates of CK-1

Peptide sequence	Ref.
Sites phosphorylated[a]	
α_{S2}-Casein: -Ser(P)-Ser(P)-Ser(P)-Glu-Glu-**Ser**-Ile Thr-[b]	112
-Ser(P)-Thr-Ser(P)-Glu-Glu-Asn-**Ser**-Lys-Lys-[b]	112
β-Casein: -Ser(P)-Ser(P)-Ser(P)-Glu-Glu-**Ser**-Ile Thr-Arg-[b]	91, 112
α_{S1}-Casein: -Val-Asn-Glu-Leu-**Ser**-Lys-Asp-Ile-	91, 112[c]
-Gly-Ser(P)-Glu-Ser(P)-**Thr**-Glu-Asp-Gln-	118
Nonphosphorylatable peptides	
Glu-Ser-Glu-Glu-Glu-Glu-Glu	unpub.
Glu-Glu-Glu-Glu-Glu-Ser	unpub.
Arg-Arg-Arg-Glu-Glu-Glu-Thr-Glu-Glu-Glu	104
Glu-Glu-Ser-Pro-Ala-Ser-Asp-Glu-Ala	unpub.
Asp-Ser-Glu-Glu-Asn-Lys	unpub.
Glu-Ser-Leu-Ser-Ser-Ser-Glu-Glu	unpub.

[a] Ser(P) denotes phosphoserines already phosphorylated before incubation with CK-1. Residues affected by CK-1 are indicated with bold type.
[b] These sites are still phosphorylatable once they are included into relatively small proteolytic fragments.[112]
[c] According to this reference this site is poorly affected by CK-1.

located on either the N-[91,112] or C-terminal side of serine,[113] respectively. Together with CK-2 and possibly with other as yet unidentified enzymes responsible, for example, for the phosphorylation of fibrinogen, pepsin, ovalbumin, and ACTH at sites bearing an acidic residue at the +2 position,[114-117] they could be ascribed to an individual group of serine- or threonine-specific, "acidic-residue-requiring" protein kinases (see Table 9).

Despite the fact that several substrates of CK-2 besides casein also are affected by CK-1 (e.g., glycogen synthase, acetyl CoA carboxylase, initiation factors, myosin light chain, spermine binding protein, etc.), the site specificities of these two enzymes, deduced from studies with model substrates, appear to be quite distinct. In particular, they affect different residues of casein fractions, with the only exception being Thr-49 of α_{s1}-casein, a modest substrate of CK-2[10] whose phosphorylation by a CK-1 preparation from yeast also has been reported.[118] Such a residue, however, is quite peculiar in that it has acidic residues on both its sides (Table 10). Actually, the finding that the other residues affected by CK-1 in casein fractions are located downstream from acidic sequences (see Table 10) suggested that the acidic determinants of this enzyme must be located on the N-terminal side of the target residues. It was not surprising, therefore, that none of the model peptides readily phosphorylated by CK-2 for having extended acidic stretches on their C-terminal side are affected at all by CK-1.

Rather unexpectedly, however, CK-1 also proved fully inactive toward peptides having one to five acidic residues on the N-terminal side of the phosphorylatable amino acid or on both sides of it, such as the ones listed in Table 10. In this respect CK-1 represents a unique example of a protein kinase still in search of synthetic peptide substrates despite being proved active on small proteolytic fragments of proteins.[10,112]

Such a high selectivity of CK-1, as outlined by experiments with artificial substrates, seems to be flatly contradicted by the ability of CK-1 to incorporate huge amounts of phosphate into some putative endogenous substrates: at least six phosphates/subunits are incorporated by CK-1 into glycogen synthase,[119] and more than 40 residues have been reported to undergo phosphorylation by CK-1 in MAP-2,[120] an astonishing performance indeed for an enzyme so fastidious about its artificial substrates. The matter is further complicated by the identification of some residues of skeletal muscle glycogen synthase which are stoichiometrically phosphorylated by CK-1 despite their lack of any acidic residues in the surrounding sequence.[121]

Unlike CK-1, the mammary gland casein kinase(s) display a site specificity partially overlapping that of CK-2. The first indication of this was provided by the ability of CK-2 to rephosphorylate some of the seryl residues already phosphorylated in native casein, as soon as they are made available by partial dephosphorylation.[92] In more detail it was shown that the Ser(P)–Ser(P)–Ser(P)–Glu–Glu sequence of β-casein, once dephosphorylated, is radiolabeled by CK-2 at its first serine.[93]

Recently, a casein kinase preparation from the Golgi membranes of rat mammary gland[122] has been assayed on several peptide substrates of CK-2:[123] while SEEEEE is also phosphorylated very well by the mammary gland casein kinase, the peptides SAAEEE and SEAEEE are phosphorylated only by CK-2. Conversely, the peptide SEEAAA, which is not a substrate for CK-2, is readily phosphorylated by the Golgi casein kinase. These data definitely prove the validity of the hypothesis that an acidic residue at position +2 is the most critical determinant of phosphorylation by "physiological" casein kinase(s); this was inferred from the observation that all sites phosphorylated in native casein fractions invariably display the sequence Ser(Thr)–X–Glu(Asp,Ser–P).[113]

Interestingly, such a phosphorylated triplet is also found in proteins other than casein, such as pepsin,[115] fibrinogen,[114] ovalbumin,[116] and ACTH.[117] Whether or not the protein kinase(s) responsible for these phosphorylations are related to the mammary gland casein kinase remains an open question.

In conclusion, the site specificity of CK-2 appears to be different from that of any other known protein kinase, being similar (but not identical) to that of the physiological mammary gland casein kinase and of possibly related protein kinases requiring an acidic residue at position +2. Ideal substrates for the specific detection of CK-2 activity would be peptides like SEAEEE and SAAEEEEE, whose C-terminal acidic cluster occupies position +3, but not +2. On the other hand, it should be mentioned that CK-2 from sources other than rat liver, like chicken liver,[102a] yeast,[31] and maize seedlings,[122a] exhibit the same specificity toward peptide substrates, although their structural properties are quite different.

ADDENDUM

In the 2 years that have elapsed since the completion of this review article (April 1987), considerable new relevant information has become available about the structure, the physiological commitments, and the specificity of CK-2.

cDNAs encoding the α- and β-subunits of *Drosophila melanogaster* CK-2 have been isolated and the complete amino acid sequences of both polypeptides determined by DNA sequencing.[123] The α-subunit contains the expected residues present in the catalytic domains of previously sequenced protein kinases, thus confirming that it represents the catalytic subunit. While it is 64% identical to the 42-kDa catalytic subunit of yeast CK-2, thus strengthening the concept that this enzyme is highly conserved over large evolutionary distances, it appears to be related only distantly to other well-characterized members of the protein kinase family.[123] The sequence of the β-subunit of *Drosophila* CK-2 is 86% and 88% identical to those of calf lung[106a] and human[124] enzymes, respectively, while not exhibiting extensive homology with any other protein whose sequence is currently known. Unlike the calf β-subunit, the *Drosophila* and human β-subunits share a potential phosphorylation site for CK-2 at their N-terminal end (M_1SSSEEV) with the motif SSSEE, which is recurrent at many phosphoacceptor sites of casein fractions. As mentioned earlier (Table 9), this motif is readily phosphorylated by CK-2 at its first serine. Thus, it is likely to represent the site of intramolecular autophosphorylation of CK-2.

The abundance of CK-2 in mouse tumor cell nucleoli with α- and β-subunits functionally and immunologically indistinguishable from extranucleolar and cytoplasmic CK-2[125] argues against the concept[29,30] that independent form(s) of nuclear CK-2 with different subunit structure may exist.

Several new members have recently joined the growing list of potential substrates of CK-2, including the β-subunit of insulin receptor,[126] a 62-kDa viral protein detected in human fibroblasts infected with pseudorabies virus,[127] presynaptic protein B-50 (GAP-43) of brain cortex,[128] the keratin matrix protein filaggrin,[129] and a number of proteins whose identities are still uncertain (for example, see References 130 to 132).

In parallel with the discovery of new potential targets for CK-2, the concept that this pleiotropic enzyme plays a central role in the regulation of cellular functions has been strengthened by a variety of observations, namely:

1. The increment of CK-2 activity in transformed[133] and proliferating[134,135] cells
2. The involvement of this enzyme in the altered phosphorylation observed upon treatment with tumor promoter okadaic acid[132]
3. Its developmental changes in lamb adrenals[136] and rat liver[137]
4. Its induction during differentiation of 3T3-L1 cells into adipocytes[132]
5. The ability of CKII microinjection into *Xenopus* oocytes to prevent progesterone-induced meiotic cell division while facilitating meiotic maturation[139]
6. The correlation of CK-2 activity with a variety of hormonal stimuli[140-143]

In this connection the finding that CK-2 is among the serine-/threonine-specific protein kinases stimulated in response to insulin receptor signaling is of special interest.[144] In accord with this, an acetyl CoA carboxylase site affected *in vitro* by CK-2 is the same as that phosphorylated upon stimulation of intact adipocytes with insulin.[142]

The synthesis and assay of new peptide substrates has resulted in additional information about the specificity and structural requirements of CK-2. Thus, using peptides with a single glutamic acid situated at different positions from the C-terminal side of serine, it could be demonstrated that an acidic residue at position $+3$ is both necessary (Table 7) and sufficient for phosphorylation,[145] the peptide SAAEAA (yet not SAEAAA) still being phosphorylated appreciably by CK-2, albeit with a 60-fold higher K_m and a 15-fold lower V_{max} than the reference peptide SEEEEE. This accounts for the finding that the phosphoacceptor site for CK-2 in the β-subunit of eIF-2 includes a single glutamic acid at position $+3$[146] instead of a more extended acidic cluster, as is the case for most physiological targets of CK-2 (Table 6). Conversely, the casein kinase of the Golgi-enriched fraction of mammary gland, GEF-CK, exhibits an absolute requirement for an acidic residue at position $+2$ rather than $+3$, and it phosphorylates SAEAAA, but not SAAEAA.[147] Basic residues, which have been shown to be deleterious if adjacent on the N-terminal side of serine (see Table 7), also reduce the phosphorylation efficiency of CK-2 when located within the C-terminal cluster at positions where neutral residues are not so harmful, the peptide SEEEKE exhibiting a threefold lower V_{max} and sixfold higher K_m than SEEEAE and SEEEVE.[145]

The possibility that aspartic acid could replace glutamic acid as a positive determinant, as suggested by its recurrence at several phosphoacceptor sites for CK-2 (Table 6), has been checked by comparing the phosphorylation efficiencies of the peptides RRRDDDSDDD and RRREEESEEE;[148] the former turned out to be an even better substrate than the latter by virtue of its threefold lower K_m. On the other hand, the capacity of phosphorylated residues to act as specificity determinants instead of carboxylic amino acids has been assessed by showing that partial dephosphorylation converts the 34-kDa component of phosvitin, lacking –S–X–X–E/D– motifs but including stretches of up to 14 consecutive phosphoserines, and the phosphopeptide AcSer(P)–Ser(P)–Ser(P)NHMe into excellent substrates for CK-2.[149,150] This finding discloses the possibility that previous phosphorylation by another protein kinase may trigger the subsequent phosphorylation by CK-2, which in turn has been shown to convert synthetic peptides into suitable substrates for glycogen synthase kinase-3 (F_A kinase).[151]

Rather unexpectedly, a type 2 casein kinase from maize seedlings (CK-IIB) displays a peptide substrate specificity similar to (but not identical with) rat liver CK-2;[152] while it shares the absolute requirement for an acidic residue at the third position from the C-terminal side of serine, it also needs an acidic residue at position +1 which is not so critical with CK-2. Thus, the peptides SAEEEEE and SAAEEEEE, which are good substrates for rat liver CK-2 (see Table 7), are affected very poorly by maize CK-IIB.[152] These findings highlight the possibility that type 2 casein kinases from different organisms may exhibit variable substrate specificity.

ACKNOWLEDGMENTS

This work was made possible by financial aid from Italian Ministero della Pubblica Istruzione and Consiglio Nazionale delle Ricerche (grants 85.00522.04 and 86.00121.04 and Progetto finalizzato Oncologia).

REFERENCES

1. **Burnett, G. and Kennedy, E. P.**, The enzymatic phosphorylation of proteins, *J. Biol. Chem.*, 211, 969, 1954.
2. **Rabinowitz, M. and Lipmann, F.**, Reversible phosphate transfer between yolk phosphoprotein and adenosine triphosphate, *J. Biol. Chem.*, 235, 1043, 1960.
3. **Krebs, E. G. and Fischer, E. H.**, The phosphorylase *b* to *a* converting enzyme of rabbit skeletal muscle, *Biochim. Biophys. Acta*, 20, 150, 1956.
4. **Walsh, D. A., Perkins, J. P., and Krebs, E. G.**, An adenosine 3′,5′-monophosphate dependent protein kinase from rabbit skeletal muscle, *J. Biol. Chem.*, 243, 3763, 1968.
5. **Rodnight, R. and Lavin, B. E.**, Phosvitin kinase from brain: activation by ions and subcellular distribution, *Biochem. J.*, 93, 84, 1964.
6. **Pinna, L. A., Baggio, B., Moret, V., and Siliprandi, N.**, Isolation and properties of a protein kinase from rat liver microsomes, *Biochim. Biophys. Acta*, 178, 199, 1969.
7. **Takeda, M., Yamamura, H., and Ohga, Y.**, Phosphoprotein kinases associated with rat liver chromatin, *Biochem. Biophys. Res. Commun.*, 42, 103, 1971.
8. **Pinna, L. A., Donella, A., and Moret, V.**, Characterization of cytosol phosphopeptides, *FEBS Lett.*, 37, 183, 1973.
9. **Baggio, B., Pinna, L. A., Moret, V., and Siliprandi, N.**, A simple procedure for the purification of rat liver phosvitin kinase, *Biochim. Biophys. Acta*, 212, 515, 1970.
10. **Pinna, L. A., Meggio, F., and Donella-Deana, A.**, Structure of the sites undergoing phosphorylation by protein kinases, with special reference to "casein kinases", in *Protein Phosphorylation and Bioregulation*, Thomas, G., Podesta, E. G., and Gordon, J., Eds., S. Karger, Basel, 1980, 8.
11. **Hathaway, G. M. and Traugh, J. A.**, Casein kinases: multipotential protein kinases, *Curr. Top. Cell. Regul.*, 21, 101, 1982.

12. **Cochet, C., Feige, J. J., Pirollet, F., Keramidas, M., and Chambaz, E. M.**, Selective inhibition of a cyclic nucleotide independent protein kinase (G type casein kinase) by quercetin and related polyphenols, *Biochem. Pharmacol.*, 31, 1357, 1982.
13. **Meggio, F., Donella-Deana, A., Brunati, A. M., and Pinna, L. A.**, Inhibition of rat liver cytosol casein kinases by heparin, *FEBS Lett.*, 141, 257, 1982.
14. **Meggio, F., Pinna, L. A., Marchiori, F., and Borin, G.**, Polyglutamyl peptides: a new class of inhibitors of type-2 casein kinases, *FEBS Lett.*, 162, 235, 1983.
15. **Meggio, F., Marchiori, F., Borin, G., Chessa, G., and Pinna, L. A.**, Synthetic peptides including acidic clusters as substrates and inhibitors of rat liver casein kinase TS (Type-2), *J. Biol. Chem.*, 259, 14576, 1984.
16. **Maenpaa, P. H.**, Effects of polyamines and polyanions on a cyclic nucleotide-independent protein kinase, *Biochim. Biophys. Acta*, 498, 294, 1977.
17. **Cochet, C., Job, D., Pirollet, F., and Chambaz, E. M.**, Adenosine 3′,5′-monophosphate-independent protein kinase activities in the bovine adrenal cortex cytosol, *Endocrinology*, 106, 750, 1980.
18. **Hathaway, G. M. and Traugh, J. A.**, Interactions of polyamines and magnesium with casein kinase II, *Arch. Biochem. Biophys.*, 233, 133, 1984.
19. **Yamamoto, M., Criss, W. E., Yoshimi, T., Yamamura, H., and Nishizuka, Y.**, A hepatic soluble cyclic nucleotide-independent protein kinase. Stimulation by basic polypeptides, *J. Biol. Chem.*, 254, 5049, 1979.
20. **Meggio, F., Brunati, A. M., and Pinna, L. A.**, Autophosphorylation of type-2 casein kinase-TS at both its α- and β-subunits, *FEBS Lett.*, 160, 203, 1983.
21. **Thornburg, W. and Lindell, T. J.**, Purification of rat liver nuclear protein kinase NII, *J. Biol. Chem.*, 252, 6660, 1977.
22. **Cochet, C. and Chambaz, E. M.**, Oligomeric structure and catalytic activity of a G type casein kinase, *J. Biol. Chem.*, 258, 1403, 1983.
23. **Glover, C. V. C., Shelton, E. R., and Brutlag, D. L.**, Purification and characterization of a type II casein kinase from *Drosophila melanogaster*, *J. Biol. Chem.*, 258, 3258, 1983.
24. **Meggio, F. and Pinna, L. A.**, Subunit structure and autophosphorylation mechanism of casein kinase-TS (type-2) from rat liver cytosol, *Eur. J. Biochem.*, 145, 593, 1984.
25. **Dahmus, G. K., Glover, C. V. C., Brutlag, D. L., and Dahmus, M. E.**, Similarities in structure and function of calf thymus and *Drosophila* casein kinase II, *J. Biol. Chem.*, 259, 9001, 1984.
26. **Hathaway, G. M., Zoller, M. H., and Traugh, J. A.**, Identification of the catalytic subunit of casein kinase II by affinity labeling with 5′-*p*-fluorosulfonylbenzoyl adenosine, *J. Biol. Chem.*, 256, 11442, 1981.
27. **Feige, J.-J., Cochet, C., Pirollet, F., and Chambaz, E. M.**, Identification of the catalytic subunit of an oligomeric casein kinase (G type). Affinity labeling of the nucleotide site using 5′-[*p*-(fluorosulfonyl)benzoyl] adenosine, *Biochemistry*, 22, 1452, 1983.
28. **Baydoun, H., Hoppe, J., Jacob, J., and Wagner, K. G.**, Purification and quaternary structure of a cyclic nucleotide-independent protein kinase (NII) from porcine liver nuclei, *FEBS Lett.*, 122, 231, 1980.
29. **Delpech, M., Levy-Favatier, F., Moisand, F., and Kruh, J.**, Rat liver nuclear protein kinases NI and NII. Purification, subunit composition, substrate specificity, possible levels of regulation, *Eur. J. Biochem.*, 160, 333, 1986.
30. **Baydoun, H., Feth, F., Hoppe, J., Erdmann, H., and Wagner, K. G.**, The acidic peptide specific type II protein kinases from the nucleus and the cytosol of porcine liver are distinct, *Arch. Biochem. Biophys.*, 245, 504, 1986.
30a. **Agostinis, P. and Pinna, L. A.**, unpublished data.
31. **Meggio, F., Grankowski, N., Kudlicki, W., Szyszka, R., Gasior, E., and Pinna, L. A.**, Structure and properties of casein kinase-2 from *Saccharomyces cerevisiae*. A comparison with the liver enzyme, *Eur. J. Biochem.*, 159, 31, 1986.
32. **Ranart, M. F., Sastre, L., and Sebastian, J.**, Purification and properties of a cAMP-independent protein kinase from *Dictyostelium discoideum*, *Eur. J. Biochem.*, 140, 47, 1984.
33. **Padmanabha, R. and Glover, C. V. C.**, Casein kinase II of yeast contains two distinct polypeptides and an unusually large β-subunit, *J. Biol. Chem.*, 262, 1829, 1987.
34. **Erdmann, H., Bocher, M., and Wagner, K. G.**, Two protein kinases from nuclei of cultured tobacco cells with properties similar to the cyclic nucleotide-independent enzymes (NI and NII) from animal tissue, *FEBS Lett.*, 137, 245, 1982.
35. **Erdmann, H., Bocher, M., and Wagner, K. G.**, The acidic peptide specific protein kinases from suspension cultured tobacco cells: a comparison of the enzymes from whole cells and isolated nuclei, *Plant Sci.*, 41, 81, 1985.
36. **Godwa, S. and Pillay, D. T. N.**, cAMP independent protein kinases from soybean cotyledons (*Glicine max* L.), *Plant Sci. Lett.*, 25, 49, 1982.
37. **Yan, T.-F. J. and Tao, M.**, Purification and characterization of a wheat germ protein kinase, *J. Biol. Chem.*, 257, 7037, 1982.

38. **Hathaway, G. M. and Traugh, J. A.**, Casein kinase II, *Methods Enzymol.*, 99, 317, 1983.
39. **Zandomeni, R. and Weinmann, R.**, Inhibitory effect of 5,6-dichloro-1-beta D-ribofuranosylbenzimidazole on a protein kinase, *J. Biol. Chem.*, 259, 14804, 1984.
40. **Hathaway, G. M. and Traugh, J. A.**, Kinetics of activation of casein kinase II by polyamines and reversal of 2,3-bisphosphoglycerate inhibition, *J. Biol. Chem.*, 259, 7011, 1984.
41. **Hathaway, G. M. and Traugh, J. A.**, Regulation of casein kinase II by 2,3-bisphosphoglycerate in erythroid cells, *J. Biol. Chem.*, 259, 2850, 1984.
42. **Glover, C. V. C.**, A filamentous form of *Drosophila* casein kinase II, *J. Biol. Chem.*, 261, 14349, 1986.
43. **Linnala-Kankkunen, A., Palvimo, J., and Maenpaa, P.**, Polyamines and heparin do not appreciably influence phosphorylation of chromatin proteins HMG 14 and HMG 17 by nuclear protein kinase II, *Biochim. Biophys. Acta,* 799, 122, 1984.
44. **Ahmed, K., Goueli, S. A., and Williams-Ashman, H. G.**, Characteristics of polyamine stimulation of cyclic nucleotide-independent protein kinase reaction, *Biochem. J.*, 232, 767, 1985.
45. **Dougherty, J. J., Rabideau, D. A., Iannotti, A. M., Sullivan, W. P., and Toft, D. O.**, Identification of the 90 kDa substrate of rat liver type II casein kinase with the heat shock protein which binds steroid receptors, *Biochim. Biophys. Acta,* 927, 74, 1987.
46. **Meggio, F., Agostinis, P., and Pinna, L. A.**, Casein kinases and their protein substrates in rat liver cytosol: evidence for their participation in multimolecular systems, *Biochim. Biophys. Acta,* 846, 248, 1985.
47. **Meggio, F., Brunati, A. M., and Pinna, L. A.**, Polycation dependent, Ca^{2+} antagonized phosphorylation of calmodulin by casein kinase-2 and a spleen tyrosine protein kinase, *FEBS Lett.*, 215, 261, 1987.
48. **Bar-Zvi, D. and Branton, D.**, Clathrin-coated vesicles contain two protein kinase activities. Phosphorylation of clathrin β-light chain by casein kinase II, *J. Biol. Chem.*, 261, 9614, 1986.
49. **Pirollet, F., Feige, J.-J., Cochet, C., Job, D., and Chambaz, E. M.**, Identification of a specific endogenous inhibitor of casein kinase (G type) in bovine adrenal cortex as a glycosaminoglycan mixture, *Biochem. Biophys. Res. Commun.*, 100, 613, 1981.
50. **Levy-Favatier, F., Delpech, M., Riffe, A., and Kruh, J.**, A 25,000 Dalton inhibitor of cAMP independent protein kinases present in rat liver HMG protein preparations, *Biochem. Biophys. Res. Commun.*, 130, 149, 1985.
51. **Tijssen, J. P. F., Dubbleman, T. M. A. R., and Van Stevenick, J.**, Isolation and characterization of polyphosphates from the yeast cell surface, *Biochim. Biophys. Acta,* 760, 143, 1983.
52. **Offenbacher, S. and Kline, E. S.**, Evidence for polyphosphate in phosphorylated non-histone nuclear proteins, *Arch. Biochem. Biophys.*, 231, 114, 1984.
53. **Qi, S.-L., Yukioka, M., Morisawa, S., and Inoue, A.**, Heterogeneity of protein kinase NII multiple subunit-polypeptides, *FEBS Lett.*, 203, 104, 1986.
53a. **Agostinis, P., Goris, J., Pinna, L. A., and Merlevede, W.**, Regulation of casein kinase-2 by phosphorylation-dephosphorylation, *Biochem. J.*, 248, 785, 1987.
54. **Traugh, J. A., Tahara, S. M., Sharp, S. B., Safer, B., and Merrick, W. C.**, Factors involved in initiation of haemoglobin synthesis can be phosphorylated *in vitro*, *Nature (London)*, 263, 163, 1976.
55. **Issinger, O.-G., Benne, R., Hershey, J. W. B., and Trout, R. R.**, Phosphorylation *in vitro* of eukariotic initiation factors IF-E2 and IF-E3 by protein kinases, *J. Biol. Chem.*, 251, 6471, 1976.
56. **Fairbanks, G., Avruch, J., Dino, J. E., and Patel, V. P.**, Phosphorylation and dephosphorylation of spectrin, *J. Supramol. Struct.*, 9, 97, 1978.
57. **De Paoli-Roach, A. A., Roach, P. J., and Larner, J.**, Multiple phosphorylation of rabbit skeletal muscle glycogen synthase, *J. Biol. Chem.*, 254, 12062, 1979.
58. **Hathaway, G. M., Lundak, T. S., Tahara, S. M., and Traugh, J. A.**, Isolation of protein kinases from reticulocytes and phosphorylation of initiation factors, *Methods Enzymol.*, 60, 495, 1979.
59. **Inoue, A., Tei, Y., Hasuma, T., Yukioka, M., and Morisawa, S.**, Phosphorylation of HMG 17 by protein kinase NII from rat liver cell nuclei, *FEBS Lett.*, 117, 68, 1980.
60. **Dahmus, M. E.**, Phosphorylation of eukariotic DNA-dependent RNA polymerase, *J. Biol. Chem.*, 256, 3332, 1981.
61. **De Paoli-Roach, A. A., Ahmad, Z., and Roach, P. J.**, Characterization of a rabbit skeletal muscle protein kinase ($PC_{0.7}$) able to phosphorylate glycogen synthase and phosvitin, *J. Biol. Chem.*, 256, 8955, 1981.
62. **Meggio, F., Donella-Deana, A., and Pinna, L. A.**, Endogenous phosphate acceptor proteins for rat liver cytosolic casein kinases, *J. Biol. Chem.*, 256, 11958, 1981.
63. **Pinna, L. A., Meggio, F., and Dediukina, M.**, Phosphorylation of troponin-T by casein kinase TS, *Biochem. Biophys. Res. Commun.*, 100, 449, 1981.
64. **Villar-Palasi, C. and Kumon, A.**, Purification and properties of a dog cardiac troponin-t kinase, *J. Biol. Chem.*, 256, 7409, 1981.
65. **Carmichael, D. F., Geahlen, R. L., Allen, S. M., and Krebs, E. G.**, Type II regulatory subunit of cAMP-dependent protein kinase. Phosphorylation by casein kinase II at a site that is also phosphorylated *in vivo*, *J. Biol. Chem.*, 257, 10440, 1982.

66. **Hemmings, B. A., Aitken, A., Cohen, P., Rymond, M., and Hofmann, F.**, Phosphorylation of the type II regulatory subunit of cAMP-dependent protein kinase by glycogen synthase kinase 3 and glycogen synthase kinase 5, *Eur. J. Biochem.*, 127, 473, 1982.
67. **Yan, T.-F. and Tao, M.**, Studies on an endogenous substrate of wheat germ protein kinase, *J. Biol. Chem.*, 257, 7044, 1982.
68. **Walton, G. M. and Gill, G. N.**, Identity of the *in vivo* phosphorylation site in high mobility group 14 protein in HeLa cells with the site phosphorylated by casein kinase II *in vitro*, *J. Biol. Chem.*, 258, 4440, 1983.
69. **Tipper, J. P., Bacon, G. W., and Witters, L. A.**, Phosphorylation of acetyl-coenzyme A carboxylase by casein kinase I and casein kinase II, *Arch. Biochem. Biophys.*, 227, 386, 1983.
70. **Matsumura, S., Murakami, N., Tashiro, Y., Yasuda, S., and Kumon, A.**, Identification of calcium-independent myosin kinase with casein kinase II, *Arch. Biochem. Biophys.*, 227, 125, 1983.
71. **Meggio, F., Flamigni, F., Caldarera, C. M., Guarnieri, C., and Pinna, L. A.**, Phosphorylation of rat heart ornithine decarboxylase by type-2 casein kinase, *Biochem. Biophys. Res. Commun.*, 122, 997, 1984.
72. **Holcomb, E. R. and Friedman, D. L.**, Phosphorylation of the c-proteins of HeLa cell hnRNP particles: involvement of a casein kinase-II type enzyme, *J. Biol. Chem.*, 259, 31, 1984.
73. **Murakami, N., Matsumura, S., and Kumon, A.**, Purification and identification of myosin heavy chain kinase from bovine brain, *J. Biochem. (Tokyo)*, 95, 651, 1984.
74. **Goueli, S. A., Holtzman, J. L., and Ahmed, K.**, Phosphorylation of the androgen receptor by a nuclear cAMP-independent protein kinase, *Biochem. Biophys. Res. Commun.*, 123, 778, 1984.
75. **Ackerman, P., Glover, C. V. C., and Osheroff, N.**, Phosphorylation of DNA topoisomerase II by casein kinase II: modulation of eukaryotic topoisomerase II activity *in vitro*, *Proc. Natl. Acad. Sci. U.S.A.*, 82, 3164, 1985.
76. **Durban, E., Doodenough, M., Mills, J., and Bush, H.**, Topoisomerase I phosphorylation *in vitro* and in rapidly growing Novikoff hepatoma cells, *EMBO J.*, 4, 2921, 1985.
77. **Goueli, S. A., Davis, A. T., Hiipakka, R. A., Liao, S., and Ahmed, K.**, Polyamine stimulated phosphorylation of prostatic spermine-binding protein is mediated only by cAMP-independent protein kinase, *Biochem. J.*, 230, 293, 1985.
78. **Clari, G. and Moret, V.**, Phosphorylation of membrane proteins by cytosolic casein kinases in human erythrocytes. Effect of monovalent ions, 2,3-bisphosphoglycerate and spermine, *Mol. Cell. Biochem.*, 68, 181, 1985.
79. **Chan, P.-K., Aldrich, M., Cook, R. G., and Busch, H.**, Amino acid sequence of protein B23 phosphorylation site, *J. Biol. Chem.*, 261, 1868, 1986.
80. **Mamrack, M. D., Olson, M. O. J., and Bush, H.**, Amino acid sequence and sites of phosphorylation in a highly acidic region of nucleolar nonhistone protein C23, *Biochemistry*, 18, 3381, 1979.
81. **Guash, M. D., Plana, M., Pena, J. M., and Itarte, E.**, Phosphorylation of fibrinogen by casein kinase 2, *Biochem. J.*, 234, 523, 1986.
82. **Kishimoto, A., Brown, M. S., Slaughter, C. A., and Goldstein, J. L.**, Phosphorylation of serine 833 in cytoplasmic domain of low density lipoprotein receptor by a high molecular weight enzyme resembling casein kinase II, *J. Biol. Chem.*, 262, 1344, 1987.
83. **Donato, N. J., Ware, C. F., and Byus, C. V.**, A rat monoclonal antibody which interacts with mammalian ornithine decarboxylase at an epitope involved in phosphorylation, *Biochim. Biophys. Acta*, 884, 370, 1986.
84. **Picton, C., Woodgett, J., Hemmings, B., and Cohen, P.**, Multisite phosphorylation of glycogen synthase from rabbit skeletal muscle. Phosphorylation of site 5 by glycogen synthase kinase-5 (casein kinase II) is a prerequisite for phosphorylation of site 3 by glycogen synthase kinase-3, *FEBS Lett.*, 150, 191, 1982.
85. **De Paoli-Roach, A. A.**, Synergistic phosphorylation and activation of ATP-Mg-dependent phosphoprotein phosphatase by F_A/GSK-3 and casein kinase II, *J. Biol. Chem.*, 259, 12144, 1984.
86. **Ingebritsen, T. S. and Cohen, P.**, The protein phosphatases involved in cellular regulation. Classification and substrate specificities, *Eur. J. Biochem.*, 132, 255, 1983.
87. **Thoen, C., Van Hove, L., Cohen, P., and Slegers, H.**, Identification of protein phosphatases dephosphorylating mRNP proteins from cryptobiotic gastrulae of the brine shrimp, *A. salina*, *Biochem. Biophys. Res. Commun.*, 131, 84, 1985.
88. **Agostinis, P., Goris, J., Vandenheede, J. R., Waelkens, E., Pinna, L. A., and Merlevede, W.**, Phosphorylation of the modulator protein of the ATP, Mg-dependent protein phosphatase by casein kinase TS. Reversal by PCS phosphatases and control by distinct phosphorylation site(s), *FEBS Lett.*, 207, 167, 1986.
89. **Meggio, F., Donella-Deana, A., and Pinna, L. A.**, Phosphorylation of α_{s2}-casein by two rat liver "casein kinases", *FEBS Lett.*, 91, 216, 1978.
90. **Pinna, L. A., Donella-Deana, A., and Meggio, F.**, Structural features determining the site specificity of a rat liver cAMP-independent protein kinase, *Biochem. Biophys. Res. Commun.*, 87, 114, 1979.
91. **Tuazon, P. T., Bingham, E., and Traugh, J. A.**, Cyclic nucleotide independent protein kinases from rabbit reticulocytes. Site-specific phosphorylation of casein variants, *Eur. J. Biochem.*, 94, 497, 1979.

92. **Donella-Deana, A., Meggio, F., and Pinna, L. A.**, Phosphorylation of threonine and serine residues of native and partially dephosphorylated caseins by a rat liver cAMP-insensitive protein kinase, *Biochem. J.*, 179, 693, 1979.
93. **Hoppe, J. and Baydoun, H.**, Substrate specificity of the nuclear protein kinase NII from porcine liver. Studies with casein variants, *Eur. J. Biochem.*, 117, 585, 1981.
94. **Meggio, F., Donella-Deana, A., and Pinna, L. A.**, The use of soybean trypsin inhibitors as phosphorylatable substrates for a rat liver protein kinase, *J. Biochem. (Tokyo)*, 86, 261, 1979.
95. **Meggio, F., Donella-Deana, A., and Pinna, L. A.**, A study with model substrates of the structure of the sites phosphorylated by rat liver casein kinase TS, *Biochim. Biophys. Acta*, 662, 1, 1981.
96. **Cohen, P., Yellowlees, D., Aitken, A., Donella-Deana, A., Hemmings, B. A., and Parker, P. J.**, Separation and characterisation of glycogen synthase kinase 3, glycogen synthase kinase 4 and glycogen synthase kinase 5 from skeletal muscle, *Eur. J. Biochem.*, 124, 21, 1982.
97. **Tashiro, Y., Matsumura, S., Murakami, N., and Kumon, A.**, The phosphorylation site for casein kinase II on 20,000 Da light chain of gizzard myosin, *Arch. Biochem. Biophys.*, 233, 540, 1984.
98. **Walton, G. M., Spiess, J., and Gill, G. N.**, Phosphorylation of high mobility group protein 14 by casein kinase II, *J. Biol. Chem.*, 260, 4745, 1985.
99. **Holmes, C. F. B., Kuret, J., Chisholm, A. K., and Cohen, P.**, Identification of the sites on skeletal muscle protein phosphatase inhibitor-2 phosphorylated by casein kinase II, *Biochim. Biophys. Acta*, 870, 408, 1986.
100. **Meggio, F., Flamigni, F., Guarnieri, C., and Pinna, L. A.**, Location of the phosphorylation site for casein kinase-2 within the amino acid sequence of ornithine decarboxylase, *Biochim. Biophys. Acta*, 929, 116, 1987.
101. **Chang, C., Salzman, A. G., Hiipakka, R. A., Huang, I.-Y., and Liao, S.**, Prostatic spermine binding protein. Cloning and nucleotide sequence of cDNA, aminoacid sequence and androgenic control of mRNA level, *J. Biol. Chem.*, 262, 2826, 1987.
102. **Byrne, B. M., van het Schip, A. D., van de Klundest, J. A. M., Arnberg, A. C., Gruber, M., and Ab, G.**, Aminoacid sequence of phosvitin derived from the nucleotide sequence of part of the chicken vitellogenin gene, *Biochemistry*, 23, 4275, 1984.
102a. **Pinna, L. A., Meggio, F., and Marchiori, F.**, unpublished data.
103. **Pinna, L. A., Meggio, F., Marchiori, F., and Borin, G.**, Opposite and mutually incompatible structural requirements of type-2 casein kinase and cAMP-dependent protein kinase as visualized with synthetic peptide substrates, *FEBS Lett.*, 171, 211, 1984.
104. **Kuenzel, E. A. and Krebs, E. G.**, A synthetic peptide substrate specific for casein kinase II, *Proc. Natl. Acad. Sci. U.S.A.*, 82, 737, 1985.
105. **Marin, O., Meggio, F., Marchiori, F., Borin, G., and Pinna, L. A.**, Site specificity of casein kinase-2 (TS) from rat liver cytosol. A study with model peptide substrates, *Eur. J. Biochem.*, 160, 239, 1986.
105a. **Pinna, L. A., Meggio, F., Marchiori, F., Perich, J. W., and Johns, R. B.**, unpublished data.
106. **Chou, P. Y. and Fasman, G. D.**, Empirical prediction of protein conformations, *Annu. Rev. Biochem.*, 47, 251, 1978.
106a. **Takio, K., Kuenzel, E. A., Walsh, K. A., and Krebs, E. G.**, Amino acid sequence of the β-subunit of bovine lung casein kinase II, *Proc. Natl. Acad. Sci. U.S.A.*, 84, 4851, 1987.
107. **Hunter, T. and Cooper, J. A.**, Protein-tyrosine kinases, *Annu. Rev. Biochem.*, 54, 897, 1985.
108. **Brunati, A. M., Marchiori, F., and Pinna, L. A.**, Isolation and partial characterization of distinct forms of tyrosine protein kinases from rat spleen, *FEBS Lett.*, 88, 321, 1985.
108a. **Pinna, L. A., Meggio, F., Marchiori, F., and Brunati, A. M.**, unpublished data.
109. **Pinna, L. A., Agostinis, P., and Ferrari, S.**, Selectivity of protein kinases and protein phosphatases: a comparative analysis, *Adv. Protein Phosphatases*, 3, 327, 1986.
110. **Meggio, F., Chessa, G., Borin, G., Pinna, L. A., and Marchiori, F.**, Synthetic fragments of protamines as model substrates for rat liver cyclic AMP dependent protein kinase, *Biochim. Biophys. Acta*, 662, 94, 1981.
111. **Purves, F., Donella-Deana, A., Marchiori, F., Leader, D. P., and Pinna, L. A.**, The substrate specificity of the protein kinase induced in cells infected with herpesviruses studies with synthetic substrates indicate structural requirements distinct from other protein kinases, *Biochim. Biophys. Acta*, 889, 208, 1986.
112. **Meggio, F., Donella-Deana, A., and Pinna, L. A.**, Studies on the structural requirements of a microsomal cAMP-independent protein kinase, *FEBS Lett.*, 106, 76, 1979.
113. **Mercier, J. C.**, Phosphorylation of caseins, present evidence for an amino acid triplet code posttranslationally recognized by specific kinases, *Biochimie*, 63, 1, 1981.
114. **Blomback, B., Blomback, M., Erdman, P., and Hessel, B.**, Amino acid sequence and the occurrence of phosphorus in human fibrinopeptides, *Nature (London)*, 193, 883, 1962.
115. **Tang, J., Sepulveda, P., Marciniszyn, J., Chen, K. C. S., Huang, W.-Y., Tao, N., Lin, D., and Lanier, J. P.**, Amino acid sequence of porcine pepsin, *Proc. Natl. Acad. Sci. U.S.A.*, 70, 3437, 1973.

116. **Henderson, J., Moir, A. J. G., Fothergill, L. A., and Fothergill, J. E.**, Sequences of sixteen phosphoserine peptides from ovalbumins of eight species, *Eur. J. Biochem.*, 114, 439, 1981.
117. **Browne, C. A., Bennett, H. P. J., and Solomon, S.**, Isolation and characterization of corticotropin and melanotropin related peptides from the neurointermediary lobe of the rat pituitary by reversed-phase liquid chromatography, *Biochemistry*, 20, 4538, 1980.
118. **Donella-Deana, A., Grankowski, N., Kudlicki, W., Szyska, R., Gasior, E., and Pinna, L. A.**, A type-1 casein kinase from yeast phosphorylates both serine and threonine residues of casein. Identification of the phosphorylation sites, *Biochim. Biophys. Acta*, 829, 180, 1985.
119. **Ahmad, Z., Camici, M., DePaoli-Roach, A. A., and Roach, P. J.**, Glycogen synthase kinases. Classification of a rabbit liver casein and glycogen synthase kinase as a distinct enzyme, *J. Biol. Chem.*, 259, 3420, 1984.
120. **Singh, T. J., Akatsuka, A., Huang, K.-P., Murthy, S. N., and Flavin, M.**, Cyclic nucleotide and Ca^{2+}-independent phosphorylation of tubulin and microtubule associated protein-2 by glycogen synthase (casein) kinase-1, *Biochim. Biophys. Res. Commun.*, 121, 19, 1984.
121. **Kuret, J., Woodgett, J. R., and Cohen, P.**, Multisite phosphorylation of glycogen synthase from rabbit skeletal muscle. Identification of the sites phosphorylated by casein kinase-1, *Eur. J. Biochem.*, 151, 39, 1985.
122. **Moore, A., Boulton, A. P., Heid, H. W., Jarasch, E. D., and Craig, R. K.**, Purification and tissue-specific expression of casein kinase from the lactating guinea-pig mammary gland, *Eur. J. Biochem.*, 152, 729, 1985.
122a. **Pinna, L. A., Meggio, F., Marchiori, F., and Dobrowolska, G.**, unpublished data.
123. **Saxena, A., Padmanabha, R., and Glover, C. V. C.**, Isolation and sequencing of cDNA clones encoding Alpha and Beta subunits of *Drosophila melanogaster* casein kinase II, *Mol. Cell. Biol.*, 7, 3049, 1987.
124. **Jakobi, R., Voss, H., and Pyorin, W.**, Human phosvitin casein kinase type II: molecular cloning and sequencing of full length cDNA encoding subunit beta, *Eur. J. Biochem.*, in press, 1989.
125. **Pfaff, M. and Anderer, F. A.**, Casein kinase II accumulation in the nucleolus and its role in nucleolar phosphorylation, *Biochim. Biophys. Acta*, 269, 100, 1988.
126. **Grande, J., Perez, M., and Itarto, E.**, Phosphorylation of hepatic insulin receptor by casein kinase. II, *FEBS Lett.*, 232, 130, 1988.
127. **Jakubowicz, T. and Leader, D. P.**, A major phospho-protein of cells infected with pseudorabies virus is phosphorylated by cellular casein kinase II, *J. Gen. Virol.*, 68, 1159, 1987.
128. **Pisano, M., Hegazy, M. G., Reimann, E. M., and Dokas, L. A.**, Phosphorylation of protein B-50 (GAP-43) from adult rat brain cortex by casein kinase II, *Biochem. Biophys. Res. Commun.*, 155, 1207, 1988.
129. **Haugen-Scofield, J., Resing, K. A., and Dale, B. A.**, Characterization of an epidermal phosphatase specific for filaggrin phosphorylated by casein kinase II, *J. Invest. Dermatol.*, 91, 553, 1988.
130. **Koike, T. and Ohtsuki, K.**, Purification and characterization of a 400 kDa nonhistone chromatin protein that serves as an effective phosphate acceptor for casein kinase II from Ehrlich ascites tumor cells, *J. Biochem. (Tokyo)*, 103, 928, 1988.
131. **Levasseur, S., Poleck, T., Friedman, Y., and Burke, C.**, Purification of a 107 kDa casein kinase G substrate from thyroid cytosol, *Mol. Cell. Biochem.*, 83, 157, 1988.
132. **Issinger, O. G., Martin, T., Richter, W. W., Olson, M. and Fujiki, H.**, Hyperphosphorylation of N 60, a protein structurally and immunologically related to nucleolin, after tumor-promoter treatments, *EMBO J.*, 7, 1621, 1988.
133. **Brunati, A. M., Saggioro, D., Chieco-Bianchi, L., and Pinna, L. A.**, Altered protein kinase activities of lymphoid cells transformed by Abelson and Moloney leukemia viruses, *FEBS Lett.*, 206, 59, 1986.
134. **Scheider, H. R., Reichert, G. H., and Issinger, O. G.**, Enhanced casein kinase II activity during mouse embryogenesis. Identification of a 110 kDa phosphoprotein as the major phosphorylation product in mouse embryos and Krebs II mouse ascites tumor cells, *Eur. J. Biochem.*, 161, 733, 1986.
135. **Perez, M., Grande, J., and Itarte, E.**, Casein kinase 2 activity increases in the prereplicative phase of liver regeneration, *FEBS Lett.*, 238, 273, 1988.
136. **Durand, P., Vilgrain, J., Chambaz, E. M., and Saez, J. M.**, Changes in protein kinase activities in lamb adrenals at late gestation and early postnatal stages, *Mol. Cell. Endocrinol.*, 53, 195, 1987.
137. **Perez, M., Grande, J., and Itarte, E.**, Developmental changes in rat hepatic casein kinases 1 and 2, *Eur. J. Biochem.*, 170, 493, 1987.
138. **Sommercorn, J. and Krebs, E. G.**, Induction of casein kinase II during differentiation of 3T3-L1 cells, *J. Biol. Chem.*, 262, 3839, 1987.
139. **Mulner-Lorrilon, O., Marot, J., Cayla, X., Pouhle, R., and Belle, R.**, Purification and characterization of a casein kinase II-type enzyme from *Xenopus laevis* ovary. Biological effects on the meiotic cell division of full-grown oocyte, *Eur. J. Biochem.*, 171, 107, 1988.
140. **Mc Manaway, M. E., Eckberg, W. R., and Anderson, W. A.**, Characterization and hormonal regulation of casein kinase II activity in heterotransplanted human breast tumors in nude mice, *Exp. Clin. Endocrinol.*, 90, 313, 1987.

141. **Corvera, S., Roach, P. J., De Paoli-Roach, A. A., and Czech, M. P.**, Insulin action inhibits insulin like growth factor II (IGF-II) receptor phosphorylation in H-35 hepatoma cells. IGF-II receptors isolated from insulin treated cells exhibit enhanced *in vitro* phosphorylation by casein kinase II, *J. Biol. Chem.*, 263, 3116, 1988.
142. **Haystead, T. A., Campbell, D. G., and Hardie, D. G.**, Analysis of sites phosphorylated on acetyl-CoA carboxylase in response to insulin in isolated adipocytes. Comparison with sites phosphorylated by casein kinase 2 and the calmodulin-dependent multiprotein kinase, *Eur. J. Biochem.*, 175, 347, 1988.
143. **Klarlund, J. K. and Czech, M. P.**, Insulin like growth factor I and insulin rapidly increase casein kinase II activity in BALB/c 3T3 fibroblasts, *J. Biol. Chem.*, 263, 15872, 1988.
144. **Czech, M. P., Klarlund, J. K., Yagaloff, K. A., Bradford, A. P., and Lewis, R. E.**, Insulin receptor signaling activation of multiple serine kinases, *J. Biol. Chem.*, 263, 11017, 1988.
145. **Marchiori, F., Meggio, F., Marin, O., Borin, G., Calderan, A., Ruzza, P., and Pinna, L. A.**, Synthetic peptide substrates for casein kinase 2. Assessment of minimum structural requirements for phosphorylation, *Biochim. Biophys. Acta,* 971, 332, 1988.
146. **Clark, S. J., Colthrust, D. R., and Proud, C. G.**, Structure and phosphorylation of eukariotic initiation factor 2. Casein kinase 2 and protein kinase C phosphorylates distinct but adjacent sites in the β-subunit, *Biochim. Biophys. Acta,* 968, 211, 1988.
147. **Meggio, F., Boulton, A. P., Marchiori, F., Borin, G., Lennon, D. P. W., Calderan, A., and Pinna, L. A.**, Substrate-specificity determinants for a membrane bound casein kinase of lactating mammary gland. A study with synthetic peptides, *Eur. J. Biochem.*, 177, 281, 1988.
148. **Kuenzel, E. A., Mulligan, J. A., Sommercorn, J., and Krebs, E. G.**, Substrate specificity determinants for casein kinase II as deduced from studies with synthetic peptides, *J. Biol. Chem.*, 262, 9136, 1987.
149. **Meggio, F. and Pinna, L. A.**, Phosphorylation of phosvitin by casein kinase 2 provides the evidence that phosphoserines can replace carboxylic amino acids as specificity determinants, *Biochim. Biophys. Acta,* 971, 227, 1988.
150. **Meggio, F., Perich, J. W., Johns, R. B., and Pinna, L. A.**, Partially dephosphorylated phosphopeptide AcSer(P)-Ser(P)-Ser(P) is an excellent substrate for casein kinase 2, *FEBS Lett.*, 237, 225, 1988.
151. **Fiol, C. J., Mahrenholz, A. M., Wang, Y., Roeske, R. W., and Roach, P. J.**, Formation of protein kinase recognition sites by covalent modification of the substrate. Molecular mechanism for the synergistic action of casein kinase II and glycogen synthase kinase 3, *J. Biol. Chem.*, 262, 14042, 1987.
152. **Dobrowolska, G., Meggio, F., Marchiori, F., and Pinna, L. A.**, Specificity determinants of Maize casein kinase IIB are related to but distinct from those of rat liver casein kinase II, *Biochim. Biophys. Acta,* 1010, 274, 1989.

Chapter 7

SUBSTRATE SPECIFICITY OF CYCLIC AMP-DEPENDENT PROTEIN KINASE

Örjan Zetterqvist, Ulf Ragnarsson, and Lorentz Engström

TABLE OF CONTENTS

I. Introduction 172

II. Protein Substrates of Cyclic AMP-Dependent Protein Kinase 172

III. Synthetic Peptides as Substrates of Cyclic AMP-Dependent Protein Kinase 173

IV. Search for Potential Phosphorylation Sites in Proteins 179

V. Conclusions 182

References 182

I. INTRODUCTION

Cyclic AMP-dependent protein kinase (PK-A) phosphorylates many proteins. Although the substrates show little apparent similarity, they obviously must have some structure in common which is recognized by the active site of PK-A. The nature of this common denominator was initially postulated by Langan[1] to be a property of the three-dimensional structure of the substrate. However, an alternative description of the nature of a common denominator emerged along several lines of evidence. Daile and Carnegie[2] showed in 1974 that peptides from enzymatic digests of the basic proteins of human myelin were substrates of PK-A. By studying genetic variants of β-casein, Kemp et al.[3] obtained evidence for the importance of the primary structure in the region of the phosphorylatable serine. Upon the identification by Engström and co-workers[4] of liver pyruvate kinase as a substrate of PK-A, Humble et al.[5] showed that both alkali-inactivated pig liver pyruvate kinase and a cyanogen bromide fragment from the same enzyme were even better substrates than the native enzyme. All these data suggested that the substrate specificity of PK-A could be determined by the primary structure of a part of the substrate protein. This idea was further supported by the finding that a synthetic hexapeptide corresponding to the region around serine-24 in chicken lysozyme[6] and a synthetic octapeptide corresponding to the amino acid residues 106 to 113 in the basic protein of human myelin[7] were substrates of PK-A. However, these peptides displayed high K_m values compared with protein substrates such as histones.

The amino acid sequence around the phosphorylated site of pyruvate kinase was determined from both pig liver pyruvate kinase[8] and the rat liver enzyme.[9] Using this information it was possible to synthesize peptides which proved to be exceedingly good substrates, with K_m values of the same order of magnitude as the *in vivo* concentration of liver pyruvate kinase. From such studies the importance of a block of two arginine residues was demonstrated and the minimum substrate of PK-A could be defined as R R X S X.[10-12]

Although the most extensive work with synthetic peptides has been performed with variants based on the amino acid sequence from the phosphorylatable site of liver pyruvate kinase, studies on synthetic peptides modeled on the phosphorylatable site of the β-subunit of phosphorylase kinase of rabbit skeletal muscle[13] have provided further information on the substrate specificity of PK-A.[14-15] From these data an alternate formulation of the minimum substrate can be obtained, namely R X K R X X S X, the lysine residue being less important than the two arginine residues.[15]

From inspection of the amino acid sequence around the phosphorylated serine of a large set of proteins phosphorylated by PK-A, it is obvious that still other arrangements of the phosphorylatable serine and the basic residues are possible, although they have not been identified so far. In this context it may be worth considering the possibility that in some cases the proper proximity of two arginine residues and the phosphorylatable serine is defined, for example, by the tertiary structure of the substrate.[16] In such a case, the original hypothesis of Langan[1] may deserve further consideration.

The present review lists most of the currently (June 1987) known substrates of PK-A in mammals. The evidence of the role of the primary structure of the phosphorylatable site is presented. By means of the structural requirements thus formulated, known amino acid sequences of mammalian proteins are screened for the occurrence of sequences of the R R X S X and R X K R X X S X type.

Various aspects of the substrate specificity of PK-A have been treated in several reviews.[12,17-21]

II. PROTEIN SUBSTRATES OF CYCLIC AMP-DEPENDENT PROTEIN KINASE

Mammalian proteins which have been shown to be phosphorylated by PK-A are listed in Table 1. When known, the effect of the phosphorylation on the biological activity of the proteins is indicated and the amino acid sequences at the phosphorylated sites are given. Krebs and Beavo[12] and Nimmo and Cohen[17] have defined criteria for the phosphorylation of enzymes to be of physiological significance. Perhaps the most important criterion is that the phosphorylation should occur *in vivo*, in response to a hormone-induced increase in cyclic AMP and at the same site as those phosphorylated by PK-A *in vitro*. The functional properties of the protein should undergo changes which correlate with the degree of phosphorylation, and the latter should correlate to the cellular level of cAMP. A low apparent K_m for a protein substrate in a phosphorylation reaction, on the order of its concentration *in vivo*, seems to be an important indication that the phosphorylation is of physiological significance.

With some enzymes (e.g., phosphorylase kinase[12,17] and liver pyruvate kinase[46]), the criteria for a physiological role of the phosphorylation are essentially fulfilled. This is not the case for most other phosphorylatable proteins. On the other hand, it seems reasonable to assume that an *in vitro* phosphorylation of native proteins, which occurs with kinetic parameters similar to those for established substrates, is of physiological significance until the reverse has been proven.

Although a considerable number of proteins are phosphorylated by PK-A, they are few in comparison with the total number of intracellular proteins. This indicates a rather high substrate specificity of the protein kinase and is further emphasized by the fact that only one or a few serine or threonine residues in each substrate polypeptide chain are phosphorylated by PK-A.

As seen in Table 1, the sequence R R X S X occurs in the phosphorylatable site of pyruvate kinase, glycogen synthase site 1a, α-chain of phosphorylase kinase, fructose-2,6-bisphosphatase/fructose-6-phosphate-2-kinase, erythrodehydroneopterin triphosphate synthetase, cytochrome P-450 LM2, tyrosine hydroxylase, and the regulatory subunit of PK-A (R^{II}). The same sequence with threonine instead of serine is present in phosphoprotein phosphatase inhibitor 1 and human chorionic gonadotropin. Phosphorylation of the sequence R X K R X X S X so far has been found only in the β-subunit of phosphorylase kinase.

In nearly every case with a known amino acid sequence at the phosphorylated site there is at least one, but generally two, arginine residues among the six amino acid residues on the N-terminal side of the phosphorylated serine or threonine residue. The third amino acid preceding the phosphorylated residue is often an arginine. In several cases this arginine is preceded or followed by a lysine. With regard to other amino acid residues surrounding the phosphate-accepting site, the first residue on the C-terminal side is frequently hydrophobic. Additional common features of the primary structures of the phosphorylated sites of different proteins are not identified easily.

III. SYNTHETIC PEPTIDES AS SUBSTRATES OF CYCLIC AMP-DEPENDENT PROTEIN KINASE

In an attempt to define the substrate specificity of PK-A, Daile et al.[7] synthesized an octapeptide fragment (peptide *1*, Table 2) corresponding to residues 106 to 113 of human myelin basic protein and found it to be a substrate of PK-A with the rather low K_m value of 240 μ*M*. This peptide was also digested with various peptidases and the resulting products further examined with respect to their function as substrates. From these experiments it was concluded that the C-terminal arginine and the N-terminal glycine could be removed without apparent effect on substrate efficiency, whereas the arginine in position −3* was essential. In addition, they drew attention to the occurrence of the amino acid sequence pattern

---

* For reasons of brevity, in this review the positions of amino acids in peptides are generally in reference to the phosphorylatable or phosphorylated serine (position 0). Amino acid positions preceding serine have negative signs and those following it have positive signs.

TABLE 1
Mammalian Protein Substrates of cAMP-Dependent Protein Kinase

Protein substrate	Effect of phosphorylation on the biological activity of the substrate	Amino acid sequence at the phosphorylated site	Ref.
Enzymes			
Acetyl-CoA carboxylase			22
Acetyl-CoA carboxylase kinase	+		23
Arylsulfatase B	+		24
ATP citrate lyase		T A <u>S</u> F S E	25
Cholesterol esterase	+		26
Cyclic nucleotide phosphodiesterase			27
Erythrodihydroneopterin triphosphate synthetase	+	Y K R R V <u>S</u> E V	28, 29
F 1-ATPase, precursor			30
Fructose-1,6-bisphosphatase	+	K A K S R P <u>S</u> L P L R E S P V H <u>S</u> I C	31
Fructose-2,6-bisphosphatase/fructose-6-phosphate-2-kinase	+ −	L Q R R R G <u>S</u> S I P Q	32
Fructose-6-phosphate-1-kinase, muscle	−	H I S R K R <u>S</u> G E A	33
Fructose-6-phosphate-1-kinase, liver	−		34
Glycogen synthase, muscle	−	Q W P R R A <u>S</u> C T S (site 1a) G S K R S N <u>S</u> V D T (site 1b) P L S R T L <u>S</u> V S S (site 2) P R P A <u>S</u> V P (site 3a) R H S <u>S</u> P H (site 4)	35 35 36, 37 38 38
Glycogen synthase, liver	−	S L <u>S</u> V T S L M Y P R P <u>S</u> <u>S</u> V P P	39
Guanylate cyclase			40
Hormone-sensitive lipase/diglyceride lipase	+		41
Myosin light-chain kinase	−		42
Na$^+$,K$^+$-ATPase, α-subunit			43
Phenylalanine hydroxylase	+	S R K L <u>S</u> B F G	44
Phospholipid methyltransferase			45
Phosphorylase kinase	(+) +	F R R L <u>S</u> I S T (α-chain) R T K R S G <u>S</u>$_\text{I}^\text{V}$ Y E (β-chain)	13
Pyruvate kinase, liver	−	L R R A <u>S</u>$_\text{L}^\text{V}$ A Z	8, 9, 46

TABLE 1 (continued)
Mammalian Protein Substrates of cAMP-Dependent Protein Kinase

Protein substrate	Effect of phosphorylation on the biological activity of the substrate	Amino acid sequence at the phosphorylated site	Ref.
Pyruvate kinase, erythrocytes	−		46
RNA polymerase	+		47
Self			
Catalytic subunit		K R V K G R T W $\underline{T}$ L C G T	48
Regulatory subunit (R^{II})	−	R F D R R V $\underline{S}$ V C A	49
Threonyl-tRNA synthetase			50
Tyrosine hydroxylase	+	G R R Q $\underline{S}$ L I E V R R V $\underline{S}$ D D V R	51
Other proteins			
Actin			52
Atrial natriuretic peptides			53
Calciductin			54
Calcium channel, dihydropyridine-sensitive			55
Choriogonadotropin		L C R R S $\underline{T}$ T D	56
Collagen, αI			68
Cytochrome P-450 LM2		L R R F $\underline{S}$ L A	57
Fibrinogen			58
Filamin			59
G-substrate		K P R R K D $\underline{T}$ P A L	60
Glicentin			61
Glucocorticoid receptor			62
Histone		G A R R K A $\underline{S}$ G P P (H1)	63, 64
		K A K T R S $\underline{S}$ R A (H2A)	63, 64
		G K K R K R $\underline{S}$ R K E S Y S (H2B)	63, 64, 65
		P A P K K G $\underline{S}$ K K	63
HMG 14		K R K V $\underline{S}$ S A E	66
Keratin proteins			67
Lens α-crystallin			69
Lipomodulin	−		70
MAP-2			71
Myelin basic protein		Q G K G R G L $\underline{S}$ L S R	7, 72, 73
		F L P R H R D $\underline{T}$ G I L	7, 72, 73
		R P S E R H G $\underline{S}$ K Y L	7, 72
Phosphatase inhibitor 1		I R R R P $\underline{T}$ P A T	74
Phospholamban			75
Prolactin			76
Ribosomal protein S 6			77
Sodium channel α-subunit			78

TABLE 1 (continued)
Mammalian Protein Substrates of cAMP-Dependent Protein Kinase

Protein substrate	Effect of phosphorylation on the biological activity of the substrate	Amino acid sequence at the phosphorylated site	Ref.
Troponin I, cardiac muscle		A P A V R R S D R A R L R V R I S A D	79, 80
Troponin I, skeletal muscle		R R R V R M S A D A	80, 81

R X X S X at the phosphorylation sites of some phosphorylatable proteins and explicitly predicted the need for a basic amino acid in close proximity to the serine to be phosphorylated. This paper was a forerunner of many others in which synthetic peptides were applied in the elucidation of the substrate specificity of the enzyme.

A step forward in this context was taken using peptides based on the sequences of phosphopeptides derived from pig and rat liver pyruvate kinase.[8,9] Thus, the heptapeptide 2 has a K_m of <10 µM, i.e., more than one order of magnitude lower than the octapeptide 1. In addition to the common feature with an arginine residue in position −3, peptide 2 has an additional arginine in position −2. The critical importance of this residue is demonstrated by the difference in the rate of phosphorylation between peptide 3 and peptide 4, the latter peptide being an extremely poor substrate. Removal of alanine in position +2 of peptide 3 results in the pentapeptide 5, which seems to be the minimum substrate as the activity with tetrapeptide 6 is insignificant. Protein kinase substrates containing arginine in both positions −3 and −2 may be referred to as being of the pyruvate kinase type, i.e., R R X S X. The shift in properties with varying X will be discussed further below.

The points made in the preceding paragraph apply to peptides 7 to 11 as well. These peptides are based on pig liver pyruvate kinase, but some of them have an additional glycine in position +2. Together with analogues based on the rat liver enzyme, they will serve as reference peptides in the following discussion.

Peptides 13 to 16 are derived from the β-subunit of rabbit skeletal muscle phosphorylase kinase. As can be seen in Table 2, the nonapeptide 13 and the octapeptide 14 are both excellent substrates, with K_m values essentially identical to those of peptides 2 and 3. Again, elimination of an arginine residue, this time in position −6, leads to a drastic increase in K_m: whereas this quantity doubles in passing from peptide 13 to 14, it increases by two orders of magnitude from peptide 14 to 15. Similarly, elimination of valine in position +1 of peptide 14 results in an increase in K_m of nearly two orders of magnitude (peptide 16). Obviously, the minimum substrate of this type is an octapeptide.

The importance of the two arginine residues in positions −3 and −2 in substrates of the pyruvate kinase type, and of the arginines in positions −6 and −3 as well as the lysine residue in position −4 in the second type of substrate, is illustrated in Table 3. Thus, the introduction of a leucine or lysine residue instead of an arginine in peptide 5 results in very poor substrates (peptides 17 to 20). Similarly, when Kemp et al.[11] introduced alanine, lysine, histidine, or homoarginine in either position −3 or −2 of peptide 7 ("Kemptide"), the K_m increased at least one order of magnitude (peptides 21 to 28).

Glycine substitutions in positions −6 and −3 of the octapeptide 14 have also been performed which demonstrated that an alternative arrangement of the two arginines is possible. Thus, the K_m values for peptides 30 and 32 are approximately two orders of magnitude larger than for 14. The lysine residue in position −4 in this type of substrate appears to play a less prominent role than the two arginine residues in the phosphorylation event. Thus,

TABLE 2
Selected Synthetic Substrates of cAMP-Dependent Protein Kinase with Particular Reference to the Minimum Partial Sequence Required

Peptide	Amino acid sequence[a]	K_m (μM)	Ref.
1	G R G L $\underline{S}$ L S R	240	7
2	L R R A $\underline{S}$ V A	<10	10
2	L R R A $\underline{S}$ V A	2[b]	82
3	R R A $\underline{S}$ V A	20	10
3	R R A $\underline{S}$ V A	3[b]	82
3	R R A $\underline{S}$ V A	14	83
4	R A $\underline{S}$ V A	—[c]	10
5	R R A $\underline{S}$ V	80	10
5	R R A $\underline{S}$ V	24[b]	82
6	R R A $\underline{S}$	—[c]	10
7	L R R A $\underline{S}$ L G	16	11
8	R R A $\underline{S}$ L G	26	11
9	R A $\underline{S}$ L G	4400	11
10	L R R A $\underline{S}$ L	57	11
11	L R R A $\underline{S}$	—[d]	11
12	V L Q R R R G $\underline{S}$ S I P Q	4	84
13	A R T K R S G $\underline{S}$ V	10	15
14	R T K R S G $\underline{S}$ V	21	15, 85
15	T K R S G $\underline{S}$ V	2100	15
16	R T K R S G $\underline{S}$	1200	15

[a] Structures of peptides are related to the phosphorylatable site of myelin basic protein (1), rat liver pyruvate kinase (2 to 6), pig liver pyruvate kinase (7 to 11), 6-phosphofructo-2-kinase/fructose-2,6-bisphosphatase (12), and the β-subunit of rabbit skeletal muscle phosphorylase kinase (13 to 16). The phosphorylatable serine residue is underlined.
[b] The differences between the K_m values obtained from Reference 82 and the values of the corresponding peptides from Reference 10 are explained by different compositions of the incubation mixtures.
[c] K_m not determined. The rate of phosphorylation of peptides 4 and 6 was 2% and <1%, respectively, of that of 5.
[d] K_m not determined since the peptide was not detectably phosphorylated.

although the K_m of peptide *31* is higher than of *14*, *31* is a much better substrate than *30* and *32*. On the basis of this evidence, we would like to propose that protein kinase substrates containing the amino acid sequence R X K R X X S X are referred to as being of phosphorylase kinase β-subunit type (or simply of β-subunit type).

Glass et al.[84] have studied synthetic peptides corresponding to the phosphorylated site of 6-phosphofructo-2-kinase/fructose-2,6-bisphosphatase. This sequence contains a block of three arginine residues at positions −2, −3, and −4 (peptide *12*). Exchange of the latter arginine residue with alanine gives a peptide with a threefold higher K_m (peptide *29*). Still, this peptide, which has a sequence of the pyruvate kinase type, is a good substrate by itself.

A large number of analogues to the minimum peptide of pyruvate kinase type have been made with replacements in position +1 (peptides *33* to *40*). The data of Table 4 indicate that neutral, hydrophobic amino acids in this position give rise to the best substrates, while glycine is less favorable. Basic amino acids such as lysine and arginine also give poorer substrates, perhaps due to some misfit in the binding of the peptides to PK-A, and an acidic amino acid such as glutamic acid seems to be particularly unfavorable in this position.

TABLE 3
Selected Synthetic Substrates of cAMP-Dependent Protein Kinase, Illustrating the Importance of the Basic Amino Acids

Peptide	Amino acid sequence[a]	K_m (µM)	Ref.[b]
5	R R A S V	80	10
5	R R A S V	16	
17	L̲ R A S V	—[c]	10
18	R L̲ A S V	—[c]	10
19	K̲ R A S V	—[c]	
20	R K̲ A S V	—[c]	
7	L R R A S L G	16	11
21	L A̲ R A S L G	4900	11
22	L R A̲ A S L G	6300	11
23	L K̲ R A S L G	1400	11
24	L R K̲ A S L G	260	11
25	L H̲ R A S L G	415	11
26	L R H̲ A S L G	1340	11
27	L X̲[d] R A S L G	350	11
28	L R X̲[d] A S L G	440	11
12	V L Q R R R G S S I P Q	4	84
29	V L Q A̲ R R G S S I P Q	11	84
14	R T K R S G S V	21	15, 85
30	G̲ T K R S G S V	1300	15
31	R T G̲ R S G S V	140	15
32	R T K G̲ S G S V	2700	15

[a] Substitutions for arginine residues are underlined.
[b] Entries lacking citations are from this review. Phosphorylation was performed as described,[82] i.e., under conditions outlined by Kemp et al.,[11] but with the catalytic subunit of rat liver PK-A.[10]
[c] The rate of phosphorylation of peptides 17 to 20 at 100 µM was <1% of that of 5.
[d] X in this table stands for homoarginine.

Fewer modifications have been made so far in position −1 (peptides 41 to 47). Neutral amino acids seem to be able to occupy this place well, with the possible exception of valine (peptide 42), which gives the highest K_m value in this series. Proline (peptides 45 and 47) and aspartic acid (peptide 43) both work surprisingly well. Peptides with β-alanine or D-alanine in position −1 of the corresponding sequence are not detectably phosphorylatable.[85] Finally, it should be pointed out that, among substrates of the β-subunit type, serine and alanine can be substituted for threonine in position −5 (peptides 48 and 49) without significant change in K_m.

When a threonine is substituted for the phosphorylatable serine in small peptides of the pyruvate kinase type, poor substrates of protein kinase are obtained, with K_m values in the 500 to 2000 µM range (peptides 51, 57, and 63, Table 5). A few remarkably good threonine peptide substrates, however, have also been made, such as 56, 61, 62, and, particularly, 60. All of them contain three or more arginines. Peptide 56 is an octapeptide with two additional arginines, and the other three are dodecapeptides. It should be pointed out that the latter all have K_m values one order of magnitude higher than the corresponding serine peptides.

Substitution of hydroxyproline for serine in the Kemptide results in a peptide with K_m = 18,000 µM. No kinetic data have been reported for the corresponding tyrosine peptide.

TABLE 4
Selected Synthetic Substrates of cAMP-Dependent Protein Kinase, Illustrating the Influence of Other Substitutions than Those for Basic Amino Acids and Serine

Peptide	Amino acid sequence[a]	K_m (μM)	Ref.[b]
5	R R A S V	80	10
5	R R A S V	24	82
5	R R A S V	16	
33	R R A S G	—[c]	10
34	R R A S F	13	
35	R R A S L	26	
36	R R A S I	7	
37	R R A S S	180	
38	R R A S E	—[d]	
39	R R A S K	—[c]	10
40	R R A S R	170	
41	R R G S V	5	
42	R R V S V	95	
43	R R D S V	56	
44	R R F S V	5	
45	R R P S V	35	82
46	R R L S I	4	
12	V L Q R R G S S I P Q	4	84
47	V L Q R R P S S I P Q	12.5	84
14	R T K R S G S V	21	15, 85
48	R S K R S G S V	26	15
49	R A K R S G S V	33	15

[a] The substitutions for A and V in peptide *5*, G in peptide *12*, and T in peptide *14* are underlined.
[b] Entries lacking citations are from this review. Phosphorylation was performed as described.[82]
[c] K_m not determined. The rate of phosphorylation of peptides *33* and *39* at 100 μM was 7 and 17%, respectively, of that of *5*.
[d] K_m not determined. No phosphorylation was detected at a 100 μM concentration of peptide *38*.

IV. SEARCH FOR POTENTIAL PHOSPHORYLATION SITES IN PROTEINS

The requirement for, and the optimal location of, the basic residues near the phosphorylatable serine in liver pyruvate kinase and the β-subunit of skeletal muscle phosphorylase kinase were established by the use of synthetic peptides, as described in Section III. Although it is apparent from Table 1 that arrangements of the basic residues other than R R X S X and R X K R X X S X exist in natural substrates of PK-A, the pyruvate kinase type is found in about one fourth of the cases. It therefore seemed appropriate to look for this pattern in a current protein sequence data base in an attempt to suggest some additional substrates of PK-A. In addition, a search was also made for the β-subunit pattern. The results are shown in Table 6.

The searches were made by means of the data base Protein Identification Resource (PIR) of the National Biomedical Research Foundation, Washington, D.C. (release 11.0, December 4, 1986). This covered 4,028 sequences and 963,031 residues. In addition, an appendix,

TABLE 5
Synthetic Substrates of cAMP-Dependent Protein Kinase, Containing Other Hydroxy Amino Acids in the Position of the Phosphorylatable Serine Residue

Peptide	Amino acid sequence[a]	K_m (μM)	Ref.
5	R R A S V	80	10
50	R R A <u>T</u> V	—[b]	10
3	R R A S V A	14	83
51	R R A <u>T</u> V A	1,950	83
52	R R A <u>T</u> P A	1,190	83
53	R R P <u>T</u> V A	2,127	83
54	R R P <u>T</u> P A	1,830	83
55	R R P S P A	5,940	83
56	R R R R P <u>T</u> P A	296—1,300	83
7	L R R A S L G	16	11
57	L R R A <u>T</u> L G	590	11
58	L R R A <u>X</u>[c] L G	18,000	86
59	L R R A <u>Y</u> L G	—[d]	87
12	V L Q R R G S S I P Q	4	84
60	V L Q R R G <u>T</u> S I P Q	39	84
29	V L Q A R R G S S I P Q	11	84
61	V L Q A R R G <u>T</u> S I P Q	139	84
47	V L Q R R R P S S I P Q	12.5	84
62	V L Q R R R P <u>T</u> S I P Q	118	84
14	R T K R S G S V	21	15
63	R T K R S G <u>T</u> V	650	15

[a] The substitutions for the phosphorylatable serine residue are underlined.
[b] The rate of phosphorylation of peptide 50 at 100 μM was <1% of that of 5.
[c] X in this table stands for hydroxyproline.
[d] This peptide has been applied as a substrate for tyrosine-specific protein kinases.[87] No data on its use as a substrate of cAMP-dependent protein kinase have been reported.

"New Entries in Preparation" (release 26.0, December 4, 1986), which covered 584 sequences and 99,118 residues, was used. The searches were performed by the Protein Sequence Query System (PSQ) program. Only sequences of mammalian proteins were included. In the case of several entries for one type of protein, only one species was chosen if the sequences of the selected peptide segments were identical. Already identified substrates of PK-A, which appeared in the search, are accounted for in Table 1 only.

With these criteria, 42 sites of the pyruvate kinase type and 3 sites of the β-subunit type are listed in Table 6. A judgment of the value of such a list certainly has to await actual phosphorylation experiments. If phosphorylation is attempted *in vitro*, it is important to select conditions which preserve the native structure of the protein. Denaturation would expose potential phosphorylation sites, which are normally inaccessible to PK-A, with a risk for artifactual phosphorylation.

If a potentially phosphorylatable site of a native protein is known to be accessible to another enzyme, such as a protease, it may be accessible to PK-A as well. In this context it is of particular interest that doublets of basic residues are located at processing sites of several proteins. Among the proteins of Table 6, proteolytic cleavage within the sequences listed has been shown, or suggested, for the complement C3 precursor, the interleukin-1 precursor, the factor IX precursor, and the β-chain precursor of nerve growth factor.

TABLE 6
RRXSX and RXKRXXSX Sequences in Mammalian Proteins[a]

Amino acid sequence	Position of serine[b] residue no.	Protein	Ref.
R R T S T	18	Cytochrome c_1, nonheme 7-kDa protein	88
R R Y S D	441	Cytochrome c oxidase, polypeptide I	89
R R Y S R	176	Glutathione peroxidase	90
R R Q S E	703	Ceruloplasmin	91
R R M S L	428	Glycogen phosphorylase	92
R R A S D	43	Kinase-related transforming protein	93
R R V S V	78	Complement subcomponent C1r, b chain	94
R R D S C	173	Complement factor D	95
R R N S W	37	Preproelastase I	96
R R N S G	57	Collagenase precursor	97
R R R S V	672	Complement C3 precursor	98
R R C S V	1545	Complement C4A	99
R R P S E	98	Histidine-rich glycoprotein precursor	100
R R S S S	280	Transforming protein (N-*myc*)	101
R R N S S	66	Pro-opiomelanocortin	102
R R T S P	1162	Thyroglobulin precursor	103
R R P S A	1949	Thyroglobulin precursor	103
R R I S L	25	Interferon α-3	104
R R I S P	48	Interferon α-4 precursor	105
R R I S P	48	Interferon α-8 precursor	106
R R L S P	48	Interferon α-1 precursor	107
R R A S Q	165	Interferon γ-precursor	108
R R L S F	90	Interleukin-1 precursor	109
R R L S L	87	Interleukin-1 α-precursor	110
R R W S W	95	T-cell receptor γ-chain precursor	111
R R T S V	159	T-cell receptor γ-chain, C region	111, 112
R R C S D	201	Ig ε-chain, C region	113
R R D S Y	62	Ig δ-chain, C region	114
R R K S S	337	HLA class I histocompatibility antigen, CW3 precursor	115
R R R S R	6	HLA class II histocompatibility antigen, γ-chain precursor	116
R R K S K	22	60S ribosomal protein L37	117
R R C S P	311	Link protein 2	118
R R S S A	218	Major prion PrP27-30 protein	119
R R K S C	236	Estrogen receptor protein	120
R R D S W	169	Sex steroid-binding protein	121
R R R S V	6	Lactotransferrin	122
R R P S R	397	Vitronectin precursor	123
R R G S E	71	Inhibin α-chain precursor	124
R R N S F	52	T-cell receptor α-chain	125
R R L S E	241	Protein kinase C	126
R R D S L	374	Myosin β heavy chain	127
R R C S C	161	Protein C precursor	128
R P K R Y N S G	49	Factor IX precursor	129
R S K R S S S H	169	Nerve growth factor, β-chain precursor	130
R T K R Q T S G	22	Coupling factor 6, mitochondrial	131

[a] Known protein substrates of PK-A emerging from this search are listed in Table 1 only. When a particular amino acid sequence is found in more than one entrance for the same protein, reference is given only to one species (human in the first place).

[b] The position of the potentially phosphorylatable serine residue refers to the numbering of the amino acid residues of the listed protein, as given in the data base.

V. CONCLUSIONS

Substrate specificity rules for PK-A have been defined by use of synthetic peptides representing the phosphorylatable site of liver pyruvate kinase in particular, but also that of the β-subunit of phosphorylase kinase. The minimal structural requirements of PK-A which emerge from these studies are of two types — namely, R R X S X (the pyruvate kinase type) and R X K R X X S X (the phosphorylase kinase β-subunit type). Therefore, the occurrence of the structure R R X S X in about one in four and R X K R X X S X in one case of identified mammalian protein substrates of PK-A seems to explain why these proteins become phosphorylated.

Still, other arrangements of the phosphorylatable serine residue and the basic residues are present in known protein substrates of PK-A, although the minimal structural requirements so far have not been identified. In a few substrates, a threonine residue is phosphorylatable instead of serine, and experiments with synthetic peptides suggest that the substrate specificity can be partly defined by a block of three arginine residues. Additional experiments are, however, required on this matter.

The specificity rules established so far merit their use in the search for additional substrates of PK-A with the help of current protein sequence data bases. Although the mere existence of a sequence matching the specificity rules does not define a protein substrate of PK-A per se, the list obtained seems to be long enough to contain at least a few potential substrates.

REFERENCES

1. **Langan, T. A.**, Protein kinases and protein kinase substrates, *Adv. Cyclic Nucleotide Res.*, 3, 99, 1973.
2. **Daile, P. and Carnegie, P. R.**, Peptides from myelin basic protein as substrates for adenosine 3',5'-cyclic monophosphate-dependent protein kinase, *Biochem. Biophys. Res. Commun.*, 61, 852, 1974.
3. **Kemp, B. E., Bylund, D. B., Huang, T.-S., and Krebs, E. G.**, Substrate specificity of the cyclic AMP-dependent protein kinase, *Proc. Natl. Acad. Sci. U.S.A.*, 72, 3448, 1975.
4. **Ljungström, O., Hjelmquist, G., and Engström, L.**, Phosphorylation of purified rat liver pyruvate kinase by cyclic 3',5'-AMP-stimulated protein kinase, *Biochim. Biophys. Acta*, 358, 289, 1974.
5. **Humble, E., Berglund, L., Titanji, V., Ljungström, O., Edlund, B., Zetterqvist, Ö., and Engström, L.**, Non-dependence on native structure of pig liver pyruvate kinase when used as a substrate for cyclic 3',5'-AMP-stimulated protein kinase, *Biochem. Biophys. Res. Commun.*, 66, 614, 1975.
6. **Bylund, D. B. and Krebs, E. G.**, Effect of denaturation on the susceptibility of proteins to enzymic phosphorylation, *J. Biol. Chem.*, 250, 6355, 1975.
7. **Daile, P., Carnegie, P. R., and Young, J. D.**, Synthetic substrate for cyclic AMP-dependent protein kinase, *Nature (London)*, 257, 416, 1975.
8. **Hjelmquist, G., Andersson, J., Edlund, B., and Engström, L.**, Amino acid sequence of a (^{32}P)phosphopeptide from pig liver pyruvate kinase phosphorylated by cyclic 3',5'-AMP-stimulated protein kinase and γ-(^{32}P)ATP, *Biochem. Biophys. Res. Commun.*, 61, 559, 1974.
9. **Edlund, B., Andersson, J., Titanji, V., Dahlqvist, U., Ekman, P., Zetterqvist, Ö., and Engström, L.**, Amino acid sequence at the phosphorylated site of rat liver pyruvate kinase, *Biochem. Biophys. Res. Commun.*, 67, 1516, 1975.
10. **Zetterqvist, Ö., Ragnarsson, U., Humble, E., Berglund, L., and Engström, L.**, The minimum substrate of cyclic AMP-stimulated protein kinase, as studied by synthetic peptides representing the phosphorylatable site of pyruvate kinase (type L) of rat liver, *Biochem. Biophys. Res. Commun.*, 70, 696, 1976.
11. **Kemp, B. E., Graves, D. J., Benjamini, E., and Krebs, E. G.**, Role of multiple basic residues in determining the substrate specificity of cyclic AMP-dependent protein kinase, *J. Biol. Chem.*, 252, 4888, 1977.
12. **Krebs, E. G. and Beavo, J. A.**, Phosphorylation-dephosphorylation of enzymes, *Annu. Rev. Biochem.*, 48, 923, 1979.

13. **Yeaman, S. J., Cohen, P., Watson, D. C., and Dixon, G. H.**, The substrate specificity of adenosine 3':5'-cyclic monophosphate-dependent protein kinase of rabbit skeletal muscle, *Biochem. J.*, 162, 411, 1977.
14. **Feramisco, J. R., Glass, D. B., and Krebs, E. G.**, Optimal spatial requirements for the location of basic residues in peptide substrates for the cyclic AMP-dependent protein kinase, *J. Biol. Chem.*, 255, 4240, 1980.
15. **Zetterqvist, Ö. and Ragnarsson, U.**, The structural requirements of substrates of cyclic AMP-dependent protein kinase, *FEBS Lett.*, 139, 287, 1982.
16. **Ekdahl, K. N.**, Rat liver fructose-1,6-bisphosphatase. Identification of serine 338 as a third major phosphorylation site for cyclic AMP-dependent protein kinase and activity changes associated with multisite phosphorylation *in vitro*, *J. Biol. Chem.*, 262, 16699, 1987.
17. **Nimmo, H. G. and Cohen, P.**, Hormonal control of protein phosphorylation, *Adv. Cyclic Nucleotide Res.*, 8, 145, 1977.
18. **Carlson, G. M., Bechtel, P. J., and Graves, D. J.**, Chemical and regulatory properties of phosphorylase kinase and cyclic AMP-dependent protein kinase, *Adv. Enzymol.*, 50, 41, 1979.
19. **Engström, L., Ekman, P., Humble, E., Ragnarsson, U., and Zetterqvist, Ö.**, Detection and identification of substrates for protein kinases: use of proteins and synthetic peptides, *Methods Enzymol.*, 107, 130, 1984.
20. **Bramson, H. N., Mildvan, A. S., and Kaiser, E. T.**, Mechanistic studies of cAMP-dependent protein kinase action, *Crit. Rev. Biochem.*, 15, 93, 1984.
21. **Engström, L., Humble, E., Ekman, P., Ragnarsson, U., and Zetterqvist, Ö.**, Structural requirements of substrates of cyclic AMP-dependent protein kinase, in *Cellular Regulation and Malignant Growth*, Ebashi, S., Ed., Springer-Verlag, Berlin, 1985, 240.
22. **Brownsey, R. W. and Hardie, D. G.**, Regulation of acetyl-CoA carboxylase: identity of sites phosphorylated in intact cells treated with adrenaline and *in vitro* by cyclic AMP-dependent protein kinase, *FEBS Lett.*, 120, 67, 1980.
23. **Lent, B. A. and Kim, K. H.**, Phosphorylation and activation of acetyl-coenzyme A carboxylase kinase by the catalytic subunit of cyclic AMP-dependent protein kinase, *Arch. Biochem. Biophys.*, 225, 972, 1983.
24. **Gasa, S., Balbaa, M., Nakamura, M., Yonemori, H., and Makita, A.**, Phosphorylation of human lysosomal arylsulfatase B by cAMP-dependent protein kinase, *J. Biol. Chem.*, 262, 1230, 1987.
25. **Pierce, M. W., Palmer, J. L., Keutmann, H. T., Hall, T. A., and Avruch, J.**, The insulin-directed phosphorylation site on ATP-citrate lyase is identical with the site phosphorylated by the cAMP-dependent protein kinase *in vitro*, *J. Biol. Chem.*, 257, 10681, 1982.
26. **Khoo, J. C., Mahoney, E. M., and Steinberg, D.**, Neutral cholesterol esterase activity in macrophages and its enhancement by cAMP-dependent protein kinase, *J. Biol. Chem.*, 256, 12659, 1981.
27. **Sharma, R. K., Wang, T. H., Wirch, E., and Wang, J. H.**, Purification and properties of bovine brain calmodulin-dependent cyclic nucleotide phosphodiesterase, *J. Biol. Chem.*, 255, 5916, 1980.
28. **Gál, E. M. and Sherman, A. D.**, Phosphorylation, a factor controlling the synthesis of L-erythrodihydrobiopterin (BH_2), *Biochem. Biophys. Res. Commun.*, 83, 593, 1978.
29. **Gál, E. M. and Sherman, A. D.**, 6-(D-Erythro-1',2',3'-trihydroxypropyl)-7,8-dihydropterin triphosphate synthetase, *Fed. Proc.*, 38, 324, 1979.
30. **Steinberg, R. A.**, Cyclic AMP-dependent phosphorylation of the precursor to beta subunit of mitochondrial F_1-ATPase: a physiological mistake?, *J. Cell Biol.*, 98, 2174, 1984.
31. **Chatterjee, T., Rittenhouse, J., Marcus, F., Reardon, I., and Heinrikson, R. L.**, Identification of the *in vivo* and *in vitro* phosphorylation sites of rat liver fructose 1,6-bisphosphatase, *J. Biol. Chem.*, 259, 3831, 1984.
32. **Murray, K. J., El-Maghrabi, M. R., Kountz, P. D., Lukas, T. J., Soderling, T. R., and Pilkis, S. J.**, Amino acid sequence of the phosphorylation site of rat liver 6-phosphofructo-2-kinase/-fructose-2,6-bisphosphatase, *J. Biol. Chem.*, 259, 7673, 1984.
33. **Kemp, R. G., Foe, L. G., Latshaw, S. P., Poorman, R. A., and Heinrikson, R. L.**, Studies on the phosphorylation of muscle phosphofructokinase, *J. Biol. Chem.*, 256, 7282, 1981.
34. **Furuya, E. and Uyeda, K.**, Regulation of phosphofructokinase by a new mechanism, *J. Biol. Chem.*, 255, 11656, 1980.
35. **Parker, P. J., Aitken, A., Bilham, T., Embi, N., and Cohen, P.**, Amino acid sequence of a region in rabbit skeletal muscle glycogen synthase phosphorylated by cyclic AMP-dependent protein kinase, *FEBS Lett.*, 123, 332, 1981.
36. **Embi, N., Parker, P. J., and Cohen, P.**, A reinvestigation of the phosphorylation of rabbit skeletal-muscle glycogen synthase by cyclic-AMP-dependent protein kinase. Identification of the third site of phosphorylation as serine-7, *Eur. J. Biochem.*, 115, 405, 1981.
37. **Embi, N., Rylatt, D. B., and Cohen, P.**, Glycogen synthase kinase-2 and phosphorylase kinase are the same enzyme, *Eur. J. Biochem.*, 100, 339, 1979.
38. **Sheorain, V. S., Corbin, J. D., and Soderling, T. R.**, Phosphorylation of sites 3 and 4 in rabbit skeletal muscle glycogen synthase by cAMP-dependent protein kinase, *J. Biol. Chem.*, 260, 1567, 1985.

39. **Wang, Y., Bell, A. W., Hermodson, M. A., and Roach, P. J.**, Liver isozyme of rabbit glycogen synthase. Amino acid sequences surrounding phosphorylation sites recognized by cyclic AMP-dependent protein kinase, *J. Biol. Chem.*, 261, 16909, 1986.
40. **Zwiller, J., Revel, M.-O., and Basset, P.**, Evidence for phosphorylation of rat brain guanylate cyclase by cyclic AMP-dependent protein kinase, *Biochem. Biophys. Res. Commun.*, 101, 1381, 1981.
41. **Nilsson, N. Ö., Strålfors, P., Fredrikson, G., and Belfrage, P.**, Regulation of adipose tissue lipolysis: effects of noradrenaline and insulin on phosphorylation of hormone-sensitive lipase and on lipolysis in intact rat adipocytes, *FEBS Lett.*, 111, 125, 1980.
42. **Conti, M. A. and Adelstein, R. S.**, The relationship between calmodulin binding and phosphorylation of smooth muscle myosin kinase by the catalytic subunit of $3':5'$ cAMP-dependent protein kinase, *J. Biol. Chem.*, 256, 3178, 1981.
43. **Mårdh, S.**, Phosphorylation of a kidney preparation of Na,K-ATPase by the catalytic subunit of cAMP-dependent protein kinase, *Curr. Top. Membr. Transf.*, 19, 999, 1983.
44. **Wretborn, M., Humble, E., Ragnarsson, U., and Engström, L.**, Amino acid sequence at the phosphorylated site of rat liver phenylalanine hydroxylase and phosphorylation of a corresponding synthetic peptide, *Biochem. Biophys. Res. Commun.*, 93, 403, 1980.
45. **Villalba, M., Varela, I., Merida, I., Pajares, M. A., Martinez del Pozo, A., and Mato, J. M.**, Modulation by the ratio S-adenosyl-methionine/S-adenosylhomocysteine of cyclic AMP-dependent phosphorylation of the 50 kDa protein of rat liver phospholipid methyltransferase, *Biochim. Biophys. Acta*, 847, 273, 1985.
46. **Engström, L., Ekman, P., Humble, E., and Zetterqvist, Ö.**, Pyruvate kinase, in *The Enzymes*, Vol. 18, Part B, Boyer, P. D. and Krebs, E. G., Eds., Academic Press, New York, 1987, 47.
47. **Kranias, E. G., Schweppe, J. S., and Jungman, R. A.**, Phosphorylative and functional modifications of nucleoplasmic RNA polymerase II by homologous adenosine $3':5'$-monophosphate-dependent protein kinase from calf thymus and by heterologous phosphatase, *J. Biol. Chem.*, 252, 6750, 1977.
48. **Shoji, S., Parmelee, D. C., Wade, R. D., Kumar, S., Ericsson, L. H., Walsh, K. A., Neurath, H., Long, G. L., Demaille, J. G., Fischer, E. H., and Titani, K.**, Complete amino acid sequence of the catalytic subunit of bovine cardiac muscle cyclic AMP-dependent protein kinase, *Proc. Natl. Acad. Sci. U.S.A.*, 78, 848, 1981.
49. **Takio, K., Smith, S. B., Krebs, E. G., Walsh, K. A., and Titani, K.**, Primary structure of the regulatory subunit of type II cAMP-dependent protein kinase from bovine cardiac muscle, *Proc. Natl. Acad. Sci. U.S.A.*, 79, 2544, 1982.
50. **Pendergast, A. M. and Traugh, J. A.**, Identification of three protein kinases which phosphorylate threonyl-tRNA synthetase from rat liver, *FEBS Lett.*, 206, 335, 1986.
51. **Campbell, D. G., Hardie, D. G., and Vulliet, P. R.**, Identification of four phosphorylation sites in the N-terminal region of tyrosine hydroxylase, *J. Biol. Chem.*, 261, 10489, 1986.
52. **Walsh, M. P., Hinkins, S., and Hartshorne, D. J.**, Phosphorylation of smooth muscle actin by the catalytic subunit of the cAMP-dependent protein kinase, *Biochem. Biophys. Res. Commun.*, 102, 149, 1981.
53. **Rittenhouse, J., Moberly, L., O'Donnell, M. E., Owen, N. E., and Marcus, F.**, Phosphorylation of atrial natriuretic peptides by cyclic AMP-dependent protein kinase, *J. Biol. Chem.*, 261, 7607, 1986.
54. **Haiech, J. and Demaille, J. G.**, Phosphorylation and the control of calcium fluxes, *Philos. Trans. R. Soc. London Ser. B*, 302, 91, 1983.
55. **Hosey, M. M., Borsotto, M., and Lazdunski, M.**, Phosphorylation and dephosphorylation of dihydropyridine-sensitive voltage-dependent Ca^{2+} channel in skeletal muscle membranes by cAMP- and Ca^{2+}-dependent processes, *Proc. Natl. Acad. Sci. U.S.A.*, 83, 3733, 1986.
56. **Keutmann, H. T., Ratanabanangkoon, K., Pierce, M. W., Kitzmann, K., and Ryan, R. J.**, Phosphorylation of human choriogonadotropin. Stoichiometry and sites of phosphate incorporation, *J. Biol. Chem.*, 258, 14521, 1983.
57. **Müller, R., Schmidt, W. E., and Stier, A.**, The site of cyclic AMP-dependent protein kinase catalyzed phosphorylation of cytochrome P-450 LM2, *FEBS Lett.*, 187, 21, 1985.
58. **Engström, L., Edlund, B., Ragnarsson, U., Dahlqvist-Edberg, U., and Humble, E.**, Phosphorylation of human fibrinogen in vitro with cyclic $3',5'$-AMP-stimulated protein kinase and (^{32}P)ATP, *Biochem. Biophys. Res. Commun.*, 96, 1503, 1980.
59. **Wallach, D., Davies, P. J. A., and Pastan, I.**, Cyclic AMP-dependent phosphorylation of filamin in mammalian smooth muscle, *J. Biol. Chem.*, 253, 4739, 1978.
60. **Aitken, A., Bilham, T., Cohen, P., Aswad, D., and Greengard, P.**, A specific substrate from rabbit cerebellum for guanosine-$3':5'$-monophosphate-dependent protein kinase. III. Amino acid sequences at the two phosphorylation sites, *J. Biol. Chem.*, 256, 3501, 1981.
61. **Conlon, J. M., Thim, L., Moody, A. J., and Söling, H. D.**, Cyclic-AMP-dependent phosphorylation of glicentin, *Biosci. Rep.*, 4, 489, 1984.

62. **Singh, V. B. and Moudgil, V. K.**, Phosphorylation of rat liver glucocorticoid receptor, *J. Biol. Chem.*, 260, 3684, 1985.
63. **Shlyapnikov, S. V., Arutyunyan, A. A., Kurochkin, S. N., Memelova, L. V., Nesterova, M. V., Sashchenko, L. P., and Severin, E. S.**, Investigation of the sites phosphorylated in lysine-rich histones by protein kinase from pig brain, *FEBS Lett.*, 53, 316, 1975.
64. **DeLange, R. J. and Smith, E. L.**, The structure of histones, *Acc. Chem. Res.*, 5, 368, 1972.
65. **Hashimoto, E., Takeda, M., Nishizuka, Y., Hamana, K., and Iwai, K.**, Phosphorylated sites of calf thymus histone H2B by adenosine 3′,5′-monophosphate-dependent protein kinase from silkworm, *Biochem. Biophys. Res. Commun.*, 66, 547, 1975.
66. **Walton, G. M., Spiess, J., and Gill, G. N.**, Phosphorylation of high mobility group 14 protein by cyclic nucleotide-dependent protein kinases, *J. Biol. Chem.*, 257, 4661, 1982.
67. **Inohard, S. and Sagami, S.**, Phosphorylation of epidermal keratin proteins by cyclic AMP-dependent protein kinase, *Arch. Dermatol. Res.*, 275, 417, 1983.
68. **Glass, D. B. and McPherson, J. M.**, In vitro phosphorylation of type I collagen by cyclic AMP-dependent protein kinase, *J. Biol. Chem.*, 261, 5674, 1986.
69. **Spector, A., Chiesa, R., Sredy, J., and Garner, W.**, cAMP-dependent phosphorylation of bovine lens alpha-crystallin, *Proc. Natl. Acad. Sci. U.S.A.*, 82, 4712, 1985.
70. **Hirata, F.**, The regulation of lipomodulin, a phospholipase inhibitory protein, in rabbit neutrophils by phosphorylation, *J. Biol. Chem.*, 256, 7730, 1981.
71. **Theurkauf, W. E. and Vallee, R. B.**, Extensive cAMP-dependent and cAMP-independent phosphorylation of microtubule-associated protein 2, *J. Biol. Chem.*, 258, 7883, 1983.
72. **Dunkley, P. R. and Carnegie, P. R.**, Amino acid sequence of the smaller basic protein from rat brain myelin, *Biochem. J.*, 141, 243, 1974.
73. **Eylar, E. H., Brostoff, S., Hashim, G., Caccam, J., and Burnett, P.**, Basic A1 protein of the myelin membrane. The complete amino acid sequence, *J. Biol. Chem.*, 246, 5770, 1971.
74. **Cohen, P., Rylatt, D. B., and Nimmo, G. A.**, The hormonal control of glycogen metabolism: the amino acid sequence at the phosphorylation site of protein phosphatase inhibitor-1, *FEBS Lett.*, 76, 182, 1977.
75. **Antipenko, A. E., Goncharov, O. G., Skvortsova, G. P., and Lyzlova, S. N.**, Phosphorylation of phospholamban in experimental myocardial infarction and proteolysis stabilization during phosphorylation, *Biull. Eksp. Biol. Med.*, 96, 42, 1983.
76. **Oetting, W. S., Tuazon, P. T., Traugh, J. A., and Walker, A. M.**, Phosphorylation of prolactin, *J. Biol. Chem.*, 261, 1649, 1986.
77. **Padel, V. and Söling, H.-D.**, Phosphorylation of the ribosomal protein S6 during agonist-induced exocytosis in exocrine glands is catalyzed by calcium-phospholipid-dependent protein kinase (protein kinase C). Experiments with guinea pig parotid glands, *Eur. J. Biochem.*, 151, 1, 1985.
78. **Costa, M. R. C. and Catterall, W. A.**, Cyclic AMP-dependent phosphorylation of the alpha subunit of the sodium channel in synaptic nerve ending particles, *J. Biol. Chem.*, 259, 8210, 1984.
79. **Grand, R. J. A., Wilkinson, J. M., and Mole, L. E.**, The amino acid sequence of rabbit cardiac troponin I, *Biochem. J.*, 159, 633, 1976.
80. **Moir, A. J. G. and Perry, S. V.**, The sites of phosphorylation of rabbit cardiac troponin I by adenosine 3′:5′-cyclic monophosphate-dependent protein kinase. Effect of interaction with troponin C, *Biochem. J.*, 167, 333, 1977.
81. **Wilkinson, J. M. and Grand, R. J. A.**, The amino acid sequence of troponin I from rabbit skeletal muscle, *Biochem. J.*, 149, 493, 1975.
82. **Hider, R. C., Ragnarsson, U., and Zetterqvist, Ö.**, The role of the phosphate group for the structure of phosphopeptide products of adenosine 3′,5′-cyclic monophosphate-dependent protein kinase, *Biochem. J.*, 229, 485, 1985.
83. **Chessa, G., Borin, G., Marchiori, F., Meggio, F., Brunati, A. M., and Pinna, L. A.**, Synthetic peptides reproducing the site phosphorylated by cAMP-dependent protein kinase in protein phosphatase inhibitor-1. Effect of structural modifications on the phosphorylation efficiency, *Eur. J. Biochem.*, 135, 609, 1983.
84. **Glass, D. B., El-Maghrabi, M. R., and Pilkis, S. J.**, Synthetic peptides corresponding to the site phosphorylated in 6-phosphofructo-2-kinase/fructose-2,6-bisphosphatase as substrates of cyclic nucleotide-dependent protein kinases, *J. Biol. Chem.*, 261, 2987, 1986.
85. **Ragnarsson, U., Zetterqvist, Ö., Titanji, V., Edlund, B., and Engström, L.**, Substrate specificity of cyclic AMP-stimulated protein kinase and some related enzymes: a synthetic approach, in *Peptides 1978*, Siemion, I. Z. and Kupryszewski, G., Eds., Wroclaw University Press, Wroclaw, Poland, 1979, 339.
86. **Feramisco, J. R., Kemp, B. E., and Krebs, E. G.**, Phosphorylation of hydroxyproline in a synthetic peptide catalyzed by cyclic AMP-dependent protein kinase, *J. Biol. Chem.*, 254, 6987, 1979.
87. **Pike, L. J., Gallis, B., Casnellie, J. E., Bornstein, P., and Krebs, E. G.**, Epidermal growth factor stimulates the phosphorylation of synthetic tyrosine-containing peptides by A431 cell membranes, *Proc. Natl. Acad. Sci. U.S.A.*, 79, 1443, 1982.

88. Schägger, H., von Jagow, G., Borchart, U., and Machleidt, W., Amino-acid sequence of the smallest protein of the cytochrome c_1 subcomplex from beef heart mitochondria, *Hoppe-Seylers Z. Physiol. Chem.*, 364, 307, 1983.
89. Anderson, S., Bankier, A. T., Barrell, B. G., de Bruijn, M. H. L., Coulson, A. R., Drouin, J., Eperon, I. C., Nierlich, D. P., Roe, B. A., Sanger, F., Schreier, P. H., Smith, A. J. H., Staden, R., and Young, I. G., Sequence and organization of the human mitochondrial genome, *Nature (London)*, 290, 457, 1981.
90. Günzler, W. A., Steffens, G. J., Grossmann, A., Kim, S.-M. A., Ötting, F., Wendel, A., and Flohe, L., The amino-acid sequence of bovine glutathione peroxidase, *Hoppe-Seylers Z. Physiol. Chem.*, 365, 195, 1984.
91. Dwulet, F. E. and Putnam, F. W., Complete amino acid sequence of a 50,000-dalton fragment of human ceruloplasmin, *Proc. Natl. Acad. Sci. U.S.A.*, 78, 790, 1981.
92. Titani, K., Koide, A., Ericsson, L. H., Kumar, S., Hermann, J., Wade, R. D., Walsh, K. A., Neurath, H., and Fischer, E. H., Sequence of the carboxyl-terminal 492 residues of rabbit muscle glycogen phosphorylase including the pyridoxal 5'-phosphate binding site, *Biochemistry*, 17, 5680, 1978.
93. Bonner, T. I., Oppermann, H., Seeburg, P., Kerby, S. B., Gunnell, M. A., Young, A. C., and Rapp, U. R., The complete coding sequence of the human *raf* oncogene and the corresponding structure of the c-*raf*-1 gene, *Nucleic Acids Res.*, 14, 1009, 1986.
94. Arlaud, G. J. and Gagnon, J., Complete amino acid sequence of the catalytic chain of human complement subcomponent C1r, *Biochemistry*, 22, 1758, 1983.
95. Niemann, M. A., Bhown, A. S., Bennett, J. C., and Volanakis, J. E., Amino acid sequence of human D of the alternative complement pathway, *Biochemistry*, 23, 2482, 1984.
96. MacDonald, R. J., Swift, G. H., Quinto, C., Swain, W., Pictet, R. L., Nikovits, W., and Rutter, W. J., Primary structure of two distinct rat pancreatic preproelastases determined by sequence analysis of the complete cloned messenger ribonucleic acid sequences, *Biochemistry*, 21, 1453, 1982.
97. Goldberg, G. I., Wilhelm, S. M., Kronberger, A., Bauer, E. A., Grant, G. A., and Eisen, A. Z., Human fibroblast collagenase. Complete primary structure and homology to an oncogene transformation-induced rat protein, *J. Biol. Chem.*, 261, 6600, 1986.
98. de Bruijn, M. H. L. and Fey, G. H., Human complement component C3: cDNA coding sequence and derived primary structure, *Proc. Natl. Acad. Sci. U.S.A.*, 82, 708, 1985.
99. Belt, K. T., Carroll, M. C., and Porter, R. R., The structural basis of the multiple forms of human complement component C4, *Cell*, 36, 907, 1984.
100. Koide, T., Foster, D., Yoshitake, S., and Davie, E. W., Amino acid sequence of human histidine-rich glycoprotein derived from the nucleotide sequence of its cDNA, *Biochemistry*, 25, 2220, 1986.
101. Stanton, L. W., Schwab, M., and Bishop, J. M., Nucleotide sequence of the human N-*myc* gene, *Proc. Natl. Acad. Sci. U.S.A.*, 83, 1772, 1986.
102. Seidah, N. G. and Chrétien, M., Complete amino acid sequence of a human pituitary glycopeptide: an important maturation product of pro-opiomelanocortin, *Proc. Natl. Acad. Sci. U.S.A.*, 78, 4236, 1981.
103. Mercken, L., Simons, M.-J., Swillens, S., Massaer, M., and Vassart, G., Primary structure of bovine thyroglobulin deduced from the sequence of its 8,431-base complementary DNA, *Nature (London)*, 316, 647, 1985.
104. Allen, G. and Fantes, K. H., A family of structural genes for human lymphoblastoid (leukocyte-type) interferon, *Nature (London)*, 287, 408, 1980.
105. Goeddel, D. V., Leung, D. W., Dull, T. J., Gross, M., Lawn, R. M., McCandliss, R., Seeburg, P. H., Ullrich, A., Yelverton, E., and Gray, P. W., The structure of eight distinct cloned human leucocyte interferon cDNAs, *Nature (London)*, 290, 20, 1981.
106. Lawn, R. M., Adelman, J., Dull, T. J., Gross, M., Goeddel, D., and Ullrich, A., DNA sequence of two closely linked human leukocyte interferon genes, *Science*, 212, 1159, 1981.
107. Shaw, G. D., Boll, W., Taira, H., Mantei, N., Lengyel, P., and Weissmann, C., Structure and expression of cloned murine IFN-α genes, *Nucleic Acids Res.*, 11, 555, 1983.
108. Gray, P. W., Leung, D. W., Pennica, D., Yelverton, E., Najarian, R., Simonsen, C. C., Derynck, R., Sherwood, P. J., Wallace, D. M., Berger, S. L., Levinson, A. D., and Goeddel, D. V., Expression of human immune interferon cDNA in *E. coli* and monkey cells, *Nature (London)*, 295, 503, 1982.
109. Lomedico, P. T., Gubler, U., Hellmann, C. P., Dukovich, M., Giri, J. G., Pan, Y.-C. E., Collier, K., Semionow, R., Chua, A. O., and Mizel, S. B., Cloning and expression of murine interleukin-1 cDNA in *Escherichia coli*, *Nature (London)*, 312, 458, 1984.
110. March, C. J., Mosley, B., Larsen, A., Cerretti, D. P., Braedt, G., Price, V., Gillis, S., Henney, C. S., Kronheim, S. R., Grabstein, K., Conlon, P. J., Hopp, T. P., and Cosman, D., Cloning, sequence and expression of two distinct human interleukin-1 complementary DNAs, *Nature (London)*, 315, 641, 1985.
111. Dialynas, D. P., Murre, C., Quertermous, T., Boss, J. M., Leiden, J. M., Seidman, J. G., and Strominger, J. L., Cloning and sequence analysis of complementary DNA encoding an aberrantly rearranged human T-cell γ-chain, *Proc. Natl. Acad. Sci. U.S.A.*, 83, 2619, 1986.

112. **Saito, H., Kranz, D. M., Takagaki, Y., Hayday, A. C., Eisen, H. N., and Tonegawa, S.**, Complete primary structure of a heterodimeric T-cell receptor deduced from cDNA sequences, *Nature (London)*, 309, 757, 1984.
113. **Hellman, L., Pettersson, U., Engström, Å., Karlsson, T., and Bennich, H.**, Structure and evolution of the heavy chain from rat immunoglobulin E, *Nucleic Acids Res.*, 10, 6041, 1982.
114. **Putnam, F. W., Takahashi, N., Tetaert, D., Debuire, B., and Lin, L.-C.**, Amino acid sequence of the first constant region domain and the hinge region of the δ heavy chain of human IgD, *Proc. Natl. Acad. Sci. U.S.A.*, 78, 6168, 1981.
115. **Sodoyer, R., Damotte, M., Delovitch, T. L., Trucy, J., Jordan, B. R., and Strachan, T.**, Complete nucleotide sequence of a gene encoding a functional human class I histocompatibility antigen (HLA — CW3), *EMBO J.*, 3, 879, 1984.
116. **Kudo, J., Chao, L.-Y., Narni, F., and Saunders, G. F.**, Structure of the human gene encoding the invariant γ-chain of class II histocompatibility antigens, *Nucleic Acids Res.*, 13, 8827, 1985.
117. **Lin, A., McNally, J., and Wool, I. G.**, The primary structure of rat liver ribosomal protein L37. Homology with yeast and bacterial ribosomal proteins, *J. Biol. Chem.*, 258, 10664, 1983.
118. **Neame, P. J., Christner, J. E., and Baker, J. R.**, The primary structure link protein from chondrosarcoma proteoglycan aggregate, *J. Biol. Chem.*, 261, 3519, 1986.
119. **Oesch, B., Westaway, D., Wälchli, M., McKinley, M. P., Kent, S. B. H., Aebersold, R., Barry, R. A., Tempst, P., Teplow, D. B., Hood, L. E., Prusiner, S. B., and Weissmann, C.**, A cellular gene encodes scrapie PrP 27-30 protein, *Cell*, 40, 735, 1985.
120. **Greene, G. L., Gilna, P., Waterfield, M., Baker, A., Hort, Y., and Shine, J.**, Sequence and expression of human estrogen receptor complementary DNA, *Science*, 231, 1150, 1986.
121. **Walsh, K. A., Titani, K., Takio, K., Kumar, S., Hayes, R., and Petra, P. H.**, Amino acid sequence of the sex steroid binding protein of human blood plasma, *Biochemistry*, 25, 7584, 1986.
122. **Metz-Boutigue, M.-H., Mazurier, J., Jolles, J., Spik, G., Montreuil, J., and Jolles, P.**, The present state of the human lactotransferrin sequence. Study and alignment of the cyanogen bromide fragments and characterization of N- and C-terminal domains, *Biochim. Biophys. Acta*, 670, 243, 1981.
123. **Jenne, D. and Stanley, K. K.**, Molecular cloning of S-protein, a link between complement, coagulation and cell-substrate adhesion, *EMBO J.*, 4, 3153, 1985.
124. **Mason, A. J., Hayflick, J. S., Ling, N., Esch, F., Ueno, N., Ying, S.-Y., Guillemin, R., Niall, H., and Seeburg, P. H.**, Complementary DNA sequences of ovarian follicular fluid inhibin show precursor structure and homology with transforming growth factor-β, *Nature (London)*, 318, 659, 1985.
125. **Sim, G. K., Yague, J., Nelson, J., Marrack, P., Palmer, E., Augustin, A., and Kappler, J.**, Primary structure of human T-cell receptor α-chain, *Nature (London)*, 312, 771, 1984.
126. **Parker, P. J., Coussens, L., Totty, N., Rhee, L., Young, S., Chen, E., Stabel, S., Waterfield, M. D., and Ullrich, A.**, The complete primary structure of protein kinase C — the major phorbol ester receptor, *Science*, 233, 853, 1986.
127. **Kavinsky, C. J., Umeda, P. K., Levin, J. E., Sinha, A. M., Nigro, J. M., Jakovcic, S., and Rabinowitz, M.**, Analysis of cloned mRNA sequences encoding subfragment 2 and part of subfragment 1 of α- and β-myosin heavy chains of rabbit heart, *J. Biol. Chem.*, 259, 2775, 1984.
128. **Foster, D. C., Yoshitake, S., and Davie, E. W.**, The nucleotide sequence of the gene for human protein C, *Proc. Natl. Acad. Sci. U.S.A.*, 82, 4673, 1985.
129. **Kurachi, K. and Davie, E. W.**, Isolation and characterization of a cDNA coding for human factor IX, *Proc. Natl. Acad. Sci. U.S.A.*, 79, 6461, 1982.
130. **Ullrich, A., Gray, A., Berman, C., and Dull, T. J.**, Human β-nerve growth factor gene sequence highly homologous to that of mouse, *Nature (London)*, 303, 821, 1983.
131. **Fang, J., Jacobs, J. W., Kanner, B. I., Racker, E., and Bradshaw, R. A.**, Amino acid sequence of bovine heart coupling factor 6, *Proc. Natl. Acad. Sci. U.S.A.*, 81, 6603, 1984.

Chapter 8

UNIQUE SPECIFICITY DETERMINANTS FOR AN S6/H4 KINASE AND PROTEIN KINASE C: PHOSPHORYLATION OF SYNTHETIC PEPTIDES DERIVED FROM THE SMOOTH MUSCLE MYOSIN LIGHT CHAIN

Ruthann A. Masaracchia, Fern E. Murdoch, and Tommy C. Hassell

TABLE OF CONTENTS

I. Introduction 190

II. Substrate Specificity of H4 Kinase 190

III. Synthetic Peptide Design 196

IV. Enzyme Preparation 198

V. Is H4 Kinase Derived from Protein Kinase C? 200

VI. Extended H4 Kinase Specificity Determinants 203

VII. Summary 205

Acknowledgment 206

References 206

I. INTRODUCTION

In 1977 the purification of a murine lymphosarcoma protein kinase which exhibited a high affinity for the protein substrate histone H4 was reported.[1] Since that time we have identified the presence of this H4 kinase in several tissues, including human placenta, 3T3 cells, 3T3-SV40 cells, rat liver and testes, and HL60 cells. The data indicate that the enzyme appears to be widely distributed in both normal and transformed cells. The enzyme does not appear to be associated with plasma membranes, nuclei, or other cellular organelles since it is readily extracted in low salt buffers. The M_r based on gel filtration is 85,000.[1]

Since the initial report of H4 kinase purification, several other protein kinases with similar properties have been reported. Traugh and co-workers[2-4] have characterized two rabbit reticulocyte protein kinases which have chromatographic properties similar to those observed with H4 kinase. Both enzymes are activated by limited proteolysis *in vitro*[2-4] and are therefore designated protease-activated kinase I (PAK-I) and protease-activated kinase II (PAK-II). As discussed below, H4 kinase can also be activated by limited proteolysis, although it is unlikely that this is the cellular regulatory mechanism for either that enzyme or PAK-I and PAK-II.

In addition to chromatographic behavior, H4 kinase and PAK-II have similar M_r values (as determined by gel filtration), Mg^{+2}/Mn^{+2} requirements, and inhibitory properties with phosphate and fluoride. The substrate specificities of the enzymes are similar, but not identical. Other protein kinases with chromatographic properties similar to those observed with H4 kinase have been purified from frog oocytes,[5] 3T3 cells,[6] and P12 cells.[7] In all cases the enzymes catalyze S6 phosphorylation in the 40S ribosomal subunit, although other properties of the purified proteins differ somewhat from the properties of H4 kinase.[8-11] Finally, the chromatographic behavior and properties of H4 kinase are somewhat similar to protein kinase C, the ubiquitous calcium/phospholipid-dependent protein kinase.[8,12] Preliminary studies in this laboratory[11] and others[13] used synthetic peptide substrates to establish that the H4 kinase (PAK-II) and protein kinase C were distinct entities. These studies have now been extended in order to establish the unique substrate specificities of H4 kinase and protein kinase C and to determine that H4 kinase is not a proteolytic product of protein kinase C. The results of those studies are presented below.

II. SUBSTRATE SPECIFICITY OF H4 KINASE

Initial kinetic studies with intact histone H4 established that H4 kinase catalyzed phosphorylation of a single serine in the protein.[1] Histone H4 contains only two serines, at residues 1 and 47. The NH_2-terminal serine is acetylated.[14] Synthetic peptides corresponding to the NH_2-terminus of H4 (Ac–Ser–Gly–Arg–Gly–Lys–Gly) did not serve as an H4 kinase substrate.[1] However, a synthetic peptide containing serine-47 (Val–Lys–Arg–Ile–Ser–Gly–Leu) had a K_m of 36 µM, compared to a K_m of 16 µM for intact H4.[1] Subsequent studies in which the site in H4 phosphorylated by the H4 kinase was determined directly[9] confirmed that serine-47 is the only residue in H4 modified by the H4 kinase.

The sequence phosphorylated in H4 by H4 kinase[9] is remarkably similar to the sequence from pyruvate kinase which is phosphorylated by the cyclic AMP-dependent protein kinase.[15] In both cases dibasic residues are located on the NH_2-terminus, one residue removed from the phosphorylated serine.

H4 sequence: `Val-Lys-Arg-Ile-Ser-Gly-Leu`
Pyruvate kinase sequence: `Leu-Arg-Arg-Ala-Ser-Leu`

The similarity in these sequences might suggest that the enzymes share similar substrate specificity determinants and/or are related protein kinases. However, further studies with peptides derived from these sequences have shown that these are not valid predictions.[9]

Synthetic peptides corresponding to the H4 sequence have been used to show that H4 kinase requires the presence of a lysine-arginine sequence one residue NH_2-terminal from the modified serine for optimal activity (Table 1). Extension of the peptide in the NH_2-terminal direction did not significantly alter the kinetics of phosphorylation by H4 kinase. Translocation of the lysine-arginine sequence to positions further removed from the modified serine or deletion of either the lysine or arginine residue markedly decreased the ability of the peptides to serve as substrates for the H4 kinase (Table 1).

Of particular interest in the series of modified H4 peptides is the sequence in which the lysine is replaced by arginine; i.e., the sequence was modified to more closely resemble the cyclic AMP-dependent specificity determinants.[15] The K_m for this peptide was 7-fold higher than the parent peptide, and the V_{max}/K_m (V/K) with H4 kinase was decreased 34-fold. The high degree of specificity of H4 kinase for the lysine-arginine sequence was confirmed in studies with the peptide Leu–Arg–Arg–Ala–Ser–Leu (Kemptide), which is the phosphorylated sequence in pyruvate kinase. This peptide was phosphorylated relatively poorly by H4 kinase (K_m = 500 μM, V/K = 33 $min^{-1}mg^{-1}$; Table 1), but substitution of an arginine to generate a lysine-arginine sequence markedly improved the phosphorylation kinetics (Table 1).

The cyclic AMP-dependent protein kinase demonstrated a comparably high degree of specificity with respect to utilization of the substrates derived from the H4 sequence (Table 1). All synthetic peptides with the lysine-arginine sequence were relatively poor substrates for the cyclic AMP-dependent protein kinase (K_m > 670 μM). However, substitution of the lysine by arginine to give an arginine-arginine sequence NH_2-terminal to the serine generated a substrate with kinetic constants comparable to those observed with the pyruvate kinase sequence (Table 1). Similarly, modification of the arginine-arginine sequence in the pyruvate kinase peptide to a lysine-arginine sequence decreased the ability of this peptide to serve as a cyclic AMP-dependent protein kinase substrate by increasing the K_m 120-fold and decreasing the V/K 265-fold.

On the basis of these data, it was concluded that although the substrate specificity determinants for H4 kinase were similar to those of the cyclic AMP-dependent protein kinase they were not identical and that the H4 kinase favored substrates containing a lysine-arginine sequence NH_2-terminal to the phosphorylated serine.

The analyses of peptides derived from H4 are valuable in establishing the *in vitro* specificity of H4 kinase-catalyzed phosphorylation, but it is unlikely that H4 is phosphorylated by this kinase *in vivo*.[16] *In vivo* phosphorylation of H4 has been reported by Allfrey and co-workers,[17] but the site that was modified was serine-1, not serine-47. The ability of other proteins to serve as H4 kinase substrates has been investigated. In contrast to several other well-characterized protein kinases, H4 kinases did not catalyze phosphorylation of histones H2B, H2A, or H3, casein, phosvitin, protamine, phosphorylase kinase, phosphorylase, or myelin basic protein. Two protein substrates of potential physiological importance have been identified. These are the S6 protein of the 40S ribosomal subunit and the regulatory light chain of smooth and nonmuscle myosin.

The 40S ribosomal subunit was phosphorylated specifically at protein S6 by the H4 kinase (Figure 1). At least three serine residues were modified in this reaction.[10] Further evidence that both S6 and H4 are phosphorylated by the same kinase is provided by the observation that H4 can inhibit the S6 phosphorylation competitively (Figure 2).

All sites in S6 which are phosphorylated in response to growth factors[18,19] or partially purified S6 kinases[5-7,20,21] have not been sequenced. The phosphorylated S6 sequence which has been determined by Wettenhall and Cohen[22] is somewhat nonspecific in that cyclic AMP-dependent protein kinase,[13] protein kinase C,[23] and protease-activated kinases[13]

TABLE 1
Kinetic Constants for the Phosphorylation by H4 Kinase and Cyclic AMP-Dependent Protein Kinase of Synthetic Peptides Derived from Histone H4 and Pyruvate Kinase

Peptide sequence	H4-PK			Cyclic AMP-dependent protein kinase		
	Apparent K_m (μM)	V_{max} (nmol min^{-1}mg^{-1})	V/K_m (min^{-1}mg^{-1})	Apparent K_m (μM)	V_{max} (μmol min^{-1}mg^{-1})	V/K_m (min^{-1}mg^{-1})
Histone H4						
Val-Lys-Arg-Ile-Ser-Gly-Leu	35 ± 5.0	93 ± 11	27	1,300 ± 280	2.2 ± 0.91	17
Val-Arg-Arg-Ile-Ser-Gly-Leu	250 ± 42	21 ± 4.0	0.84	88 ± 16	12 ± 2.1	1,360
Val-Lys-Arg-Gly-Ser-Gly-Leu	121 ± 19	102 ± 9.1	8.4	670 ± 110	2.7 ± 0.33	40
Val-Arg-Lys-Ile-Ser-Gly-Leu	470 ± 57	16 ± 1.8	0.34	1,600 ± 310	1.2 ± 0.41	
Val-Lys-Arg-Gly-Ile-Ser-Gly-Leu	680 ± 32	27 ± 3.0	0.40	4,300 ± 370	4.2 ± 0.30	9.8
Pyruvate kinase						
Leu-Arg-Arg-Ala-Ser-Leu-Gly	500 ± 55	33 ± 6.1	0.67	7.2 ± 1.8	19 ± 1.4	26,000
Leu-Lys-Arg-Ala-Ser-Leu-Gly	143 ± 31	250 ± 18	17	915 ± 31	9 ± 1.1	98

Note: Synthetic peptides were assayed according to the procedures of Glass et al.[33] with H4 kinase purified from murine lymphosarcoma[8] or the catalytic subunit of cyclic AMP-dependent protein kinase purified from beef heart[32].

From Eckols, T. K., Thompson, R. E., and Masaracchia, R. A., *Eur. J. Biochem.*, 134, 249, 1983. With permission.

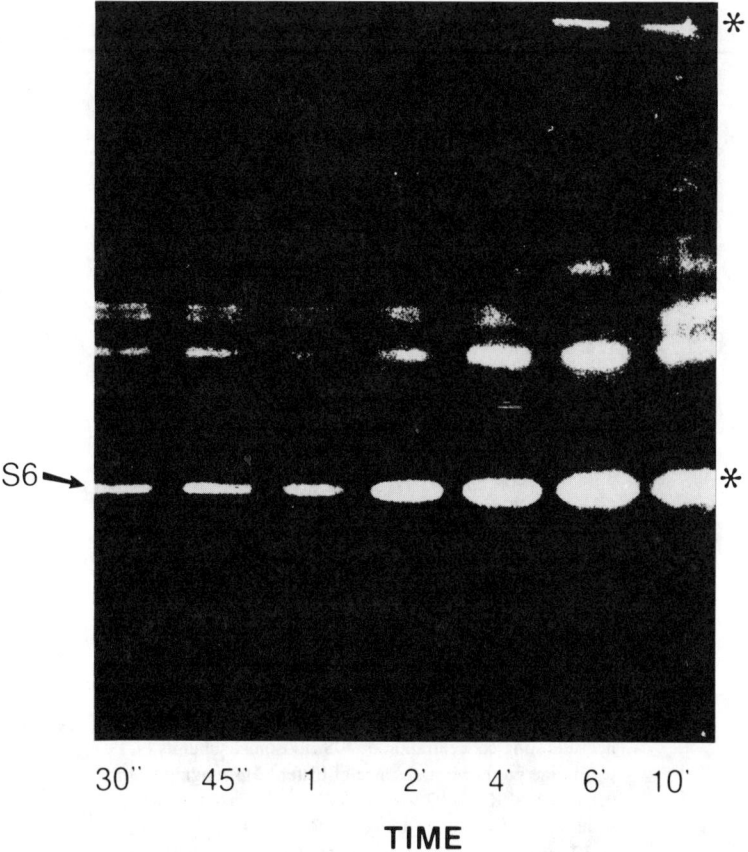

FIGURE 1. Phosphorylation of 40S ribosomal subunits by H4 kinase. Ribosomal subunits (total absorbance units = 7.5) were incubated with 267 μM [γ-^{32}P]ATP (195 dpm/pmol) and activated H4 kinase (4 μg) in a total volume of 1.0 ml. At the designated time intervals, aliquots (50 μl) were transferred to sodium dodecyl sulfate (SDS) sample buffer (50 μl) containing 6% SDS and 0.1 M dithiothreitol. Samples were prepared and electrophoresed in a 10% polyacrylamide gel for 12 h at 70 V. The autoradiogram (shown) was prepared from the dried, stained gel. Asterisk (*) denotes bands present only in samples containing H4 kinase. (From Donahue, M. J. and Masaracchia, R. A., *J. Biol. Chem.*, 259, 435, 1984. With permission.)

catalyze phosphorylation of the synthetic peptides derived from the sequence Arg–Arg–Leu–Ser–Ser–Leu–Arg–Ala.

Data obtained from the study of H4 peptides do not predict that this sequence is a favorable substrate for H4 kinase, although the role of basic residues in the COOH-terminal portion of a synthetic substrate was not investigated previously. The kinetic properties observed when H4 kinase activity was measured with the S6 peptide Arg–Arg–Leu–Ser–Ser–Leu–Arg–Ala–Gly were consistent with the predicted specificity determinants derived from previous data. The apparent K_m was 81 μM and the V/K was 3.1 min^{-1}mg^{-1}. This V/K is 56-fold less than that observed with H4. As the data with the H4 peptides and Kemptide predict, the diarginine sequence NH$_2$-terminal to the modified serine does contain sufficient specificity determinants to support phosphorylation by H4 kinase, but the kinetics are less favorable than those observed with a lysine-arginine-X-serine sequence.

The apparent inconsistency of the poor phosphorylation kinetics for the synthetic peptide and the efficient phosphorylation of the intact 40S ribosomal subunit may be attributed to

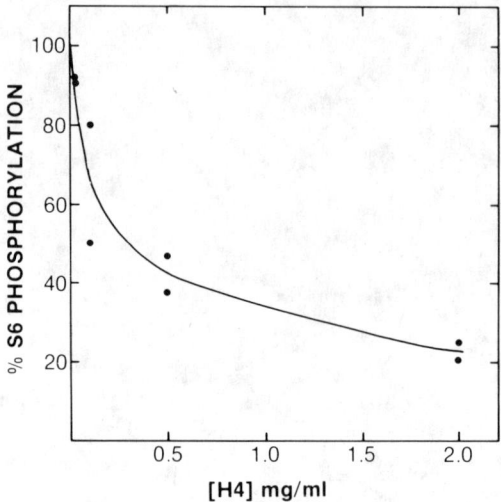

FIGURE 2. Competitive inhibition of placenta H4/S6 kinase-catalyzed S6 phosphorylation by histone H4.[4] H4/S6 kinase was trypsin activated for 3 min. The activated S6 kinase (5 µg) was preincubated with 20 mM MES, pH 6.8, 10 mM $MgCl_2$, 0.2 mM [γ-^{32}P]ATP (300 dpm/pmol), and 2 mM β-glycerophosphate for 10 min at 30°C. The phosphotransferase reaction was initiated by the addition of a nonsaturating concentration of 40S ribosomal subunits (7.1 µg) in the presence or absence of histone H4 ranging from 0.1 to 10 × K_m (0.02 to 2 mg/ml). Reaction was stopped by addition of an equal volume of SDS-PAGE sample buffer at 10 min. Proteins were separated by 10% SDS-PAGE, and S6 phosphorylation was quantitated by liquid scintillation counting of the S6 protein band cut from the dried gel. See Table 3 for further data.

several factors. Among these are (1) the site reproduced in the synthetic peptide may not represent the sites modified by H4 kinase in the intact ribosome, (2) contributions from other basic residues in S6 or other ribosomal proteins may enhance the ability to serve as an H4 kinase substrate in the ribosome, and (3) this site may not be modified stoichiometrically by H4 kinase *in vivo*.

The second protein phosphorylated by H4 kinase of potential physiological significance is the regulatory light chain of smooth muscle and nonmuscle myosin (LC-1). H4 kinase catalyzed LC-1 phosphorylation of intact thymus myosin and isolated mixed myosin light chains (Figure 3). Similar to results obtained with smooth muscle myosin,[24,25] protein kinase C and myosin light chain kinase (MLCK) also catalyzed phosphorylation of the thymus myosin and light chains (Figure 3).

When comparative rates of substrate phosphorylation with H4 kinase, protein kinase C, and MLCK were studied with myosin, light chains, and a peptide containing the MLCK phosphorylation site [MLC(1—23); Figure 4], significant differences in the reactivity of the three enzymes were observed (Table 2). MLCK catalyzed phosphorylation of all substrates at approximately equivalent rates. In contrast, protein kinase C-catalyzed phosphorylation of the peptide MLC(1—23) was approximately 100 times faster than myosin or light chain phosphorylation. The rates at which H4 kinase catalyzed phosphorylation of the isolated light chain and MLC(1—23) were 8- and 400-fold faster, respectively, than the intact myosin phosphorylation.

The data illustrate two important points. First, the significant differences in enzyme

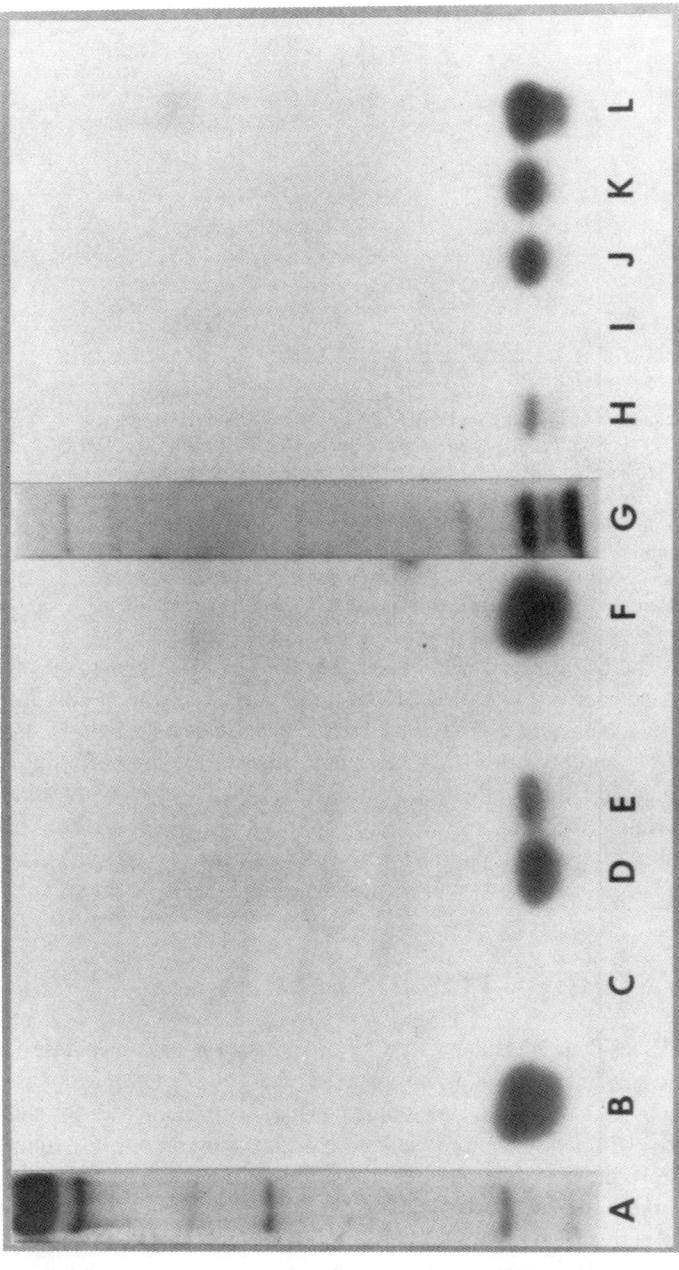

FIGURE 3. Phosphorylation of thymus myosin and isolated light chains by calcium-dependent and calcium-independent protein kinases. Purified thymus myosin (A to F; 70 μg) or myosin light chains (G to K; 20 μg) were incubated for 10 min with Mg[γ-^{32}P]ATP and bovine trachea MLCK in the presence (B,H) or absence (C,I) of calcium/calmodulin, protein kinase C in the presence (D,J) or absence (E,K) of calcium/phosphatidylserine, or H4/S6 kinase (F,L). The assay was terminated by the addition of SDS-PAGE buffer. The proteins were separated by 10% SDS-PAGE. Coomassie brilliant blue-stained samples (A,G) and autoradiographs (B to F; H to L) of the SDS-PAGE analysis are shown. (From Hassell, T. C., Kemp, B. E., and Masaracchia, R. A., *Biochem. Biophys. Res. Commun.*, 134, 240, 1986. With permission.)

NAME	SEQUENCE
MLC(1-23)	S-S-K-R-A-K-A-K-T-T-K-K-R-P-Q-R-A-T-S-N-V-F-G
MLC(1-13)	S-S-K-R-A-K-A-K-T-T-K-K-R-G
MLC(1-8)	S-S-K-R-A-K-A-K-G
MLC(3-13)	K-R-A-K-A-K-T-T-K-K-R-G
MLC(3-13, A10)	K-R-A-K-A-K-T-A-K-K-R-G
MLC(11-23)	K-K-R-P-Q-R-A-T-S-N-V-F-G
MLC(11-23, A14)	K-K-R-A-Q-R-A-T-S-N-V-F-G
MLC(14-23)	P-Q-R-A-T-S-N-V-F-G
MLC(14-23, K15)	P-K-R-A-T-S-N-V-F-G
MLC(14-23, K15, des T)	L-R-A-S-N-V-F-G

FIGURE 4. Amino acid sequences of synthetic peptides derived from the NH_2-terminal sequence of smooth muscle regulatory light chain. Peptides were prepared by the Merrifield method[34] adapted to a BioSearch SAM-II solid-phase peptide synthesizer. Peptides were cleaved from resin and deprotected with HF. Peptides were purified by HPLC on a preparative C-18 reverse-phase column (BioRad, Richmond, CA) with a linear gradient of 0 to 60% acetonitrile in 0.1% trifluoroacetic acid. All sequences were confirmed with an Applied Biosystems 470A gas-phase sequencer with a Model 120A on-line PTH analyzer. The peptides were used for the determination of specificity determinants of protein kinase C and 56/H4 kinase. Phosphorylated residues are underlined.

affinity for myosin and light chains observed with H4 kinase, but not MLCK, support the hypothesis that H4 kinase is not derived from MLCK by a proteolytic reaction. This hypothesis is further supported by the enzymes' different substrate specificities; i.e., H4 kinase catalyzes phosphorylation of H4 and S6, but MLCK does not. Second, the high affinity of protein kinase C and H4 kinase for the synthetic peptide MLC(1—23) suggests that this sequence may be useful in defining the unique substrate specificities of these enzymes. The results obtained from the study of synthetic peptides derived from MLC(1—23) and their reactivity with H4 kinase and protein kinase C will be presented in the subsequent sections of this chapter.

III. SYNTHETIC PEPTIDE DESIGN

Analysis of MLC(1—23) residues phosphorylated by H4 kinase or protein kinase C demonstrated that H4 kinase catalyzed serine phosphorylation exclusively, but protein kinase C catalyzed phosphorylation of both serine and threonine.[26] This observation was not sufficient to identify the respective H4 kinase and protein kinase C phosphorylation sites, since the peptide contains multiple serine and threonine residues. The serine residues are at the NH_2-terminus and in an internal basic sequence, which is the MLCK phosphorylation site. Earlier studies with peptides derived from the NH_2-terminus of histone H4[1] demonstrated that H4 kinase did not utilize the NH_2-terminal serine in that sequence as a phosphorylation site. The NH_2-terminal sequence

Peptide	NH_2-terminal sequence
H4	Ac-Ser-Gly-Arg-Gly-Lys-Gly
MLC(1—23)	Ac-Ser-Ser-Lys-Arg-Ala-Lys-Ala

TABLE 2
Comparative Rates of Myosin, Light Chain, and Peptide Phosphorylation by MLCK, Protein Kinase C, and H4 Kinase

Enzyme	Protein kinase activity (pmol/10 min)		
	Myosin	Light chain	MLC(1—23)
MLCK + Ca-Calmodulin	1.3	<1.0	<1.0
	5.8	3.0	11.3
Protein kinase C + Ca^{+2}, PS, OAG	3.5	6.0	<1.0
	7.0	5.5	597
H4 kinase	6.4	49	2900

Note: Enzymes (MLCK, 0.9 μg; PKC, 0.4 μg; H4PK, 10 μg) were incubated with myosin (100 μg), light chains (23 μg), or MLC(1—23) (435 μM) and Mg[γ-^{32}P]ATP (200 dpm/pmol) for 10 min. Enzyme concentrations were selected to give comparable rates of phosphorylation with myosin. Effector concentrations were as follows: CaCl$_2$ = 0.2 mM, calmodulin = 5 μM, phosphatidylserine (PS) = 100 μg/ml, 1-oleoyl-2-acetyl-sn-glycerol (OAG) = 9 μg/ml.

From Hassell, T. C., Kemp, B. E., and Masaracchia, R. A., *Biochem. Biophys. Res. Commun.*, 134, 240, 1986. With permission.

of MLC(1—23) is somewhat homologous to the NH$_2$-terminus of H4 and would not be predicted to be an H4 kinase phosphorylation site on the basis of these data. Furthermore, H4 kinase phosphorylation of myosin activated MgATPase activity, suggesting that H4 kinase catalyzed phosphorylation of LC-1 at the regulatory, or MLCK, site.[26] On the basis of these data, serine-19 was assigned as the most probable site of H4 kinase-catalyzed phosphorylation of MLC(1—23).

Prediction of the specific MLC(1—23) threonine and serine residues which are modified by protein kinase C is more difficult. All the potential threonine phosphorylation sites contain basic residues NH$_2$-terminal to the threonine. Using several synthetic peptides derived from a variety of proteins, Woodgett et al.[27] and Turner et al.[28] reported that arginine residues NH$_2$-terminal to a protein kinase C phosphorylation site were required for favorable kinetics. The exception to this result was the observation that the NH$_2$-terminal serine in lactate dehydrogenase was phosphorylated by protein kinase C with a K$_m$ of 10 μM.[27]

Woodgett et al.[27] and Turner et al.[28] also demonstrated that an arginine residue COOH-terminal to the phosphorylation site improved reaction kinetics with protein kinase C, although a peptide which was derived from glycogen synthase and contained no COOH-terminal basic residue was phosphorylated by protein kinase C with a K$_m$ of 40 μM.[23]

The threonines at positions 9 and 10 are flanked by both an arginine and lysines in the COOH-sequence, whereas the threonine at position 18 does not have a basic residue in the COOH-terminal sequence (Figure 4). On this basis, the threonines at positions 9 and 10 may be predicted to be the favored phosphorylation site. To test this hypothesis, MLC(1—23) phosphorylated by protein kinase C was digested with thermolysin. The specificity of thermolysin predicts that at pH 6.4 the NH$_2$-terminal seryl residues will occur in a short (n = 5) basic peptide, that the threonine-9 and threonine-10 residues will occur in a large (n = 12) basic peptide, and that threonine-18 and serine-19 (the MLCK phosphorylation sites) will occur in a short (n = 4 or 6) neutral peptide. Results of the thermolysin digestion demonstrated that the major portion of the protein kinase C-catalyzed phosphorylation occurred at threonine-9 or -10.[26] Recent data obtained by sequencing the smooth muscle myosin light chain phosphorylated by protein kinase C have confirmed that threonine-9 is the protein kinase C phosphorylation site.[29] This is consistent with an earlier report which established that the threonine modified by protein kinase C was not the threonine at the MLCK site.[30]

Collectively, the data support the hypothesis that two phosphorylation domains exist in the synthetic peptide MLC(1—23). One domain includes the threonine modified by protein kinase C (Figure 4). In addition, this domain includes the NH_2-terminal serine residues which are phosphorylated by protein kinase C in both heavy meromyosin and isolated LC-1 from platelets.[29] The second domain contains serine-19, which is the MLCK phosphorylation site. This domain differs from the first domain significantly in that there are no basic residues COOH-terminal to the modified amino acid.

Synthetic peptides utilized to probe the specificity determinants of protein kinase C, H4 kinase, and MLCK in the MLC(1—23) sequence were designed to focus on the significance of basic residues NH_2-terminal to the phosphorylation site. By design, therefore, the contribution of COOH-terminal amino acids to protein kinase C specificity determinants[27,28] was not addressed in this study. The peptides studied are shown in Figure 4.

IV. ENZYME PREPARATION

H4 kinase was purified from human placenta by a modification of the method described for lymphosarcoma.[1,8] Two modifications of the protocol were made. First, the placental extract was prepared in 20 mM Tris$^+$Cl$^-$, pH 7.5, containing 30 mM 2-mercaptoethanol, 0.5 mM EDTA, 2 µM leupeptin, 0.2 mM phenylmethylsulfonyl fluoride (PMSF), and 6 mM benzamidine. All other buffers were also modified to contain these additions. Second, after the ammonium sulfate precipitation the sample was applied to a column of phosphocellulose (40 ml) equilibrated with 20 mM 2-(N-morpholino)ethanesulfonic acid (MES), pH 6.8, containing 30 mM 2-mercaptoethanol, 0.5 mM EDTA, 2 µM leupeptin, 0.2 mM PMSF, and 6 mM benzamidine. The proteins were eluted with a 200-ml linear gradient containing 0 to 0.6 M KCl. The S6 kinase activity eluted at 25 to 45 mmho (approximately 0.2 to 0.4 M KCl) The fractions containing enzyme activity were pooled, dialyzed, and chromatographed sequentially on CM-Sephadex® and Sephacryl® S200 as described previously.[1,8]

The enzyme was eluted from the Sephacryl® S200 column in two forms. An active form of the enzyme eluted with an apparent M_r of 85,000 (Figure 5). The enzyme catalyzed phosphorylation of H4, S6, S6-peptide, LC-1, Kemptide, and MLC(1—23) (Table 3). Little activity was observed with casein, protamine, phosvitin, and H1. The specific activity of the enzyme with H4 was 200 to 400 nmol P_i transferred per minute-milligram. The enzyme is analogous to the lymphosarcoma enzyme which was shown to catalyze multisite S6 phosphorylation. This enzyme is designated H4/S6 PK-II.

When the Sephacryl® S200 fractions were briefly preactivated with trypsin[2,13] a second peak of protein kinase activity was observed (Figure 5). This enzyme eluted with an apparent M_r of 158,000. This enzyme catalyzed phosphorylation of precisely the same substrates as H4/S6 PK-II (Table 3) and on this basis was designated H4/S6 PK-I. The specific activity of this enzyme with H4 was 400 to 600 nmol P_i transferred per minute-milligram.

The proteins in the Sephacryl® S200 fractions (Figure 5) were analyzed by sodium dodecyl sulfate polyacrylamide gel electrophoresis (SDS-PAGE) in order to determine if the activity of the H4/S6 PK coincided with the elution of specific proteins. Laser scanning densitometry of the Coomassie blue-stained SDS-PAGE gels revealed two proteins which coincided with the protein kinase activity (Figure 6). The major protein in the active H4 kinase peak (H4/S6 PK-II) migrated with an apparent M_r of 66,000 in the SDS-PAGE. The protein correlated precisely with the occurrence of the protein kinase activity (Figures 5 and 6).

The distribution of the M_r 66,000 protein was biphasic, and a second peak of M_r 66,000 protein in the SDS-PAGE coincided with the trypsin-activatable H4 kinase (H4/S6 PK-I). In the fractions containing H4/S6 PK-I activity the M_r 66,000 protein was one of two major

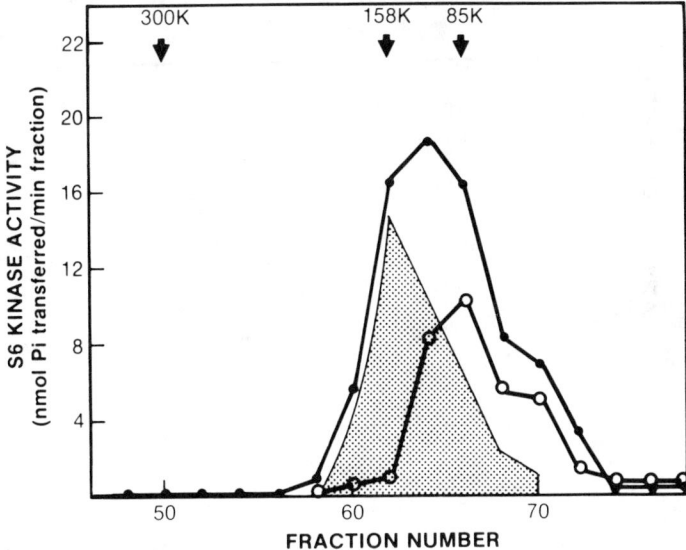

FIGURE 5. Elution of H4/S6 kinases from Sephacryl® S200. H4 kinase purified from 283 g of human placenta was applied to a column (2.5 × 95 cm) of Sephacryl® S200 equilibrated in 20 mM MES, pH 6.8, 1 mM EDTA, 1 μM leupeptin, and 1 mM dithiothreitol. Proteins were eluted in 3.2-ml fractions. Protein kinase activity was assayed with S6 peptide (RRLSSLRA) after activation by MgATP (−○−) or limited trypsin digestion (−●−). The shaded area is net trypsin-dependent activity.

TABLE 3
Substrate Specificity of H4/S6 Kinases from Human Placenta

Substrate	Protein kinase activity (% H4 activity)	
	H4/S6 kinase I	H4/S6 kinase II
H4 (1 mg/ml)	100	100
H1 (1 mg/ml)	8.2	10
40S ribosomes (10 μg)	47	57
Myosin light chains (0.5 mg/ml)	9.7	8.9
V–K–R–I–S–G–L[a] (100 μM)	36	44
R–R–L–S–S–L–R–A[b] (100 μM)	9.2	16
MLC(1—23)[c] (100 μM)	57	63
L–R–R–A–S–L–G[d] (100 μM)	1.1	3.0

Note: H4/S6 kinase I (0.69 μg) or H4/S6 kinase II (0.91 μg) purified from human placenta was incubated with substrates and Mg[γ-32P]ATP for 10 min. Protein phosphorylation was quantitated according to Glass et al.[33] except for the phosphorylation of S6 in the 40S ribosome which was quantitated by SDS-PAGE and autoradiography.[10] The H4/S6 kinase I was activated by incubation of enzyme (approximately 2 μg) in 10 mM MES, pH 6.8, 24 mM KCl, and 2 mM 2-mercaptoethanol (10 μl) with 60 ng trypsin in 20 mM TRIS-Cl, pH 8.0, and 50 mM 2-mercaptoethanol containing 40 μg bovine serum albumin (5 μl) for 10 min at 30°C. The reaction was stopped by the addition of 120 ng soybean trypsin inhibitor in 10 mM TRIS-Cl, pH 7.5 (5 μl).

[a] Phosphorylation sequence in histone H4 (see Table 1).
[b] Phosphorylation sequence in S6 (see text).
[c] Phosphorylation sequence in the myosin light chain (see Figure 4).
[d] Phosphorylation sequence in pyruvate kinase (Kemptide; see Table 1).

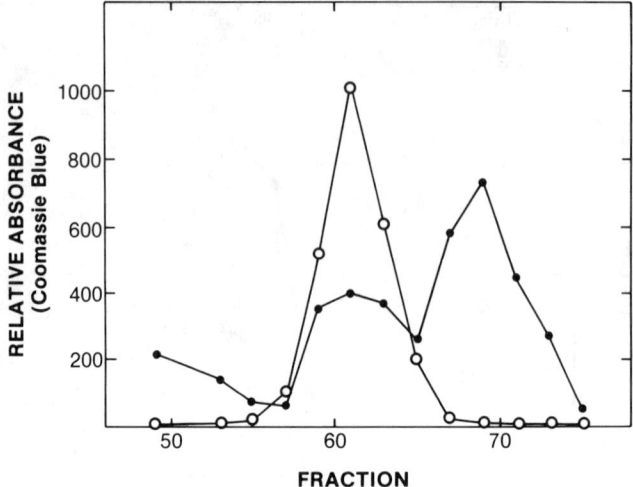

FIGURE 6. Quantitation of proteins in the H4/S6 kinase preparation from human placenta. Aliquots (20 µl) from alternate fractions of the Sephacryl® S200 chromatography shown in Figure 5 were added to SDS-PAGE buffer and analyzed by electrophoresis in a 12.5% polyacrylamide gel containing SDS. Proteins were stained with Coomassie brilliant blue, destained electrophoretically, and quantitated by lasar scanning densitometry of the dried gel (shown). The absorbance of proteins migrating with $M_r = 95,000$ ($-\circ-$) and 66,000 ($-\bullet-$) is shown.

proteins whose occurrence coincided with the protein kinase activity. The second major protein, which comigrated in the gel filtration column with enzyme activity and the M_r 66,000 protein, migrated in the SDS-PAGE with an M_r of 95,000 (Figure 6). The occurrence of this protein is of particular interest since Erikson and Maller[5] have reported purification of an M_r 95,000 S6 kinase in frog oocytes. These data suggest that H4 kinase may contain multiple subunits that can be dissociated to generate the active enzyme.

The hypothesis that the M_r 66,000 protein is the H4 kinase has been further substantiated by binding experiments with 5-fluorosulfonylbenzoyladenosine (Table 4). The M_r 66,000 protein bound the ATP analogue, and this binding was decreased in the presence of MgATP. No other protein in the preparation demonstrated significant 5-fluorobenzoyladenosine binding. The active H4 kinase was used as the enzyme source in all experiments in this report, although comparable results have been obtained with the trypsin-activated enzyme.

The protein kinase C used in this report was prepared from the lymphosarcoma P1798 by a modification of published procedures.[11,31] This tissue is particularly rich in protein kinase C. The published procedures were modified to include Sephacryl® S200 chromatography subsequent to the CM-Sephadex® chromatography[11] and prior to phenyl-Sepharose® chromatography.[31] The purified enzyme had a specific activity of 3 µmol P_i transferred per minute-milligram with histone H1, and activity was 94% dependent on the presence of effectors (0.2 mM $CaCl_2$, 60 µg/ml phosphatidylserine, and 10 µg/ml 1-oleoyl-2-acetyl-sn-glycerol). The catalytic subunit of the cyclic AMP-dependent protein kinase was prepared from bovine heart.[32]

V. IS H4 KINASE DERIVED FROM PROTEIN KINASE C?

Although initial data indicated that H4 kinase and protein kinase C were highly similar in chromatographic properties and substrate specificity,[8,12] studies with the H4 peptides[9,11] and the S6 peptides[13] determined that the two enzymes were unique entities. However, the

TABLE 4
Binding of [^{14}C]FSBA to Proteins in the H4/S6 Kinase Preparation

Protein M_r	^{14}C-FSBA (pmol/band)		[Protein] (Absorbance/M_r)
	A	B	
95,000	<0.2	<0.2	8
83,000	<0.2	<0.2	5
72,000	0.46	<0.2	12
66,000	4.1	0.87	60
38,000	0.91	0.36	21

Note: H4/S6 kinase (188 μg) was incubated with 50 μM [8-adenosine-^{14}C]5'-p-fluorosulfonylbenzoyladenosine (FSBA; 48 mCi/mmol) in the absence (A) or presence (B) of 2 mM ATP. The incubation was carried out at 30°C for 120 min, and radiolabeled products were detected by SDS-PAGE and autoradiography. Labeled protein bands were cut from the dried gel, solubilized in Protosolve, and quantitated by liquid scintillation counting. [Protein] was estimated by laser scanning densitometry of a duplicate SDS-PAGE stained with Coomassie blue dye. Dye binding was normalized by the ratio of relative absorbance to M_r.

data in these studies did not establish conclusively that a proteolyzed product of protein kinase C (protein kinase M)[31] was unrelated to H4 kinase. Since both protein kinase C and H4 kinase catalyzed myosin light chain phosphorylation as well as MLC(1—23) phosphorylation, this sequence was used to determine what relationship, if any, there was between the two enzymes.

The synthetic peptide MLC(1—23), which contains both the threonine phosphorylated by protein kinase C and the serine phosphorylated by MLCK, was phosphorylated by H4 kinase and protein kinase C with comparable K_m values (Table 5). The V_{max} are difficult to compare in this experiment, since protein kinase C is more pure than the H4 kinase, but it can be noted that the V_{max} observed with MLC(1—23) and H4 kinase is approximately 30% of the V_{max} observed with the H4 peptide shown in Table 1. The V_{max} observed with MLC(1—23) and protein kinase C is approximately 14% of the V_{max} observed with that enzyme and histone H1.

When the parent peptide was truncated into an NH$_2$-terminal domain MLC(1—23) and a COOH-terminal domain MLC(11—23), substantial differences in the enzymes' reactivities were observed (Table 5). All peptides derived from the NH$_2$-terminal sequence of the smooth muscle myosin light chain were poor substrates for H4 kinase.

No reactivity of these peptides with H4 kinase was detectable with peptide concentrations as high as 2 mM. This is not surprising since threonine residues were previously noted to be poor H4 kinase phosphorylation sites.[9] In addition, the MLC(1—13) sequence lacks a dibasic sequence NH$_2$-terminal to the threonines. As discussed previously, the NH$_2$-terminal serines would not be predicted to be H4 kinase phosphorylation sites.

In contrast, the truncated peptide MLC(1—13) was a much better substrate for protein kinase C than was the full-length peptide MLC(1—23) (Table 5). The K_m was slightly higher with the truncated peptide (25 vs. 15 μM). The V_{max} was increased sevenfold, yielding an overall increase in the pseudo-first-order rate constant V/K of fourfold. These data suggest that the COOH-terminal domain (MLCK domain) of MLC(1—23) may interact with protein kinase C and regulate the reactivity of that enzyme with the threonine site in the NH$_2$-terminal domain.

TABLE 5
Kinetic Constants for the Phosphorylation by H4 Kinase and Protein Kinase C of Synthetic Peptides Derived from the NH_2-Terminal Domain of the Regulatory Myosin Light Chain

Peptide sequence[b]	H4-PK[a]			Protein kinase C		
	Apparent K_m (μM)	V_{max} (nmol min^{-1}mg^{-1})	V/K_m (min^{-1}mg^{-1})	Apparent K_m (μM)	V_{max} (μmol min^{-1}mg^{-1})	V/K_m (min^{-1}mg^{-1})
MLC(1—23)	26	34	26	15	0.43	0.54
MLC(1—13)	N.D.	N.D.	N.D.	25	3.0	2.4
MLC(11—23)	54	56	21	91	0.13	0.028
MLC(3—13)	N.D.	N.D.	N.D.	20	3.0	3.0
MLC(1—8)	N.D.	N.D.	N.D.	400	0.41	0.02
MLC(3—13, A10)	N.D.	N.D.	N.D.	90	0.36	0.075

[a] N.D.: no phosphorylation was detected with 5 mM peptide.
[b] Sequences for all peptides are shown in Figure 4.

With both MLC(1—23) and MLC(1—13), protein kinase C catalyzed phosphorylation of both serine and threonine, as determined by partial acid hydrolysis and thin layer electrophoresis (data not shown). These data indicate that both the NH_2-terminal serines and threonines 9 and 10 can be phosphorylation sites. Based on the distribution of ^{32}P label in the phosphoamino acids recovered from the partial acid hydrolysis, the phosphorylation of threonine accounted for approximately 80% of the modified residues. This is in agreement with results obtained by others with the intact light chain.[26]

When the NH_2-terminal peptide MLC(1—13) was modified to contain only the threonine phosphorylation site [MLC(3—13); Figure 4], little change in the kinetic constants with protein kinase C was observed (Table 5). These results, in conjunction with the analysis of phosphorylated residues, suggest that the threonine site is the favored substrate of protein kinase C. This was further substantiated by the synthesis of a peptide containing the NH_2-terminal serine residues, but lacking the threonine site [MLC(1—8); Figure 4]. This peptide contains all the basic residues COOH-terminal to the serine that are predicted to determine protein kinase C reactivity based on the results of others.[23,27] Nevertheless, the peptide was a poor substrate for protein kinase C, demonstrating a pseudo-first-order rate constant V/K which was at least 100-fold lower than that observed with any of the peptides which contained the threonine sequence. The data confirm that the kinetic constants observed for protein kinase C and MLC(1—13) principally reflect the phosphorylation of threonine-9 or -10.

A peptide which modified the dithreonine sequence was synthesized in order to confirm the results of Bengur et al.[29] demonstrating that threonine-9 is phosphorylated in the intact light chain. In this peptide [MLC(3—13, A-10)] the threonine in position 10 was substituted with an alanine (Figure 4). The observed K_m and V_{max} with protein kinase C were 90 μM and 360 nmol/min-mg, respectively. The V/K was 0.075 $min^{-1}mg^{-1}$. These kinetic parameters are far less favorable for protein kinase C-catalyzed phosphorylation than are those with the dithreonine sequence. The data suggest that threonine-10 also may be a phosphorylation site in the peptide. Alternatively, threonine-10 may be an important specificity determinant for protein kinase C in this sequence.

The failure of H4 kinase to catalyze phosphorylation of any of these peptides derived from the NH_2-terminal domain of MLC(1—23) provides strong evidence that H4 kinase is different from protein kinase C. To establish this distinction conclusively, protein kinase C was digested briefly with trypsin to generate an effector-independent enzyme (protein kinase "M"; Figure 7). Protein kinase "M" catalyzed phosphorylation of intact light chain, MLC(1—23), MLC(1—13), and MLC(3—13) in addition to histone H1 and protamine (data not shown). The kinetic constants for protein kinase "M" were determined with MLC(3—13). The apparent K_m of MLC(3—13) with protein kinase "M" was 45 μM as compared to a K_m of 20 μM observed with that peptide and protein kinase C. The V_{max} was decreased approximately tenfold to 0.38 nmol/min-mg. It was noted in these experiments that protein kinase "M", generated by limited digestion with trypsin, was unstable and activity was lost at a rate of approximately 35% per hour when the enzyme was stored on ice and assayed for 10 min at 30°C. The significant decrease in V_{max} may reflect enzyme instability. Nevertheless, the proteolysis clearly did not alter the substrate specificity of protein kinase C, and the data do not support the hypothesis that H4 kinase is derived from protein kinase C, since H4 kinase does not catalyze phosphorylation of these peptides.

VI. EXTENDED H4 KINASE SPECIFICITY DETERMINANTS

Phosphorylation of synthetic peptides derived from H4 established that the sequence Lys–Arg–X–Ser was a favored sequence for H4 kinase,[9] but not cyclic AMP-dependent kinase[9] or protein kinase C.[11] The phosphorylation of the MLC(1—23) peptide by H4 kinase is not predicted by those results, since the Lys–Arg sequence in MLC(1—23) is five residues

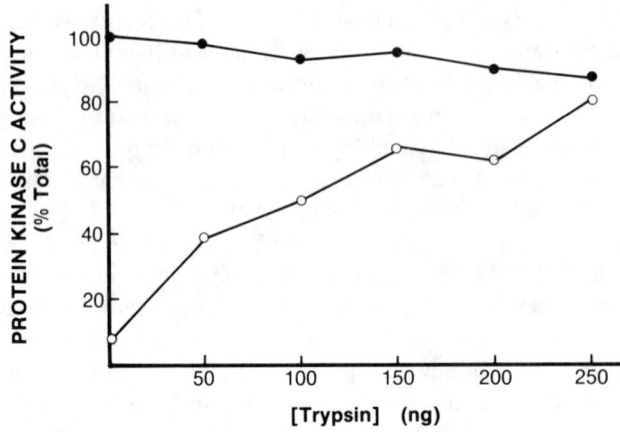

FIGURE 7. Conversion of protein kinase C to protein kinase M by limited trypsin digestion. Protein kinase C (2.4 μg) was incubated with trypsin at the indicated concentrations for 3 min at 30°C in the presence of 20 mM Tris, pH 7.2, and 40 μg bovine serum albumin. The reaction was terminated by the addition of trypsin inhibitor, and aliquots containing 0.4 μg protein kinase C were assayed for protein kinase activity with 200 μM MLC(3—13) in the presence (—○—) or absence (—●—) of 0.2 M CaCl$_2$, 100 μg/ml phosphatidylserine, and 9 μg/ml 1-oleoyl-2-acetyl-sn-glycerol.

removed from the modified serine. Previous studies demonstrated that displacement of the lysine-arginine sequence from the serine by even one residue significantly decreased the K_m and the V_{max} for H4 kinase (Table 1).

When MLC(1—23) was truncated so that the synthetic peptide contained only the COOH-terminal MLCK domain [MLC(11—23); Figure 4], H4 kinase catalyzed phosphorylation of serine at the MLCK phosphorylation site with kinetics comparable to those observed with the histone H4-derived peptides (Table 6). The apparent K_m with MLC(11—23) was 54 μM as compared to 36 μM with the H4 peptide. In addition, the V/K for MLC(11—23) was comparable to that observed with the H4 peptide (Table 6). The data predict that either the single arginine, perhaps in combination with other determinants such as the adjacent threonine, contributes the necessary determinants to the peptide to permit H4 kinase phosphorylation, or the cluster of remote basic residues can serve as specificity determinants for H4 kinase.

To test the possibility that the single arginine proximal to the serine alone fulfills the H4 kinase requirement for basic residues in the substrate sequence, a peptide [MLC(14—23)] which lacked the remote cluster of basic amino acids was prepared and kinetic constants with H4 kinase determined. With this peptide the apparent K_m increased approximately 7-fold and the V_{max} decreased 3-fold, resulting in an overall decrease of V/K of 20-fold (Table 6). The data clearly indicate that the proximal arginine in the sequence Arg–Ala–Thr–Ser contributes only a small portion of the specificity determinants required for the phosphorylation of MLC(11—23). The data support the hypothesis that the remote basic sequence Lys–Lys–Arg–Pro must make a significant contribution to the reactivity of the peptide.

The amino acid sequence in the myosin light chain peptides differs from the major portion of synthetic peptides utilized in studies of protein phosphorylation in that a proline occurs between the phosphorylated residue and the remote cluster of basic amino acids which apparently determine the substrate specificity (Figure 4). A kink in the peptide chain, introduced by this proline, may bring the remote basic residues into closer proximity with the serine, thereby resulting in a spatial arrangement comparable to that of the H4 peptide.

TABLE 6
Kinetic Constants for the Phosphorylation by H4 Kinase of Synthetic Peptides Derived from the MLCK-Phosphorylation Site of the Regulatory Myosin Light Chain

Peptide sequence	Apparent K_m (μM)	V_{max} (nmol min^{-1}mg^{-1})	V/K (min^{-1}mg^{-1})
MLC(1—23)[a]	26	34	26
MLC(11—23)	54	56	21
MLC(11—23, A14)	94	9	1.9
MLC(14—23)	350	9	0.5
MLC(14—23, K15)	220	16	1.5
MLC(14—23, K15, des T)	42	86	41
V–K–R–I–S–G–L[b]	35	93	27
V–R–R–I–S–G–L	250	21	0.84

[a] Sequences for all MLC peptides are shown in Figure 4.
[b] Sequences derived from H4 as shown in Table 1.

To test this hypothesis, the proline was replaced by alanine and the kinetic constants of the peptide with H4 kinase were determined (Table 6).

The removal of the proline from the MLC(11—23) sequence increased the K_m 4-fold and decreased the V_{max} 3-fold, resulting in an overall 14-fold decrease in the V/K. These data demonstrate that the remote basic residues contribute to the specificity determinants of the sequence and that the proline is required to achieve the appropriate orientation of those residues.

To confirm that the Lys–Arg sequence established as an important determinant in the H4 studies could also confer the appropriate specificity determinants to the myosin light chain sequence, the MLC(14—23) peptide was modified to contain a proximal Lys–Arg sequence (Figure 4). The MLC(14—23), which was a poor substrate for H4 kinase (Table 6), was modified to contain the sequence Lys–Arg–Ala–Thr–Ser in place of Gln–Arg–Ala–Thr–Ser. The kinetic parameters obtained with this peptide [MLC(14—23, K15)] were slightly better than those observed with MLC(14—23), but the peptide was still utilized relatively poorly compared to the H4 peptide (Table 6). In a further attempt to improve the reaction kinetics, the MLC(14—23) sequence was modified to position the dibasic sequence more closely to the phosphorylated residue. This peptide [MLC(14—23, K15, des T)] was highly analogous to the histone H4 peptide:

> H4 peptide: Lys–Arg–Ile–Ser–Gly
> MLC(14—23, K15, des T): Lys–Arg–Ala–Ser–Asn

This peptide was the best H4 kinase substrate derived from the myosin light chain sequence (Table 6). Both the K_m and V_{max} were comparable to those observed with the histone H4 peptide. These data confirm that a lysine-arginine sequence within one or two residues NH$_2$-terminal to the phosphorylated serine is a principal specificity determinant for H4 kinase.

VII. SUMMARY

The NH$_2$-terminal amino acid sequence of the smooth muscle myosin light chain has been used to establish that the H4 kinase is not related to protein kinase C, although both enzymes catalyze phosphorylation of several common substrates. Synthetic peptides derived from the protein kinase C phosphorylation site in the light chain clearly establish that the specificity determinants for protein kinase C and H4 kinase differ. In the course of these studies a synthetic peptide which may prove to be a specific substrate for protein kinase C and protein kinase "M" has been developed, i.e., K–R–A–K–A–K–T–T–K–K–R.

In addition, studies on peptides derived from the myosin light chain kinase phosphorylation site in the smooth muscle sequence (serine-19) have confirmed that the most favored sequence for H4 kinase-catalyzed phosphorylation is a lysine-arginine sequence one of two residues NH_2-terminal to the modified serine.

ACKNOWLEDGMENT

This work was supported by Grant B864 from The Robert A. Welch Foundation. We gratefully acknowledge the expert technical assistance of Peggy Magnino and Becky A. de la Houssaye.

REFERENCES

1. **Masaracchia, R. A., Kemp, B. E., and Walsh, D. A.**, Histone 4 phosphotransferase activities in proliferating lymphocytes. Partial purification and characterization of an enzyme specific for serine-47, *J. Biol. Chem.*, 252, 7109, 1977.
2. **Tahara, S. M. and Traugh, J. A.**, Differential activation of two protease-activated protein kinases from reticulocytes by a Ca^{+2}-stimulated protease and identification of phosphorylated translational components, *Eur. J. Biochem.*, 126, 1395, 1982.
3. **Lubben, T. H. and Traugh, J. A.**, Cyclic nucleotide-independent protein kinases from rabbit reticulocytes. Purification and characterization of protease-activated kinase II, *J. Biol. Chem.*, 258, 13992, 1983.
4. **Perisic, O. and Traugh, J. A.**, Protease-activated kinase II as the potential mediator of insulin-stimulated phosphorylation of ribosomal protein S6, *J. Biol. Chem.*, 258, 9589, 1983.
5. **Erikson, E. and Maller, J. L.**, Purification and characterization of a protein kinase from *Xenopus* eggs highly specific for ribosomal protein S6, *J. Biol. Chem.*, 261, 350, 1986.
6. **Cobb, M. H.**, An insulin-stimulated ribosomal protein S6 kinase in 3T3-L1 cells, *J. Biol. Chem.*, 261, 12994, 1986.
7. **Rowland, E. A., Muller, T. H., Goldstein, M., and Greene, L. A.**, Cell-free detection and characterization of a novel nerve growth factor-activated protein kinase in PC12 cells, *J. Biol. Chem.*, 262, 7504, 1987.
8. **de la Houssaye, B. A., Eckols, T. K., and Masaracchia, R. A.**, Activation of a cyclic AMP-independent protein kinase by an endogenous ATP-requiring protease from lymphosarcoma cells, *J. Biol. Chem.*, 258, 4272, 1983.
9. **Eckols, T. K., Thompson, R. E., and Masaracchia, R. A.**, Primary substrate specificity determinants for H4-specific protease-activated protein phosphotransferase, *Eur. J. Biochem.*, 134, 249, 1983.
10. **Donahue, M. J. and Masaracchia, R. A.**, Phosphorylation of ribosomal protein S6 at multiple sites by a cyclic AMP-independent protein kinase from lymphoid cells, *J. Biol. Chem.*, 259, 435, 1984.
11. **Magnino, P. E., de la Houssaye, B. A., and Masaracchia, R. A.**, Resolution and characterization of calcium/phospholipid-dependent protein kinase and H4 protease-activated protein kinase activities in lymphoid cells, *Biochem. Biophys. Res. Commun.*, 116, 675, 1983.
12. **Ogawa, Y., Takai, Y., Kawahara, Y., Kimura, S., and Nishizuka, Y.**, A new possible regulatory system for protein phosphorylation in human peripheral lymphocytes. I. Characterization of a calcium activated, phospholipid-dependent protein kinase. *J. Immunol.*, 127, 1369, 1981.
13. **Gabrielli, B., Wettenhall, R. E. H., Kemp, B. E., Quinn, M., and Bizonova, L.**, Phosphorylation of ribosomal protein S6 and a peptide analogue of S6 by a protease-activated kinase isolated from rat liver, *FEBS Lett.*, 175, 219, 1984.
14. **DeLange, R. J., Fambrough, D. M., Smith, E. L., and Bonner, J.**, Calf and pea histone IV. II. The complete amino acid sequence of calf thymus histone IV: presence of ϵ-N-acetyl lysine, *J. Biol. Chem.*, 244, 319, 1969.
15. **Kemp, B. E., Graves, D. J., Benjamini, E., and Krebs, E. G.**, Role of multiple basic residues in determining the substrate specificity of cyclic AMP-dependent protein kinase, *J. Biol. Chem.*, 252, 4888, 1977.

16. **Masaracchia, R. A. and Walsh, D. A.**, Protein phosphotransferase activities and cyclic nucleotide action in proliferating lymphocytes, *Cancer Res.*, 36, 3227, 1976.
17. **Ruiz-Carrillo A., Waugh, L., and Allfrey, V.**, Processing of newly synthesized histone molecules. Nascent histone H4 chains are reversibly phosphorylated and acetylated, *Science,* 190, 117, 1975.
18. **Thomas, G., Martin-Perez, J., Siegmann, M., and Otto, A. M.**, The effect of serum, EGF, $PGF_{2\alpha}$ and insulin on S6 phosphorylation and the initiation of protein and DNA synthesis, *Cell,* 30, 235, 1982.
19. **Smith, C. J., Rubin, C. S., and Rosen, O. M.**, Insulin-treated 3T3-L1 adipocytes and cell-free extracts derived from them incorporate ^{32}P into ribosomal protein S6, *Proc. Natl. Acad. Sci. U.S.A.,* 77, 2641, 1980.
20. **Blenis, J. and Erikson, R. L.**, Regulation of a ribosomal protein S6 kinase activity by the Rous sarcoma virus transforming protein, serum or phorbol ester, *Proc. Natl. Acad. Sci. U.S.A.,* 82, 7621, 1985.
21. **Novak-Hofer, I. and Thomas, G.**, Epidermal growth factor-mediated activation of an S6 kinase in Swiss mouse 3T3 cells, *J. Biol. Chem.,* 260, 10314, 1985.
22. **Wettenhall, R. E. H. and Cohen, P.**, Isolation and characterization of cyclic AMP-dependent phosphorylation sites from rat liver ribosomal protein S6, *FEBS Lett.,* 140, 263, 1982.
23. **House, C., Wettenhall, R. E. H., and Kemp, B. E.**, The influence of basic residues on the substrate specificity of protein kinase C, *J. Biol. Chem.,* 262, 772, 1987.
24. **Adelstein, R. S. and Eisenberg, E.**, Regulation and kinetics of the actin-myosin-ATP interaction, *Annu. Rev. Biochem.,* 49, 921, 1980.
25. **Nishikawa, M., Sellers, J. R., Adelstein, R. S., and Hidaka, H.**, Protein kinase C modulates *in vitro* phosphorylation of the smooth muscle meromyosin by myosin light chain kinase, *J. Biol. Chem.,* 259, 8808, 1984.
26. **Hassell, T. C., Kemp, B. E., and Masaracchia, R. A.**, Nonmuscle myosin phosphorylation sites for calcium-dependent and calcium-independent protein kinases, *Biochem. Biophys. Res. Commun.,* 134, 240, 1986.
27. **Woodgett, J. R., Gould, K. L., and Hunter, T.**, Substrate specificity of protein kinase C. Use of synthetic peptides corresponding to physiological sites as probes for substrate recognition requirements, *Eur. J. Biochem.,* 160, 177, 1986.
28. **Turner, R. S., Kemp, B. E., Su, H.-I., and Kuo, J. F.**, Substrate specificity of phospholipid/Ca^{+2}-dependent protein kinase probed with synthetic peptide fragments of the bovine myelin basic protein, *J. Biol. Chem.,* 260, 11503, 1985.
29. **Bengur, A. R., Robinson, E. A., Appella, E., and Sellers, J. R.**, Sequence of the sites phosphorylated by protein kinase C in the smooth muscle myosin light chain, *J. Biol. Chem.,* 262, 7613, 1987.
30. **Nishikawa, M., Hidaka, H., and Adelstein, R. S.**, Phosphorylation of smooth muscle heavy meromyosin by calcium-activated phospholipid-dependent protein kinase, *J. Biol. Chem.,* 258, 14069, 1983.
31. **Takai, Y., Kishimoto, A., Inoue, M., and Nishizuka, Y.**, Studies on cyclic nucleotide-independent protein kinase and its proenzyme in mammalian tissues. I. Purification and characterization of an active enzyme from bovine cerebellum, *J. Biol. Chem.,* 252, 7603, 1977.
32. **Beavo, J. A., Bechtel, P. J., and Krebs, E. G.**, Preparation of homogeneous cyclic AMP-dependent protein kinase(s) and its subunits from rabbit skeletal muscle, *Methods Enzymol.,* 38C, 299, 1974.
33. **Glass, D. B., Masaracchia, R. A., Feramisco, J., and Kemp, B. E.**, Isolation of phosphorylated peptides and proteins on ion exchange papers, *Anal. Biochem.,* 87, 566, 1978.
34. **Merrifield, R. B.**, Solid-phase peptide synthesis. III. An improved synthesis of bradykinin, *Biochemistry,* 3, 1385, 1964.

Chapter 9

SUBSTRATE SPECIFICITY OF THE CYCLIC GMP-DEPENDENT PROTEIN KINASE

David B. Glass

TABLE OF CONTENTS

I.	Introduction	210
II.	Homology between Cyclic Nucleotide-Dependent Protein Kinases	211
	A. Regulatory Domains	211
	B. Catalytic Domains	213
III.	Autophosphorylation of cGMP-Dependent Protein Kinase	215
	A. Site Specificity and Kinetics of Autophosphorylation	215
	B. Modeling Autophosphorylation with Synthetic Peptides	216
	C. Functional Significance of Autophosphorylation	217
IV.	*In Vitro* Substrate Specificities of Protein Kinases	217
	A. Protein Substrates	217
	1. Relatively Selective for cAMP-Dependent Protein Kinase	217
	2. Relatively Selective for cGMP-Dependent Protein Kinase	218
	B. Synthetic Peptides as Model Substrates	221
V.	Endogenous Protein Substrates of cGMP-Dependent Protein Kinase	226
	A. Proteins Phosphorylated in Vascular Smooth Muscle	226
	B. Proteins Phosphorylated in Other Tissues	227
VI.	Peptide Inhibitors of cGMP-Dependent Protein Kinase	228
VII.	Concluding Remarks	230
	Acknowledgments	230
	References	230

I. INTRODUCTION

Cyclic GMP-dependent protein kinase is one of the intracellular targets of the second messenger cGMP.[1,2] When the cGMP-dependent protein kinase was first discovered by Kuo and Greengard,[3,4] they proposed that all of the physiological actions of cGMP would be mediated by this enzyme. It is now known that, in addition to activating the cGMP-dependent protein kinase, cGMP also affects the activity of several types of cyclic nucleotide phosphodiesterase[5] and directly regulates a specific cation channel in photoreceptors[6,7] and possibly some other tissues.[8] Several of the physiological roles of cGMP can be explained, at least in part, by these latter actions. However, there have been few biological systems in which the action of cGMP mediated by the cGMP-dependent protein kinase is fully understood. The recent discovery of atrial natriuretic factor[9,10] and its ability to regulate cGMP metabolism in target tissues[11-14] by stimulating particulate guanylate cyclase[15-17] has refocused attention on the intracellular action of this cyclic nucleotide to activate the cGMP-dependent protein kinase, particularly in vascular smooth muscle.[18-21] This question has been emphasized further by the fact that other vasodilator substances act through endothelial-derived relaxing factor[22] and by the recent discovery that endothelial-derived relaxing factor is nitric oxide,[23] a potent stimulator of the soluble form of guanylate cyclase.[24-26]

Vascular smooth muscle is one of the tissues that contains relatively high concentrations of the cGMP-dependent protein kinase.[27,28] The well-known cAMP-dependent protein kinase is also present in this tissue, as it is expressed in rather uniform amounts in almost every cell type. The tissue distributions and relative levels of cGMP-dependent protein kinase are more restricted as compared to those of the cAMP-dependent enzyme.[28,29] However, the cGMP-dependent protein kinase is expressed and probably plays important physiological roles in a number of tissues other than vascular smooth muscle. These include cerebellar Purkinje cells[28-30] and other neurons,[31] nonvascular smooth muscle,[32] platelets,[33] lymphocytes,[34] and intestinal epithelial cells.[35,36]

The two cyclic nucleotide-dependent protein kinases are homologous proteins[37] and are thought to have evolved from a common ancestor.[38] Therefore, although these protein kinases differ in that each is activated rather selectively by its cognate cyclic nucleotide effector,[39] they have many properties in common. Foremost among these is that each enzyme catalyzes the same general phosphotransferase reaction. In addressing the role(s) of cGMP-dependent protein kinase, an understanding of the physiologically relevant protein substrates of the enzyme and their biochemical identities and functions is most important. With several notable exceptions,[40] the natural protein substrates of the cGMP-dependent protein kinase are not well known. The particular substrates that are phosphorylated by a protein kinase in a given cell type depend on the expression of substrate proteins by that differentiated cell, the subcellular localization of substrate proteins and the protein kinase, and the inherent substrate specificity of the protein kinase. One approach to the latter question has been to investigate the protein substrate specificity and active site configuration of the protein kinase at the biochemical level. This method has been particularly useful in delineating the recognition of potential substrates by cGMP- and cAMP-dependent protein kinases. Both of these enzymes fall into the broad class of protein kinases that recognize basic amino acid residues near the sites of phosphorylation in their substrate proteins. Also, as a superfamily of protein kinases is being described,[41] it is of interest to compare and contrast their protein substrate specificities.

The purpose of this chapter is to summarize recent findings concerning the substrate specificity of cGMP-dependent protein kinase and to compare this specificity to that of the cAMP-dependent enzyme. A number of previous reviews have covered the structure, function, and potential physiological roles of cGMP-dependent protein kinase extensively,[42-47] and this earlier literature will not be covered exhaustively here. The reader is referred to these reviews as well as to recent coverage of the cAMP-dependent protein kinase, other protein kinases, and protein phosphorylation in general.[48-51]

II. HOMOLOGY BETWEEN CYCLIC NUCLEOTIDE-DEPENDENT PROTEIN KINASES

The cGMP-dependent and cAMP-dependent protein kinases have each been purified to homogeneity,[52-56] and their physicochemical properties and subunit structures are well characterized. The cGMP-dependent protein kinase is a homodimer of 76-kDa subunits.[52,53] Each polypeptide chain contains two cGMP binding sites[56,57] and a catalytic site. The holoenzyme is relatively inactive in the absence of cGMP and is activated allosterically by the binding of the effector cyclic nucleotide.[52,53] The subunits of the enzyme are not dissociated during this activation of the phosphotransferase activity. The inactive holoenzyme of cAMP-dependent protein kinase has an overall size, shape, and amino acid composition similar to those of the cGMP-dependent enzyme,[58] but it is a heterotetramer composed of two cAMP-binding regulatory subunits and two catalytic subunits.[59-61] Cyclic AMP activates the protein kinase by binding to the regulatory subunit and causing the dissociation of the holoenzyme into a regulatory subunit dimer and two free, active catalytic subunits.[59-61] Therefore, one important difference between the activated forms of the two protein kinases is their relative sizes. The catalytic subunit of cAMP-dependent protein kinase is a rather small, globular protein of 41 kDa, while the cGMP-dependent holoenzyme is a much larger protein of 153 kDa with an extended structure.[53] These size differences may present important steric constraints in the abilities of their active sites to recognize large protein substrates or even smaller heat-stable inhibitor proteins like the inhibitor protein of cAMP-dependent protein kinase.[62] The relative sizes and structural similarities of the subunits of the cyclic nucleotide-dependent protein kinases are depicted in Figure 1.

Although the two protein kinases have different subunit structures, modes of activation by cyclic nucleotides, and immunological properties,[63] their overall holoenzyme structures and general functions are similar. Each has the ability to bind cyclic nucleotides and catalyze the transfer of the γ-phosphate of ATP to seryl or threonyl residues in proteins. Also, as is generally characteristic of protein kinases, each enzyme undergoes an autophosphorylation reaction.[64-67] These similarities in function between the two protein kinases were thought to be due to an underlying amino acid sequence homology.[38,58,68] These predictions have been confirmed by the complete sequencing of both protein kinases.[37,69-72] From this work, the cGMP-dependent protein kinase is known to be composed of 670 amino acids. In addition, the regulatory and catalytic subunits of the cAMP-dependent protein kinase have recently been cloned.[73-76]

A. REGULATORY DOMAINS

The amino-terminal half of each subunit of cGMP-dependent protein kinase contains the regulatory domain that is homologous to the regulatory subunits of the cAMP-dependent enzyme. Approximately the first 50 residues are involved in the dimerization of the subunits in each enzyme.[77-79] However, the degree of structural similarity between the two protein kinases in this region is quite low.[37]

The sites of autophosphorylation in each enzyme are in their regulatory domains or subunits between amino acid residues 50 and 100. These regions of cGMP- and cAMP-dependent protein kinases also have little similarity in primary structure.[37] The autophosphorylation site in the cAMP-dependent protein kinase is serine-95 of the type II regulatory subunit.[80,81] Multiple sites are autophosphorylated in the cGMP-dependent enzyme under some conditions; however, the major autophosphorylation site is a threonyl residue, threonine-58.[82] Autophosphorylation of the cGMP-dependent kinase will be considered in detail below.

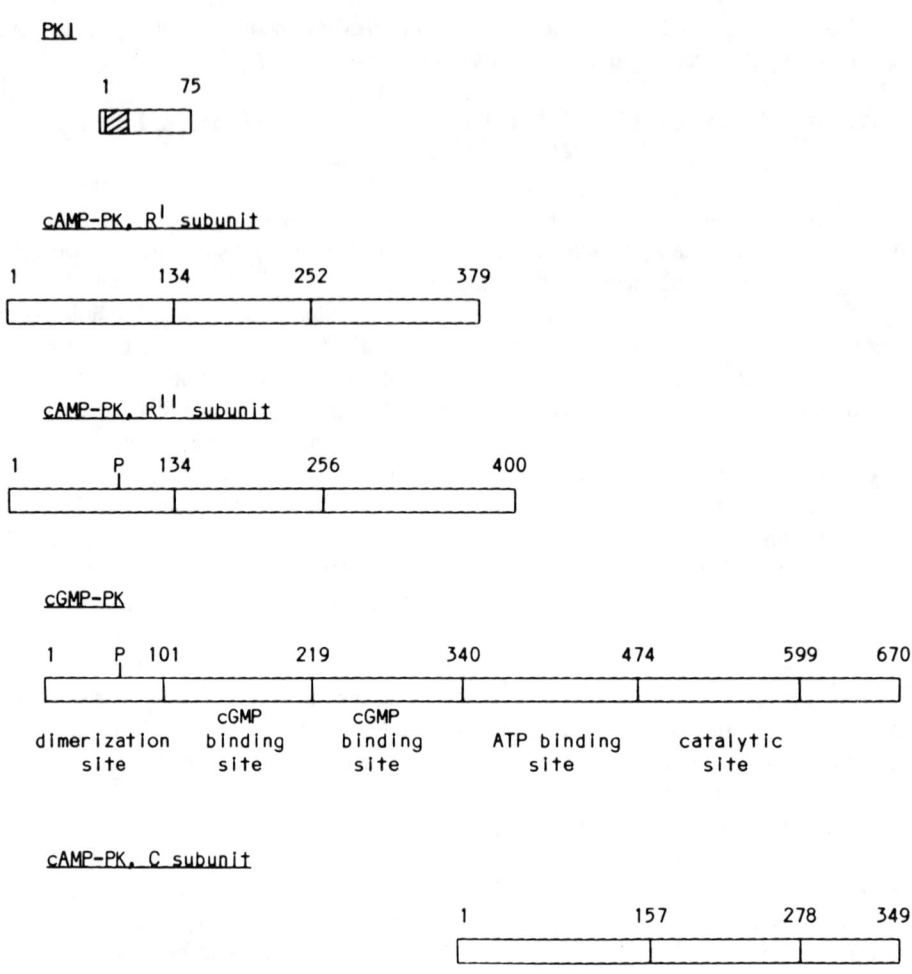

FIGURE 1. Structural homologies among the regulatory and catalytic subunits of the cAMP-dependent protein kinase and the regulatory and catalytic domains of the subunit of cGMP-dependent protein kinase. The domain designations are as proposed by Takio et al.[37] The numbers above the subunits refer to their lengths and the amino acids in their sequences that limit each domain. PKI, the heat-stable inhibitor protein of cAMP-dependent protein kinase; cAMP-PK, cAMP-dependent protein kinase; C subunit, catalytic subunit; R^I subunit, type I regulatory subunit; R^{II} subunit, type II regulatory subunit; cGMP-PK, cGMP-dependent protein kinase. P indicates the sites of autophosphorylation in cGMP-dependent and type II cAMP-dependent protein kinases. The hatched area in the heat-stable inhibitor protein is the biologically active portion.[62]

Both of the cyclic nucleotide-dependent protein kinases bind 2 mol of cyclic nucleotide per mole of subunit monomer.[56,57] Residues 101 through 340 of cGMP-dependent protein kinase contain two similarly sized cGMP binding domains. These are highly homologous to one another and to the twin cAMP binding domains in the regulatory subunits of the cAMP-dependent protein kinase, as well as to the cAMP binding domain of the catabolite gene activator protein of *Escherichia coli*.[37] The binding of cyclic nucleotides to the two types of cGMP binding sites of cGMP-dependent protein kinase has been studied extensively by several laboratories, but this topic will not be covered here as it is not directly related to the substrate specificity of the activated enzyme. Additional details concerning the structure and function of the regulatory subunits of cAMP-dependent protein kinase can be found in another chapter in this volume.[83]

Mild treatment of the cGMP-dependent protein kinase with trypsin removes the first 77 amino acids of each subunit, yielding a 65-kDa fragment that still contains the two cGMP binding sites and the entire catalytic domain of the enzyme, but lacks the dimerization and autophosphorylation sites.[84] This proteolytic fragment is fully active catalytically in the absence or presence of cGMP. This indicates that the dimerization region and autophosphorylation sites in the amino-terminal portion of the regulatory domain as well as the cGMP binding sites are involved in controlling the activation of the holoenzyme. The autophosphorylation sites are analogues of protein substrates. As such, in the inactive native enzyme, they are probably oriented within the protein/peptide binding portion of the active site, making it unavailable for binding of exogenous substrate proteins.

B. CATALYTIC DOMAINS

The catalytic domain of cGMP-dependent protein kinase is comprised of residues 341 through the carboxy-terminus.[37] This catalytic domain is homologous to the free catalytic subunit of the cAMP-dependent protein kinase. Areas of particularly high sequence homology in the catalytic domains between the two protein kinases are shown in Table 1. Each enzyme has an ATP binding domain that is characteristic of the nucleotide-fold structure described by Rossmann and colleagues.[85,86] These domains contain a homologous lysyl residue that has been labeled covalently in each enzyme by the chemically reactive ATP analogue p-fluorosulfonylbenzoyladenosine. In the catalytic subunit of cAMP-dependent protein kinase this is lysine-72,[87,88] and in the cGMP-dependent protein kinase this is lysine-389.[89,90] Closely amino-terminal to these lysyl residues in each enzyme is a region of high homology, containing the sequence –Gly–X–Gly–X–Phe–Gly–, which is present in the ATP binding domains of all known protein kinases.[91,92]

Carboxy-terminal to the ATP binding sites in each enzyme are the catalytic sites.[37] As depicted in Table 1, these regions include stretches of sequence in which 19 out of 28 and 17 out of 20 residues are identical. These domains are thought to contain the protein substrate binding portions of the active sites. This has been demonstrated clearly for the catalytic subunit of cAMP-dependent protein kinase by the covalent labeling of cysteine-199 with an analogue of the peptide substrate Kemptide.[93] The corresponding cysteinyl residue in the cGMP-dependent protein kinase is cysteine-518, although its proximity to the Kemptide binding site has not been demonstrated directly. These same cysteines also have been labeled during the inactivation of each protein kinase by o-phthalaldehyde.[94,95] Treatment of either protein kinase with this reagent inactivates phosphotransferase activity, and this is correlated with the formation of an isoindole moiety involving o-phthalaldehyde and unique lysyl and cysteinyl residues. In the catalytic subunit of cAMP-dependent protein kinase, lysine-72 and cysteine-199 are close in space and are presumably the residues cross-linked by o-phthalaldehyde.[94] In the cGMP-dependent enzyme, lysine-389 and cysteine-518 may be cross-linked in a similar fashion,[95] although the exact residues participating in the isoindole formation are not known. In the latter enzyme, the regulatory domains also contain an appropriately located lysyl and cysteinyl residue because o-phthalaldehyde forms an extra equivalent of isoindole adduct and inactivates one of the two cGMP binding sites.[95] Immediately amino-terminal to the cysteinyl residues in the catalytic domains of each enzyme is another highly homologous region. In the cAMP-dependent protein kinase catalytic subunit, this sequence is modified covalently during inactivation of the enzyme with dicyclohexylcarbodiimide.[96] Parallel studies have not been conducted yet with the cGMP-dependent protein kinase. Several additional carboxy-terminal regions of the catalytic domains of each enzyme contain highly homologous sequences,[37] but functions have not been assigned directly to these areas.

Both activated protein kinases bind ATP with similar affinities. The inactive holoenzyme of both enzymes can bind ATP readily, as evidenced by autophosphorylation occurring in the absence of cyclic nucleotides. The homologous nucleotide-fold regions of the two enzymes underlie the binding of ATP, and it is thought that minor differences in the primary

TABLE 1
Selected Sequence Homologies between the Catalytic Domain of cGMP-Dependent Protein Kinase and the Catalytic Subunit of cAMP-Dependent Protein Kinase[a]

cGMP-dependent protein kinase	-Ile^{362}-Asp-Thr-Leu-Gly-Val-Gly-Gly-Phe-Gly-Arg-Val-Glu-Leu-Val^{376}-
cAMP-dependent protein kinase, catalytic subunit	-Ile^{46} -Lys-Thr-Leu-Gly-Thr-Gly-Ser-Phe-Gly-Arg-Val-Met-Leu-Val^{60}-

-Phe^{386}-Ala-Met-Lys^{389}-Ile-Leu-Lys-Lys^{393}-

-Tyr^{69}-Ala-Met-Lys^{72} -Ile-Leu-Asp-Lys^{76}-

-Ile^{480}-Tyr-Arg-Asp-Leu-Lys-Pro-Glu-Asn-Leu-Ile-Leu-Asp-His-Arg-Gly-Tyr-Ala-Lys-Leu-Val-Asp-Phe-Gly-Phe-Ala-Lys-Lys^{507}-

-Ile^{163}-Tyr-Arg-Asp-Leu-Lys-Pro-Glu-Asn-Leu-Ile-Asp-Gln-Gln-Gly-Tyr-Ile-Gln-Val-Thr-Asp-Phe-Gly-Phe-Ala-Lys-Arg^{190}-

-Thr^{514}-Trp-Thr-Phe-Cys^{518}-Gly-Thr-Pro-Glu-Tyr-Val-Ala-Pro-Glu-Ile-Ile-Leu-Asn-Lys-Gly^{533}-

-Thr^{195}-Trp-Thr-Leu-Cys^{199}-Gly-Thr-Pro-Glu-Tyr-Leu-Ala-Pro-Glu-Ile-Ile-Leu-Ser-Lys-Gly^{214}-

[a] The cGMP-dependent and cAMP-dependent protein kinases are listed on the top and bottom, respectively, of each set of sequences.[37,72] Identical amino acids are represented by a vertical line, and conserved residues are indicated by a dotted line. Chemically modified residues and selected regions of homology are described in the text.

structures of the two enzymes in this domain are of little importance to this function. However, it is anticipated that differences in the amino acid sequences or in higher orders of structure of the two enzymes in the protein/peptide binding portions of their catalytic domains may underlie the observed differences in their protein/peptide substrate specificities.

III. AUTOPHOSPHORYLATION OF cGMP-DEPENDENT PROTEIN KINASE

One protein of known primary structure which is a substrate for phosphorylation by the cGMP-dependent protein kinase is the enzyme itself. The regulatory portion of the cGMP-dependent protein kinase cannot be considered a true exogenous substrate for the enzyme's active site because the interaction between these two domains probably is influenced by other forces, due to the fact that they are on the same polypeptide chain. However, the sites of autophosphorylation do reveal some characteristics of the substrate specificity of the enzyme. The rate of the autophosphorylation reaction for cGMP-dependent protein kinase is slow compared to that of the cAMP-dependent holoenzyme. The first-order rate constant for the former reaction is 0.07 min^{-1}.[67] The corresponding value for the cAMP-dependent protein kinase under similar conditions is at least tenfold higher.[97,98]

A. SITE SPECIFICITY AND KINETICS OF AUTOPHOSPHORYLATION

The cGMP-dependent protein kinase undergoes autophosphorylation in the absence of bound cyclic nucleotide or in the presence of either cGMP or cAMP.[65-67,99,100] The rate of autophosphorylation is stimulated more by cAMP than by cGMP.[101] Autophosphorylation of the enzyme is stimulated by occupancy of only one of the two cGMP binding sites.[102] In the presence of cGMP, approximately 1 mol of phosphate is incorporated per mole of enzyme subunit.[82,99-101] In the presence of cAMP, as much as a fourfold greater amount of phosphate is incorporated.[99-101] This is not due to the presence of contaminating cAMP-dependent protein kinase because that enzyme does not phosphorylate the cGMP-dependent protein kinase.[101]

The amino acid sequences at the various autophosphorylation sites in cGMP-dependent protein kinase have been determined.[82,100] In the presence of cGMP, threonine-58 is the only autophosphorylated residue.[82,100] In the presence of cAMP, threonine-58 is still the major site of autophosphorylation, but serine-50, serine-72, and threonine-84 also are phosphorylated.[100] In the latter situation, additional phosphorylation occurs to a minor extent on serine-1 and serine-64. The amino acid sequences of the major sites of autophosphorylation as well as the sequence at the autophosphorylation site in the type II regulatory subunit of the cAMP-dependent protein kinases are shown in Table 2. For cGMP-dependent protein kinase, all of the major autophosphorylation sites except one contain one or more basic amino acids residues closely amino-terminal to the phosphorylatable hydroxyamino acid. Serine-50 is the exception.[100] It is preceded by a rather hydrophobic sequence of eight residues, and the rate of phosphorylation of this site is quite slow. The autophosphorylation site in the type II regulatory subunit contains the classical Arg–X–X–Arg–Arg–X–Ser–X– sequence that is typical of substrates of cAMP-dependent protein kinase.[44,48,49,103] If histidine is included, all four of the autophosphorylation sites in cGMP-dependent protein kinase are followed closely on their carboxy-terminal side by a basic amino acid residue. The two threonine autophosphorylation sites contain a basic residue immediately adjacent in the carboxy-terminal direction. The presence of a basic amino acid residue closely carboxy-terminal to sites of phosphorylation has been observed in several exogenous protein or peptide substrates of the cGMP-dependent protein kinase (see Sections IV.A and B). The proximity of the four autophosphorylation sites in the primary sequence of cGMP-dependent protein kinase suggests that multisite interactions might occur, but there is little data to

TABLE 2
Amino Acid Sequences at Autophosphorylation Sites in cGMP-Dependent Protein Kinase and Type II Regulatory Subunit of cAMP-Dependent Protein Kinase

Protein kinase	Autophosphorylation site sequence[a]	Ref.
cGMP-dependent protein kinase[b] (regulatory domain)	-Val-Leu-Pro-Val-Pro-<u>Ser</u>^{50}-Thr-**His**-Ile-Gly-Pro-	100
	-Ile-Gly-Pro-**Arg**-Thr-<u>Thr</u>^{58}-**Arg**-Ala-Gln-Gly-Ile-	82, 100
	-Pro-Gln-Thr-Tyr-**Arg**-<u>Ser</u>^{72}-Phe-**His**-Asp-Leu-**Arg**-	100
	-Ala-Phe-**Arg**-**Lys**-Phe-<u>Thr</u>^{84}-**Lys**-Ser-Glu-**Arg**-Ser-	100
cAMP-dependent protein kinase (regulatory subunit, type II)	-Phe-Asp-**Arg**-**Arg**-Val-<u>Ser</u>^{95}-Val-Cys-Ala-Glu-Thr-	80, 81

[a] The underlined residues are the autophosphorylated amino acids. They are numbered according to the amino acid sequences of the protein kinase subunits. Basic amino acids are in bold print.

[b] All four sites are autophosphorylated in the presence of cAMP. Under this condition, the site with the fastest initial rate of autophosphorylation is threonine-58. In the presence of cGMP, only threonine-58 is autophosphorylated.

indicate that phosphorylation of one site has an influence on the subsequent phosphorylation of another site. Precise data on the initial rates of autophosphorylation of all these sites are also lacking, other than that threonine-58 is the first residue to be phosphorylated.

B. MODELING AUTOPHOSPHORYLATION WITH SYNTHETIC PEPTIDES

The autophosphorylation of the threonine-58 site in cGMP-dependent protein kinase has been modeled using synthetic peptides corresponding to this sequence.[104] The peptide Ile–Gly–Pro–Arg–Thr–Thr^{58}–Arg–Ala–Gln–Gly–Ile is phosphorylated to 1 mol of phosphate per mole of peptide, and nearly all of this phosphate is on threonine-58. This is the same residue specificity among the two adjacent threonines as the autophosphorylation reaction in the native enzyme. Phosphorylation of the peptide is similar whether the cGMP-dependent protein kinase is activated by a saturating concentration of cGMP or cAMP. This model peptide has a low affinity for the enzyme (K_m of 580 μM) and is phosphorylated at a very low velocity relative to other exogenous peptide or protein substrates of the cGMP-dependent protein kinase.[105-107] However, the autophosphorylation site peptide is kinetically a better substrate for cGMP-dependent protein kinase than for the catalytic subunit of the cAMP-dependent enzyme.[104] An analogue peptide that has serine in place of threonine-58 is phosphorylated exclusively on that seryl residue with a 70-fold higher maximal velocity than the parent peptide.[104] Therefore, in this peptide sequence, serine is a better phosphate-accepting residue than is threonine. This result suggests that if the native cGMP-dependent protein kinase contained serine instead of threonine at position 58 (or could be made so by site-directed mutagenesis), its rate of autophosphorylation might rival that of the cAMP-dependent protein kinase.

Both arginyl residues in the synthetic autophosphorylation site peptide are crucial to its ability to serve as a substrate for cGMP-dependent protein kinase. Replacement of arginine-56 by alanine results in an analogue that is essentially not a substrate, while replacement of the carboxy-terminal arginine-59 produces a substrate peptide having an affinity 40-fold lower than the parent peptide.[104] The high K_m value of the parent autophosphorylation site peptide may not reflect the true binding affinity of the regulatory domain of the enzyme to its active site if forces other than the primary sequence also promote this interaction. Thus, the degree of autophosphorylation in the native enzyme might depend more on the velocity of the catalytic reaction than on the affinity between the two domains of the protein kinase.

C. FUNCTIONAL SIGNIFICANCE OF AUTOPHOSPHORYLATION

The physiological significance of the autophosphorylation of cGMP-dependent protein kinase is unclear. Autophosphorylation has no significant effect on the kinetics by which the protein kinase phosphorylates exogenous substrate proteins or peptides.[99,108] However, autophosphorylation clearly alters the kinetics of activation of the enzyme by cGMP or cAMP. Autophosphorylation of cGMP-dependent protein kinase affects the function of one of the two cGMP binding sites.[108,109] Autophosphorylation of the enzyme abolishes the positively cooperative binding of cGMP that is typically observed at site 1.[109] In addition, the ability of MgATP to inhibit binding of cGMP to site 1 is lost in the autophosphorylated enzyme.[99,109] Autophosphorylation of the protein kinase in the presence of cAMP results in a tenfold increase in the affinity of binding site 1 for cAMP, but no change in its affinity for cGMP.[99,108] This increased ability of cAMP to activate the enzyme must be due to the autophosphorylation of serine-50, serine-72, and/or threonine-84 because autophosphorylation in the presence of cGMP (which results in phosphorylation of only threonine-58) does not increase the binding affinity for cAMP.[99,100] Through these several effects, autophosphorylation of the cGMP-dependent protein kinase influences the interactions between the cyclic nucleotide binding sites and the catalytic site to generally facilitate activation of the enzyme.[110]

IV. *IN VITRO* SUBSTRATE SPECIFICITIES OF PROTEIN KINASES

A. PROTEIN SUBSTRATES
1. Relatively Selective for cAMP-Dependent Protein Kinase

The *in vitro* protein substrate specificities of the two protein kinases are grossly similar. That is, most purified proteins that are substrates for the catalytic subunit of cAMP-dependent protein kinase are also phosphorylated *in vitro* by the cGMP-dependent protein kinase. These substrate proteins include phosphorylase *b* kinase, glycogen synthase, L-type pyruvate kinase, fructose-1,6-bisphosphatase, hormone-sensitive lipase (cholesterol ester hydrolase), various histone fractions, and the inhibitory subunit of cardiac troponin.[44,58] The activities of those protein substrates that are enzymes generally are regulated in the same manner by phosphorylations catalyzed by either of the protein kinases,[58,111] suggesting that the two enzymes phosphorylate the same functional sites. In most cases, the cGMP-dependent protein kinase is a poorer catalyst for phosphorylation of these proteins than is the catalytic subunit of the cAMP-dependent enzyme. That conclusion results from experiments in which the phosphorylatable protein was used at a single fixed concentration as substrate for either of the two protein kinases. For the most part, equimolar amounts of each activated enzyme were used as catalysts in these studies. Because of its dimeric subunit structure, the cGMP-dependent protein kinase provides twice as many active sites as an equimolar amount of the free catalytic subunit of cAMP-dependent protein kinase. Therefore, if the reaction velocities with these substrates are expressed as a turnover number per catalytic site, the activity of the former enzyme is clearly less than that of the latter. In many of these studies, a direct correlation between the actual kinetic constants for phosphorylation by either protein kinase or the phosphorylation site specificity of each enzyme is lacking. More recent studies that provide these data are investigations with purified troponin,[112,113] glycogen synthase,[114] ATP citrate lyase,[115] hormone-sensitive lipase,[116] and tyrosine hydroxylase.[117] These studies are consistent with the favorable *in vitro* phosphorylation of the protein substrates by cAMP-dependent protein kinase. Reaction velocities for the cGMP-dependent protein kinase were $1/10$ to $1/100$ of those catalyzed by the cAMP-dependent enzyme. The one exception to this was tyrosine hydroxylase, which was phosphorylated and activated by cGMP-dependent kinase at one half the rate that it was by cAMP-dependent protein kinase. Other recently identified substrates of cAMP-dependent protein kinase that are also phosphorylated by the cGMP-dependent enzyme are nonhelical α-chains of type I collagen,[118] cardiac C-protein,[119,120] and atrial natriuretic peptides.[121-124]

A most interesting exception to the overlapping protein substrate specificities of the two cyclic nucleotide-dependent protein kinases involves the bifunctional enzyme 6-phosphofructo-2-kinase/fructose-2,6-bisphosphatase. This liver protein is a physiologically relevant substrate for the cAMP-dependent protein kinase.[125] *In vitro*, 6-phosphofructo-2-kinase/fructose-2,6-bisphosphatase is an excellent substrate for cAMP-dependent protein kinase, but it is not phosphorylated at all by the cGMP-dependent enzyme.[107,125] The site of cAMP-dependent kinase-catalyzed phosphorylation in this substrate protein is serine-32,[126,127] which is close to its amino terminus and therefore might well be accessible to either enzyme. In addition, small synthetic peptides corresponding to the amino acid sequence at this phosphorylation site are phosphorylated readily by the cGMP-dependent protein kinase (see Section IV.B).[107] The reason why the native bifunctional enzyme is not a substrate for cGMP-dependent protein kinase is unclear, but the explanation presumably involves higher orders of structure in the substrate protein and potential steric constraints in the interactions between this protein and the two kinases.

A similar situation exists with respect to the heat-stable inhibitor protein of the cAMP-dependent protein kinase.[62] While it is an inhibitor and not a substrate, it is a pseudosubstrate that interacts with the protein binding site of this enzyme.[62] However, the inhibitor protein, even at high concentrations, does not affect the catalytic activity of cGMP-dependent protein kinase.[62,128] This is particularly surprising since the protein substrate specificities of the two protein kinases are similar and small synthetic peptides that correspond to the active portion of the heat-stable inhibitor protein are able to fully inhibit both protein kinases (although with different potencies).[62,128]

2. Relatively Selective for cGMP-Dependent Protein Kinase

Although the cGMP- and cAMP-dependent protein kinases are homologous proteins with similar protein substrate specificities, numerous examples of differences in the kinetic details of their interactions with protein or peptide substrates have been reported. Several proteins tested as *in vitro* substrates are phosphorylated equally well by each protein kinase or are "selectively" phosphorylated by the cGMP-dependent compared to cAMP-dependent protein kinase. This selective substrate specificity is based on favorable kinetic constants, a unique site or subunit of phosphorylation by cGMP-dependent protein kinase, or, in the case of one substrate, an apparently absolute specificity by this enzyme. The purified protein substrates that are phosphorylated favorably by the cGMP-dependent protein kinase are listed in Table 3.

Each protein kinase phosphorylates skeletal muscle phosphorylase *b* kinase at the same unique sites on both its α- and β-subunits. Although the initial rate of overall phosphorylation of this substrate by cGMP-dependent protein is slower than by cAMP-dependent protein kinase, the relative rates of phosphorylation of the individual subunits of native phosphorylase kinase differ between the two enzymes.[129] The cGMP-dependent protein kinase phosphorylates the α-subunit threefold faster than the β-subunit, while the cAMP-dependent enzyme phosphorylates the β-subunit fivefold faster than the α-subunit.[129] It is unlikely that phos-

TABLE 3
Protein Substrates of cGMP-Dependent Protein Kinase which are Phosphorylated *In Vitro* with Favorable Kinetics or Site Specificity

Protein substrate	Source	Specificity basis[a]	Ref.
Phosphorylase b kinase	Rabbit skeletal muscle	Different subunits	129
Histone H1	Calf thymus	Unique site, selective kinetics	130
Histone H2B	Calf thymus	Selective site kinetics	131—133
Type I regulatory subunit of cAMP-dependent protein kinase	Bovine skeletal muscle	Absolute specificity	134
G-substrate	Rabbit cerebellum	Selective kinetics	138
DARPP-32	Bovine caudate nucleus	Equal kinetics	142
Phosphatase inhibitor-1	Rabbit skeletal muscle	Equal kinetics	142
High-mobility group 14 protein	Calf thymus	Selective kinetics	143
Myosin light chain kinase	Human platelet	Unique site	152
Ribosomal proteins	Krebs II ascites tumor and rabbit reticulocytes	Selective sites or kinetics	153, 154

[a] Specificity of cGMP-dependent protein kinase for phosphorylation of substrate proteins is compared to the catalytic subunit of cAMP-dependent protein kinase. In cases where each protein kinase phosphorylates the substrate protein, the relative specificity of cGMP-dependent protein kinase is based on phosphorylation of selective sites or on a comparison of initial phosphorylation rates or actual K_m and k_{cat} values.

phorylation of phosphorylase kinase by cGMP-dependent protein kinase occurs *in vivo*.[129] The phosphorylation site sequence in the α-subunit of phosphorylase kinase and those in other protein substrates of the cGMP-dependent protein kinase are listed in Table 4.

Certain sites in histones H1 and H2B are selectively phosphorylated by cGMP-dependent protein kinase. Both protein kinases phosphorylate serine-37 in histone H1, but the cGMP-dependent enzyme also phosphorylates a unique seryl residue in the carboxy terminus of this protein.[130] The selectivity of cGMP-dependent protein kinase for this site as compared to cAMP-dependent kinase is indicated by the ratios of their kinetic parameters, k_{cat}/K_m. This value was 520-fold higher for the former enzyme when using the carboxy-terminal fragment of histone H1 as a substrate.[130] The exact location and amino acid sequence around this phosphorylated residue, which is in the microheterogeneous portion of histone H1, have not been determined yet. Histone H2B is phosphorylated on both serine-32 and serine-36 by each of the cyclic nucleotide-dependent protein kinases.[131] However, the former site is phosphorylated more rapidly by cGMP-dependent protein kinase,[131,132] while the latter site is phosphorylated at a greater initial rate by the cAMP-dependent enzyme.[133] The phosphorylation of these sites has been studied using synthetic peptide substrates (see Section IV.B).[105]

The cGMP-dependent protein kinase catalyzes the *in vitro* phosphorylation of the purified type I regulatory subunit of the cAMP-dependent protein kinase when cAMP is bound to this substrate.[134] The nucleotide-free type I regulatory subunit also interacts with the cGMP-dependent enzyme, but as an inhibitor. This isozyme of the regulatory subunit does not undergo autophosphorylation by the catalytic subunit in the type I holoenzyme of cAMP-dependent protein kinase.[135] Therefore, the cGMP-dependent protein kinase has an absolute specificity for this substrate protein. Serine-99 is the residue in the regulatory subunit that is phosphorylated.[136] The sequence around this site is listed in Table 4. The physiological significance of this phosphorylation reaction is unclear.

The best characterized physiological substrate of the cGMP-dependent protein kinase is termed G-substrate.[30,40] It is a low molecular weight protein purified from the cerebellum.[137] Both cyclic nucleotide-dependent protein kinases phosphorylate G-substrate on threonine residues to a stoichiometry of 2 mol of phosphate per mole of protein.[138] The two protein kinases have similar V_{max} values for G-substrate; however, the cGMP-dependent protein

TABLE 4
Amino Acid Sequences at Phosphorylation Sites in Selected *In Vitro* or *In Vivo* Protein Substrates of cGMP-Dependent Protein Kinase

Protein substrate	Phosphorylation site sequence[a]	Ref.
Phosphorylase kinase (α-subunit)	Phe-**Arg-Arg**-Leu-Ser^5-Ile-Ser-Thr-Glu-	129
Histone H2B	-Asp-Gly-**Lys-Lys-Arg-Lys-Arg**-Ser^{32}-**Arg-Lys**-Glu-Ser-	131
Regulatory subunit (type I)	-**Arg-Arg-Arg-Arg**-Gly-Ala-Ile-Ser^{99}-Ala-Glu-Val-Tyr-	136
G-substrate (site 1)	-**Lys-Lys**-Pro-**Arg-Arg-Lys**-Asp-Thr-Pro-Ala-Leu-His-	139
G-substrate (site 2)	-Gln-**Lys**-Pro-**Arg-Arg-Lys**-Asp-Thr-Pro-Ala-Leu-His-	139
DARPP-32	-**Arg-Arg-Arg**-Pro-Thr-Pro-Ala-Met-Leu-	140
Phosphatase inhibitor-1	-Gln-Ile-**Arg-Arg-Arg-Arg**-Pro-Thr^{35}-Pro-Ala-Thr-Leu-	140
High-mobility group 14 protein	Pro-**Lys-Arg-Lys**-Val-Ser^6-Ser-Ala-Glu-Gly-	143
	-**Lys-Arg-Arg**-Ser-Ala-**Arg**-Leu-Ser^{24}-Ala-**Lys**-Pro-Ala-	143

[a] The underlined residues are the phosphorylated amino acids. They are numbered according to the amino acid sequences of the native proteins. Basic amino acids are in bold print.

kinase has a 28-fold higher affinity for this substrate based on a favorable K_m value.[138] In G-substrate the primary structures around the two phosphorylated threonyl residues are homologous to one another.[139] These amino acid sequences are shown in Table 4. Two other endogenous proteins related to G-substrate in structure and probably in function are DARPP-32 and phosphoprotein phosphatase inhibitor-1.[140] DARPP-32 and phosphatase inhibitor-1 are regulated in intact cells via phosphorylation by cAMP-dependent protein kinase.[141] *In vitro*, both cAMP-dependent and cGMP-dependent protein kinases phosphorylate these two proteins on threonyl residues.[140] The sequences at the phosphorylation sites in DARPP-32 and phosphatase inhibitor-1 are homologous to one another and to those in G-substrate (Table 4). These sites are rather unusual in that they contain multiple basic residues amino-terminal to the phosphorylated threonines, which in turn are flanked by prolyl residues. DARPP-32 is phosphorylated with nearly identical kinetic constants by cGMP-dependent protein kinase and the catalytic subunit of cAMP-dependent kinase.[142] Both protein kinases exhibit similar k_{cat} values with phosphatase inhibitor-1, but the cAMP-dependent protein kinase has a fivefold higher affinity than cGMP-dependent protein kinase for this substrate.[142] The phosphorylated forms of all of these proteins are able to inhibit phosphoprotein phosphatases.[140]

High-mobility group 14 protein is a highly basic, chromatin-associated protein that is phosphorylated on the same sites by both cyclic nucleotide-dependent protein kinases.[143,144] Phosphorylation of this substrate on serine-6 is kinetically selective for cGMP-dependent protein kinase, based mainly on a lower K_m value and a slightly higher V_{max}.[143] However, studies in thyroid slices suggest that the thyrotropin-stimulated phosphorylation of high-mobility group 14 in intact cells is mediated by the cAMP-dependent protein kinase.[145]

Myosin light chain kinase from several tissues is phosphorylated by cAMP-dependent protein kinase.[146-149] In smooth muscle and nonmuscle cells, myosin light chain kinase is thought to be regulated by this covalent modification.[150,151] The phosphorylation by cAMP- and cGMP-dependent protein kinases of myosin light chain kinase purified from tracheal smooth muscle or blood platelets has been studied in the presence and absence of calmodulin, an effector of this enzyme.[152] The number of sites of phosphorylation in myosin light chain kinase from the two tissues is different with the two protein kinases. The cGMP-dependent protein kinase phosphorylates a unique site in the platelet myosin light chain kinase,[152] but the functional effect and physiological relevance of this reaction are not clear.

Several eukaryotic ribosomal proteins are *in vitro* substrates for the cAMP- and cGMP-dependent protein kinases. Ribosomal protein S6 is phosphorylated by both protein kinases, while several other ribosomal proteins, notably S2, S10, and L5, appear to be phosphorylated preferentially by the cGMP-dependent enzyme.[153,154]

Finally, the $(Ca^{2+}-Mg^{2+})$ATPase from sarcolemma of smooth muscle has recently been investigated as a potential physiologically relevant substrate of the cGMP-dependent protein kinase. One report indicates that purified $(Ca^{2+}-Mg^{2+})$ATPase which is reconstituted into liposomes is a substrate for the cGMP-dependent protein kinase.[155] In this and other studies on partially purified particulate $(Ca^{2+}-Mg^{2+})$ATPase from smooth muscle cells, incubation with activated cGMP-dependent protein kinase stimulated either the ATPase activity or the rate of Ca^{2+} uptake into microsomes.[155-158] However, the mechanism of this effect is unclear, since several investigators have since shown that the highly purified $(Ca^{2+}-Mg^{2+})$ATPase is not a substrate for this protein kinase.[159,160]

As can be seen in Table 4, there is not one consensus amino acid sequence that occurs invariably around all of the sites of phosphorylation in protein substrates of the cGMP-dependent protein kinase. However, most of them have multiple basic amino acids amino-terminal to their phosphorylation sites, and a number of them also have one basic amino acid on the carboxy-terminal side of the phosphorylated residue. Threonine (rather than serine) residues are phosphorylated in some cases, and in many of these sequences the threonine is preceded or followed by a prolyl residue.

B. SYNTHETIC PEPTIDES AS MODEL SUBSTRATES

Synthetic peptides that correspond to amino acid sequences at phosphorylation sites in proteins are useful models for determining the kinetics of substrate specificity for protein kinases. Synthetic peptides are available in amounts that allow use at saturating concentrations, they usually contain only one phosphorylation site (or two sites that can be mapped readily), and analogue peptides with altered sequences can be produced easily. The latter strategy allows conclusions to be made concerning the importance of specific amino acid residues as determinants of specificity. However, one must always keep in mind the possibility that a single amino acid substitution in a peptide substrate changes the kinetics of phosphorylation, not because that residue itself interacts with the active site of the protein kinase, but because it alters the overall conformation of the peptide.

The synthetic peptides that have been used to study the substrate specificity of the cGMP-dependent protein kinase include those modeled after the two phosphorylation sites in histone H2B, as well as the phosphorylation sites in 6-phosphofructo-2-kinase/fructose-2,6-bis-phosphatase, DARPP-32, and L-type pyruvate kinase and the major autophosphorylation site in the cGMP-dependent protein kinase itself. Studies with the latter peptide and its analogues[104] have been covered above in Section III. All of the literature on synthetic peptides employed as model substrates for the cAMP-dependent protein kinase will not be covered here, especially if the peptides have not been compared as substrates for the cGMP-dependent enzyme. Many of these peptides are compiled in a review by Bramson et al.[48]

One of the first studies on the substrate specificity of cGMP-dependent protein kinase to use synthetic peptides involved heptapeptides corresponding to sequences around serine-32 and serine-36 in histone H2B.[105] The former peptide, Arg–Lys–Arg–Ser^{32}–Arg–Lys–Glu, is probably the best studied model substrate for this enzyme.[106,161,162] The serine-32 site peptide is a kinetically favorable substrate for cGMP-dependent as compared with cAMP-dependent protein kinase, based both on a higher affinity and greater maximal velocity.[105] The serine-36 peptide is preferentially phosphorylated by the cAMP-dependent protein kinase, based mainly on a higher affinity.[105] The k_{cat}/K_m values of both cyclic nucleotide-dependent protein kinases for these histone peptides and selected other synthetic peptides are listed in Table 5. The ratios of these apparent first-order rate constants for the cGMP- and cAMP-dependent enzymes are calculated as a measure of selectivity of phosphorylation by the former enzyme (Table 5). Arg–Lys–Arg–Ser^{32}–Arg–Lys–Glu is the most selective peptide substrate for cGMP-dependent protein kinase of those that have reasonably high k_{cat} and low K_m values for this enzyme. Arg–Lys–Arg–Ser^{32}–Arg–Lys–Glu is rather unusual among protein kinase substrate peptides in that every residue except the phosphoryl-accepting serine is charged. This may explain the effect of this peptide and its analogues to alter the binding affinity of the cGMP-dependent protein kinase for the nucleotide substrate as measured with the fluorescent ATP analogue lin-benzo ADP.[163,164] On the other hand, these synthetic peptides do not affect the ATP binding affinity of the catalytic subunit of cAMP-dependent protein kinase.[165,166] This observation, then, is another difference in the substrate interactions between the two protein kinases.

Reasons for the excellent ability of Arg–Lys–Arg–Ser^{32}–Arg–Lys–Glu to serve as a substrate of cGMP-dependent protein kinase were investigated with a series of analog peptides in which alanine was substituted for selected amino acid residues.[106] In this sequence, the three basic residues amino-terminal to the serine are important determinants of specificity for both protein kinases. The lysine at position 34 is a negative determinant of specificity for each enzyme; its replacement by alanine causes appreciable increases in the velocities. The selectivity of cGMP-dependent protein kinase for this peptide is mainly due to the arginyl residue at position 33, immediately carboxy-terminal to the serine. Replacement of this residue results in the loss of almost all selectivity as a substrate for the cGMP-dependent protein kinase (Table 5).[106] Attempts to make kinetically better and more selective substrates for cGMP-dependent protein kinase based on substitutions at position 34 or amino-terminal acetylation of the Arg–Lys–Arg–Ser^{32}–Arg–Lys–Glu sequence have not produced analogues more selective than the parent peptide (Table 5).[167] In this parent histone peptide, serine is kinetically a better phosphate acceptor than is threonine (Table 5).[106] Inhibitor peptides that are somewhat selective for cGMP-dependent protein kinase have been made from the histone H2B substrate peptide by replacement of serine-32 with alanine (see Section VI).[161,162]

Because 6-phosphofructo-2-kinase/fructose-2,6-bisphosphatase is phosphorylated readily by cAMP-dependent protein kinase, but not at all by the cGMP-dependent kinase,[107,125] an 11-residue synthetic peptide corresponding to its site of phosphorylation was tested as a substrate for each enzyme. This peptide is phosphorylated readily by both protein kinases, although it and several analogues are slightly better substrates for the catalytic subunit of cAMP-dependent protein kinase (Table 5).[107] Therefore, structures other than the primary sequence at the site of phosphorylation in 6-phosphofructo-2-kinase/fructose-2,6-bisphosphatase must be responsible for the inability of the native enzyme to be phosphorylated by cGMP-dependent protein kinase.

Another reason to investigate analogues of the 6-phosphofructo-2-kinase/fructose-2,6-bisphosphatase peptide substrate is that the amino acid sequence at its phosphorylation site resembles those in G-substrate, DARPP-32, and phosphatase inhibitor-1 in that they all

TABLE 5
Kinetics of Phosphorylation of Selected Synthetic Peptide Substrates by cGMP-Dependent and cAMP-Dependent Protein Kinases[a]

Protein	Model peptide and analogue sequences[b]	k_{cat}/K_m cGMP kinase ($\mu M^{-1} min^{-1}$)	k_{cat}/K_m cAMP kinase ($\mu M^{-1} min^{-1}$)	$\dfrac{k_{cat}/K_m \text{ cGMP kinase}}{k_{cat}/K_m \text{ cAMP kinase}}$	Ref.
Histone H2B (site 32)	Arg-Lys-Arg-Ser-Arg-Lys-Glu	31.0	0.43	72	105,106
	Arg-Lys-Arg-Ser-**Ala**-Lys-Glu	0.62	0.26	2.5	
	Arg-Lys-Arg-**Thr**-Arg-Lys-Glu	0.25	0.022	11	
	Arg-Lys-Arg-Ser-Arg-**Ala**-Glu	106	9.7	11	
	acetyl-Arg-lys-Arg-Ser-Arg-**Ala**-Glu	204	13	16	
Histone H2B (site 36)	Arg-lys-Glu-Ser-Tyr-Ser-Val	2.5	6.7	0.37	105
cGMP-dependent protein kinase (autophosphorylation)	Ile-Gly-Pro-Arg-Thr-Thr-Arg-Ala-Gln-Gly-Ile	0.018	0.00057	31.6	104
	Ile-Gly-Pro-Arg-Thr-**Ser**-Arg-Ala-Gln-Gly-Ile	0.69	—	—	
6-Phoshofructo-2-kinase/fructose-2,6-bisphosphatase	Val-Leu-Gln-Arg-Arg-Arg-Gly-Ser-Ser-Ile-Pro-Gln	4.2	147	0.029	107
	Val-Leu-Gln-Arg-Arg-Arg-**Pro**-Ser-Ser-Ile-Pro-Gln	33	48	0.69	
	Val-Leu-Gln-Arg-Arg-Arg-Gly-**Thr**-Ser-Ile-Pro-Gln	1.1	4.8	0.23	
	Val-Leu-Gln-Arg-Arg-Arg-**Pro**-**Thr**-Ser-Ile-Pro-Gln	2.1	2.3	0.91	
DARPP-32	Arg-Arg-Pro-Thr-Pro-Ala-Met-Leu-Phe	0.21	0.012	18	142

TABLE 5 (continued)
Kinetics of Phosphorylation of Selected Synthetic Peptide Substrates by cGMP-Dependent and cAMP-Dependent Protein Kinases[a]

Protein	Model peptide and analogue sequences[b]	k_{cat}/K_m cGMP kinase ($\mu M^{-1} \cdot min^{-1}$)	k_{cat}/K_m cAMP kinase ($\mu M^{-1} \cdot min^{-1}$)	k_{cat}/K_m cGMP kinase / k_{cat}/K_m cAMP kinase	Ref.
L-type pyruvate kinase	Leu-Arg-Arg-Ala-<u>Ser</u>-Leu-Gly (Kemptide)	10.0	68.2	0.15	172—174
	Leu-Arg-Arg-Ala-**MeSer**-Leu-Gly[c]	0.010	<0.000015	667	
	Leu-Arg-Arg-**Gly**-**MeSer**-Leu-Gly[c]	0.061	0.00055	111	

[a] The k_{cat} values are based on the holoenzyme of cGMP-dependent protein kinase ($M_r = 153,000$) and the free catalytic subunit of cAMP-dependent protein kinase ($M_r = 41,000$). Ratios of k_{cat}/K_m were calculated from the V_{max} and K_m values in the indicated references.

[b] The underlined residues are the phosphorylated amino acids. Amino acids that are substituted for a residue in the parent peptide are in bold print.

[c] MeSer, N-methyl seryl residue.

contain three or more arginyl residues amino-terminal to the phosphorylatable hydroxyamino acid, rather than just two (Table 4). Also, the phosphorylation site in 6-phosphofructo-2-kinase/fructose-2,6-bisphosphatase contains two adjacent seryl residues, of which only serine-32 (the first serine) is phosphorylated in the native protein.[125] Studies with analogues of this sequence indicate that peptides containing three arginines are better substrates for either protein kinase than are those with only two arginines.[107] Substitution of threonine for serine-32 in the parent peptide results in a kinetically poorer peptide substrate, but increases the selectivity for cGMP-dependent kinase (Table 5). The cAMP-dependent protein kinase predominantly phosphorylates the first of the two adjacent hydroxyamino acids in either the parent peptide or its analogues. However, the cGMP-dependent protein kinase phosphorylates both of the hydroxyamino acid residues in a mutually exclusive fashion.[107] Replacement of glycine-31 with proline yields peptides containing the sequence −Arg−Arg−Arg−Pro−Ser−Ser− or −Arg−Arg−Arg−Pro−Thr−Ser−, similar to the phosphorylation site sequences in DARPP-32 and its related proteins. The prolyl residue immediately amino-terminal to the phosphorylatable amino acids enhances the selectivity of these analogues as substrates for the cGMP-dependent protein kinase (Table 5). In addition, proline in this position causes the cGMP-dependent enzyme to phosphorylate the threonyl residue at position 32 preferentially.[107] These results are consistent with a role for a prolyl residue in allowing the facile phosphorylation of an adjacent threonine, a residue that is usually a poor phosphate acceptor in the cyclic nucleotide-dependent protein kinase-catalyzed phosphotransferase reaction. A synthetic nonapeptide corresponding to the amino acid sequence in DARPP-32 is a rather poor substrate for both protein kinases, having K_m values of greater than 1 mM.[142] However, this peptide had only two arginyl residues amino-terminal to the phosphorylatable threonine.

The prototypical synthetic peptide substrate for cyclic nucleotide-dependent protein kinases is Leu−Arg−Arg−Ala−Ser−Leu−Gly, or Kemptide,[103] which corresponds (other than the carboxy-terminal glycyl residue) to the sequence at the phosphorylation site in pig liver L-type pyruvate kinase.[168] Initial studies by Kemp et al.[103] and Zetterqvist et al.[169] demonstrated the importance of multiple arginyl residues in Kemptide as determinants of substrate specificity for cAMP-dependent protein kinase. The distance of the arginyl residues from the phosphorylatable seryl residue in the peptide is also an important determinant of specificity for this enzyme.[170] Kemptide and its analogues also are phosphorylated by the cGMP-dependent protein kinase,[58,171] but kinetically it is a poorer substrate for this enzyme than for the cAMP-dependent protein kinase.[105,162] This is mainly due to a much higher K_m value with the cGMP-dependent enzyme; Kemptide has similar high maximal velocities with either kinase. Based on analogues of the Kemptide sequence the cGMP-dependent protein kinase, like the cAMP-dependent enzyme, has a strict preference for substrates in which only one amino acid residue separates the two amino-terminal arginines from the phosphorylatable hydroxyamino acid.[162]

Kemptide has served as a model substrate or inhibitor peptide for numerous studies of the active site of cAMP-dependent protein kinase. Recently, Thomas and colleagues[172,173] used a series of N-methylated or depsipeptide analogues of Kemptide to investigate the role of hydrogen bonding in the peptide-enzyme interaction. When these conformationally restricted peptides were tested as substrates for the cGMP-dependent protein kinase, several analogues that contain an N-methylated seryl residue were shown to be 100- to nearly 700-fold more selective for this enzyme as compared to the cAMP-dependent protein kinase (Table 5).[174] Unfortunately, both of these peptides kinetically were quite poor substrates of the cGMP-dependent protein kinase. Their selectivity for this enzyme was due to the fact that they were almost not substrates for the cAMP-dependent protein kinase at all. Nonetheless, the former enzyme seems better able to accommodate backbone methylation of a peptide substrate bound to its active site than the latter. This may be relevant to the ability of the cGMP-dependent protein kinase to phosphorylate threonine residues more selectively in some peptide and protein substrates. The β-methyl group of threonine might assume an orientation similar to that of the methyl group in an N-methylated amino acid residue.

V. ENDOGENOUS PROTEIN SUBSTRATES OF cGMP-DEPENDENT PROTEIN KINASE

A. PROTEINS PHOSPHORYLATED IN VASCULAR SMOOTH MUSCLE

Because of the evidence for cGMP and the cGMP-dependent protein kinase mediating the relaxant effects of various vasodilator substances,[2,11-13,18-21] potential endogenous protein substrates of this enzyme have been studied extensively in vascular smooth muscle. These studies have been performed either using cell-free extracts to which cyclic nucleotides have been added or using intact cells prelabeled with radioactive phosphorus which were then exposed to drugs known to elevate cGMP levels. Additional studies have used fractionated subcellular extracts with addition of purified protein kinases to catalyze phosphorylation of endogenous proteins. Although these studies have spanned at least 14 years, only marginal progress has been made in identifying physiologically relevant protein substrates for cGMP-dependent protein kinase for which the function of the protein is known. In most cases, a radiolabeled polypeptide simply has been identified as a band on an autoradiogram of cellular proteins separated by polyacrylamide gel electrophoresis in the presence of sodium dodecyl sulfate.

Both the cGMP-dependent protein kinase and its substrate proteins are present in the medial layer of rabbit aorta. Photoaffinity labeling with a radiolabeled, photoreactive cyclic nucleotide analogue was used to localize the enzyme.[175,176] Approximately 25% of the cGMP-dependent protein kinase is present in the particulate fraction.[176] The protein kinase is also present in primary cultures of smooth muscle cells isolated from this tissue. Substrate proteins for the cGMP-dependent protein kinase are not detected in the soluble fraction from vascular smooth muscle. However, in particulate fractions from both medial tissue and the smooth muscle cells in culture, four proteins are present whose phosphorylation is stimulated selectively by addition of cGMP.[176] These substrate proteins are 240 to 250, 130, 85, and 75 kDa in size. The phosphorylation of the 130-kDa protein occurs extremely rapidly, suggesting that it is a kinetically favorable substrate or is associated closely with the endogenous protein kinase.[21] Subcellular fractionation of the medial layer of aorta by sucrose density-gradient centrifugation indicates that the four substrate proteins comigrate with a fraction rich in plasma membrane, sarcoplasmic reticulum, and Golgi complex.[18] In addition, these substrate proteins appear to be integral membrane proteins. The cGMP-dependent protein kinase appears in both the soluble fraction and in the same subcellular fraction that contains the substrate proteins.[18] Unlike the substrate proteins, however, the particle-associated protein kinase is a peripheral membrane protein.

The phosphorylation of the 240- to 250-, 130-, and 85-kDa substrate proteins in particulate fractions of vascular smooth muscle has been investigated further by Jamieson and his colleagues.[177] The K_a values for stimulation of phosphorylation of these substrates by cGMP and cAMP are 0.01 and 0.2 μM, respectively. Use of the heat-stable inhibitor protein of cAMP-dependent protein kinase and antibodies against cGMP-dependent protein kinase showed that phosphorylation of the substrate proteins by either cyclic nucleotide is mediated by the activation of the endogenous, membrane-associated cGMP-dependent protein kinase.[21,177] If this endogenous protein kinase is removed from the membrane preparation, addition of purified exogenous cGMP-dependent protein kinase results in the phosphorylation of the same protein substrates. Peptide mapping of the phosphorylated protein substrates indicates that both the soluble and particulate forms of the cGMP-dependent protein kinase have the same site specificity. There is almost no cAMP-dependent protein kinase associated with the washed particulate fractions that contain the substrate proteins for cGMP-dependent protein kinase. However, either the endogenous cAMP-dependent protein kinase in homogenates or the exogenously added catalytic subunit of this enzyme also phosphorylates these same substrate proteins. Peptide mapping of the 130-kDa substrate phosphorylated by cAMP-dependent protein kinase indicates that this enzyme recognizes a subset of the sites phosphorylated by the cGMP-dependent protein kinase.[177] Thus, the phosphorylation of the cGMP-dependent protein kinase substrate proteins could be promoted in vascular smooth muscle via activation of the membrane-associated cGMP-dependent protein kinase by either cGMP or cAMP, or possibly by activation of the cytosolic cAMP-dependent protein kinase by cAMP.

One other study of substrate proteins in a sarcolemmal fraction from porcine aorta reported that cGMP and cGMP-dependent protein kinase stimulated phosphorylation of a 35-kDa polypeptide.[158]

Murad and co-workers[20,178] studied the phosphorylation of cellular proteins in intact strips of rat aorta exposed to agents that are known to raise cellular cGMP levels and relax vascular smooth muscle. Tissue was treated with sodium nitroprusside or other relaxant agents and then homogenized and separated into soluble and particulate fractions under conditions that minimized protein phosphorylation and dephosphorylation. Sodium nitroprusside produces a concentration-dependent increase in the phosphorylation of nine proteins and a decrease in the phosphorylation of two proteins. These patterns of protein phosphorylation are mimicked by exposing the tissue to 8-bromo-cGMP and are unaffected by removal of the endothelium. Acetylcholine, an endothelial-dependent vasodilator, causes the same pattern of protein phosphorylation in intact aorta as does sodium nitroprusside.[20] Removal of the endothelium from tissues prior to stimulation with acetylcholine prevents the phosphorylation of protein substrates. Derivatives of cAMP have only small effects on the phosphorylation of these same cellular proteins, indicating that the changes are mediated rather selectively by cGMP.

All of the phosphoproteins studied above are within the size range of 20 to 50 kDa. The two-dimensional gel electrophoresis conditions used gave poor resolution in the higher molecular weight areas.[20,178] Therefore, these experiments with intact tissue are not readily comparable to the results observed in cell-free extracts, in which substrate proteins of 240 to 250, 130, and 85 kDa are phosphorylated routinely. The two 22-kDa proteins whose degree of phosphorylation is decreased by vasodilators are identified as myosin light chains.[20] Thus, agents that act through cGMP may produce relaxation by ultimately causing dephosphorylation of myosin light chains.

The enhanced phosphorylation of proteins in intact tissues can result from a change in the activities of either a protein kinase or a phosphoprotein phosphatase. While the phosphorylation of proteins in intact smooth muscle in the above studies is apparently cGMP dependent, it is not possible to be sure that this result is directly due to an activation of the cGMP-dependent protein kinase. However, as measured by activity ratios of protein kinase assayed in the absence and presence of cGMP,[179] the cGMP-dependent protein kinase is indeed activated in intact aorta on exposure to either endothelium-dependent or -independent vasodilators.[19] Cholinergic agents and sodium nitroprusside produce a similar activation of the cGMP-dependent protein kinase in tracheal smooth muscle.[32]

All of the data from both intact tissues and cell-free tissue extracts strongly suggest a physiological role for cGMP in the relaxation of vascular smooth muscle. The action of this cyclic nucleotide most probably is mediated by the activation of cGMP-dependent protein kinase resulting in the phosphorylation of endogenous protein substrates.

B. PROTEINS PHOSPHORYLATED IN OTHER TISSUES

Several studies have reported the cGMP-dependent phosphorylation of endogenous substrates in tissues other than vascular smooth muscle. All of these studies employed cell-free extracts and *in vitro* phosphorylation. The functional identity of these phosphorylated proteins remains unknown.

The proteins selectively phosphorylated by cGMP in other tissues rich in smooth muscle are similar to some of those phosphorylated in vascular smooth muscle. The molecular weights as well as the subcellular localization of the protein substrates from various smooth muscle are alike. Polypeptides of 240, 130, and 100 kDa are phosphorylated in the presence of low concentrations of cGMP in membrane fractions from guinea pig vas deferens and uterus and from rabbit small intestine.[180,181] In membrane fractions of tracheal smooth muscle, phosphorylation of 130- and 80-kDa proteins is stimulated by cGMP and also by calcium and calmodulin.[182] These substrates appear to be integral membrane proteins. The K_a values for stimulation of phosphorylation by cGMP are at least tenfold lower than those for cAMP.

In brush border fractions of isolated villous cells from rat small intestine, cGMP promotes the selective phosphorylation of an 86-kDa integral membrane protein.[35] This cyclic nucleotide is 30-fold more potent than cAMP in stimulating the phosphorylation.

Various proteins in membrane fractions isolated from human placenta are phosphorylated *in vitro* in the presence of low concentrations of cGMP, but not cAMP.[183] These polypeptides have apparent sizes of approximately 160 to 170, 123, 94, 85, 75, 61, and 44 kDa. The largest of these proteins is also phosphorylated in placental membranes by the addition of atrial natriuretic factor.[183] It is not yet clear whether this 160- to 170-kDa protein substrate is identical to the similarly sized membrane receptor for atrial natriuretic factor.

Cytosol from rat heart contains a 70-kDa protein whose phosphorylation is stimulated by cGMP alone or by the combination of cGMP and added cGMP-dependent protein kinase.[184] The substrate protein also is phosphorylated readily after partial purification from the soluble cell extract. This phosphoprotein does not represent autophosphorylation of the cGMP-dependent protein kinase.[184] The 70-kDa substrate is different from other endogenous substrates in this tissue whose phosphorylation is stimulated by the cAMP-dependent, calcium/calmodulin-dependent, or calcium/phospholipid-dependent (protein kinase C) protein kinases.

The selective substrate of the cGMP-dependent protein kinase in cerebellum has been mentioned previously because it is the one substrate protein that has been purified and characterized and whose function is beginning to be understood.[51] Unlike most of the endogenous substrates described above, G-substrate is a soluble protein that is present in the cytosol of cerebellar Purkinje cells.[29,30,40] G-substrate is a 23-kDa protein whose use as an *in vitro* substrate of cGMP-dependent protein kinase has been covered above in Section IV.A.2.[137-139]

VI. PEPTIDE INHIBITORS OF cGMP-DEPENDENT PROTEIN KINASE

General approaches to inhibition of the cyclic nucleotide-dependent protein kinases include compounds that interfere with their activation by cyclic nucleotides[110,185] and inhibitors based on the structures of either of the two substrates, MgATP or the protein/peptide phosphorylation site sequence. Additionally, although it has not been done for the protein kinases, inhibitors could include components of both substrate structures, nucleotide and peptide, in the same molecule. Such compounds might mimic the "transition state" of the substrates at the active site or, at the least, provide additional bonding interactions to enhance the potency of inhibition. The first of these general approaches, cyclic nucleotide antagonists, will not be covered here. The use of ATP analogues as protein kinase inhibitors has been useful in the *in vitro* affinity labeling of active sites of various protein kinases.[87-89] However, ATP analogues would not necessarily allow specific inhibition of different types of protein kinases and most certainly would inhibit other ATP-utilizing enzymes in intact cells. Compounds with nonnucleotide structures recently have been discovered that apparently interact at the nucleotide binding sites of various protein kinases, including the cyclic nucleotide-dependent enzymes.[186-188] Because of the potential lack of specificity of nucleotide analogues as protein kinase inhibitors and the wealth of information on the peptide/protein substrate specificities of these enzymes, peptide analogues are being developed as protein kinase inhibitors. The synthetic peptide inhibitors may exhibit some degree of specificity for certain protein kinases because of variability in the structures of the protein binding sites among the different protein kinases.

Kemp et al.[189] first synthesized a peptide inhibitor of cAMP-dependent protein kinase based on a phosphorylatable sequence in denatured lysozyme. Feramisco and Krebs[190] replaced the seryl residue in Kemptide with an alanine. The resulting peptide contains the determinants of specificity required for recognition by the catalytic subunit of cAMP-dependent protein kinase, but cannot be phosphorylated. Thus, it is a competitive inhibitor of the enzyme vs. peptide or protein substrates. However, although Kemptide has a low micromolar K_m value, its true K_d value as well as the K_i value of the inhibitor analogue are 300 to 500 μM.[161,190]

A similar strategy of replacing serine with alanine has been undertaken with the histone H2B peptide substrate of the cGMP-dependent protein kinase.[161] The synthetic peptide Arg–Lys–Arg–Ala–Arg–Lys–Glu is a competitive inhibitor of the cGMP-dependent protein kinase, with a K_i value of approximately 80 μM vs. other synthetic peptides or various nonhistone protein substrates.[161,162] This peptide inhibits the autophosphorylation of the enzyme with a similar potency.[162] Interestingly, although the inhibitor peptide is based on a sequence from a phosphorylation site in histone, it will not inhibit the cGMP-dependent protein kinase when various histones are used as substrates.[161] Arg–Lys–Arg–Ala–Arg–Lys–Glu inhibits the phosphorylation of peptides and histones by cAMP-dependent protein kinase equally well.[161] The reason for the lack of inhibition of cGMP-dependent protein kinase with histone substrates is not known, but it probably involves the subunit structure of this protein kinase and the fact that histones bind extremely tightly to the enzyme.[191,192]

Arg–Lys–Arg–Ala–Arg–Lys–Glu has about a sevenfold selectivity in its potency for inhibition of cGMP-dependent protein kinase as compared to the cAMP-dependent enzyme.[161] The Kemptide inhibitor analogue, on the other hand, is slightly more selective for inhibition of the catalytic subunit of cAMP-dependent protein kinase.[161] The only other peptides that inhibit the cGMP-dependent protein kinase with appreciable potency are fragments of the active portion of the heat-stable inhibitor of cAMP-dependent protein kinase.[62,128] One of these is protein kinase inhibitor (5-22)amide, with the sequence Thr–Thr–Tyr–Ala–Asp–Phe–Ile–Ala–Ser–Gly–Arg–Thr–Gly–Arg–Arg–Asn–Ala–Ile–NH$_2$.[128] These inhibitor peptides are pseudosubstrates that have multiple basic amino acids amino-terminal to an alanine residue. Also, there is some sequence similarity between these protein kinase inhibitor peptides and the autophosphorylation site in the regulatory domain of cGMP-dependent protein kinase.[193] However, these peptides are as much as three or four orders of magnitude more potent as inhibitors of the cAMP-dependent protein kinase than they are of the cGMP-dependent enzyme.[128] The structure-function relationships of protein kinase inhibitor (5-22)amide and its analogues have been studied intensively with respect to inhibition of the cAMP-dependent protein kinase,[62] and many of the analogues are just now being tested as inhibitors of cGMP-dependent protein kinase. Therefore, at this time the most selective inhibitor of the cGMP-dependent protein kinase remains Arg–Lys–Arg–Ala–Arg–Lys–Glu, the histone H2B peptide analogue.[161]

VII. CONCLUDING REMARKS

The main aspects of the substrate specificity of cyclic nucleotide-dependent protein kinases that have been dealt with above are sites of phosphorylation in substrate proteins and the kinetics of phosphorylation of these sites or their peptide models. In nearly all cases, these studies addressed the substrate specificities of the protein kinases by assessing the primary amino acid sequences in the substrate proteins or peptides. From these data it is clear that the cGMP-dependent protein kinase is one of the protein kinases that recognize basic amino acid residues as important determinants of specificity. The substrate specificity of this enzyme is similar, but clearly not identical, to that of the homologous cAMP-dependent protein kinase. Each of these protein kinases has a rather broad substrate specificity. One of the differences between their specificities is that the cGMP-dependent protein kinase will favor phosphorylation of a site that contains basic amino acids not only amino-terminal but also immediately carboxy-terminal to the phosphorylated residue. Such a carboxy-terminal charged residue in a substrate is not at all well tolerated by the cAMP-dependent protein kinase. However, not all of the substrates of the cGMP-dependent protein kinase have a basic amino acid located carboxy-terminal to the serine or threonine residue at their phosphorylation sites, so such a sequence is not a required determinant of specificity. While the cyclic nucleotide-dependent protein kinases clearly recognize determinants of specificity in primary sequences, additional experimental work must now address the role of higher orders of structure in the interactions of substrates and inhibitors with the active sites of these protein kinases.

In addition, to gain a full understanding of the physiological relevance of the substrate specificities of the cyclic nucleotide-dependent protein kinases, details of the expression, cellular localization, and functions of their substrate proteins will continue to be an important research endeavor.

ACKNOWLEDGMENTS

The work from the author's laboratory was supported by U.S. Public Health Service Grant GM28144.

REFERENCES

1. **Goldberg, N. D. and Haddox, M. K.**, Cyclic GMP metabolism and involvement in biological regulation, *Annu. Rev. Biochem.*, 46, 823, 1977.
2. **Waldman, S. A. and Murad, F.**, Cyclic GMP synthesis and function, *Pharmacol. Rev.*, 39, 163, 1987.
3. **Kuo, J. F. and Greengard, P.**, Cyclic nucleotide-dependent protein kinases. VI. Isolation and partial purification of a protein kinase activated by guanosine 3′,5′-monophosphate, *J. Biol. Chem.*, 245, 2493, 1970.
4. **Kuo, J. F.**, Guanosine 3′,5′-monophosphate-dependent protein kinase in mammalian tissues, *Proc. Natl. Acad. Sci. U.S.A.*, 71, 4037, 1974.
5. **Beavo, J. A.**, Multiple isozymes of cyclic nucleotide phosphodiesterase, *Adv. Second Messenger Phosphoprotein Res.*, 22, 1, 1988.
6. **Cook, N. J., Zeilinger, C., Koch, K.-W., and Kaupp, B.**, Solubilization and functional reconstitution of the cGMP-dependent cation channel from bovine rod outer segments, *J. Biol. Chem.*, 261, 17033, 1986.
7. **Hanke, W., Cook, N. J., and Kaupp, U. B.**, cGMP-dependent channel protein photoreceptor membranes: single channel activity of the purified and reconstituted protein, *Proc. Natl. Acad. Sci. U.S.A.*, 85, 94, 1988.
8. **Nakamura, T. and Gold, G. H.**, A cyclic nucleotide-gated conductance in olfactory receptor cilia, *Nature (London)*, 325, 442, 1987.

9. **de Bold, A. J.**, Atrial natriuretic factor: a hormone produced by the heart, *Science,* 230, 767, 1985.
10. **Cantin, M. and Genest, J.**, The heart and the atrial natriuretic factor, *Endocrine Rev.,* 6, 107, 1985.
11. **Rapoport, R. M., Winquist, R. J., Baskin, E. P., Faison, E. P., Waldman, S. A., and Murad, F.**, Effects of atriopeptins on relaxation and cyclic GMP levels in rat and rabbit aortas, *Eur. J. Pharmacol.,* 120, 123, 1986.
12. **Ignarro, L. J., Wood, K. S., Harbison, R. G., and Kadowitz, P. J.**, Atriopeptin II relaxes and elevates cGMP in bovine pulmonary artery but not vein, *J. Appl. Physiol.,* 60, 1128, 1986.
13. **Huang, C.-L., Ives, H. E., and Cogan, M. G.**, In vivo evidence that cGMP is the second messenger for atrial natriuretic factor, *Proc. Natl. Acad. Sci. U.S.A.,* 83, 8015, 1986.
14. **Fiscus, R., Robles, B. T., Waldman, S. A., and Murad, F.**, Atrial natriuretic factors stimulate accumulation and efflux of cyclic GMP in C6-2B rat glioma and PC12 rat pheochromocytoma cell cultures, *J. Neurochem.,* 48, 522, 1987.
15. **Tremblay, J., Gerzer, R., Vinay, P., Pang, S. C., Beliveau, R., and Hamet, P.**, The increase in cGMP by atrial natriuretic factor correlates with the distribution of particulate guanylate cyclase, *FEBS Lett.,* 181, 17, 1985.
16. **Kuno, T., Andresen, J. W., Kamisaki, Y., Waldman, S. A., Chang, L. Y., Saheki, S., Leitman, D. C., Nakane, M., and Murad, F.**, Co-purification of an atrial natriuretic factor receptor and particulate guanylate cyclase from rat lung, *J. Biol. Chem.,* 261, 5817, 1986.
17. **Tremblay, J., Gerzer, R., Pang, S. C., Cantin, M., Genest, J., and Hamet, P.**, ANF stimulation of detergent-dispersed particulate guanylate cyclase from bovine adrenal cortex, *FEBS Lett.,* 194, 210, 1986.
18. **Ives, H. E., Casnellie, J. E., Greengard, P., and Jamieson, J. D.**, Subcellular localization of cyclic GMP-dependent protein kinase and its substrates in vascular smooth muscle, *J. Biol. Chem.,* 255, 3777, 1980.
19. **Fiscus, R. R., Rapoport, R. M., and Murad, F.**, Endothelium-dependent and nitrovasodilator-induced activation of cyclic GMP-dependent protein kinase in rat aorta, *J. Cyclic Nucleotide Protein Phosphorylation Res.,* 9, 415, 1984.
20. **Rapoport, R. M., Draznin, M. B., and Murad, F.**, Endothelium-dependent relaxation in rat aorta may be mediated through cyclic GMP-dependent protein phosphorylation, *Nature (London),* 306, 174, 1983.
21. **Lincoln, T. M. and Johnson, R. M.**, Possible role of cyclic GMP-dependent protein kinase in vascular smooth muscle function, *Adv. Cyclic Nucleotide Protein Phosphorylation Res.,* 17, 285, 1984.
22. **Furchgott, R. F.**, The role of endothelium in the response of vascular smooth muscle to drugs, *Annu. Rev. Pharmacol. Toxicol.,* 24, 175, 1984.
23. **Palmer, R. M. J., Ferrige, A. G., and Moncada, S.**, Nitric oxide release accounts for the biological activity of endothelium-derived relaxing factor, *Nature (London),* 327, 524, 1987.
24. **Ignarro, L. J., Adams, J. B., Horwitz, P. M., and Woods, K. S.**, Activation of soluble guanylate cyclase by the NO-hemoproteins involves NO-heme exchange. Comparison of heme-containing and heme-deficient enzyme forms, *J. Biol. Chem.,* 261, 4997, 1986.
25. **Ignarro, L. J., Harbison, R. G., Woods, K. S., and Kadowitz, P. J.**, Activation of purified soluble guanylate cyclase by endothelial-derived relaxing factor from intrapulmonary artery and vein — stimulation by acetylcholine, bradykinin, and arachidonic acid, *J. Pharmacol. Exp. Ther.,* 237, 893, 1986.
26. **Mulsch, A., Bohme, E., and Busse, R.**, Stimulation of soluble guanylate cyclase by endothelium-derived relaxing factor from cultured endothelial cells, *Eur. J. Pharmacol.,* 135, 247, 1987.
27. **Joyce, N. C., DeCamilli, P., Lohmann, S. M., and Walter, U.**, cGMP-dependent protein kinase is present in high concentrations in contractile cells of the kidney vasculature, *J. Cyclic Nucleotide Protein Phosphorylation Res.,* 11, 191, 1986.
28. **Walter, U.**, Distribution of cyclic GMP-dependent protein kinase in various rat tissues and cell lines determined by a sensitive and specific radioimmunoassay, *Eur. J. Biochem.,* 118, 339, 1981.
29. **Lohmann, S. M., Walter, U., Miller, P. E., Greengard, P., and DeCamilli, P.**, Immunohistochemical localization of cyclic GMP-dependent protein kinase in mammalian brain, *Proc. Natl. Acad. Sci. U.S.A.,* 78, 653, 1981.
30. **Schlichter, D. J., Detre, J. A., Aswad, D. W., Chehrazi, B., and Greengard, P.**, Localization of cyclic GMP-dependent protein kinase and substrate in mammalian cerebellum, *Proc. Natl. Acad. Sci. U.S.A.,* 77, 5537, 1980.
31. **Paupardin-Tritch, D., Hammond, C., Gerschenfeld, H. M., Nairn, A. C., and Greengard, P.**, cGMP-dependent protein kinase enhances Ca^{+2} current and potentiates the serotonin-induced Ca^{+2} current increase in snail neurones, *Nature (London),* 323, 812, 1986.
32. **Fiscus, R. R., Torphy, T. J., and Mayer, S. E.**, Cyclic GMP-dependent protein kinase activation in canine tracheal smooth muscle by methacholine and sodium nitroprusside, *Biochim. Biophys. Acta,* 805, 382, 1984.
33. **Waldmann, R., Nieberding, M., and Walter, U.**, Vasodilator-stimulated protein phosphorylation in platelets is mediated by cAMP- and cGMP-dependent protein kinases, *Eur. J. Biochem.,* 167, 441, 1987.

34. **Largen, M. T. and Votta, B.**, Immunocytochemical evidence for 3',5'-cGMP and 3',5'-cGMP-dependent protein kinase involvement in lymphocyte proliferation, *J. Cyclic Nucleotide Protein Phosphorylation Res.*, 9, 231, 1983.
35. **de Jonge, H. R.**, Cyclic nucleotide-dependent phosphorylation of intestinal epithelium proteins, *Nature (London)*, 262, 590, 1976.
36. **van Dommelen, F. S. and de Jonge, H. R.**, Local changes in fractional saturation of cGMP- and cAMP-receptors in intestinal microvilli in response to cholera toxin and heat-stable *Escherichia coli* toxin, *Biochim. Biophys. Acta*, 886, 135, 1986.
37. **Takio, K., Wade, R. D., Smith, S. B., Krebs, E. G., Walsh, K. A., and Titani, K.**, Guanosine cyclic 3',5'-phosphate-dependent protein kinase, a chimeric protein homologous with two separate protein families, *Biochemistry*, 23, 4207, 1984.
38. **Lincoln, T. M. and Corbin, J. D.**, On the role of the cAMP and cGMP-dependent protein kinases in cell function, *J. Cyclic Nucleotide Res.*, 4, 3, 1978.
39. **Khoo, J. C. and Gill, G. N.**, Comparison of cyclic nucleotide specificity of guanosine 3',5'-monophosphate-dependent protein kinase and adenosine 3',5'-monophosphate-dependent protein kinase, *Biochim. Biophys. Acta*, 584, 21, 1979.
40. **Schlichter, D. J., Casnellie, J. E., and Greengard, P.**, An endogenous substrate for cGMP-dependent protein kinase in mammalian cerebellum, *Nature (London)*, 273, 61, 1978.
41. **Hunter, T.**, A thousand and one protein kinases, *Cell*, 50, 823, 1987.
42. **Kuo, J. F., Shoji, M., and Kuo, W.-N.**, Molecular and physiopathological aspects of mammalian cyclic GMP-dependent protein kinase, *Annu. Rev. Pharmacol. Toxicol.*, 18, 341, 1978.
43. **Gill, G. N. and McCune, R. W.**, Guanosine 3',5'-monophosphate-dependent protein kinase, *Curr. Top. Cell. Regul.*, 15, 1, 1979.
44. **Glass, D. B. and Krebs, E. G.**, Protein phosphorylation catalyzed by cyclic AMP-dependent and cyclic GMP-dependent protein kinases, *Annu. Rev. Pharmacol. Toxicol.*, 20, 363, 1980.
45. **Kuo, J. F. and Shoji, M.**, Cyclic GMP-dependent protein phosphorylation, in *Handbook of Experimental Pharmacology*, Vol. 58, Nathanson, J. A. and Kebabian, J. W., Eds., Springer-Verlag, Berlin, 1982, chap. 12.
46. **Lincoln, T. M. and Corbin, J. D**, Characterization and biological role of the cGMP-dependent protein kinase, *Adv. Cyclic Nucleotide Res.*, 15, 139, 1983.
47. **Walter, U.**, Cyclic GMP-regulated enzymes and their possible physiological functions, *Adv. Cyclic Nucleotide Protein Phosphorylation Res.*, 17, 249, 1984.
48. **Bramson, H. N., Kaiser, E. T., and Mildvan, A. S.**, Mechanistic studies of cAMP-dependent protein kinase action, *Crit. Rev. Biochem.*, 15, 93, 1984.
49. **Krebs, E. G. and Beavo, J. A.**, Phosphorylation-dephosphorylation of enzymes, *Annu. Rev. Biochem.*, 48, 923, 1979.
50. **Flockhart, D. A. and Corbin, J. D.**, Regulatory mechanisms in the control of protein kinases, *Crit. Rev. Biochem.*, 12, 133, 1982.
51. **Nestler, E. J. and Greengard, P.**, *Protein Phosphorylation in the Nervous System*, John Wiley & Sons, New York, 1984.
52. **Lincoln, T. M., Dills, W. L., Jr., and Corbin, J. D.**, Purification and subunit composition of guanosine 3':5'-monophosphate-dependent protein kinase from bovine lung, *J. Biol. Chem.*, 252, 4269, 1977.
53. **Gill, G. N., Walton, G. M., and Sperry, P.**, Guanosine 3':5'-monophosphate-dependent protein kinase from bovine lung: subunit structure and characterization of the purified enzyme, *J. Biol. Chem.*, 252, 6443, 1977.
54. **Bechtel, P. J., Beavo, J. A., and Krebs, E. G.**, Purification and characterization of catalytic subunit of skeletal muscle adenosine 3':5'-monophosphate-dependent protein kinase, *J. Biol. Chem.*, 252, 2691, 1977.
55. **Dills, W. L., Goodwin, C. D., Lincoln, T. M., Beavo, J. A., Bechtel, P. J., Corbin, J. D., and Krebs, E. G.**, Purification of cyclic nucleotide receptor proteins by cyclic nucleotide affinity chromatography, *Adv. Cyclic Nucleotide Res.*, 10, 199, 1979.
56. **Corbin, J. D., Sugden, P. H., West, L., Flockhart, D. A., Lincoln, T. M., and McCarthy, D.**, Studies on the properties and mode of action of the purified regulatory subunit of bovine heart adenosine 3':5'-monophosphate-dependent protein kinase, *J. Biol. Chem.*, 253, 3997, 1978.
57. **Corbin, J. D., Øgreid, D., Miller, J. P., Suva, R. H., Jastorff, B., and Døskeland, S. O.**, Studies of cGMP analog specificity and function of the two intrasubunit binding sites of cGMP-dependent protein kinase, *J. Biol. Chem.*, 261, 1208, 1986.
58. **Lincoln, T. M. and Corbin, J. D.**, Adenosine 3':5'-cyclic monophosphate- and guanosine 3':5'-cyclic monophosphate-dependent protein kinases: possible homologous proteins, *Proc. Natl. Acad. Sci. U.S.A.*, 74, 3239, 1977.
59. **Tao, M., Salas, M. L., and Lipmann, F.**, Mechanism of activation by adenosine 3':5'-cyclic monophosphate of a protein phosphokinase from rabbit reticulocytes, *Proc. Natl. Acad. Sci. U.S.A.*, 67, 408, 1970.

60. **Reiman, E. M., Walsh, D. A., and Krebs, E. G.,** Purification and properties of rabbit skeletal muscle adenosine 3',5'-monophosphate-dependent protein kinases, *J. Biol. Chem.*, 246, 1986, 1971.
61. **Reiman, E. M., Brostrom, C. O., Corbin, J. D., King, C. A., and Krebs, E. G.,** Separation of regulatory and catalytic subunits of the cyclic 3',5'-adenosine monophosphate-dependent protein kinase(s) of rabbit skeletal muscle, *Biochem. Biophys. Res. Commun.*, 42, 187, 1971.
62. **Walsh, D. A., Angelos, K. L., Van Patten, S. M., Glass, D. B., and Garetto, L. P.,** The inhibitor protein of the cAMP-dependent protein kinase, in *Peptides and Protein Phosphorylation*, Kemp, B. E., Ed., CRC Press, Boca Raton, FL, 1990, chap. 2.
63. **Walter, U., Miller, P., Wilson, F., Menkes, D., and Greengard, P.,** Immunological distinction between guanosine 3':5'-monophosphate-dependent and adenosine 3':5'-monophosphate-dependent protein kinases, *J. Biol. Chem.*, 255, 3757, 1980.
64. **Erlichman, J., Rosenfeld, R., and Rose, O. M.,** Phosphorylation of a cyclic adenosine 3':5'-monophosphate-dependent protein kinase from bovine cardiac muscle, *J. Biol. Chem.*, 249, 5000, 1974.
65. **de Jonge, H. R. and Rosen, O. M.,** Self-phosphorylation of cyclic guanosine 3':5'-monophosphate-dependent protein kinase from bovine lung, *J. Biol. Chem.*, 252, 2780, 1977.
66. **Lincoln, T. M., Flockhart, D. A., and Corbin, J. D.,** Studies on the structure and mechanism of activation of the guanosine 3':5'-monophosphate-dependent protein kinase, *J. Biol. Chem.*, 253, 6002, 1978.
67. **Foster, J. L., Guttmann, J., and Rosen, O. M.,** Autophosphorylation of cGMP-dependent protein kinase, *J. Biol. Chem.*, 256, 5029, 1981.
68. **Gill, G. N.,** A hypothesis concerning the structure of cAMP- and cGMP-dependent protein kinases, *J. Cyclic Nucleotide Res.*, 3, 153, 1977.
69. **Titani, K., Sasagawa, T., Ericsson, L. H., Kumar, S., Smith, S. B., Krebs, E. G., and Walsh, K. A.,** Amino acid sequence of the regulatory subunit of bovine type I adenosine cyclic 3',5'-phosphate dependent protein kinase, *Biochemistry*, 23, 3193, 1984.
70. **Takio, K., Smith, S. B., Krebs, E. G., Walsh, K. A., and Titani, K.,** Primary structure of the regulatory subunit of type II cAMP-dependent protein kinase from bovine cardiac muscle, *Proc. Natl. Acad. Sci. U.S.A.*, 79, 2544, 1982.
71. **Takio, K., Smith, S. B., Krebs, E. G., Walsh, K. A., and Titani, K.,** Amino acid sequence of the regulatory subunit of bovine type II adenosine cyclic 3',5'-phosphate dependent protein kinase, *Biochemistry*, 23, 4200, 1984.
72. **Shoji, S., Parmelee, D. C., Wade, R. D., Kumar, S., Ericsson, L. H., Walsh, K. A., Neurath, H., Long, G. L., Demaille, J. G., Fischer, E. H., and Titani, K.,** Complete amino acid sequence of the catalytic subunit of bovine cardiac muscle cyclic AMP-dependent protein kinase, *Proc. Natl. Acad. Sci. U.S.A.*, 78, 848, 1981.
73. **Uhler, M. D., Carmichael, D. F., Lee, D. C., Chrivia, J. C., Krebs, E. G., and McKnight, G. S.,** Isolation of cDNA clones coding for the catalytic subunit of mouse cAMP-dependent protein kinase, *Proc. Natl. Acad. Sci. U.S.A.*, 83, 1300, 1986.
74. **Showers, M. O. and Maurer, R. A.,** A cloned bovine cDNA encodes an alternate form of the catalytic subunit of cAMP-dependent protein kinase, *J. Biol. Chem.*, 261, 16288, 1986.
75. **Kuno, T., Ono, Y., Hirai, M., Hashimoto, S., Shuntoh, H., and Tanaka, C.,** Molecular cloning and cDNA structure of the regulatory subunit of type I cAMP-dependent protein kinase from rat brain, *Biochem. Biophys. Res. Commun.*, 146, 878, 1987.
76. **Clegg, C. H., Correll, L. A., Cadd, G. G., and McKnight, G. S.,** Inhibition of intracellular cAMP-dependent protein kinase using mutant genes of the regulatory type I subunit, *J. Biol. Chem.*, 262, 13111, 1987.
77. **Monken, C. E. and Gill, G. N.,** Structural analysis of cGMP-dependent protein kinase using limited proteolysis, *J. Biol. Chem.*, 255, 7067, 1980.
78. **Monken, C. E. and Gill, G. N.,** A comparison of the cyclic nucleotide-dependent protein kinases using chemical cleavage at tryptophan and cysteine, *Arch. Biochem. Biophys.*, 240, 888, 1985.
79. **Reimann, E. M.,** Conversion of bovine cardiac adenosine cyclic 3',5'-phosphate dependent protein kinase to a heterodimer by removal of 45 residues at the N-terminus of the regulatory subunit, *Biochemistry*, 25, 119, 1986.
80. **Huang, T.-S., Feramisco, J. R., Glass, D. B., and Krebs, E. G.,** Specificity considerations relevant to protein kinase activation and function, *Proc. Miami Winter Symp.*, 16, 449, 1979.
81. **Takio, K., Walsh, K. A., Neurath, H., Smith, S. B., Krebs, E. G., and Titani, K.,** The amino acid sequence of a hinge region in the regulatory subunit of bovine cardiac muscle cyclic AMP-dependent protein kinase II, *FEBS Lett.*, 114, 83, 1980.
82. **Takio, K., Smith, S. B., Walsh, K. A., Krebs, E. G., and Titani, K.,** Amino acid sequence around a hinge region and its autophosphorylation site in bovine lung cGMP-dependent protein kinase, *J. Biol. Chem.*, 258, 5531, 1983.
83. **Taylor, S. S., Buechler, J. A., and Knighton, D. R.,** cAMP-dependent protein kinase: mechanism for ATP:protein phosphotransfer, in *Peptides and Protein Phosphorylation*, Kemp, B. E., Ed., CRC Press, Boca Raton, FL, 1990, chap. 1.

84. **Heil, W. G., Landgraf, W., and Hofmann, F.**, A catalytically active fragment of cGMP-dependent protein kinase — occupation of its cGMP-binding sites does not affect its phosphotransferase activity, *Eur. J. Biochem.*, 168, 117, 1987.
85. **Rossmann, M. G., Moras, D., and Olsen, K. W.**, Chemical and biological evolution of a nucleotide-binding protein, *Nature (London)*, 250, 194, 1974.
86. **Rossmann, M. G., Liljas, A., Branden, C.-I., and Banaszak, L. J.**, Evolution and structural relationships among dehydrogenases, in *The Enzymes*, Vol. 11, Part A, Boyer, P.D., Ed., Academic Press, New York, 1975, chap. 2.
87. **Hixson, C. S. and Krebs, E. G.**, Affinity labeling of catalytic subunit of bovine heart muscle cyclic AMP-dependent protein kinase by 5'-p-fluorosulfonylbenzoyladenosine, *J. Biol. Chem.*, 254, 7509, 1979.
88. **Zoller, M. J. and Taylor, S. S.**, Affinity labeling of the nucleotide binding site of the catalytic subunit of cAMP-dependent protein kinase using p-fluorosulfonyl-[^{14}C]benzoyl-5'-adenosine, *J. Biol. Chem.*, 254, 8363, 1979.
89. **Hixson, C. S. and Krebs, E. G.**, Affinity labeling of the ATP binding site of bovine lung cyclic GMP-dependent protein kinase with 5'-p-fluorosulfonylbenzoyladenosine, *J. Biol. Chem.*, 256, 1122, 1981.
90. **Hashimoto, E., Takio, K., and Krebs, E. G.**, Amino acid sequence at the ATP-binding site of cGMP-dependent protein kinase, *J. Biol. Chem.*, 257, 727, 1982.
91. **Barker, W. C. and Dayhoff, M. O.**, Viral *src* gene products are related to the catalytic chain of mammalian cAMP-dependent protein kinase, *Proc. Natl. Acad. Sci. U.S.A.*, 79, 2836, 1982.
92. **Reimann, E. M., Titani, K., Ericsson, L. H., Wade, R. D., Fischer, E. H., and Walsh, K. A.**, Homology of the γ subunit of phosphorylse *b* kinase with cAMP-dependent protein kinase, *Biochemistry*, 23, 4185, 1984.
93. **Bramson, H. N., Thomas, N., Matsueda, R., Nelson, N. C., Taylor, S. S., and Kaiser, E. T.**, Modification of the catalytic subunit of bovine heart cAMP-dependent protein kinase with affinity labels related to peptide substrates, *J. Biol. Chem.*, 257, 10575, 1982.
94. **Puri, R. N., Bhatnagar, D., and Roskoski, R., Jr.**, Adenosine cyclic 3',5'-monophosphate dependent protein kinase: fluorescent affinity labeling of the catalytic subunit from bovine skeletal muscle with *o*-phthalaldehyde, *Biochemistry*, 24, 6499, 1985.
95. **Puri, R. N., Bhatnagar, D., Glass, D. B., and Roskoski, R., Jr.**, Inactivation of guanosine cyclic 3',5'-monophosphate dependent protein kinase from bovine lung by *o*-phthalaldehyde, *Biochemistry*, 24, 6508, 1985.
96. **Toner-Webb, J. and Taylor, S. S.**, Inhibition of the catalytic subunit of cAMP-dependent protein kinase by dicyclohexylcarbodiimide, *Biochemistry*, 26, 7371, 1987.
97. **Rosen, O. M. and Erlichman, J.**, Reversible autophosphorylation of a cyclic 3':5'-AMP-dependent protein kinase from bovine cardiac muscle, *J. Biol. Chem.*, 250, 7788, 1975.
98. **Rangel-Aldao, R. and Rosen, O. M.**, Mechanism of self-phosphorylation of adenosine 3':5'-monophosphate-dependent protein kinase from bovine cardiac muscle, *J. Biol. Chem.*, 251, 7526, 1976.
99. **Hofmann, F. and Flockerzi, V.**, Characterization of phosphorylated and native cGMP-dependent protein kinase, *Eur. J. Biochem.*, 130, 599, 1983.
100. **Aitken, A., Hemmings, B. A., and Hofmann, F.**, Identification of the residues on cyclic GMP-dependent protein kinase that are autophosphorylated in the presence of cyclic AMP and cyclic GMP, *Biochim. Biophys. Acta*, 790, 219, 1984.
101. **Hofmann, F. and Gensheimer, H.-P.**, Cyclic AMP-dependent protein kinase does not phosphorylate cyclic GMP-dependent protein kinase in vitro, *FEBS Lett.*, 151, 71, 1983.
102. **Hofmann, F., Gensheimer, H.-P., and Gobel, C.**, Autophosphorylation of cGMP-dependent protein kinase is stimulated only by occupancy of one of the two cGMP binding sites, *FEBS Lett.*, 164, 350, 1983.
103. **Kemp, B. E., Graves, D. J., Benjamini, E., and Krebs, E. G.**, Role of multiple basic residues in determining the substrate specificity of cyclic AMP-dependent protein kinase, *J. Biol. Chem.*, 252, 4888, 1977.
104. **Glass, D. B. and Smith, S. B.**, Phosphorylation by cyclic GMP-dependent protein kinase of a synthetic peptide corresponding to the autophosphorylation site in the enzyme, *J. Biol. Chem.*, 258, 14797, 1983.
105. **Glass, D. B. and Krebs, E. G.**, Comparison of the substrate specificity of adenosine 3':5'-monophosphate- and guanosine 3':5'-monophosphate-dependent protein kinases. Kinetic studies using synthetic peptides corresponding to phosphorylation sites in histone H2B, *J. Biol. Chem.*, 254, 9728, 1979.
106. **Glass, D. B. and Krebs, E. G.**, Phosphorylation by guanosine 3':5'-monophosphate-dependent protein kinase of synthetic peptide analogs of a site phosphorylated in histone H2B, *J. Biol. Chem.*, 257, 1196, 1982.
107. **Glass, D. B., El-Maghrabi, M. R., and Pilkis, S. J.**, Synthetic peptides corresponding to the site phosphorylated in 6-phosphofructo-2-kinase/fructose-2,6-bisphosphatase as substrates of cyclic nucleotide-dependent protein kinases, *J. Biol. Chem.*, 261, 2987, 1986.
108. **Landgraf, W., Hullin, R., Gobel, G., and Hofmann, F.**, Phosphorylation of cGMP-dependent protein kinase increases the affinity for cyclic AMP, *Eur. J. Biochem.*, 154, 113, 1986.

109. **Hofmann, F., Gensheimer, H.-P., and Gobel, C.**, cGMP-dependent protein kinase. Autophosphorylation changes the characteristics of binding site 1, *Eur. J. Biochem.*, 147, 361, 1985.
110. **Hofmann, F., Gensheimer, H.-P., Landgraf, W., Hullin, R., and Jastorff, B.**, Diastereomers of adenosine 3′,5′-monothionophosphate (cAMP[S]) antagonize the activation of cGMP-dependent protein kinase, *Eur. J. Biochem.*, 150, 85, 1985.
111. **Khoo, J. C., Sperry, P. J., Gill, G. N., and Steinberg, D.**, Activation of hormone-sensitive lipase and phosphorylase kinase by purified cyclic GMP-dependent protein kinase, *Proc. Natl. Acad. Sci. U.S.A.*, 74, 4843, 1977.
112. **Blumenthal, D. K., Stull, J. T., and Gill, G. N.**, Phosphorylation of cardiac troponin by guanosine 3′:5′-monophosphate-dependent protein kinase, *J. Biol. Chem.*, 253, 334, 1978.
113. **Lincoln, T. M. and Corbin, J. D.**, Purified cyclic GMP-dependent protein kinase catalyzes the phosphorylation of cardiac troponin inhibitory subunit (TN-I), *J. Biol. Chem.*, 253, 337, 1978.
114. **Embi, N., Parker, P. J., and Cohen, P.**, A reinvestigation of the phosphorylation of rabbit skeletal-muscle glycogen synthase by cyclic AMP-dependent protein kinase, *Eur. J. Biochem.*, 115, 405, 1981.
115. **Guy, P. S., Cohen, P., and Hardie, D. G.**, Purification and physicochemical properties of ATP citrate (pro-3S) lyase from lactating rat mammary gland and studies of its reversible phosphorylation, *Eur. J. Biochem.*, 114, 399, 1981.
116. **Stralfors, P. and Belfrage, P.**, Phosphorylation of hormone-sensitive lipase by cyclic GMP-dependent protein kinase, *FEBS Lett.*, 180, 280, 1985.
117. **Roskoski, R., Jr., Vulliet, P. R., and Glass, D. B.**, Phosphorylation of tyrosine hydroxylase by cyclic GMP-dependent protein kinase, *J. Neurochem.*, 48, 840, 1987.
118. **Glass, D. B. and McPherson, J. M.**, *In vitro* phosphorylation of type I collagen by cyclic AMP-dependent protein kinase, *J. Biol. Chem.*, 261, 5674, 1986.
119. **Hartzell, H. C. and Glass, D. B.**, Phosphorylation of purified cardiac muscle C-protein by purified cAMP-dependent and endogenous Ca^{2+}-calmodulin-dependent protein kinases, *J. Biol. Chem.*, 259, 15587, 1984.
120. **Glass, D. B. and Hartzell, H. C.**, unpublished data, 1986.
121. **Rittenhouse, J., Moberly, L., O'Donnell, M. E., Owen, N.E., and Marcus, F.**, Phosphorylation of atrial natriuretic peptides by cyclic AMP-dependent protein kinase, *J. Biol. Chem.*, 261, 7607, 1986.
122. **Olins, G. M., Mehta, P. P., Blehm, D. J., Patton, D. R., Zupec, M. E., Whipple, D. E., Tjoeng, F. S., Adams, S. P., Olins, P. O., and Gierse, J. K.**, Phosphorylation of high- and low-molecular-mass atrial natriuretic peptide analogs by cyclic AMP-dependent protein kinase, *FEBS Lett.*, 224, 325, 1987.
123. **Gagelmann, M., Hock, D., and Forssmann, W. G.**, Relaxation of smooth muscle by cardiodilantin/ atrial natriuretic peptide is inhibited by cAMP-dependent phosphorylation, *FEBS Lett.*, 225, 251, 1987.
124. **Marcus, F. and Rittenhouse, J.**, personal communication, 1987.
125. **Murray, K. J., El-Maghrabi, M. R., Kountz, P. D., Lukas, T. J., Soderling, T. R., and Pilkis, S. J.**, Amino acid sequence at the phosphorylation site of rat liver 6-phosphofructo-2-kinase/fructose-2,6-bisphosphatase, *J. Biol. Chem.*, 259, 7673, 1984.
126. **Lively, M. O., El-Maghrabi, M. R., Pilkis, J., D'Angelo, G., Colosia, A. D., Ciavola, J.-A., Fraser, B. A., and Pilkis, S. J.**, Complete amino acid sequence of rat liver 6-phosphofructo-2-kinase/fructose-2,6-bisphosphatase, *J. Biol. Chem.*, 263, 839, 1988.
127. **Darville, M. I., Crepin, K. M., Vandekerckhove, J., Van Damme, J., Octave, J. N., Rider, M. H., Marchand, M. J., Hue, L., and Rousseau, G. G**, Complete nucleotide sequence coding for rat liver 6-phosphofructo-2-kinase/fructose-2,6-bisphosphatase derived from a cDNA clone, *FEBS Lett.*, 224, 317, 1987.
128. **Glass, D. B., Cheng, H.-C., Kemp, B. E., and Walsh, D. A.**, Differential and common recognition of the catalytic sites of the cGMP-dependent and cAMP-dependent protein kinases by inhibitory peptides derived from the heat-stable inhibitor protein, *J. Biol. Chem.*, 261, 12166, 1986.
129. **Cohen, P.**, Phosphorylation of rabbit skeletal muscle phosphorylase kinase by cyclic GMP-dependent protein kinase, *FEBS Lett.*, 119, 301, 1980.
130. **Zeilig, C. E., Langan, T. A., and Glass, D. B.**, Sites in histone H1 selectively phosphorylated by guanosine 3′:5′-monophosphate-dependent protein kinase, *J. Biol. Chem.*, 256, 994, 1981.
131. **Hashimoto, E., Takeda, M., Nishizuka, Y., Hamana, K., and Iwai, K.**, Studies on the sites in histones phosphorylated by adenosine 3′:5′-monophosphate-dependent and guanosine 3′:5′-monophosphate-dependent protein kinases, *J. Biol. Chem.*, 251, 6287, 1976.
132. **Yamamoto, M., Takai, Y., Hashimoto, E., and Nishizuka, Y.**, Intrinsic activity of guanosine 3′,5′-monophosphate-dependent protein kinase similar to adenosine 3′,5′-monophosphate-dependent protein kinase, *J. Biochem.*, 81, 1857, 1977.
133. **Yeaman, S. J., Cohen, P., Watson, D. C., and Dixon, G. H.**, The substrate specificity of adenosine 3′:5′-cyclic monophosphate-dependent protein kinase of rabbit skeletal muscle, *Biochem. J.*, 162, 411, 1977.
134. **Geahlen, R. L. and Krebs, E. G.**, Regulatory subunit of type I cAMP-dependent protein kinase as an inhibitor and substrate of the cGMP-dependent protein kinase, *J. Biol. Chem.*, 255, 1164, 1980.

135. **Hofmann, F., Beavo, J. A., Bechtel, P. J., and Krebs, E. G.**, Comparison of adenosine 3':5'-monophosphate-dependent protein kinases from rabbit skeletal and bovine heart muscle, *J. Biol. Chem.*, 250, 7795, 1975.
136. **Hashimoto, E., Takio, K., and Krebs, E. G.**, Studies on the site in the regulatory subunit of type I cAMP-dependent protein kinase phosphorylated by cGMP-dependent protein kinase, *J. Biol. Chem.*, 256, 5604, 1981.
137. **Aswad, D. W. and Greengard, P.**, A specific substrate from rabbit cerebellum for guanosine 3':5'-monophosphate-dependent protein kinase. I. Purification and characterization, *J. Biol. Chem.*, 256, 3487, 1981.
138. **Aswad, D. W. and Greengard, P.**, A specific substrate from rabbit cerebellum for guanosine 3':5'-monophosphate-dependent protein kinase. II. Kinetic studies on its phosphorylation by guanosine 3':5'-monophosphate-dependent and adenosine 3':5'-monophosphate-dependent protein kinases, *J. Biol. Chem.* 256, 3494, 1981.
139. **Aitken, A., Bilham, T., Cohen, P., Aswad, D., and Greengard, P.**, A specific substrate from rabbit cerebellum for guanosine 3':5'-monophosphate-dependent protein kinase. III. Amino acid sequences at the two phosphorylation sites, *J. Biol. Chem.*, 256, 3501, 1981.
140. **Hemmings, H. C., Jr., Williams, K. R., Konigsberg, W. H., and Greengard, P.**, DARPP-32, a dopamine- and adenosine 3':5'-monophosphate-regulated neuronal phosphoprotein. I. Amino acid sequence around the phosphorylated threonine, *J. Biol. Chem.*, 259, 14486, 1984.
141. **Foulkes, J. G. and Cohen, P.**, The hormonal control of glycogen metabolism. Phosphorylation of protein phosphatase inhibitor-1 *in vivo* in response to adrenaline, *Eur. J. Biochem.*, 97, 251, 1979.
142. **Hemmings, H. C., Jr., Nairn, A. C., and Greengard, P.**, DARPP-32, a dopamine- and adenosine 3':5'-monophosphate-regulated neuronal phosphoprotein. II. Comparison of the kinetics of phosphorylation of DARPP-32 and phosphatase inhibitor 1, *J. Biol. Chem.*, 259, 14491, 1984.
143. **Walton, G. M., Spiess, J., and Gill, G. N.**, Phosphorylation of high mobility group 14 protein by cyclic nucleotide-dependent protein kinases, *J. Biol. Chem.*, 257, 4661, 1982.
144. **Palvimo, J., Kinnala-Kankkunen, A., and Maenpaa, P. H.**, Differential phosphorylation of high mobility group protein HMG 14 from calf thymus and avian erythrocytes by a cyclic GMP-dependent protein kinase, *Biochem. Biophys. Res. Commun.*, 110, 378, 1983.
145. **Walton, G. M., Gill, G. N., Cooper, E., and Spaulding, S. W.**, Thyrotropin-stimulated phosphorylation of high mobility group 14 protein at the site catalyzed by cyclic nucleotide-dependent protein kinases *in vitro*, *J. Biol. Chem.*, 259, 601, 1984.
146. **Adelstein, R. S., Conti, M. A., and Hathaway, D. R.**, Phosphorylation of smooth muscle myosin light chain kinase by the catalytic subunit of adenosine 3':5'-monophosphate-dependent protein kinase, *J. Biol. Chem.*, 253, 8347, 1978.
147. **Conti, M. A. and Adelstein, R. S.**, The relationship between calmodulin binding and phosphorylation of smooth muscle myosin kinase by the catalytic subunit of 3':5' cAMP-dependent protein kinase, *J. Biol. Chem.*, 256, 3178, 1981.
148. **Hathaway, D. R., Eaton, C. R., and Adelstein, R. S.**, Regulation of human platelet myosin light chain kinase by the catalytic subunit of cyclic AMP-dependent protein kinase, *Nature (London)*, 291, 252, 1981.
149. **Edelman, A. M. and Krebs, E. G.**, Phosphorylation of skeletal muscle myosin light chain kinase by the catalytic subunit of cAMP-dependent protein kinase, *FEBS Lett.*, 138, 293, 1982.
150. **de Lanerolle, P., Nishikawa, M., Yost, D. A., and Adelstein, R. S.**, Increased phosphorylation of myosin light chain kinase after an increase in cyclic AMP in intact smooth muscle, *Science*, 223, 1415, 1984.
151. **Lamb, N. J. C., Fernandez, A., Conti, M. A., Adelstein, R., Glass, D. B., Welch, W. J., and Feramisco, J. R.**, Regulation of actin microfilament integrity in living nonmuscle cells by the cAMP-dependent protein kinase and the myosin light chain kinase, *J. Cell Biol.*, 106, 1955, 1988.
152. **Nishikawa, M., de Lanerolle, P., Lincoln, T. M., and Adelstein, R. S.**, Phosphorylation of mammalian myosin light chain kinases by the catalytic subunit of cyclic AMP-dependent protein kinase and by cyclic GMP-dependent protein kinase, *J. Biol. Chem.*, 259, 8429, 1984.
153. **Issinger, O.-G., Beier, H., Speichermann, N., Flokerzi, V., and Hofmann, F.**, Comparison of phosphorylation of ribosomal proteins from HeLa and Krebs II ascites-tumor cells by cyclic AMP-dependent and cyclic GMP-dependent protein kinases, *Biochem. J.*, 185, 89, 1980.
154. **Del Grande, R. W. and Traugh, J. A.**, Phosphorylation of 40-S ribosomal subunits by cAMP-dependent, cGMP-dependent and protease-activated protein kinases, *Eur. J. Biochem.*, 123, 421, 1982.
155. **Furukawa, K.-I. and Nakamura, H.**, Cyclic GMP regulation of the plasma membrane (Ca^{2+}-Mg^{2+})ATPase in vascular smooth muscle, *J. Biochem.*, 101, 287, 1987.
156. **Popescu, L. M., Panoiu, C., Hinescu, M., and Nutu, O.**, The mechanism of cGMP-induced relaxation in vascular smooth muscle, *Eur. J. Pharmacol.*, 107, 393, 1985.
157. **Rashatwar, S. S., Cornwell, T. L., and Lincoln, T. M.**, Effects of 8-bromo-cGMP on Ca^{2+} levels in vascular smooth muscle cells: possible regulation of Ca^{2+}-ATPase by cGMP-dependent protein kinase, *Proc. Natl. Acad. Sci. U.S.A.*, 84, 5685, 1987.

158. **Suematsu, E., Hirata, M., and Kuriyama, H.**, Effects of cAMP- and cGMP-dependent protein kinases and calmodulin on Ca^{2+} uptake by highly purified sarcolemmal vesicles of vascular smooth muscle, *Biochim. Biophys. Acta,* 773, 83, 1984.
159. **Lincoln, T. M.**, personal communication, 1987.
160. **Hofmann, F.**, personal communication, 1987.
161. **Glass, D. B.**, Differential responses of cyclic GMP-dependent and cyclic AMP-dependent protein kinases to synthetic peptide inhibitors, *Biochem. J.,* 213, 159, 1983.
162. **Glass, D. B., McFann, L. J., Miller, M. D., and Zeilig, C. E.**, Interactions of cyclic GMP-dependent protein kinase with phosphate-accepting proteins and peptides, in *Protein Phosphorylation, Cold Spring Harbor Conferences on Cell Proliferation,* Vol. 8, Rosen, O. M. and Krebs, E. G., Eds., Cold Spring Harbor Laboratory, Cold Spring Harbor, New York, 1981, 267.
163. **Bhatnagar, D., Glass, D. B., Roskoski, R., Jr., Lessor, R. A., and Leonard, N. J.**, Interaction of guanosine cyclic 3',5'-monophosphate-dependent protein kinase with *lin*-benzoadenine nucleotides, *Biochemistry,* 24, 1122, 1985.
164. **Bhatnagar, D., Glass, D. B., Roskoski, R., Jr., Lessor, R. A., and Leonard, N. J.**, Synthetic peptide analogs differentially alter the binding affinities of cyclic nucleotide-dependent protein kinases for nucleotide substrates, *Biochemistry,* 27, 1988, 1988.
165. **Hartl, F. T., Roskoski, R., Jr., Rosendahl, M. S., and Leonard, N. J.**, Adenosine cyclic 3',5'-monophosphate dependent protein kinase: Interactions of the catalytic subunit and holoenzyme with *lin*-benzoadenine nucleotides, *Biochemistry,* 22, 2347, 1983.
166. **Bhatnagar, D., Roskoski, R., Jr., Rosendahl, M. S., and Leonard, N. J.**, Adenosine cyclic 3',5'-monophosphate dependent protein kinase: a new fluorescent displacement titration technique for characterizing the nucleotide binding site on the catalytic subunit, *Biochemistry,* 22, 6310, 1983.
167. **Glass, D. B., Horton, T. M., and Williams, V. L.**, unpublished data, 1987.
168. **Humble, E.**, Amino acid sequence around the phosphorylated sites of porcine and rat liver pyruvate kinase, type L, *Biochim. Biophys. Acta,* 626, 179, 1980.
169. **Zetterqvist, O., Ragnarsson, U., Humble, E., Berglund, L., and Engstrom, L.**, The minimum substrate of cyclic AMP-stimulated protein kinase, as studied by synthetic peptides representing the phosphorylatable site of pyruvate kinase (L type) of rat liver, *Biochem. Biophys. Res. Commun.,* 70, 696, 1976.
170. **Feramisco, J. R., Glass, D. B., and Krebs, E. G.**, Optimal spatial requirements for the location of basic residues in peptide substrates for the cyclic AMP-dependent protein kinase, *J. Biol. Chem.,* 255, 4240, 1980.
171. **Edlund, B., Zetterqvist, O., Ragnarsson, U., and Engstrom, L.**, Phosphorylation of synthetic peptides by (^{32}P)ATP and cyclic GMP-stimulated protein kinase, *Biochem. Biophys. Res. Commun.,* 79, 139, 1977.
172. **Thomas, N. E., Bramson, H. N., Miller, W. T., and Kaiser, E. T.**, Role of enzyme-peptide substrate backbone hydrogen bonding in determining protein kinase substrate specificities, *Biochemistry,* 26, 4461, 1987.
173. **Bramson, H. N., Thomas, N. E., Miller, W. T., Fry, D. C., Mildvan, A. S., and Kaiser, E. T.**, Conformation of Leu-Arg-Arg-Ala-Ser-Leu-Gly bound in the active site of adenosine cyclic 3',5'-phosphate dependent protein kinase, *Biochemistry,* 26, 4466, 1987.
174. **Thomas, N. E., Bramson, H. N., Nairn, A. C., Greengard, P., Fry, D. C., Mildvan, A. S., and Kaiser, E. T.**, Distinguishing among protein kinases by substrate specificities, *Biochemistry,* 26, 4471, 1987.
175. **Casnellie, J. E., Schlichter, D. J., Walter, U., and Greengard, P.**, Photoaffinity labeling of a guanosine 3':5'-monophosphate-dependent protein kinase from vascular smooth muscle, *J. Biol. Chem.,* 253, 4771, 1978.
176. **Casnellie, J. E., Ives, H. E., Jamieson, J. D., and Greengard, P.**, Cyclic GMP-dependent protein phosphorylation in intact medial tissue and isolated cells from vascular smooth muscle, *J. Biol. Chem.,* 255, 3770, 1980.
177. **Parks, T. P., Nairn, A. C., Greengard, P., and Jamieson, J. D.**, The cyclic nucleotide-dependent phosphorylation of aortic smooth muscle membrane proteins, *Arch. Biochem. Biophys.,* 255, 361, 1987.
178. **Rapoport, R. M., Draznin, M. B., and Murad, F.**, Sodium nitroprusside-induced protein phosphorylation in intact rat aorta is mimicked by 8-bromo cyclic GMP, *Proc. Natl. Acad. Sci. U.S.A.,* 79, 6470, 1982.
179. **Lincoln, T. M. and Keely, S. L.**, Regulation of cardiac cyclic GMP-dependent protein kinase, *Biochim. Biophys. Acta,* 676, 230, 1981.
180. **Casnellie, J. E. and Greengard, P.**, Guanosine 3':5'-cyclic monophosphate-dependent phosphorylation of endogenous substrate proteins in membranes of mammalian smooth muscle, *Proc. Natl. Acad. Sci. U.S.A.,* 71, 1891, 1974.
181. **Wallach, D., Davies, P. J. A., and Pastan, I.**, Cyclic AMP-dependent phosphorylation of filamin in mammalian smooth muscle, *J. Biol. Chem.,* 253, 4739, 1978.
182. **Hogaboom, G. K., Emler, C. A., Butcher, F. R., and Fedan, J. S.**, Concerted phosphorylation of endogenous tracheal smooth muscle membrane proteins by Ca^{2+}·calmodulin-, cyclic GMP-, and cyclic AMP-dependent protein kinases, *FEBS Lett.,* 139, 309, 1982.

183. **Sen, I. and Roy, P.**, Atrial natriuretic factor induced phosphorylation of human placental membrane protein: an effect mimicked by guanosine 3':5'-cyclic monophosphate, *Biochem. Biophys. Res. Commun.*, 139, 431, 1986.
184. **Wrenn, R. W. and Kuo, J. F.**, Cyclic GMP-dependent phosphorylation of an endogenous protein from rat heart, *Biochem. Biophys. Res. Commun.*, 101, 1274, 1981.
185. **Connelly, P. A., Botelho, L. H. P., Sisk, R. B., and Garrison, J. C.**, A study of the mechanism of glucagon-induced protein phosphorylation in isolated rat hepatocytes using (S_p)-cAMPS and (R_p)-cAMPS, the stimulatory and inhibitory diastereomers of adenosine cyclic 3',5'-phosphorothioate, *J. Biol. Chem.*, 262, 4324, 1987.
186. **Kase, H., Iwahashi, K., Nakanishi, S., Matsuda, Y., Yamada, K., Takahashi, M., Murakata, C., Sato, A., and Kaneko, M.**, K-252 compounds, novel and potent inhibitors of protein kinase C and cyclic nucleotide-dependent protein kinases, *Biochem. Biophys. Res. Commun.*, 142, 436, 1987.
187. **Inagaki, M., Kawamoto, S., Itoh, H., Saitoh, M., Hagiwara, M., Takahashi, J., and Hidaka, H.**, Naphthalenesulfonamides as calmodulin antagonists and protein kinase inhibitors, *Mol. Pharmacol.*, 29, 577, 1986.
188. **Hidaka, H., Inagaki, M., Kawamoto, S., and Sasaki, Y.**, Isoquinolinesulfonamides, novel and potent inhibitors of cyclic nucleotide dependent protein kinase and protein kinase C, *Biochemistry*, 23, 5036, 1984.
189. **Kemp, B. E., Benjamini, E., and Krebs, E. G.**, Synthetic hexapeptide substrates and inhibitors of 3':5'-cyclic AMP-dependent protein kinase, *Proc. Natl. Acad. Sci. U.S.A.*, 73, 1038, 1976.
190. **Feramisco, J. R. and Krebs, E. G**, Inhibition of cyclic AMP-dependent protein kinase by analogues of a synthetic peptide substrate, *J. Biol. Chem.*, 253, 8968, 1978.
191. **Walton, G. M. and Gill, G. N.**, Protein effects on the activity of guanosine 3':5'-monophosphate-dependent protein kinase, *J. Biol. Chem.*, 255, 1603, 1980.
192. **Walton, G. M. and Gill, G. N.**, Regulation of cyclic nucleotide-dependent protein kinase activity by histones and poly(L-arginine), *J. Biol. Chem.*, 256, 1681, 1981.
193. **Scott, J. D., Fischer, E. H., Demaille, J. G., and Krebs, E. G.**, Identification of an inhibitory region of the heat-stable protein inhibitor of the cAMP-dependent protein kinase, *Proc. Natl. Acad. Sci. U.S.A.*, 82, 4379, 1985.

Chapter 10

PROTEIN-TYROSINE KINASES

Robert L. Geahlen and Marietta L. Harrison

TABLE OF CONTENTS

I. Introduction 240

II. Studies on the Substrate Specificity of Protein-Tyrosine Kinases 240
   A. Techniques for the Assay of Peptide Phosphorylation 241
   B. Results of Studies on Substrate Specificity 241
   C. Sites of Phosphorylation on Protein Substrates 245

III. Use of Synthetic Peptides for the Purification and Characterization of
    Protein-Tyrosine Kinases 247
   A. Detection of Normal Cellular Protein-Tyrosine Kinases 247
   B. Characterization of Protein-Tyrosine Kinases 248

IV. Summary 249

References 250

I. INTRODUCTION

The phosphorylation of proteins is an important and common mechanism for the intracellular regulation of enzymic activity. Numerous key regulatory enzymes are phosphorylated on specific residues at specific sites by the action of protein kinases in events that often alter the kinetic properties of the modified enzymes.[1,2] There has been much interest in determining the mechanisms by which protein kinases discriminate between different protein substrates and different potential phosphorylation sites. The use of synthetic peptides of determined sequence has revolutionized studies on the substrate specificity of these enzymes. By varying the sequence of amino acids surrounding the phosphorylated residue, it has been possible to decipher the primary structural determinants that specify phosphorylation sites for a number of specific kinases. Based on such studies, investigators have often been able to predict, solely on the basis of sequence information, the identity of the kinase responsible for the phosphorylation of a specific substrate.

Based on the relative selectivity of different protein kinases for specific amino acid sequences, peptide probes can often be fashioned for use in detecting the activity of specific kinases in, for example, subcellular fractions that contain multiple kinase activities. This is seldom feasible with protein substrates that more often than not can serve as substrates for more than one protein kinase. In this manner, purification schemes can be developed for the isolation of protein kinases based on their ability to phosphorylate peptides of a certain sequence. Such peptides can also serve as valuable tools for the quantification of protein kinase activity, for the comparison of activities from different sources, and for the determination of kinetic mechanisms.

The use of peptides as substrates originated with studies on the substrate specificity of protein-serine and -threonine kinases. In recent years, the use of synthetic peptide substrates has also been applied energetically to the study of a new class of protein kinase: those that specifically recognize and phosphorylate substrates on tyrosine residues. Interest in these enzymes has been intense due to their probable involvement in the regulation of cellular growth processes. Numbering among the protein-tyrosine kinases are the oncogene products of several acute transforming retroviruses and their cellular homologues, and the receptors for a variety of polypeptide growth factors.[3] The association of tyrosine phosphorylation with both viral oncogenesis and growth factor-dependent mitogenesis has suggested an important role for these enzymes and their substrates in the control of cell proliferation.

II. STUDIES ON THE SUBSTRATE SPECIFICITY OF PROTEIN-TYROSINE KINASES

The amino acid sequences of peptides selected for the study of protein-serine and -threonine kinases typically were based on the sequences surrounding the phosphorylation sites of known, physiologically relevant substrates. The first protein-tyrosine kinase to be described (pp60^{v-src}, the protein product of the Rous sarcoma virus *src* gene) was identified based on its fortuitous ability to phosphorylate the antibody employed for its immunoprecipitation.[4,5] Further analyses indicated that the kinase was also phosphorylated on a tyrosine residue *in vitro*.[5,6] Since the identities of the physiologically important substrates for protein-tyrosine kinases were uncertain, the first peptides to be synthesized for use as substrates[7] were based on the primary structure of the phosphorylation site on pp60^{v-src}.[8-10] A second series of peptides that have received widespread use are based on the structures of angiotensin I and II, which are tyrosine-containing peptides that have been shown to serve as substrates for a number of protein-tyrosine kinases, but are not directly related to known protein substrates.[11] The original impetus for the use of both of these classes of peptides was the need for convenient assay systems for the detection, purification, and characterization of protein-tyrosine kinases.

A. TECHNIQUES FOR THE ASSAY OF PEPTIDE PHOSPHORYLATION

The reaction mixture for a typical peptide phosphorylation assay includes the protein-tyrosine kinase, present as a purified or partially purified enzyme, an immunocomplex, or a subcellular fraction; [γ-^{32}P]ATP; a divalent cation, normally Mg^{2+} or Mn^{2+} (for many protein-tyrosine kinases the preferred cation is Mn^{2+}); and the peptide substrate. A suitable method is required for the separation of radiolabeled phosphopeptide from other reaction products and substrates. A number of procedures have been developed for the separation and quantification of [^{32}P]phosphopeptides formed in protein-tyrosine kinase assays. Examples of separation techniques include high pressure liquid chromatography (HPLC),[7] high-voltage paper electrophoresis,[11] thin-layer electrophoresis,[12] anion exchange chromatography,[13] hydrolysis of [γ-^{32}P]ATP and extraction of [^{32}P]orthophosphate,[14] and adsorption to ion exchange filter paper.[7] The most commonly employed of these procedures is the isolation of phosphopeptides on small squares of phosphocellulose paper, which are subsequently washed in dilute acetic or phosphoric acid to remove unreacted [γ-^{32}P]ATP. The binding of the phosphopeptide to phosphocellulose paper requires that the phosphopeptide carry a net positive charge. To utilize this assay procedure, some investigators have incorporated additional basic amino acid residues into the design of their synthetic peptides.[7,15]

B. RESULTS OF STUDIES ON SUBSTRATE SPECIFICITY

The regions on protein substrates that are recognized by protein kinases have been shown in many cases to depend on the number and position of charged amino acids in the neighborhood of the phosphorylated residue. For example, the recognition of substrates by both the cAMP-dependent protein kinase and the Ca^{2+}/phospholipid-dependent protein kinase (protein kinase C) is dependent on the presence of basic amino acids at an appropriate distance from the phosphorylated amino acid residue.[16-18] Casein kinases I and II, on the other hand, prefer to phosphorylate residues that are located either on the N-terminal (casein kinase II) or C-terminal (casein kinase I) side of acidic amino acids.[19] Based on these observations and the observation that the sites of phosphorylation of a number of viral protein-tyrosine kinases are located near acidic amino acid residues,[8,20] most investigations of substrate specificity have centered on the role of charged residues in the recognition of peptides as substrates by protein-tyrosine kinases.

Tables 1 to 3 summarize the kinetic data obtained by several investigators using synthetic peptides as substrates for a number of different protein-tyrosine kinases. Table 1 includes the data for peptides with sequences based on the *in vitro* site of autophosphorylation of pp60^{v-src}. These peptides have been used for the majority of the systematic studies that have been carried out on the effects of various amino acid substitutions on the kinetics of peptide phosphorylation. Table 2 includes data for analogues of angiotensin I and II, which have received widespread use as substrates for a number of different kinases. Table 3 includes data for the phosphorylation of a wide range of additional peptide substrates. Due to the different kinases employed and the different states of purity of each kinase, it is not possible to directly compare the V_{max} values obtained in different studies.

The apparent K_m values of protein-tyrosine kinases for various peptide substrates are generally in the millimolar range, which is considerably higher than the range of values obtained for the best substrates for the cAMP-dependent protein kinase[16,17] or protein kinase C.[18] The dependency of protein-tyrosine kinases on the presence and location of specific charged amino acid residues was also much less than that seen for these other kinases. The systematic replacement of any single basic amino acid residue in peptides based on the structure of the pp60^{v-src} autophosphorylation site (Table 1) had only a relatively small effect on the kinetic constants. Deletion of the arginine residue three amino acids on the C-terminal side of the tyrosine decreased the K_m 4-fold for the epidermal growth factor (EGF) receptor kinase[13] and 2.5-fold for the tyrosine kinase P90,[12] suggesting that a basic residue in this

TABLE 1
Phosphorylation of pp60[v-src]-Related Peptides by Protein-Tyrosine Kinases

Peptide[a]	Kinase[b]	K_m (mM)	V_{max} (nmol/min/mg)	Ref.
R L I E D N E Y T A R Q G	EGF-R	3.3	3.2	13
R L I E D N E Y T A R K G	EGF-R	3.5	3.9	13
R L I E D A E Y T A	EGF-R	0.84	4.8	13
R L I E D A L Y T A	EGF-R	3.0	—[c]	13
L I E D A E Y T A	EGF-R	0.44	5.7	13
L I E D A L Y T A	EGF-R	1.2	—[c]	13
I E E A A Y T A	EGF-R	4.1	—[c]	13
I D D A A Y T A	EGF-R	7.0	—[c]	13
R R L I E D N E Y T A R G	EGF-R	0.28	7.5	21
R R L I E D A E Y A A R G	EGF-R	0.5	9.8	21
R R L I A D A E Y A A R G	EGF-R	1.1	8.5	21
R R L I E A A E Y A A R G	EGF-R	1.8	9.2	21
I E D N E Y T A R Q G	EGF-R	1.1	11.5	21
L E D A E Y A A R R R G	EGF-R	1.5	11.2	22
E D N E Y T A R Q G	pp60[v-src]	6.25	0.5	11
E D N E Y V A R Q G	pp60[v-src]	5.89	0.67	11
R R L I E D A E Y A A R G	p56[lck]	2.8	1.4	23
I E D N E Y T A R Q G	p56[lck]	5.0	0.62	23
R R L I E D A E Y A A R G	p56[lck]	2.0	8.0	24
R R L I A D A E Y A A R G	p56[lck]	1.3	5.5	24
R R L I E A A E Y A A R G	p56[lck]	2.7	5.5	24
R R L I E D A A Y A A R G	p56[lck]	4.6	9.1	24
E D A E Y A A R R G	p56[lck]	3.4	9.0	24
G G A G Y A A R R G	p56[lck]	3.2	1.1	24
K L I E D N E Y T A R	Y73 P90	5	1.0[d]	12
K L I E D N E Y T A	Y73 P90	2	0.9[d]	12
L I E D N E Y T A R	Y73 P90	7	0.6[d]	12
K L I D N E Y T A R	Y73 P90	37	2.3[d]	12
K L D N E Y T A R	Y73 P90	40	2.5[d]	12
K L I E D N E Y	Y73 P90	—[e]	—	12
Y T A R	Y73 P90	—[e]	—	12
R R L I E D A E Y A A R G	IR	1.2	11.8	25
R R L I E D A E Y A R G	IR	1.2	2.2	26
R R L I E D A E Y A R G	IR	0.96	18[f]	27
E D A E Y A A R R R G	Spleen	2.2	—	15
R R L I E D A E Y A A R G	PBLs	2.2	2.3	28
E D A E Y A A R R R G	PBLs	7.1	3.7	28

Note: Sequence surrounding site of autophosphorylation on pp60[v-src] is R L I E D N E Y T A R Q G.

[a] The one-letter symbols for the amino acids are: A, alanine; C, cysteine; D, aspartic acid; E, glutamic acid; F, phenylalanine; G, glycine; H, histidine; I, isoleucine; K, lysine; L, leucine; M, methionine; N, asparagine; P, proline; Q, glutamine; R, arginine; S, serine; T, threonine; V, valine; W, tryptophan; Y, tyrosine.

[b] Abbreviations used are: EGF-R, receptor for the epidermal growth factor, usually constitutes membranes prepared from A431 cells, which overexpress the receptor; pp60[v-src], the 60-kDa product of the Rous sarcoma virus *src* gene; p56[lck], 56-kDa product of the *lck* gene, usually constitutes membranes prepared from LSTRA cells, which overexpress p56[lck]; Y73 P90, 90-kDa transforming protein of the Y73 virus; IR, insulin receptor; spleen, detergent-solubilized membranes from rat spleen; PBLs, particulate fraction from human peripheral blood lymphocytes.

[c] Apparent values for V_{max} varied from 3 to 6 nmol/min/mg for these peptides.

[d] Values represent relative maximal velocities normalized to 1.0 for the peptide K L I E D N E Y T A R.

[e] Rates of phosphorylation were too low for accurate determinations.

[f] V_{max} value expressed as pmol/min.

TABLE 2
Phosphorylation of Angiotensin Analogues by Protein-Tyrosine Kinases

Peptide[a]	Kinase[b]	K_m (mM)	V_{max} (nmol/min/mg)	Ref.
D R V Y I H P F H L	pp60[v-src]	1.54	2.6	11
D R V Y I H P F	pp60[v-src]	2.0	10	11
Sar R V Y I H P F	pp60[v-src]	4.2	2	11
D R V Y V H P F	pp60[v-src]	0.87	6.7	11
N R V Y V H P F	pp60[v-src]	1.33	1.7	11
R V Y I H P F	pp60[v-src]	5.0	2.2	11
R G Y S L G	pp60[v-src]	—[c]	—	11
D R V Y I H P F	EGF-R	0.8	11	29
D R V Y V H P F	P75 liver	0.27	1.3	30
D R V Y I H P F	IR	2.6	2.2[d]	26
R V Y V H P F	IR	8.0	0.05[d]	26
D R V Y I H P F	ptabl50	3.7	1250	31
D R V Y I H P F	P60[v-abl]	3.8	9.2	32
D R V Y I H P F H L	B cells	3.6	0.8	33
D R V Y I H P F H L	p56[lck]	2.3	4.7	33
D R V Y I H P F H L	p40 thymus	2.5	—	34

Note: Angiotensin sequence, D R V Y I H P F H L; angiotensin II sequence, D R V Y I H P F.

[a] One-letter amino acid codes are given in Table 1; Sar, sarcosine.
[b] Abbreviations used in addition to those defined in Table 1: P75 liver, soluble 75-kDa kinase isolated from rat liver; ptabl50, fusion protein carrying the kinase domain of the v-*abl* gene product expressed in *Escherichia coli;* B cells, membranes prepared from murine B lymphocytes; p40 thymus, soluble 40-kDa kinase isolated from bovine thymus.
[c] Not phosphorylated.
[d] V_{max} values expressed as pmol/min.

position has a slight negative effect on the phosphorylation of the peptide. However, the addition of extra arginine residues near the C-terminus does not tend to increase the K_m values for synthetic peptides dramatically.[15,22,24]

The acidic amino acids N-terminal to the tyrosine residue have attracted much attention due to their presence in a number of known phosphorylation sites.[8,20] In general, peptides bearing acidic residues proximal to the phosphorylated tyrosine were much better substrates than those possessing basic residues. Peptides similar in sequence to the cAMP-dependent protein kinase substrate L R R A S L G (Kemptide), but containing a tyrosine residue in place of the serine (Table 2), were poor substrates for a number of different protein-tyrosine kinases, with V_{max} values 8- to 30-fold lower than those obtained for the pp60[v-src]-related peptides.

Replacement of any single acidic amino acid residue with a neutral amino acid in the pp60[v-src]-peptide series, however, had only small effects on the kinetics of peptide phosphorylation (Table 1). The glutamate residue four amino acids N-terminal to the tyrosine appeared to play little or no role in the recognition of peptides by the EGF receptor kinase[21] or the p56[lck] kinase,[24] but its elimination caused a sevenfold increase in the K_m of the P90 kinase for the peptide.[12] Replacement of the aspartate residue three amino acids to the N-terminal side of the tyrosine increased the K_m of the EGF receptor kinase for the peptide threefold,[21] but had little effect on the p56[lck] kinase.[24] Finally, the glutamate residue adjacent to the tyrosine generally appeared to be a positive determinant, and its replacement increased K_m values two- to fourfold, with little effect on the V_{max}.[13,21,24]

TABLE 3
Phosphorylation of Additional Peptides by Protein-Tyrosine Kinases

Peptide source[a]	Peptide[b]	Kinase[c]	K_m (mM)	V_{max} (nmol/min/mg)	Ref.
Kemptide analogues	L R R A Y L G	EGF-R	6.0	1.2	21
	L R R A Y L G	p56[lck]	6.7	0.6	24
	L R R A Y L G	IR	1.3	0.41	25
	I E E A Y L G	EGF-R	3.2	—	13
	I E E Y Y L G	EGF-R	1.9	—	13
Insulin β-chain	R G F F Y T P K T	EGF-R	NP	—	13
p21[ras]	R R L D T T G Q E E E Y S A	EGF-R	0.24	1.65	23
Gastrin	R R L E E E E E A Y G	EGF-R	0.15	1.7	23
	R R L E E E E E A Y G	EGF-R	0.11	1.7	35
	pE G P W L E E E E E A Y G W M D F–NH₂	EGF-R	0.057	0.1	36
Polyoma virus middle T	E E E E Y M P M E	Y73 P90	3	—	12
LHRF	pE H W S Y G L R P G–NH₂	Y73 P90	—[d]	—	12
—	R G Y A L G	Y73 P90	—[d]	—	12
Proctolin	R Y L P T	IR	>25	—	26
β-lipotropin	R Y G G F M	IR	>50	—	26
—	Y R	IR	>80	—	26
—	K Y K	ptabl50	12	110	31
—	R Y	ptabl50	36	100	31
IR-960	L F A S S N P E Y L S A R R	IR	1.20	25	27
IR-1150	T R D I Y E T D Y Y R K	IR	0.24	5	27
IR-1316	K R S Y E E H I P Y T H M N G G K	IR	1.13	33	27
Enolase	V P S G A S T G I Y E A L E L R	P140[gag-fps]	0.15	—	37

[a] Kemptide, commonly employed substrate for cAMP-dependent protein kinase, has the sequence L R R A S L G; p21[ras], 21-kDa protein encoded by the *ras* oncogene; LHRF, luteinizing hormone releasing factor; IR, insulin receptor.
[b] One-letter amino acid codes are as given in Table 1; pE, pyroglutamate.
[c] P140[gag-fps], transforming protein of the Fujinami sarcoma virus.
[d] Phosphorylated, but extremely poorly.

The N-terminal aspartate residue of the angiotensin analogues was also a positive determinant. However, its replacement with an asparagine residue did not greatly reduce the ability of the peptide to be phosphorylated by pp60[v-src], indicating that peptides totally lacking acidic amino acid residues could still serve as substrates.[11]

The best peptide substrates that have been described for protein-tyrosine kinases contain acidic amino acid residues near the phosphorylated tyrosine. If these residues play a role in determining substrate specificity, however, then they must be able to function when positioned on either side of the phosphorylated tyrosine, since the peptide based on the phosphorylation site of enolase has acidic residues only on the C-terminal side.[37] In systematic studies, the substitution of any charged amino acid with a neutral residue resulted in modest changes in kinetic parameters compared to those seen with peptide substrates for certain protein-serine and -threonine kinases.[16-18] It is probable that ionic interactions do not play as important a role in substrate recognition by protein-tyrosine kinases as they do for these other enzymes. For example, the cAMP-dependent protein kinase, which recognizes sites C-terminal to arginine residues, is strongly inhibited by polyarginine.[38] The EGF receptor kinase, however, is largely unaffected by either polyglutamic or polyaspartic acid.[13] Likewise, the association of angiotensin analogues with a number of protein-tyrosine kinases is not inhibited by substantial changes in the ionic strength of the reaction medium.[30,34,39]

It is also possible that amino acids other than basic or acidic ones play an important role in substrate recognition. It has been suggested that an aromatic amino acid adjacent to the phosphorylated tyrosine residue may be a positive structural determinant for the phosphorylation of substrates by the insulin receptor and the v-ros gene product.[27] Also, replacement of the isoleucine residue in angiotensin II with a valine decreased the K_m of pp60[v-src] for peptide phosphorylation twofold.[11] However, a similar modification had no effect on the phosphorylation of a pp60[v-src]-related peptide.[11]

Observations drawn from the study of synthetic peptides have frustrated efforts to clearly define primary structural determinants that are involved in the recognition of substrates by protein-tyrosine kinases. It is perhaps of some consolation to investigators of protein-tyrosine kinase substrate specificity that not all peptides containing tyrosine residues are substrates (Table 3).

C. SITES OF PHOSPHORYLATION ON PROTEIN SUBSTRATES

The relative lack of specificity of protein-tyrosine kinases for small peptide substrates has led to the conjecture that secondary and tertiary structure play an important role in the recognition of protein substrates.[12,21,24] This idea is supported by the fact that the K_m values for the phosphorylation of peptide substrates by protein-tyrosine kinases are much higher than those for the phosphorylation of protein substrates.[33,37,40,41] A direct comparison of peptide and protein substrates has been made for enolase, where the intact protein is phosphorylated with a K_m value 100 times lower than that of the corresponding peptide.[37]

An examination of sites of phosphorylation on several protein substrates does not reveal an absolute requirement for a specific amino acid sequence (Table 4). Many of the sites do, however, share a number of common features. The sites of *in vitro* phosphorylation of the products of the *src*, *yes*, *lck*, *fps*, *fes*, and *abl* genes and the insulin receptor all have arginine residues seven amino acids N-terminal to the phosphorylated tyrosine.[20] These same proteins also share a glycine residue in common located five residues on the C-terminal side of the tyrosine.[20] In addition, of the 23 sites listed, 13 have basic amino acid residues (lysine or arginine) located 5 to 7 amino acids downstream from the tyrosine. This is not a strict structural requirement for exogenous substrates, however, since erythrocyte band 3, which is an excellent substrate for a number of protein-tyrosine kinases,[33,34,41] is phosphorylated on amino acid number 8 and the nearest basic amino acid is not encountered until amino acid 75.[53]

TABLE 4
Sites of Phosphorylation on Protein Substrates

Substrate	Kinase	Site	Ref.
pp60$^{v\text{-}src}$	pp60$^{v\text{-}src}$	R L I E D N E Y T A R Q G A K	10
P90$^{gag\text{-}yes}$	P90$^{gag\text{-}yes}$	R L I E D N E Y T A R Q G A K	42
p56^{lck}	p56^{lck}	R L I E D N E Y T A R E G A K	43, 44
P140$^{gag\text{-}fps}$	P140$^{gag\text{-}fps}$	R Q E E D G V Y A S T G G M K	45
P85$^{gag\text{-}fes}$	P85$^{gag\text{-}fes}$	R E E A D G V Y A A S G G L R	46
P110$^{gag\text{-}fes}$	P110$^{gag\text{-}fes}$	R E A A D G I Y A A S G G L R	46
P120$^{gag\text{-}abl}$	P120$^{gag\text{-}abl}$	R L M T G D T Y T A H A G A G	47
Insulin receptor[a]	IR	R D I Y E T D Y Y R K G G K G	27, 48
EGF-R-1173[b]	EGF-R	S T R E N A E Y L R V A P Q S	49
EGF-R-1148[c]	EGF-R	I S L D N P D Y Q Q D F F P K	49
EGF-R-1068[c]	EGF-R	T F L P V P E Y I N Q S V P K	49
LDH[d]	P140$^{gag\text{-}fps}$	K Q V V D S A Y E V I K L K G	37
Enolase	P140$^{gag\text{-}fps}$	S G A S T G I Y E A L E L R	37
Myosin light chain	EGF-R	D E E V D E M Y R E A P I D K	50, 51
Myosin light chain	EGF-R	D V K G N F N Y V E F T R I L	50, 51
Band 3	Endogenous kinase	M E E L Q D D Y E D D M E E N	52, 53
35-kDa substrate[e]	EGF-R	I D N E E Q E Y I K T V K G S	54
HLA-B7	pp60$^{v\text{-}src}$	S G G K G G S Y S Q A A C S D	55
HLA-A2	pp60$^{v\text{-}src}$	S D R K G G S Y S Q A A S S D	55
p36[e]	pp60$^{v\text{-}src}$	H S T P P S A Y G S V K A Y T	56
pp60$^{c\text{-}src}$	Unknown	F T S T E P Q Y Q P G E N L	57
Polyoma virus middle T	Unknown[f]	L E E E E E Y M P M E D L Y	58
Polyoma virus middle T	Unknown[f]	S L L S N P T Y S V M R S H S	58
Protein kinase inhibitor protein	EGF-R	T D V E T T Y A D F I A S G	59

Note: One-letter amino acid abbreviations as given in Table 1.

[a] Major site of autophosphorylation of the insulin receptor (IR) *in vitro*, is also phosphorylated at this site *in vivo*.
[b] Major site of autophosphorylation of epidermal growth factor receptor (EGF-R) *in vitro* and *in vivo*.
[c] Sites of autophosphorylation of EGF-R *in vitro*.
[d] Lactate dehydrogenase
[e] The 35-kDa substrates from A431 cells and p36 have recently been identified as lipocortin I and II, respectively.[55,61]
[f] Phosphorylation may be catalyzed by pp60$^{c\text{-}src}$, which forms a stable complex with middle T.[60]

Most of the phosphorylation sites have acidic amino acid residues located proximal to the phosphorylated tyrosine. In fact, some reported sites have a prolonged stretch of acidic amino acids preceding or surrounding the phosphorylated residue (e.g., band 3, polyoma middle T, myosin light chain site 1). Since proteins bearing these sequences are excellent substrates, such sequences may be of value in predicting potential phosphorylation sites on protein substrates. Other sites, however, are notably lacking in acidic amino acids (e.g., HLA antigens, p36), and it would be difficult, based solely on primary structural information, to predict these as potential sites of phosphorylation. Such observations support the idea that higher-order structure is important in substrate recognition for protein-tyrosine kinases.

The disparity in phosphorylation site sequences shown in Table 4 may also partially reflect differences in the substrate specificity of different protein-tyrosine kinases. The identification of the sequences surrounding sites of tyrosine phosphorylation on protein substrates should provide investigators with new leads for the further design of synthetic peptide substrates. One important goal would be the design of peptide substrates that are specific for a particular type or class of protein-tyrosine kinase.

III. USE OF SYNTHETIC PEPTIDES FOR THE PURIFICATION AND CHARACTERIZATION OF PROTEIN-TYROSINE KINASE

The discovery of protein-tyrosine kinases preceded the identification of physiologically relevant substrates. This lack of specific, readily obtainable protein substrates provided the original impetus for the design of synthetic peptides as substrates.[7] By using peptides containing tyrosine as the sole phosphorylatable residue, investigators could detect and measure the activity of protein-tyrosine kinases even in crude subcellular fractions that contained multiple kinase activities. Thus, synthetic peptides have found their greatest utility as substrates for the detection and measurement of protein-tyrosine kinase activity in cells and tissues, for monitoring the purification of protein-tyrosine kinases, and for the characterization of the purified or partially purified enzymes.

A. DETECTION OF NORMAL CELLULAR PROTEIN-TYROSINE KINASES

The protein-tyrosine kinases that are present in nontransformed cells have been categorized into two main classes.[3] The first includes those whose genes have at one time been captured by retroviruses, and the second includes the receptors for polypeptide growth factors. In addition to the two main classes, however, there are a growing number of enzymes being described that are not growth factor receptors and for which there is no known viral homologue. Synthetic peptide substrates have been particularly useful for the characterization of these enzymes, since other convenient means of identification, such as specific antisera and specific activators, have not been available. Examples of such enzymes have now been described from a variety of sources based on their ability to phosphorylate synthetic peptides (Table 5). In addition to synthetic peptides, synthetic polymers of glutamate and tyrosine have also received widespread use as artificial substrates.[71]

The availability of peptides as substrates has also permitted the design of purification protocols for the isolation of protein-tyrosine kinases by providing a means of detecting the activity of the enzymes following the various steps used for protein fractionation. For example, by monitoring the ability of protein fractions to catalyze the phosphorylation of various angiotensin analogues, investigators have recently purified enzymes from rat liver (p75 kinase),[30] calf thymus (p40 kinase),[34] and bovine spleen (a 50-kDa enzyme) to near homogeneity.[64] (Angiotensin II has also been used to follow the isolation of an active fragment of the v-*abl* protein expressed in *Escherichia coli*.[31]) Using a peptide related to the site of autophosphorylation on pp60[v-src], investigators have purified a 56-kDa kinase over 200-fold from rat spleen.[62] The purification and characterization of protein-tyrosine kinases from normal tissues should prove useful for deciphering the role of these enzymes in cell growth and metabolism.

The phosphorylation of angiotensin analogues by the enzymes from liver and thymus was found to be stimulated several fold by high concentrations of NaCl.[30,34,39] Elevated ionic strength increased the V_{max} of the enzymes, but had little or no effect on the K_m values of the enzymes for their substrates. Our laboratory has used this observation to aid in the purification of p40 from bovine thymus.[34] At NaCl concentrations approaching 2 M, the activity of endogenous protein-serine and -threonine kinases as well as the activity of phosphotyrosine phosphatases found in thymocyte extracts was strongly inhibited while the activity of p40 was stimulated. We have found this to be a property of protein-tyrosine kinases expressed in numerous cell types and tissues. The use of elevated concentrations of NaCl in peptide phosphorylation assays can be exploited not only to aid in the purification

TABLE 5
Detection of Protein-Tyrosine Kinases in Normal Tissues

Tissue source	Peptide[a]	Ref.
Rat spleen	E D A E Y A A R R R G	15, 62, 63
Bovine spleen	D R V Y V H P F	64
Murine T lymphocytes	R R L I E D A E Y A A R G	65
Murine B lymphocytes	D R V Y I H P F H L	33
Bovine thymus	D R V Y I H P F H L	34, 39
Human peripheral blood lymphocytes	R R L I E D A E Y A A R G	28
Human platelets	E D A E Y A A R R R G	66
Rat brain	E D A E Y A A R R R G	67
Rat liver	D R V Y V H P F	30, 68
Human serum	D R V Y I H P F	69
Sea urchin embryo	E D A E Y A A R R R G	70

[a] One-letter amino acid codes as given in Table 1.

of protein-tyrosine kinases, but also to allow the detection of many protein-tyrosine kinases that are normally expressed at low levels which are difficult to measure.[39]

B. CHARACTERIZATION OF PROTEIN-TYROSINE KINASES

Synthetic tyrosine-containing peptides can be phosphorylated *in vitro* by a wide variety of different protein-tyrosine kinases ranging from growth factor receptors to viral oncogene products (as illustrated in Tables 1 to 5). The overall utility of peptides arises from the relative ease with which their phosphorylation can be monitored. They can be and have been used extensively for multiple purposes such as comparing the activities of different kinases, monitoring changes in kinase activity in response to different treatments, characterizing purified enzymes, and determining kinetic mechanisms. For example, the first synthetic peptides were created as substrates for a protein-tyrosine kinase ($p56^{lck}$) that was expressed at high levels in a murine lymphoma cell line (LSTRA).[7] These same peptides and other peptides have been used subsequently to (1) identify the site of autophosphorylation on $p56^{lck}$;[72] (2) show that normal spleen T lymphocytes express a kinetically similar enzyme;[65] (3) compare levels of $p56^{lck}$ kinase activity between T cells, LSTRA cells, and other T leukemia and lymphoma cell lines;[65,73] (4) compare and contrast the kinetics of T cell and B cell kinases;[33] (5) characterize antibodies to $P56^{lck}$;[74] and (6) determine the effects of phorbol esters on $p56^{lck}$ kinase activity.[75]

Synthetic peptides have played an equally important role in the characterization of growth factor receptor kinases. Early work showed that a series of peptides based on the sequence of the autophosphorylation site of $pp60^{v-src}$ could be phosphorylated in an EGF-dependent manner by membranes from cells (A431) that overexpress the EGF receptor.[21] Subsequently, peptides have been shown to serve as substrates for both the insulin[26] and platelet-derived growth factor (PDGF) receptor kinases.[76] Table 6 illustrates a number of examples of how synthetic peptides have been used for the characterization of the properties of growth factor receptors. In addition to the growth factor receptors, numerous other protein-tyrosine kinases, including the products of several of the viral oncogenes, have been shown to use synthetic peptides as substrates *in vitro*.[11,12,14,31,32,37]

Studies such as those described above illustrate some of the wide variety of uses for synthetic peptides in the characterization of protein-tyrosine kinases. Their use has been sufficiently widespread that their phosphorylation is used routinely as a basis of comparison of different preparations of protein-tyrosine kinases. It is clear that their use as substrates

TABLE 6
Use of Synthetic Peptides for the Characterization of the Protein-Tyrosine Kinase Activity of Growth Factor Receptors

Analysis[a]	Ref.
Activation of EGF-R kinase by EGF and TGF	77
Comparison of EGF-R autophosphorylation and peptide phosphorylation	78
Kinetic mechanism of EGF-R kinase	22
Comparison of EGF-R and IR from placenta	40
Desensitization of EGF-R by EGF	79
Characterization of purified EGF-R	29
Characterization of proteolytic fragments of EGF-R	80
Characterization of anti-EGF-R antibodies	81
Characterization of purified PDGF-R	76
Kinase activity of purified IR	82
Activation of IR by autophosphorylation	83
Effect of dithiothreitol on IR kinase	84, 85
Characterization of purified IR	25
Identification of sites of autophosphorylation of IR	27

[a] Abbreviations used are: EGF-R, epidermal growth factor receptor; EGF, epidermal growth factor; TGF, transforming growth factor; IR, insulin receptor; PDGF-R, platelet-derived growth factor receptor.

is playing an important part in the rapid development of information regarding the characteristics of protein-tyrosine kinases.

IV. SUMMARY

Synthetic peptides have received widespread use as substrates for the detection and characterization of protein-tyrosine kinases. They have been particularly useful for studies of this general class of protein kinase due, in part, to a lack of well-defined, physiologically relevant protein substrates. Thus, synthetic peptides have helped fill the need for specific substrates in assays of protein-tyrosine kinase activity. Studies on the substrate specificity of protein-tyrosine kinases using peptides of defined sequence have failed to uncover rigid requirements for primary structure. The best substrates generally possess acidic amino acids proximal to the phosphorylated tyrosine, but exceptions to this rule are known. An examination of sites of tyrosine phosphorylation on protein substrates suggests that a wide variety of primary sequences are phosphorylated by various members of the tyrosine kinase family. The apparent K_m values of protein-tyrosine kinases for protein substrates are generally one to two orders of magnitude lower than those for peptides. These observations suggest that secondary and/or tertiary structure is important in the recognition of protein substrates by protein-tyrosine kinases.

Despite the relative lack of specificity and affinity displayed by protein-tyrosine kinases for peptide substrates, the phosphorylation of peptides has proven to be a powerful technique for the study of protein-tyrosine kinases. A number of enzymes now have been detected in a wide range of tissues based on their ability to catalyze the phosphorylation of synthetic peptides, and several of these have been purified to near homogeneity. In addition, peptides have played important roles in the characterization of the kinase activity of both growth factor receptors and oncogene products.

REFERENCES

1. **Krebs, E. G. and Beavo, J. A.**, Phosphorylation-dephosphorylation of enzymes, *Annu. Rev. Biochem.*, 48, 923, 1979.
2. **Cohen, P.**, The role of protein phosphorylation in neural and hormonal control of cellular activity, *Nature (London)*, 296, 613, 1982.
3. **Hunter, T. and Cooper, J. A.**, Protein-tyrosine kinases, *Annu. Rev. Biochem.*, 54, 897, 1985.
4. **Collett, M. S. and Erikson, R. L.**, Protein kinase activity associated with the avian sarcoma virus *src* gene product, *Proc. Natl. Acad. Sci. U.S.A.*, 75, 2021, 1978.
5. **Hunter, T. and Sefton, B. M.**, The transforming gene product of Rous sarcoma virus phosphorylates tyrosine, *Proc. Natl. Acad. Sci. U.S.A.*, 77, 1311, 1980.
6. **Collett, M. S., Erikson, E., and Erikson, R. L.**, Structural analysis of the avian sarcoma virus transforming protein: sites of phosphorylation, *J. Virol.*, 29, 770, 1979.
7. **Casnellie, J. E., Harrison, M. L., Pike, L. J., Hellstrom, K. E., and Krebs, E. G.**, Phosphorylation of synthetic peptides by a tyrosine protein kinase from the particulate fraction of a lymphoma cell line, *Proc. Natl. Acad. Sci. U.S.A.*, 79, 282, 1982.
8. **Neil, J. C., Ghysdael, J., Vogt, P. K., and Smart, J. E.**, Homologous tyrosine phosphorylation sites in transformation-specific gene products of distinct avian sarcoma viruses, *Nature (London).*, 291, 675, 1981.
9. **Smart, J. E., Oppermann, H., Czernilofsky, A. P., Purrchio, A. F., Erikson, R. L., and Bishop, J. M.**, Characterization of sites for tyrosine phosphorylation in the transforming protein of Rous sarcoma virus (p60^{v-src}) and its normal cellular homologue (p60^{c-src}), *Proc. Natl. Acad. Sci. U.S.A.*, 78, 6013, 1981.
10. **Czernilofsky, A. P., Levinson, A. D., Varmus, H. E., Bishop, J. M., Tischer, E., and Goodman, H. M.**, Nucleotide sequence of an avian sarcoma virus oncogene (*src*) and proposed amino acid sequence for gene product, *Nature (London)*, 287, 193, 1980.
11. **Wong, T. W. and Goldberg, A. R.**, *In vitro* phosphorylation of angiotensin analogs by tyrosyl protein kinases, *J. Biol. Chem.*, 258, 1022, 1983.
12. **Hunter, T.**, Synthetic peptide substrates for a tyrosine protein kinase, *J. Biol. Chem.*, 257, 4843, 1982.
13. **House, C., Baldwin, G. S., and Kemp, B. E.**, Synthetic peptide substrates for the membrane tyrosine protein kinase stimulated by epidermal growth factor, *Eur. J. Biochem.*, 140, 363, 1984.
14. **Braun, S., Ghany, M. A., and Racker, E.**, A rapid assay for protein kinases phosphorylating small polypeptides and other substrates, *Anal. Biochem.*, 135, 369, 1983.
15. **Swarup, G., Dasgupta, J. D., and Garbers, D. L.**, Tyrosine protein kinase activity of rat spleen and other tissues, *J. Biol. Chem.*, 258, 10341, 1983.
16. **Kemp, B. E., Graves, D. J., Benjamini, E., and Krebs, E. G.**, Role of multiple basic residues in determining the substrate specificity of cyclic AMP-dependent protein kinase, *J. Biol. Chem.*, 252, 4888, 1977.
17. **Feramisco, J. R., Glass, D. B., and Krebs, E. G.**, Optimal spatial requirements for the location of basic residues of peptide substrates for the cyclic AMP-dependent protein kinase, *J. Biol. Chem.*, 235, 4240, 1980.
18. **House, C., Wettenhall, R. E. H., and Kemp, B. E.**, The influence of basic residues on the substrate specificity of protein kinase C., *J. Biol. Chem.*, 262, 772, 1987.
19. **Hathaway, G. M. and Traugh, J. A.**, Casein kinases — multipotential protein kinases, in *Current Topics in Cellular Regulation*, Vol. 21, Horecker, B. L. and Stadtman, E. R., Eds., Academic Press, New York, 1982, 101.
20. **Patschinsky, T., Hunter, T., Esch, F. S., Cooper, J. A., and Sefton, B. M.**, Analysis of the sequence of amino acids surrounding sites of tyrosine phosphorylation, *Proc. Natl. Acad. Sci. U.S.A.*, 79, 973, 1982.
21. **Pike, L. J., Gallis, B., Casnellie, J. E., Bornstein, P., and Krebs, E. G.**, Epidermal growth factor stimulates the phosphorylation of synthetic tyrosine-containing peptides by A431 cell membranes, *Proc. Natl. Acad. Sci. U.S.A.*, 79, 1443, 1982.
22. **Erneux, C., Cohen, S., and Garbers, D. L.**, The kinetics of tyrosine phosphorylation by the purified epidermal growth factor receptor kinase of A431 cells, *J. Biol. Chem.*, 258, 4137, 1983.
23. **Baldwin, G. S., Stanley, I. J., and Nice, E. C.**, A synthetic peptide containing the autophosphorylation site of the transforming protein of Harvey sarcoma virus is phosphorylated by the EGF-stimulated tyrosine kinase, *FEBS Lett.*, 153, 257, 1983.
24. **Casnellie, J. E. and Krebs, E. G.**, The use of synthetic peptides for defining the specificity of tyrosine protein kinases, *Adv. Enzyme Regul.*, 22, 501, 1984.
25. **Kasuga, M., Fujita-Yamaguchi, Y., Blithe, D. L., White, M. F., and Kahn, C. R.**, Characterization of the insulin receptor kinase purified from human placental membranes, *J. Biol. Chem.*, 258, 10973, 1983.
26. **Stadtmauer, L. A. and Rosen, O. M.**, Phosphorylation of exogenous substrates by the insulin receptor-associated protein kinase, *J. Biol. Chem.*, 258, 6682, 1983.

27. **Stadtmauer, L. and Rosen, O. M.**, Phosphorylation of synthetic insulin receptor peptides by the insulin receptor kinase and evidence that the preferred sequence containing Tyr-1150 is phosphorylated *in vivo*, *J. Biol. Chem.*, 261, 10000, 1986.
28. **Trevillyan, J. M., Nordstrom, A., and Linna, T. J.**, High tyrosine-specific protein kinase activity in normal human peripheral blood lymphocytes, *Biochim. Biophys. Acta*, 845, 1, 1985.
29. **Weber, W., Bertics, P. J., and Gill, G.**, Immunoaffinity purification of the epidermal growth factor receptor. Stoichiometry of binding and kinetics of self-phosphorylation, *J. Biol. Chem.*, 259, 14631, 1984.
30. **Wong, T. W. and Goldberg, A. R.**, Purification and characterization of the major species of tyrosine protein kinase in rat liver, *J. Biol. Chem.*, 259, 8505, 1984.
31. **Foulkes, J. G., Chow, M., Gorka, C., Frackelton, A. R., Jr., and Baltimore, D.**, Purification and characterization of a protein-tyrosine kinase encoded by the Abelson murine leukemia virus, *J. Biol. Chem.*, 260, 8070, 1985.
32. **Ferguson, B., Pritchard, M. L., Feild, J., Rieman, D., Greig, R. G., Poste, G., and Rosenberg, M.**, Isolation and analysis of an Abelson murine leukemia virus-encoded tyrosine-specific kinase produced in *Escherichia coli*, *J. Biol. Chem.*, 260, 3652, 1985.
33. **Harrison, M. L., Low, P. S., and Geahlen, R. L.**, T and B lymphocytes express distinct tyrosine protein kinases, *J. Biol. Chem.*, 259, 9348, 1984.
34. **Zioncheck, T. F., Harrison, M. L., and Geahlen, R. L.**, Purification and characterization of a protein-tyrosine kinase from bovine thymus, *J. Biol. Chem.*, 261, 15637, 1986.
35. **Baldwin, G. S., Burgess, A. W., and Kemp, B. E.**, Phosphorylation of a synthetic gastrin peptide by the tyrosine kinase of A431 cell membranes, *Biochem. Biophys. Res. Commun.*, 109, 656, 1982.
36. **Baldwin, G. S., Knesel, J., and Monckton, J. M.**, Phosphorylation of gastrin-17 by epidermal growth factor-stimulated tyrosine kinase, *Nature (London)*, 301, 435, 1983.
37. **Cooper, J. A., Esch, F. S., Taylor, S. S., and Hunter, T.**, Phosphorylation sites in enolase and lactate dehydrogenase utilized by tyrosine protein kinases *in vivo* and *in vitro*, *J. Biol. Chem.*, 259, 7835, 1984.
38. **Matsuo, M., Huang, C.-H., and Huang, L. C.**, Evidence for an essential arginine recognition site on adenosine 3':5'-cyclic monophosphate-dependent protein kinase of rabbit muscle, *Biochem. J.*, 173, 441, 1978.
39. **Mason, D. L., Harrison, M. L., and Geahlen, R. L.**, Properties of a tyrosine protein kinase from calf thymus. Response to ionic strength and divalent cations, *Biochim. Biophys. Acta*, 829, 221, 1985.
40. **Pike, L. J., Kuenzel, E. A., Casnellie, J. E., and Krebs, E. G.**, A comparison of the insulin- and epidermal growth factor-stimulated protein kinases from human placenta, *J. Biol. Chem.*, 259, 9913, 1984.
41. **Foulkes, J. G., Mathey-Prevot, B., Guild, B. C., Prywes, R., and Baltimore, D.**, A comparison of the protein-tyrosine kinases encoded by Abelson murine leukemia virus and Rous sarcoma virus, in *Cancer Cells/3, Growth Factors and Transformation*, Feramisco, J., Ozanne, B., and Stiles, C., Eds., Cold Spring Harbor Laboratories, Cold Spring Harbor, NY, 1985, 329.
42. **Kitamura, N., Kitamura, A., Yoyoshima, K., Hirayama, Y., and Yoshida, M.**, Avian sarcoma virus Y73 genome sequence and structural similarity of its transforming gene product to that of Rous sarcoma virus, *Nature (London)*, 297, 205, 1982.
43. **Marth, J. D., Peet, R., Krebs, E. G., and Perlmutter, R. M.**, A lymphocyte-specific protein-tyrosine kinase gene is rearranged and overexpressed in the murine T cell lymphoma LSTRA, *Cell*, 43, 393, 1985.
44. **Voronova, A. F. and Sefton, B. M.**, Expression of a new tyrosine protein kinase is stimulated by retrovirus promoter insertion, *Nature (London)*, 319, 682, 1986.
45. **Shibuya, M. and Hanafusa, H.**, Nucleotide sequence of Fujinami sarcoma virus: evolutionary relationship of its transforming gene with transforming genes of other sarcoma viruses, *Cell*, 30, 787, 1982.
46. **Hampi, A., Laprevotte, I., Galibert, F., Fedele, L. A., and Sherr, C. J.**, Nucleotide sequences of feline retroviral oncogenes (v-*fes*) provide evidence for a family of tyrosine-specific protein kinase genes, *Cell*, 30, 775, 1982.
47. **Reddy, E. P., Smith, M. J., and Srinivasan, A.**, Nucleotide sequence of Abelson murine leukemia virus genome: structural similarity of its transforming gene product to other *onc* gene products with tyrosine-specific kinase activity, *Proc. Natl. Acad. Sci. U.S.A.*, 80, 3623, 1983.
48. **Ullrich, A., Bell, J. R., Chen. E. Y., Herrera, R., Petruzzelli, L. M., Dull, T. J., Gray, A., Coussens, L., Liao, Y.-C., Tsubokawa, M., Mason, A., Seeburg, P. H., Grunfeld, C., Rosen, O. M., and Ramachandran, J.**, Human insulin receptor and its relationship to the tyrosine kinase family of oncogenes, *Nature (London)*, 313, 756, 1985.
49. **Downward, J., Parker, P., and Waterfield, M. D.**, Autophosphorylation sites on the epidermal growth factor receptor, *Nature (London)*, 311, 483, 1984.
50. **Gallis, B., Edelman, A. M., Casnellie, J. E., and Krebs, E. G.**, Epidermal growth factor stimulates tyrosine phosphorylation of the myosin regulatory light chain from smooth muscle, *J. Biol. Chem.*, 258, 13089, 1983.
51. **Maita, T., Chen, J.-T., and Matsuda, G.**, Amino-acid sequence of the 20000-molecular-weight light chain of chicken gizzard-muscle myosin, *Eur. J. Biochem.*, 117, 417, 1981.

52. **Dekowski, S. A., Rybicki, A., and Drickamer, K.,** A tyrosine kinase associated with the red cell membrane phosphorylates band 3, *J. Biol. Chem.,* 258, 2750, 1983.
53. **Kaul, R. K., Murthy, S. N. P., Reddy, A. G., Steck, T. L., and Kohler, H.,** Amino acid sequence of the Nα-terminal 201 residues of human erythrocyte membrane band 3, *J. Biol. Chem.,* 258, 7981, 1983.
54. **De, B. K., Misono, K. S., Lukas, T. J., Mroczkowski, B., and Cohen, S.,** Calcium-dependent 35-kilodalton substrate for epidermal growth factor receptor/kinase isolated from normal tissue, *J. Biol. Chem.,* 261, 13784, 1986.
55. **Guild, B. C., Erikson, R. L., and Strominger, J. L.,** HLA-A2 and HLA-B7 antigens are phosphorylated *in vitro* by Rous sarcoma virus kinase (pp60^{v-src}) at a tyrosine residue encoded in a highly conserved exon of the intracellular domain, *Proc. Natl. Acad. Sci. U.S.A.,* 80, 2894, 1983.
56. **Glenney, J. R., Jr. and Tack, B. F.,** Amino-terminal sequence of p36 and associated p10: identification of the site of tyrosine phosphorylation and homology with S-100, *Proc. Natl. Acad. Sci. U.S.A.,* 82, 7884, 1985.
57. **Cooper, J. A., Gould, K. L., Cartwright, C. A., and Hunter, T.,** Tyr^{527} is phosphorylated in pp60^{c-src}: implications for regulation, *Science,* 231, 1431, 1986.
58. **Hunter, T., Hutchinson, M. A., and Eckhart, W.,** Polyoma middle-sized T antigen can be phosphorylated on tyrosine at multiple sites *in vitro*, *EMBO J.,* 3, 73, 1984.
59. **Van Patten, S. M., Heisermann, G. J., Cheng, H.-C., and Walsh, D. A.,** Tyrosine kinase catalyzed phosphorylation and inactivation of the inhibitor protein of the cAMP-dependent protein kinase, *J. Biol. Chem.,* 262, 3398, 1987.
60. **Courtneidge, S. A. and Smith, A. E.,** Polyoma virus transforming protein associates with the product of the c-*src* cellular gene, *Nature (London),* 303, 435, 1983.
61. **Huang, K.-S., Wallner, B. P., Mattaliano, R. J., Tizard, R., Burne, C., Frey, A., Hession, C., McGray, P., Sinclair, L. K., Chow, E. P., Browning, J. L., Ramachandran, K. L., Tang, J., Smart, J. E., and Pepinsky, R. B.,** Two human 35 kd inhibitors of phospholipase A$_2$ are related to substrates of pp60^{v-src} and of the epidermal growth factor receptor/kinase, *Cell,* 46, 191, 1986.
62. **Swarup, G. and Subrahmanyam, G.,** Activation of a cellular tyrosine-specific protein kinase by phosphorylation, *FEBS Lett.,* 188, 131, 1985.
63. **Tokuda, M., Khanna, N. C., Arora, A. K., and Waisman, D. M.,** Identification and characterization of the tyrosine protein kinases of rat spleen, *Biochem. Biophys. Res. Commun.,* 139, 910, 1986.
64. **Kong, S.-K. and Wang, J. H.,** Purification and characterization of a protein tyrosine kinase from bovine spleen, *J. Biol. Chem.,* 262, 2597, 1987.
65. **Casnellie, J. E., Harrison, M. L., Hellstrom, K. E., and Krebs, E. G.,** A lymphoma cell line expressing elevated levels of tyrosine protein kinase activity, *J. Biol. Chem.,* 258, 10738, 1983.
66. **Nakamura, S.-I., Takeuchi, F., Tomizawa, T., Takasaki, N., Kondo, H., and Yamamura, H.,** Two separate tyrosine protein kinases in human platelets, *FEBS Lett.,* 184, 56, 1985.
67. **Dasgupta, J. D., Swarup, G., and Garbers, D. L.,** Tyrosine protein kinase activity in normal rat tissues: brain, *Adv. Cyclic Nucleotide Protein Phosphoryl. Res.,* 17, 461, 1984.
68. **Wong, T. W. and Goldberg, A. R.,** Tyrosyl protein kinases in normal rat liver: identification and partial characterization, *Proc. Natl. Acad. Sci. U.S.A.,* 80, 2529, 1983.
69. **Lin, M.-F., Lee, P. L., and Clinton, G. M.,** Characterization of tyrosyl kinase activity in human serum, *J. Biol. Chem.,* 260, 1582, 1985.
70. **Dasgupta, J. D. and Garbers, D. L.,** Tyrosine protein kinase activity during embryogenesis, *J. Biol. Chem.,* 258, 6174, 1983.
71. **Braun, S., Raymond, W. E., and Racker, E.,** Synthetic tyrosine polymers as substrates and inhibitors of tyrosine-specific protein kinases, *J. Biol. Chem.,* 259, 2051, 1984.
72. **Casnellie, J. E., Harrison, M. L., Hellstrom, K. E., and Krebs, E. G.,** A lymphoma protein with an *in vitro* site of tyrosine phosphorylation homologous to that in pp60^{src}, *J. Biol. Chem.,* 257, 13877, 1982.
73. **Voronova, A. F., Buss, J. E., Patschinsky, T., Hunter, T., and Sefton, B. M.,** Characterization of the protein apparently responsible for the elevated tyrosine protein kinase activity in LSTRA cells, *Mol. Cell. Biol.,* 4, 2705, 1984.
74. **Casnellie, J. E., Gentry, L. E., Rohrschneider, L. R., and Krebs, E. G.,** Identification of the tyrosine protein kinase from LSTRA cells by use of site-specific antibodies, *Proc. Natl. Acad. Sci. U.S.A.,* 81, 6676, 1984.
75. **Casnellie, J. E. and Lamberts, R. J.,** Tumor promoters cause changes in the state of phosphorylation and apparent molecular weight of a tyrosine protein kinase in T lymphocytes, *J. Biol. Chem.,* 261, 4921, 1986.
76. **Bishayee, S., Ross, A. H., Womer, R., and Scher, C. D.,** Purified human platelet-derived growth factor receptor has ligand-stimulated tyrosine kinase activity, *Proc. Natl. Acad. Sci. U.S.A.,* 83, 6756, 1986.
77. **Pike, L. J., Marquardt, H., Todaro, G. J., Gallis, B., Casnellie, J. E., Bornstein, P., and Krebs, E. G.,** Transforming growth factor and epidermal growth factor stimulate the phosphorylation of a synthetic, tyrosine-containing peptide in a similar manner, *J. Biol. Chem.,* 257, 14628, 1982.

78. **Cassel, D., Pike, L. J., Grant, G. A., Krebs, E. G., and Glaser, L.**, Interaction of epidermal growth factor-dependent protein kinase with endogenous membrane proteins and soluble peptide substrate, *J. Biol. Chem.*, 258, 2945, 1983.
79. **Chinkers, M. and Garbers, D. L.**, Suppression of protein tyrosine kinase activity of the epidermal growth factor receptor by epidermal growth factor, *J. Biol. Chem.*, 261, 8295, 1986.
80. **Basu, M., Biswas, R., and Das, M.**, 42,000-molecular weight EGF receptor has protein kinase activity, *Nature (London)*, 311, 477, 1984.
81. **Gill, G. N., Kawamoto, T., Cochet, C., Le, A., Sato, J. D., Masui, H., McLeod, C., and Mendelsohn, J.**, Monoclonal anti-epidermal growth factor receptor antibodies which are inhibitors of epidermal growth factor binding and antagonists of epidermal growth factor-stimulated tyrosine protein kinase activity, *J. Biol. Chem.*, 259, 7755, 1984.
82. **Petruzzelli, L., Herrera, R., and Rosen, O. M.**, Insulin receptor is an insulin-dependent tyrosine protein kinase: copurification of insulin-binding activity and protein kinase activity to homogeneity from human placenta, *Proc. Natl. Acad. Sci. U.S.A.*, 81, 3327, 1984.
83. **Rosen, O. M., Herrera, R., Olowe, Y., Petruzzelli, L. M., and Cobb, M. H.**, Phosphorylation activates the insulin receptor tyrosine protein kinase, *Proc. Natl. Acad. Sci. U.S.A.*, 80, 3237, 1983.
84. **Pike, L. J., Eakes, A. T., and Krebs, E. G.**, Characterization of affinity-purified insulin receptor/kinase. Effects of dithiothreitol on receptor/kinase function, *J. Biol. Chem.*, 261, 3782, 1986.
85. **Fujita-Yamaguchi, Y. and Kathuria, S.**, The monomeric $\alpha\beta$ form of the insulin receptor exhibits much higher insulin-dependent tyrosine-specific protein kinase activity than the intact $\alpha_2\beta_2$ form of the receptor, *Proc. Natl. Acad. Sci. U.S.A.*, 82, 6095, 1985.

Chapter 11

ANTIBODIES AGAINST SYNTHETIC PHOSPHORYLATION SITE PEPTIDES

David B. Glass and Laura Anne Uphouse

TABLE OF CONTENTS

I. Introduction 256

II. Applications of Antipeptide Antibodies to Protein Kinases 257
    A. Tyrosine-Protein Kinases 258
    B. Serine/Threonine Protein Kinases....................................... 259
    C. Phosphorylation Site in a Substrate of cAMP-Dependent Protein Kinase 260

III. Choice of Peptide Sequence 260

IV. Peptide Synthetic Strategies 262

V. Preparation of Antipeptide Antibodies 265
    A. Peptide-Protein Conjugate Formation and Immunization 265
    B. Screening of Antisera 267

VI. Purification and Characterization of Antibodies 268
    A. Immunoaffinity Chromatography 268
    B. Antibody Characterization 270

VII. Specificity of Anti-Phosphorylation Site Antibodies 272
    A. Antipeptide Antibodies that Recognize the Phosphorylation Site Epitope 272
    B. Anti-Phosphorylation Site Antibodies from Anti-Pyruvate Kinase Antisera 276

VIII. Discussion 278

IX. Concluding Remarks 280

Acknowledgments 280

References 280

I. INTRODUCTION

It is now clear that relatively small synthetic peptides corresponding to nearly any accessible portion of a protein can elicit antibodies that will react with the same antigenic determinant in the native protein.[1-9] Antibodies to such synthetic peptides have an inherent predetermined specificity that can be controlled by selection of the peptide sequence. The cloning and sequencing of a great number of complementary DNAs (cDNAs) have made available the deduced amino acid sequences of many proteins. Appropriately chosen regions of these proteins can be synthesized chemically for use as immunogens to raise protein-reactive antibodies. Antipeptide antibodies have been used extensively in recent years in developing synthetic vaccines,[10-16] for detecting and localizing various predicted gene products,[17-23] and as structural and functional probes of numerous proteins.[24-30] A general approach to the production and use of antipeptide antibodies is outlined in Figure 1.

Synthetic peptides are useful vaccines against infections produced by various viruses, bacteria, and protozoans.[10] The peptide immunogen approach induces neutralizing antibodies against capsid proteins of foot-and-mouth virus[7,11,31,32] and is being actively studied for other viruses including hepatitis B,[12,33] Epstein-Barr,[34,35] polio,[13,36] and human immunodeficiency virus.[16,37,38] Among bacteria, synthetic peptides have been applied as vaccines for some strains of streptococcus and gonococcus.[39,40] Also, peptides induce antibodies that recognize either the circumsporozoite or merozoite forms of *Plasmodium falciparum*, one of the malaria organisms.[14,15,41,42]

Antipeptide antibodies are useful in the analysis of the protein products of several oncogenes. These include both viral oncogenes and their cellular homologues. Peptides generally are chosen from predicted amino acid sequences based on complete nucleic acid sequences. The *myc* gene product has been identified in normal and malignant cells with antipeptide antibodies.[18] Similar immunologic reagents that recognize the products of the viral and cellular *fos* oncogenes were generated with synthetic peptides.[43] Antibodies to a peptide sequence in the viral *sis* gene product react with human platelet-derived growth factor (PDGF), the cellular homologue of *sis*.[17,44] Nearly 60% of the antipeptide antibodies generated against the Ha-*ras*, Ki-*ras*, *myb*, *fes*, *myc*, and *src* encoded oncogene products reacted with their respective targeted proteins.[45] The protein product of the transforming gene of Rous sarcoma virus, *src*, is a tyrosine-specific protein kinase, catalyzing the transfer of the γ-phosphate of ATP to the phenolic hydroxyl group of tyrosyl residues in protein substrates. As discussed in greater detail below, *src* and numerous other oncogene products and cellular proteins that belong to this class of protein kinase have been identified and functionally characterized with specific antipeptide or antityrosine antibodies.[45-49]

The structures and functions of a wide variety of other proteins also can be studied with antipeptide antibodies. These proteins include previously uncharacterized gene products predicted from cloned cDNA sequences,[19,20] gene products expressed in a tissue-specific manner,[22] propeptides of neurohormones or growth factors,[23,24] cell surface receptors, channels, and carrier proteins,[24-29,50-55] regulatory G proteins involved in signal transduction,[56,57] various effector or mediator proteins such as calmodulin and interferons,[30,58-63] and extracellular proteins including components of the extracellular matrix.[64-70] Some of the protein receptors that have been characterized with antipeptide antibodies happen to be either substrates for protein kinases or are protein kinases themselves.

Antipeptide antibodies have been applied to the study of various protein kinases and their substrates. Sites of phosphorylation in a substrate must be on the protein's solvent-exposed surface and available to the appropriate protein kinase. This is certainly the case for the one protein kinase substrate whose three-dimensional structure is known.[71] Also, sites of phosphorylation in many proteins occur near the NH_2-terminus,[71-75] a region that in many cases is exposed to solvent. Because many kinases recognize charged amino acids as

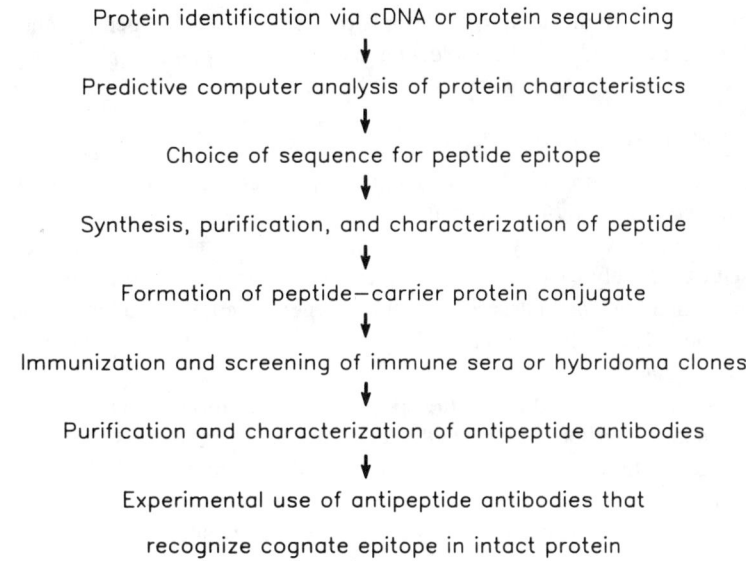

FIGURE 1. General strategy for production and use of antipeptide antibodies. Refer to text for details.

determinants of specificity,[76-83] the primary sequences around sites of phosphorylation are relatively polar and hydrophilic. Kemptide,[84] the model peptide substrate for the cAMP-dependent protein kinase, binds to the active site of that enzyme in an extended coil conformation.[85] Corresponding phosphorylation sites in proteins probably have low energy barriers between different conformations and would be expected to have a reasonable degree of flexibility. These characteristics of surface location, exposure to solvent, polarity, and segmental mobility all tend to be found in regions of proteins that are highly antigenic.[1-9,86-94] Thus, synthetic phosphorylation site peptides should promote formation of antipeptide antibodies that recognize the same phosphorylation sites in intact protein substrates. The same can be said of autophosphorylation sites on protein kinases themselves.

II. APPLICATIONS OF ANTIPEPTIDE ANTIBODIES TO PROTEIN KINASES

The protein kinases and their substrates that have been studied with antipeptide antibodies include the tyrosine-specific[95,96] as well as the serine/threonine-specific classes of protein kinases.[76-78] While the uses of antipeptide antibodies for the study of protein kinases are limited only by the imaginations of individual investigators, several approaches have been employed and several others are obvious. Antibodies have been generated toward either the phospho- or dephospho-forms of phosphorylation sites in substrates and the autophosphorylation sites in protein kinases. Also, antipeptide antibodies have been made to other regulatory regions in protein kinases. These antipeptide antibodies are useful for identification, purification, and assay of protein kinases or their substrates, for analysis of the state of phosphorylation of proteins in cells and tissues, to inhibit protein kinase activities, and for structure-function studies of phosphoproteins. Antipeptide antibodies are applied in various standard immunochemical techniques, including both liquid- and solid-phase radioimmunoassays (RIAs), enzyme-linked immunosorbent assays (ELISAs), subcellular localization by immunocytochemistry, as probes in Western blotting protocols,[97] and in immunoaffinity chromatography.[98] In theory, antipeptide antibodies recognizing a protein phosphorylation site could be used as immunogens to generate anti-idiotype antibodies that might be directed toward determinants within the active site of a protein kinase. Antipeptide antibodies also could prove useful in studying the molecular mechanisms by which phosphorylation of a protein alters its function.

A. TYROSINE-PROTEIN KINASES

The tyrosine-specific protein kinases include retroviral gene products with this activity and the normal cellular homologues of these viral oncogene products, including a number of cell surface receptors for polypeptide growth factors.[95,96] The immunological approach has been used extensively to study tyrosine-protein kinases because these enzymes are rare cellular proteins that have been described first at the genomic level and are difficult to purify in large quantities. The cellular proteins that are substrates for the tyrosine-protein kinases are also trace cellular proteins.[95,96,99]

Antibodies that recognize the tyrosine-specific protein kinases or their substrate proteins have been generated using either synthetic peptides or phosphotyrosine as immunogens. Antipeptide antibodies that are reactive toward the *src* gene product, pp60^{src}, have been made with synthetic peptides corresponding to various portions of that protein's sequence,[45-49] including one of the tyrosine autophosphorylation sites.[46] Antipeptide antibodies against regions of the *fps/fes* oncogene-encoded tyrosine-protein kinase reacted with that viral protein, its cellular homologue, and another widely distributed cellular tyrosine-protein kinase.[45,100] Synthetic peptides corresponding to the intracellular COOH-terminal domain of the human epidermal growth factor (EGF) receptor elicited antibodies that reacted with the EGF receptor as well as with the gene product of the avian v-*erb*B oncogene.[101] Antibodies made against several peptide sequences within the v-*erb*B gene product indicated that the v-*erb*B protein is a truncated homologue of the EGF receptor that retains an unregulated tyrosine-protein kinase activity.[102,103] Peptides corresponding to the major autophosphorylation site in the tyrosine-protein kinase of polyoma virus middle T antigen as well as to the COOH-terminus of this protein elicited antipeptide antibodies.[104,105] These antibodies were specific for middle T antigen and demonstrated that it does not possess phosphatidylinositol kinase activity. The protein kinase present in the lymphoma cell line LSTRA also has been identified using antibodies against a synthetic peptide corresponding to a site of tyrosine autophosphorylation in that enzyme.[106] Antibodies against a peptide sequence predicted from a cloned cDNA of a human T cell leukemia identified the tyrosine-protein kinase product of that message as well as a similar tyrosine-protein kinase from normal human and murine peripheral T lymphocytes.[107]

Finally, antipeptide antibodies have been used in structure-function studies of the human insulin receptor, whose β-subunit is a tyrosine-specific protein kinase that is itself phosphorylated on tyrosyl residues in an insulin-dependent manner.[99,108-112] After the cDNA for the human insulin proreceptor was cloned and sequenced, Rosen and colleagues developed a panel of antipeptide antibodies against selected regions of the predicted amino acid sequence of the β-subunit. Antibodies P4 (made to residues 952 to 967 in the proreceptor sequence) and P5 (made to COOH-terminal residues 1328 to 1343) both recognized the β-subunit of the receptor, but not other protein kinases.[108] Antibody P4 inhibited insulin-stimulated autophosphorylation and tyrosine-protein kinase activity toward exogenous substrates. Antipeptide antibody P2 (made to residues 1143 to 1162) recognized the phospho- and dephospho- forms of the denatured insulin receptor, but only precipitated the phospho-form of the native insulin receptor.[109,110] Thus, autophosphorylation of the insulin receptor changes its conformation, making an epitope more accessible to this antibody. Antipeptide antibody P2 also was used to demonstrate that tyrosine-1150 is a major site of insulin-dependent autophosphorylation both *in vivo* and *in vitro*.[109,110] Antibody P5 identified tyrosine-1316 as another site of autophosphorylation *in vitro*.[110] The rate of phosphorylation of tyrosine-1150 correlated best with activation of the tyrosine kinase activity.[110] Antipeptide antibodies P4 and P5 delineated conserved and nonconserved domains of the insulin receptor and the related insulin-like growth factor-I (IGF-I) receptor.[111] Antibody P4, which is directed against a cytoplasmic domain close to the membrane spanning region of the insulin receptor, also reacted with the IGF-I receptor. Antibody P5, which recognizes the COOH-terminus of the insulin receptor, did not react with the IGF-I receptor. Additionally, antipeptide antibody P2 was used to detect the β-subunit of the insulin receptor from *Drosophila* after it was autophosphorylated in an insulin-dependent manner.[112]

A more general immunologic approach to the identification of tyrosine-protein kinases and their substrate proteins has employed antiphosphotyrosine antibodies. These reagents preferentially recognize the phosphorylated forms of kinases or their substrate proteins and cross-react with various phosphotyrosine-containing proteins. Such antibodies have been generated using immunogens of phosphotyrosine or phosphotyramine covalently coupled to carrier proteins.[113-116] These conjugates contain a phosphoester bond that conceivably could be hydrolyzed *in vivo* prior to recognition of the antigen by the immune system. Therefore, other investigators have generated phosphotyrosine-reactive antibodies by immunization with hapten-protein conjugates containing the structurally related phosphonate azobenzylphosphonate.[117-121] This hapten contains a carbon-phosphorus bond that is resistant to phosphohydrolases. The antiphosphotyrosine or antibenzylphosphonate antibodies have allowed identification or isolation of phosphotyrosine-containing proteins in cells transformed by the Abelson murine leukemia virus and by the Rous sarcoma virus.[117-120] These antibodies have also identified phosphotyrosine-containing proteins resulting from stimulation of fibroblasts by PDGF[113,122] and from stimulation of epidermal carcinoma cells by EGF.[118] Additionally, the phosphorylated forms of the placental insulin receptor,[114] the fibroblast PDGF receptor,[116,121,122] and a uterine estradiol receptor[115] have been purified using antibodies that react with phosphotyrosine.

B. SERINE/THREONINE PROTEIN KINASES

Antipeptide antibodies have also been applied to serine/threonine-specific protein kinases and their substrates. One of the earliest studies described antipeptide antibodies directed toward the NH$_2$-terminal 18 residues of glycogen phosphorylase, the substrate of phosphorylase *b* kinase.[73,74] The site of phosphorylation is contained within this sequence. The antipeptide antibodies reacted better with phosphorylase *a* than with phosphorylase *b*. This specificity was not due to recognition of the phosphoseryl residue per se in the former protein, but probably was due to better recognition of the conformation of this epitope that was induced by the phosphorylation reaction. The antipeptide antibodies interfered with the regulation of phosphorylase activity by the NH$_2$-terminus, but did not affect the catalytic site of the enzyme directly.

In another study, antibodies were generated that distinguished between the phospho- and dephospho-forms of G-substrate, a neuron-specific substrate protein of the cGMP-dependent protein kinase.[123,124] Antibodies against the phosphorylation site in G-substrate were elicited with the stoichiometrically phosphorylated form of the native protein, while the antibodies directed toward the dephospho-form of the protein were raised against a synthetic peptide corresponding to the amino acid sequence around the site of phosphorylation. These antibodies were used to develop an RIA to assess the phosphorylation state of G-substrate in cerebellum under various experimental conditions.[123-125]

Antipeptide antibodies were elicited against a sequence around the proposed site in the nicotinic acetylcholine receptor that is phosphorylated by the cAMP-dependent protein kinase.[126] These antibodies preferentially recognized the nonphosphorylated form of this protein. In a separate study, a monoclonal antibody was found to selectively recognize the phosphorylated forms of phenylalanine hydroxylase as well as a synthetic phosphopeptide corresponding to the site of phosphorylation in this enzyme.[127] Finally, antipeptide antibodies have been used to study some of the phosphoprotein phosphatases that act on phosphorylated protein substrates, particularly the phospho-form of phosphatase inhibitor-1.[128]

Most of the antipeptide and antiphosphotyrosine antibodies mentioned above were used to immunoprecipitate their target proteins or to identify them on gel electrophoretograms by Western blotting techniques. As mentioned above, some anti-phosphorylation site antibodies have been used to determine the phosphorylation states of several proteins.[108,110,123-127] Some investigators have covalently coupled antibodies to agarose and have then used these immobilized antibodies to purify protein kinases or their substrates by immunoaffinity chromatography.[105,114-118,122] Elution of phosphotyrosine-containing proteins from columns of antiphosphotyrosine antibodies has been effected by the ligands phenylphosphate or nitrophenylphosphate. Antipeptide antibodies have also been used as immunocytochemical reagents to determine regional distributions of kinases and substrates in tissues or the subcellular localization of these proteins in various cells.[47,49,101,102,119,125,128] Several antipeptide or antiphosphotyrosine antibodies have been used to inhibit the activities of various protein kinases.[46,49,74,103,106,108,111] These inhibition studies have helped delineate autophosphorylation sites and other regulatory regions in several protein kinases.

C. PHOSPHORYLATION SITE IN A SUBSTRATE OF cAMP-DEPENDENT PROTEIN KINASE

The work from our laboratory described here involves the generation and characterization of anti-phosphorylation site antibodies directed against the site in a characteristic protein substrate that is phosphorylated by cAMP-dependent protein kinase. L-type pyruvate kinase was chosen as the model substrate for this purpose because there is only a single phosphorylation site in the enzyme (one in each identical subunit of the homotetramer)[129] and because its phosphorylation by cAMP-dependent protein kinase is well characterized as to stoichiometry, kinetics, and consequence of the covalent modification on enzymatic activity.[130-132] The amino acid sequence around this phosphorylation site in rat L-type pyruvate kinase is Glu–Gly–Pro–Ala–Gly–Tyr–Leu–Arg–Arg–Ala–Ser–Val–Ala–Gln–Leu–.[133-135] This structure is representative of one class of substrates of cAMP-dependent protein kinase that contain a similar sequence around their phosphorylation sites. The phosphorylatable serine (or threonine) is separated by one amino acid from two adjacent arginyl residues located toward the NH$_2$-terminus.[84,136] These features of primary structure are recognized as determinants of substrate specificity by the cAMP-dependent protein kinase.[76-78,84] Our goal was to generate antibodies that recognized the structure –Arg–Arg–X–Ser– within the dephosphoform of the phosphorylation site in the native protein substrate, L-type pyruvate kinase.

III. CHOICE OF PEPTIDE SEQUENCE

With knowledge of the partial or complete amino acid sequence of a protein, one must select regions for synthesis of peptide antigens. These peptides should be immunogenic when coupled with a carrier protein and should have a high probability of eliciting protein-reactive antibodies. Historically, it was thought that antipeptide antibodies would recognize a protein only if they corresponded to an immunogenic determinant contained within that protein. Most globular proteins contain a limited number of antigenic sites.[9,137] In addition, some of these antigenic epitopes are assembled from amino acids in discontinuous portions of a protein's primary sequence rather than being composed of linear stretches of residues.[1,86,137] Thus, sequential epitopes in small, flexible peptide immunogens would poorly mimic these conformational determinants in a native protein. However, recent experimental evidence indicates that useful antipeptide antibodies can be made with synthetic peptides corresponding to nearly any sequence of a protein.[1,3-5,7-10,45,86] Such synthetic peptides can induce antibody specificities that are not attainable by immunization with intact proteins.

These antipeptide antibodies will react with a high frequency toward the same region of an intact protein if that protein is denatured. Denatured proteins are exposed more totally to solvent and are less restricted conformationally, thus resembling the original peptide immunogen to a greater degree. However, antipeptide antibodies probably will react with the intact protein in its native form only if the peptide sequence chosen as the immunogen is present on the surface of the protein, where it is accessible for interaction with the antibody.[138,139]

Globular proteins in aqueous solution are folded so that their hydrophilic and charged amino acids are on the surface, where they contact solvent, and their nonpolar residues generally are oriented toward the interior of the molecule.[140] Thus, one major criterion used to choose synthetic peptide sequences as effective immunogens for generation of antipeptide antibodies is that they contain a relative abundance of polar amino acids.[5,9,86] Several algorithms have been developed to analyze hydrophilicity profiles along linear sequences of proteins by assigning experimentally determined hydrophilicity or hydrophobicity values to the side chains of each amino acid.[89,90] One recent study derived hydrophilicity parameters from retention times of model peptides on reverse-phase high performance liquid chromatography.[91] Usually the mean hydrophilicity value is calculated for a moving segment of approximately six amino acid residues along the primary sequence. Computer programs employing these algorithms are used to predict those regions located on the surface of the native protein. In many cases, the NH_2- or COOH-terminal portions of proteins are near their surfaces. These domains are also likely to be less constrained conformationally and therefore more like a small peptide immunogen. Thus, peptides are synthesized to correspond to hydrophilic sequences in the NH_2- or COOH-terminal regions of a target protein.[9]

Another method to choose protein sequences useful as synthetic peptide immunogens is prediction of secondary protein structure. A number of known antigenic epitopes of proteins include turn or loop regions between two domains with defined secondary structures. Proline residues have been suggested to contribute to immunogenicity, as they occur in β-turns at the surface of many proteins.[4,37] In fact, one immunodominant site of a synthetic peptide immunogen preferentially assumes a β-turn structure in aqueous solution,[88] although this conformation is different from that of the cognate site in the intact protein. Proline residues also may provide conformational constraint in an otherwise highly flexible peptide immunogen.[3,88] These concerns have led to the use of well-known algorithms for prediction of secondary structure from amino acid sequence as aids in narrowing the choices of sequences for peptide synthesis.[141,142] However, the rather low accuracy of prediction of protein secondary structure has limited the usefulness of this parameter for the generation of antipeptide antibodies.[143]

Recent studies indicate that known antigenic epitopes in several proteins correspond well to regions of high atomic mobility or chain flexibility.[92-94] Peptides corresponding to highly mobile regions of one protein were much better in eliciting protein-reactive antibodies than were peptides from well-ordered regions.[92] A predictive algorithm has been developed that assigns flexibility coefficients to each amino acid based on atomic temperature factors derived from the mean square displacement of main chain carbon atoms in crystal structures of 31 proteins.[144] Lysine, serine, glycine, and proline have the highest flexibility coefficients, followed by the other hydrophilic amino acids. The degree of usefulness of flexibility parameters for predicting amino acid sequences for antipeptide antibodies will require further study. However, the high frequency of recognition of intact proteins by antibodies generated against peptides of unordered structure may relate well to the relaxation of conformation in local mobile segments of the proteins.[3,86,88,93] As expected, newer algorithms to predict antigenic surface residues on native proteins combine both hydrophilicity and flexibility parameters.[91]

One other important concern in choosing a synthetic peptide immunogen for antipeptide antibodies is peptide length. Most known antigenic epitopes contain at least six to eight amino acids.[86,137] The three-dimensional crystal structure of the antigen-antibody complex between lysozyme and the F_{ab} fragment of an antilysozyme monoclonal antibody indicates that a total of 16 residues from two discontinuous portions of lysozyme make up the antigenic site.[145] Synthetic peptides from approximately 8 to 40 residues in length have been used to elicit protein-reactive antibodies.[45,146] Many investigators prefer peptides of about 10 to 15 residues.[1] Longer peptides may confound screening of antibodies because these peptides can contain multiple epitopes. Particularly long synthetic peptides also can present a problem because in solution they might assume a preferred conformation that differs from that same sequence in the target protein.

Peptides corresponding to the middle of protein sequences as well as to the NH_2- or COOH-termini have all been used successfully to generated protein-reactive antipeptide antibodies, as long as the peptide sequences have met most of the previous requirements. The frequency of recognition of native proteins by antipeptide antibodies appears to be somewhat higher for those antibodies made against NH_2- or COOH-terminal regions than for those made toward internal portions of the sequence.[9]

Synthetic peptides representing a site of phosphorylation in a substrate of one of the protein kinases would be expected to be consistent with most of the above "rules" for choice of amino acid sequence for generation of antipeptide antibodies. Sites of phosphorylation are on the surfaces of substrate proteins and are generally hydrophilic, containing charged amino acids around a phosphorylatable seryl, threonyl, or tyrosyl residue. Lysine and serine, which are present at phosphorylation sites of many serine-specific protein kinases, are the residues with the highest flexibility parameters. Aspartic and glutamic acids, which are determinants for tyrosine-specific protein kinases,[83,104] are also predictors of relatively high chain flexibility. In addition, many sites of phosphorylation are near the NH_2-termini of substrate proteins.[71-76]

The phosphorylation site in L-type pyruvate kinase is thought to be in the NH_2-terminal region of the protein.[135] The sequence that contains the phosphorylatable seryl residue in L-type pyruvate kinase is missing in the homologous muscle (M-type) isozyme.[135] The latter form of the enzyme is not regulated by phosphorylation and dephosphorylation.[129] The recent cloning and sequencing of the cDNA for rat L-type pyruvate kinase proved that the phosphorylation site is indeed quite close to the NH_2-terminus.[147] If the initiator methionine is excluded, the phosphorylated residue is serine-11. The hydrophilicity and predicted flexibility profiles of the NH_2-terminal region of rat liver L-type pyruvate kinase are shown in Figure 2.[89,144]

IV. PEPTIDE SYNTHETIC STRATEGIES

There are two approaches to the production of antibodies of predetermined specificity by use of synthetic peptides. Antipeptide antibodies can be produced directly, based on the choice of an amino acid sequence that is highly likely to elicit protein-reactive antibodies (as discussed in Section I). Alternatively, synthetic peptides can be used to purify a specific subpopulation of antibodies from an antiprotein antiserum if the epitope of interest is immunogenic in the native protein. We used both techniques to elicit antibodies against the phosphorylation site in L-type pyruvate kinase. In the first approach, antipeptide antibodies were produced against a synthetic phosphorylation site decapeptide as described below. In the alternate approach, native L-type pyruvate kinase was purified from rat liver by the method of Riou et al.[148] and used to immunize rabbits. The purified pyruvate kinase was predominantly in the dephosphorylated form as shown by the fact that it could be phosphorylated by the cAMP-dependent protein kinase to the extent of 0.85 mol phosphate per mole of subunit monomer. A synthetic phosphorylation site peptide then was used to purify a

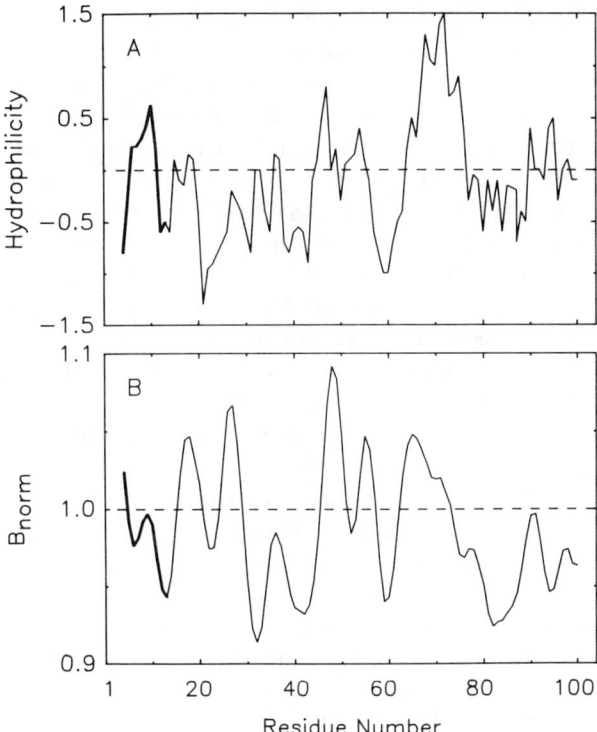

FIGURE 2. Hydrophilicity and predicted flexibility profiles for the first 100 amino acid residues at the NH_2-terminus of rat liver L-type pyruvate kinase.[147] Panel A: hydrophilicity parameters calculated as described by Hopp and Woods.[89] Panel B: flexibility parameters (B_{norm} values) calculated by the method of Karplus and Schulz.[144] Bold lines indicate the phosphorylation site sequence in pyruvate kinase corresponding to the decapeptide that was used to generate antidecapeptide antibodies.

subpopulation of anti-phosphorylation site antibodies from the resulting anti-pyruvate kinase antiserum.

Antibodies can be raised to either the dephosphorylated or phosphorylated form of a phosphorylation site sequence. Generation of the former is straightforward, and that approach has been taken for our studies of the cAMP-dependent protein kinase phosphorylation site in L-type pyruvate kinase. Production of antibodies against the phosphorylated form of a protein is somewhat more capricious. A native protein or synthetic phosphorylation site peptide can be phosphorylated stoichiometrically *in vitro* by the appropriate protein kinase, or the peptide can be phosphorylated chemically.[149] The phosphoprotein or the phosphopeptide is used for immunization after conjugation to a carrier protein. This simple approach has been successful in several instances;[123,127,128] however, there is no guarantee that the peptide will not be substantially dephosphorylated before or during recognition by the immune system *in vivo*. If the peptide is thiophosphorylated prior to immunization, endogenous phosphatases will be less likely to remove the thiophosphate group than the normal phosphate ester.[150] However, the antibodies will be directed against this altered structure. Probably the best approach would be to use an immunogen that contains a stable phosphonate bond which cannot be cleaved by a hydrolase. This approach has been employed to generate antibodies directed against phosphotyrosine (see Section II.A).[117-121] Such an analogue of phosphoserine would be L-2-amino-4-phosphonobutyric acid. This compound is used as a ligand for excitatory amino acid receptors,[151] but to our knowledge has not yet been incorporated into a synthetic peptide as a phosphoserine analogue.

The sequence of the peptide immunogen representing the desired epitope will dictate to a large degree the strategy of coupling to a carrier protein prior to immunization. The coupling strategy should first be determined and then the peptide synthesis performed to facilitate carrying out the plan. If the peptide is to be coupled via an amino group, it can be synthesized on a benzhydrylamine resin as a COOH-terminal amide so that this functional group is blocked. Alternatively, the COOH-terminus of the peptide can be blocked after its synthesis by formation of the methyl ester, although this is a less stable bond. If one wished to couple through the α-carboxyl group, the NH_2-terminus of the peptide can be blocked by acetylation during synthesis. In many cases, particularly with internal sequences, these strategies are desired because the native protein does not have free amino or carboxyl groups at these positions. On the other hand, peptide immunogens representing NH_2- or COOH-terminal protein sequences are left as the corresponding free, ionized functional groups to best mimic those structures in the target protein.

Specific amino acids that are useful in covalent coupling, such as cysteine, lysine, or tyrosine, can be included for such purposes at one end or the other of the peptide if they do not either interrupt the antigenic epitope of interest or replicate identical residues within the peptide. It is optimal to have only one functional group in the peptide that is reactive under the chosen covalent coupling conditions for formation of the peptide-carrier protein conjugate. This is to avoid any ambiguity in the coupling chemistry which would yield a population of immunogenic molecules of mixed structures. Of course, tyrosine also is added to some peptide sequences for later radioiodination and use as a ligand for liquid-phase RIAs.

During synthesis, a radiolabeled amino acid can be incorporated into an aliquot of the peptide immunogen so that coupling densities of the peptide-carrier protein conjugate can be determined conveniently. Usually a tritium- or ^{14}C-labeled aliphatic amino acid such as alanine or valine is used for this purpose. Commercially available radiolabeled amino acids can be added to carrier amino acid and converted to their N^α-tert-butyloxycarbonyl (Boc) derivates in good yield using 2-tert-butyloxycarbonyloxyimino-2-phenylacetonitrile.[152] After incorporation of the radiolabeled Boc-amino acid in a single coupling during solid-phase peptide synthesis, a second coupling of the residue is performed with an excess of nonradiolabeled Boc-amino acid to ensure completed step yield.

Peptides corresponding to protein phosphorylation sites generally are synthesized by solid-phase techniques.[153] Modern methods of solid-phase peptide synthesis have been reviewed recently.[154,155] The newly developed techniques of multiple peptide synthesis and cleavage are of potential use in the characterization of the final antibodies.[156,157] While the normal scale of the multiple synthesis method may not be suitable for synthesis of the amounts of peptide needed for complete characterization, conjugation and immunization, screening, and synthesis of immunoadsorbent columns, this approach is useful for making a series of peptide analogues to determine the epitope specificity of the generated antipeptide antibodies. Peptides of varying length or with exhaustive amino acid substitutions at each position in the sequence can be prepared readily by the multiple synthetic method. A series of such peptide analogues allows analysis of antigenicity by epitope mapping to the level of single amino acid residues.[6,7,45,156,158,159]

Our synthesis and coupling approaches to the generation of a peptide immunogen representing the phosphorylation site in L-type pyruvate kinase were formulated before the exact position of this sequence at the NH_2-terminus of the protein was known. In an attempt to generate antipeptide antibodies recognizing antigenic determinants within the sequence –Arg–Arg–Ala–Ser–Val–, the decapeptide Ala–Gly–Tyr–Leu–Arg–Arg–Ala–Ser–Val–Ala was synthesized on a standard Merrifield resin.[153] This peptide corresponds to the sequence of amino acids around the phosphorylatable seryl residue in L-type pyruvate kinase from rat liver.[134] After synthesis of the first nine residues of the peptide, a small aliquot of peptidyl resin was removed and used to make the decapeptide containing [³H]alanine at its NH₂-terminus. Boc-[³H]alanine was prepared at a specific activity of 41 mCi/mmol for this purpose.[152] The bulk of the peptidyl resin was used to complete the synthesis of the nonradiolabeled decapeptide. After cleavage, deprotection, and purification of the peptides, 1 part of tritium-labeled decapeptide was mixed with 24 parts of nonradiolabeled decapeptide. The COOH-terminus of the decapeptide was then blocked by conversion to the methyl ester.[160] This preparation of the pyruvate kinase decapeptide methyl ester was covalently bound to derivatized carrier protein via its NH₂-terminus using a carbodiimide as the coupling agent (see Section V.A).

V. PREPARATION OF ANTIPEPTIDE ANTIBODIES

A. PEPTIDE-PROTEIN CONJUGATE FORMATION AND IMMUNIZATION

Small peptides are antigenic, but usually are not highly immunogenic. In some instances, however, antibodies have been generated successfully against free peptides as short as six or seven residues.[161] Most investigators still choose to covalently couple small synthetic peptides to large carrier molecules to produce consistently immunogenic conjugates.[4,162,163] In most cases, synthetic peptide haptens are conjugated to a carrier protein. However, newer adjuvants include synthetic lipids and the immunomodulatory glycopeptide N-acetylmuramyl-L-alanyl-D-isoglutamine.[10,162-164] In addition, some investigators have immunized animals with deprotected peptides that are still covalently coupled to the polymeric beads of the solid-phase resin on which they were synthesized.[37]

Numerous proteins have been used as carriers for conjugate formation with synthetic peptides.[10] These proteins are usually chosen to be immunogenic per se in the animal species (typically rabbit or mouse) used for antibody production. In our own work and that of others, keyhole limpet hemocyanin, bovine thyroglobulin, and bovine serum albumin have proved useful. Other commonly used carrier proteins include immunoglobulins, ovalbumin, β-galactosidase, and synthetic amino acid polymers.[10] Bovine serum albumin is not recommended as a carrier if the antipeptide antibodies will be used in cultured cell systems because antibodies directed against serum proteins may be present and confound experimental results.[1]

Either peptides can be coupled directly to carrier proteins or the carrier proteins can be derivatized with linkers or spacers prior to coupling. We have derivatized the amino groups of carrier proteins using succinic anhydride[165] so that they contain only free carboxyl groups which are available for coupling to peptide amino groups. This procedure tends to reduce polymerization of carrier proteins during coupling reactions catalyzed by carbodiimides. Other approaches are available for derivatization of carrier proteins. Adipic dihydrazide-derivatized proteins provide a carrier with a nucleophilic spacer to form peptide-protein conjugates readily.[21,29]

The synthetic peptide should be coupled to the carrier protein by a stable bond in a chemically well-defined manner. Coupling of the peptide to carrier protein alters its chemical structure at the site of attachment and also potentially sterically hinders it. A coupling strategy should be chosen that does not alter or obscure important determinants of the desired immunogenic epitope. A well-conceived coupling strategy may increase the specificity of antibodies toward one epitope in the synthetic peptide. Numerous protein carriers and coupling chemistries have been used successfully to produce antipeptide antibodies.[1] Table 1 lists chemical reagents commonly used for coupling of peptides to carrier proteins and indicates the reactive functional groups involved.[5,163] In our studies the water-soluble carbodiimide 1-ethyl-3-(3-dimethylaminopropyl) carbodiimide was used to couple decapeptide methyl ester to either succinylated keyhole limpet hemocyanin, succinylated bovine thyro-

TABLE 1
Reagents for Covalent Attachment of Synthetic Peptides to Carrier Proteins[a]

Coupling reagent	Functional groups bonded
Glutaraldehyde or glyoxal	Amino to amino
Carbodiimides	Amino to carboxyl
Maleimidobenzoyl-N-hydroxysuccinimide	Sulfhydryl to amino
bis-Diazobenzidine	Tyrosine to tyrosine

[a] Refer to reviews by Walter and Doolittle[5] and Benoit et al.[163] for details.

TABLE 2
Coupling Densities of Decapeptide-Carrier Protein Conjugates

Conjugate[a]	Coupling density	
	mol peptide/mol protein	mol peptide/10^4 Da
Succinylated keyhole limpet hemocyanin decapeptide methyl ester	141	0.47
Succinylated bovine thyroglobulin decapeptide methyl ester	40	0.60
Bovine serum albumin decapeptide methyl ester	3.8	0.56

[a] [^{3}H]Ala–Gly–Tyr–Leu–Arg–Arg–Ala–Ser–Val–Ala methyl ester (specific activity 1.6 mCi/mmol) was coupled to the indicated carrier proteins at concentrations of 10 to 45 mg/ml in 10 mM sodium phosphate buffer, pH 7.0, and 150 mM NaCl. The ratios of carrier protein to decapeptide were 1:1000 for succinylated keyhole limpet hemocyanin, 1:200 for succinylated bovine thyroglobulin, and 1:13 for bovine serum albumin. An excess of 1-ethyl-3-(3-dimethylaminopropyl)carbodimide was added and the reaction allowed to proceed for 30 min at room temperature and then overnight at 4°C. After extensive dialysis of the reaction mixture, protein and radioactivity were quantitated.

globulin, or underivatized bovine serum albumin. Other less commonly employed coupling methods are also available. These include using toluene diisocyanate or difluorodinitrobenzene as coupling reagents,[166,167] coupling by the mixed anhydride procedure,[168] and formation of hydrazone bonds.[169] A wide range of coupling densities has been used in forming peptide-carrier protein conjugates. The coupling densities that we obtained are listed in Table 2 and are typical of literature values. Because carrier proteins differ in size, the coupling densities of different conjugates vary greatly when expressed as moles of peptide per mole of carrier protein. When normalized to mass of carrier protein, most conjugates have coupling densities of approximately 1 mol of peptide per 10,000 Da. The pyruvate kinase phosphorylation site decapeptide has a mass of about 1,000 Da. Therefore, these decapeptide immunogenic conjugates present a relatively high density of peptide on the surface of the carrier protein for induction of antipeptide antibodies.

It is sometimes useful to prepare several conjugates of the same peptide using different carrier proteins. In many cases, after immunizing an animal with one peptide-carrier protein conjugate, a boosting immunization with the same peptide conjugated to another carrier protein will enhance the formation of peptide-specific antibodies rather than an immune response against the carrier proteins.[4] Also, several different conjugates of the same peptide are useful for screening antipeptide sera (see Section V.B).

Antipeptide antibodies that react with the same sequence in a target protein have been generated successfully as either polyclonal antisera in rabbits or as monoclonal antibodies after immunization of mice. Each approach has advantages and disadvantages. The former technique is easy and does not require the large amount of screening sometimes necessary for the monoclonal method. Also, polyclonal antisera may have higher average affinities for binding of antigen than do corresponding monoclonal antibodies. Polyclonal antisera usually contain a mixed population of antibodies against different conformations of the same epitope or against several epitopes on the synthetic peptide immunogen. However, monospecific antibodies many times can be purified from polyclonal antisera by immunoaffinity chromatography using appropriate synthetic peptides. The monoclonal antibody method directly supplies antibodies of a pure immunoglobulin population directed against a single conformation of an antigenic epitope.[170] Of course, these antibodies are obtained in larger amounts than similar immunoglobulins in polyclonal antisera. Also, production of specific monoclonal antibodies does not require highly purified immunogen if a suitable antigen is available for screening. In the present study, polyclonal antisera against the synthetic phosphorylation site decapeptide sequence or native L-type pyruvate kinase (dephosphorylated form) were generated in rabbits. The desired antibodies were purified from the resulting antisera by immunoaffinity chromatography as described in Section VI.A.

Peptide-carrier protein conjugates usually are available in quantities to allow immunization and boosting of several animals by any of the wide variety of available techniques.[5,163] It is wise to immunize more than one animal with each peptide conjugate to avoid false negatives if an individual animal is a poor responder. Sampling of blood to obtain preimmune serum controls must be remembered. The absolute titers of antipeptide antibodies in immune sera, the times to peak titer, and the boosting needs vary.[1,4,5] Some peptide-protein conjugates are highly immunogenic; others need a number of booster injections over several months to generate suitable titers. Titers of antipeptide antibodies can be greater than or less than titers for antibodies generated against intact proteins.[1]

B. SCREENING OF ANTISERA

Screening of antipeptide sera or hybridoma clones is facilitated by preparing conjugates of a single peptide antigen with at least two different carrier proteins. Antisera are then cross-screened against the conjugate containing the heterologous carrier protein so that only antipeptide and not anti-carrier protein antibodies are detected. In our studies, sera from animals immunized with one decapeptide-carrier protein conjugate (succinylated keyhole limpet hemocyanin decapeptide) were screened against the other conjugate (succinylated bovine thyroglobulin decapeptide). Since some antisera contained antisuccinate antibodies, the underivatized bovine serum albumin-decapeptide conjugate was also useful in screening. Alternatively, anti-carrier protein antibodies can be removed from antiserum, prior to screening, by batchwise treatment with carrier protein that has been immobilized on agarose beads. A high titer of antibodies directed against carrier proteins or derivatized carrier proteins can partially mask the presence of the desired antipeptide antibodies in some immune assays.

It is most important to screen antipeptide antibodies against the protein that contains that peptide sequence if what is really desired are antibodies that react with that epitope in the intact protein. As discussed above, antipeptide antibodies probably will display different protein-reactive activities depending on whether the protein used in screening is in its native conformation or is denatured. This in turn depends on the type of immunoassay employed.

Solid-phase immune assays are rapid and sensitive, and they allow quantitation of immunoreactivity of antisera or of antibodies secreted from hydridoma clones. Standard ELISAs or solid-phase assays using [^{125}I]protein A as an antibody probe are used for initial screening.[171,172] Either the peptide-carrier protein conjugate or the intact protein containing the peptide sequence is adsorbed nonspecifically to the plastic wells of microtiter plates. Typical controls in solid-phase screening assays are either preimmune serum or antiserum pretreated with an excess of the peptide originally used for immunization. Antipeptide antisera should be screened initially at different dilutions in case the antibodies are more reactive against the peptide than against the protein antigen. Antisera from individual bleedings of the same animal should be screened separately to assess titers, which may vary appreciably. Antisera with similar titers from different bleedings can be pooled after screening if larger amounts are needed for purification of the desired antipeptide antibodies.

Other useful immune assays include immunoprecipitation of protein antigen from solution followed by sodium dodecyl sulfate polyacrylamide gel electrophoresis (SDS-PAGE) of the immune complexes, Ouchterlony immunodiffusion, and immune overlay of protein antigens blotted onto solid supports after separation by gel electrophoresis (Western blotting).[97,173] These methods are not particularly well suited for screening large numbers of samples. The former methods detect immunoreactivity against a protein antigen in its native conformation, while the latter technique detects protein that is denatured by SDS and heating. Liquid-phase RIAs in which peptide or protein antigen displaces a radioiodinated peptide from the antibody are probably the best approach for accurately quantitating the reactivity of antipeptide antibodies against native proteins that contain the targeted epitope. Antigen-antibody complexes formed in the RIA are readily separated from free antigen with preparations of heat- or chemically inactivated *Staphylococcus aureus* rich in protein A.[174] We have used this method to assess reactivities of the purified antidecapeptide antibodies against the site of phosphorylation in L-type pyruvate kinase. The liquid-phase RIA also has been used with a series of synthetic peptides to map the epitope recognized by the antidecapeptide antibodies.

VI. PURIFICATION AND CHARACTERIZATION OF ANTIBODIES

A. IMMUNOAFFINITY CHROMATOGRAPHY

Antibodies against either synthetic peptide conjugates or native substrate proteins can be purified conveniently from antisera or hybridoma clones by immunoaffinity chromatography on columns of immobilized haptogenic peptides.[98] This procedure has the advantages of producing a substantial purification in a single step, suitability to both small and large scales, and enrichment of the antibodies of desired specificity. Antibodies from an antipeptide immune serum sometimes can be subfractionated on columns containing smaller peptides corresponding to defined portions of the larger peptide antigen that was initially used for immunization. Immunoaffinity chromatography is also useful in identifying and isolating cellular proteins that are the targets of purified epitope-specific antibodies.[105,114,116-118] In this case, the purified antibodies are covalently linked to an immobilized support on which cellular or tissue extracts are then chromatographed.

For purification of antipeptide antibodies, synthetic peptides must be immobilized on agarose or other solid supports in such a way that their antigenic epitopes are not functionally altered or sterically hindered. In most cases, the original synthetic peptide is simply coupled in the identical orientation through the same chemically reactive group as it was in forming the peptide-carrier protein conjugate used for immunization. Numerous coupling chemistries are commercially available for coupling of peptides or antibodies to immobilized supports.[98,175] Agarose beads have been derivatized with a wide variety of functional groups or reactive moieties. These include cyanogen bromide-, epoxy-, sulfonate-, and hydrazide-activated agarose as well as agarose containing hydroxysuccinimide esters. These groups react covalently with nucleophilic groups of peptides or proteins. Agarose preparations containing either amino or carboxyl groups are available for coupling reactions catalyzed by carbodiimides.[69] Also, agarose and other solid supports that contain certain sulfhydryl functionalities will react with cysteine residues in peptides and antibodies. A variety of spacer arms are available to place the peptide or protein at different distances from the polymeric support. Our synthetic decapeptide methyl ester was coupled through its amino

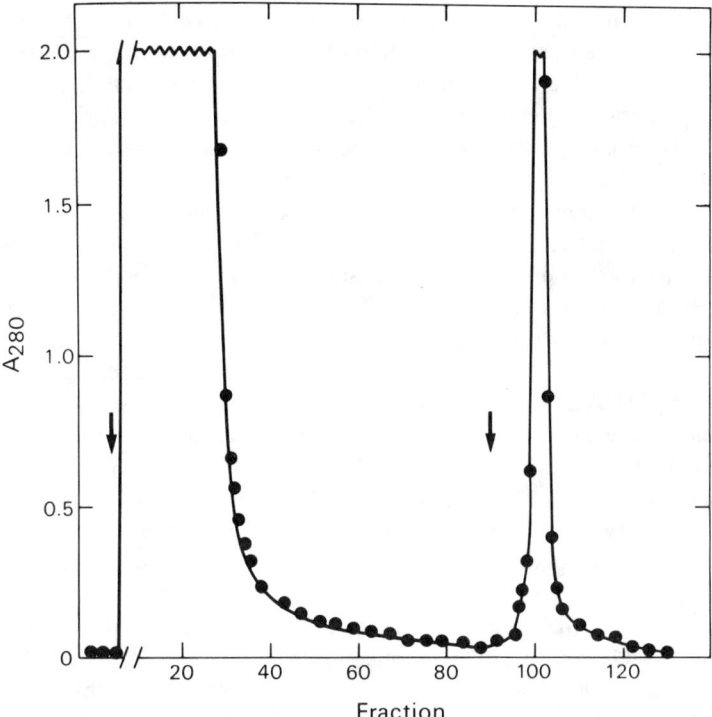

FIGURE 3. Purification of antidecapeptide antibodies from unfractionated antiserum by immunoaffinity chromatography. At the first arrow, antiserum from a rabbit immunized with succinylated bovine thyroglobulin-decapeptide conjugate was applied to a column of decapeptide covalently attached to CH-Sepharose® 4B. The column was washed with phosphate-buffered saline, pH 7.3, until the absorbance at 280 nm returned to baseline. At the second arrow, adsorbed protein was eluted with 100 mM citrate buffer, pH 2.2, and subsequently was neutralized, pooled, concentrated, and dialyzed.

group to the carboxyl groups on the ends of the six-carbon spacer arms of CH-Sepharose® 4B by a carbodiimide-catalyzed reaction.

Antibodies that become adsorbed specifically to immobilized peptide antigens can be eluted with chaotropic salts, changes in pH, or by denaturation with reagents such as guanidine hydrochloride. Care must be taken to remove the eluting agent rapidly from the purified antibodies and to concentrate and stabilize the antibodies as soon as possible. Antipeptide antibodies also can be eluted from peptide immunoadsorbent columns with an excess of free peptide.[1] The ease of removal of bound peptide from the antibodies depends on their affinities for the peptide. Peptide-antibody complexes may have to be denatured transiently and dialyzed to effect complete removal. Free peptide in solution also can be used under mild conditions to elute cellular proteins from immunoaffinity supports made of antipeptide antibodies.

The purification of antidecapeptide antibodies directed against the site of phosphorylation in L-type pyruvate kinase was accomplished by immunoaffinity chromatography on a column of immobilized decapeptide, as shown in Figure 3. Pooled antisera from one rabbit immunized with succinylated bovine thyroglubulin decapeptide was recirculated through a decapeptide-agarose column overnight, and nonspecifically bound serum proteins were removed by washing with saline buffered at pH 7.3. Specific antidecapeptide antibodies were eluted with citrate buffer, pH 2.2, and fractions of the eluate containing protein were immediately pooled, neutralized, dialyzed against buffered saline, and concentrated. The immunoreactivities of the initial pooled antiserum, the flow-through serum from the column, and the affinity-purified antibodies were tested in a solid-phase RIA against both decapeptide-carrier protein and L-type pyruvate kinase. The results indicated that antibodies directed against the decapeptide and cross-reactive toward L-type pyruvate kinase were removed from the antiserum nearly quantitatively by the immunoaffinity column and recovered as purified immunoglobulin. On the other hand, anti-carrier protein and antisuccinate antibodies were not retained by the column at all. In the alternate experimental approach, a subpopulation of anti-pyruvate kinase antibodies directed against the site of phosphorylation in that protein was purified from a crude antiserum of an animal immunized with L-type pyruvate kinase by a similar immunoaffinity chromatography protocol.

B. ANTIBODY CHARACTERIZATION

Standard immunological and protein chemistry techniques are used to assess the purity of the monospecific antibodies, to determine their immunoglobulin type and subtype, and to quantitate their immune reactivities. As controls for affinity-purified antipeptide antibodies, the corresponding types of immunoglobulins are purified from preimmune sera.

The purified antidecapeptide antibodies that also recognized determinants in the dephosphorylated form of L-type pyruvate kinase were highly purified immunoglobulins as determined by SDS-PAGE (data not shown). As depicted in Figure 4, these antibodies were immunoglobulin G (IgG) as assessed by immunodiffusion. The yield of purified antipeptide antibodies was 15 mg of IgG from 34 ml of pooled antidecapeptide antiserum. The recovery of immune reactivities of the purified antibodies toward decapeptide and L-type pyruvate kinase was compared to those present in the pooled antiserum by solid-phase RIA of appropriate dilutions of the crude and purified materials. Recoveries of reactivities against the decapeptide and against L-type pyruvate kinase were 76 and 67%, respectively, indicating that the same population of antibodies in the immune serum recognized both of these antigens. As determined by Ouchterlony immunodiffusion (Figure 4), neither the antidecapeptide antiserum nor the affinity-purified antidecapeptide antibodies immunoprecipitated L-type pyruvate kinase, although both recognized it in other immune assays. Therefore, these antibodies are nonprecipitating. Also, the purified antidecapeptide antibodies completely prevented the phosphorylation of free decapeptide by the catalytic subunit of the cAMP-dependent protein kinase. In a standard protein kinase assay, 18 μg of pure antidecapeptide IgG caused 50% inhibition of decapeptide phosphorylation; this corresponds to approximately 240 pmol of antibody combining sites preventing 160 pmol of decapeptide from being phosphorylated. This result and the recovery data above indicate that the antidecapeptide antibodies were not irreversibly denatured during purification and that they did retain the majority of their reactivity. These antibodies were less potent in inhibiting the phosphorylation of L-type pyruvate kinase by the cAMP-dependent protein kinase, probably because their affinity for native pyruvate kinase is similar to that of the protein kinase for this protein as a substrate (see Section VII.A).

When the antiserum from one rabbit that was immunized with L-type pyruvate kinase was subjected to immunoaffinity chromatography on immobilized peptide, the flow-through antiserum retained 90% of its reactivity toward L-type pyruvate kinase, but less than 20% of its reactivity toward the decapeptide. Therefore, the majority of those anti-pyruvate kinase antibodies that recognized antigenic determinants in the phosphorylation site decapeptide were bound to the column, and these were a small subpopulation of the total anti-pyruvate kinase antibodies. The recovery of reactivities toward pyruvate kinase in the purified antibodies eluted from the column (relative to the reactivity of the crude antiserum) was 11%. Recovery of reactivity toward the decapeptide, however, was 150%. These data indicated that immunoaffinity chromatography resulted in the selective purification of a subpopulation

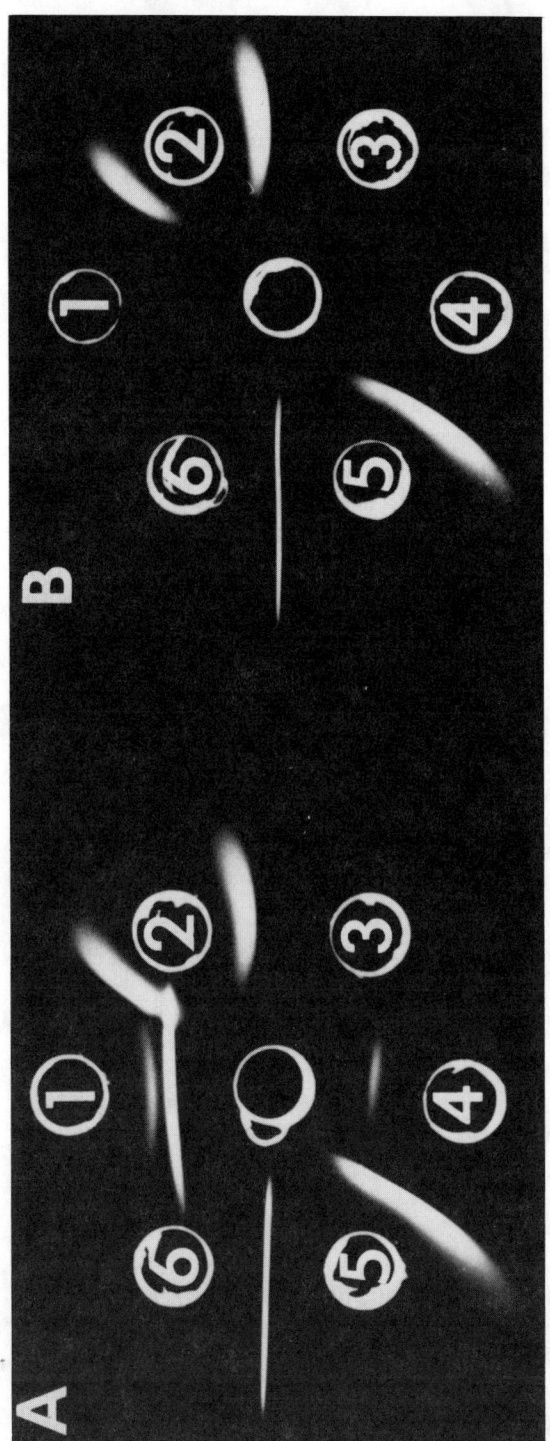

FIGURE 4. Reactivities of antidecapeptide immune serum and purified antidecapeptide antibodies as analyzed by Ouchterlony immunodiffusion. Panel A: the central well contained succinylated bovine thyroglobulin-decapeptide conjugate. The peripheral wells contained unfractionated immune serum (well 1), goat anti-rabbit IgG (wells 2 and 5), unfractionated preimmune serum (well 3), affinity-purified antidecapeptide antibodies (well 4), and IgG purified from preimmune serum (well 6). Panel B: the central well contained L-type pyruvate kinase purified from rat liver, and the peripheral wells contained the same reagents as in panel A.

of anti-pyruvate kinase antibodies, namely antibodies specific for the phosphorylation site in the enzyme. The yield of these anti-phosphorylation site antibodies was much lower than the antipeptide antibodies, about 0.6 mg total protein, and they were also IgG. Although the unfractionated immune serum precipitated L-type pyruvate kinase, the purified anti-phosphorylation site antibodies were nonprecipitating, as were the purified antidecapeptide antibodies.

VII. SPECIFICITY OF ANTI-PHOSPHORYLATION SITE ANTIBODIES

A. ANTIPEPTIDE ANTIBODIES THAT RECOGNIZE THE PHOSPHORYLATION SITE EPITOPE

All eight rabbits immunized with either of the decapeptide-carrier protein conjugates produced antidecapeptide antibodies. The antisera of four out of eight animals also recognized intact L-type pyruvate kinase in initial solid-phase RIA screens at antisera dilutions of 1:1000. The reactivities of each of these antipeptide antisera against L-type pyruvate kinase were less than against the decapeptide antigen, indicating that the titers of protein-reactive antibodies were somewhat lower than those antibodies that recognized decapeptide. Therefore, the four animals whose antipeptide antisera did not react with pyruvate kinase at dilutions of 1:1000 still could have produced protein-reactive antipeptide antibodies, but at a titer too low to be detected in the screening assay. Antidecapeptide antibodies from the antisera that reacted best against L-type pyruvate kinase were immunoaffinity purified as described in Section VI.A.

The reactivities of the purified antidecapeptide antibodies against three different antigens in the solid-phase RIA are illustrated in Figure 5. These purified antibodies were concentrated about sevenfold with respect to the volume of antiserum from which they were isolated. Therefore, a 1:690 dilution of purified IgG corresponds to a 1:100 dilution of unfractionated antidecapeptide antiserum. Reactivity against L-type pyruvate kinase was less than against the decapeptide antigen. The M-type isozyme of rat pyruvate kinase is not phosphorylated by cAMP-dependent protein kinase because it structurally lacks the L-type enzyme's NH_2-terminal 12 amino acids that contain the phosphorylation site.[135] Because the antidecapeptide antibodies are directed against antigenic determinants within the phosphorylation site in L-type pyruvate kinase, it was of interest to determine if these antibodies displayed cross-reactivity toward the M-type isozyme. As shown in Figure 5, the purified antidecapeptide antibodies did not recognize M-type pyruvate kinase at dilutions with which they reacted well with the L-type isozyme. IgG purified from preimmune serum displayed no reactivity with any of the antigens. Panel C of Figure 6 is a Western blot in which purified antidecapeptide antibodies were reacted in an immune overlay assay with L- and M-type pyruvate kinases blotted from an SDS-polyacrylamide gel electrophoretogram. Again, the antidecapeptide antibodies only recognized the L-type isozyme. These results indicate that the antidecapeptide antibodies do indeed react with the phosphorylation site epitope that is present only in L-type pyruvate kinase.

A liquid-phase RIA with [^{125}I]decapeptide was used to analyze the reactivities of the affinity-purified antidecapeptide antibodies more quantitatively. Displacement of radioiodinated decapeptide from the antibody by nonradiolabeled decapeptide in either its dephosphorylated or stoichiometrically phosphorylated form is shown in Figure 7. The 100-fold greater reactivity of the decapeptide vs. the phosphodecapeptide indicates that the antidecapeptide antibodies are quite selective for the dephosphorylated form of this antigen and that an unmodified seryl residue is a major determinant of antigenicity in the peptide. The reactivities of the purified antidecapeptide antibodies against L-type pyruvate kinase and a series of peptide analogues of the decapeptide antigen were determined in the liquid-phase

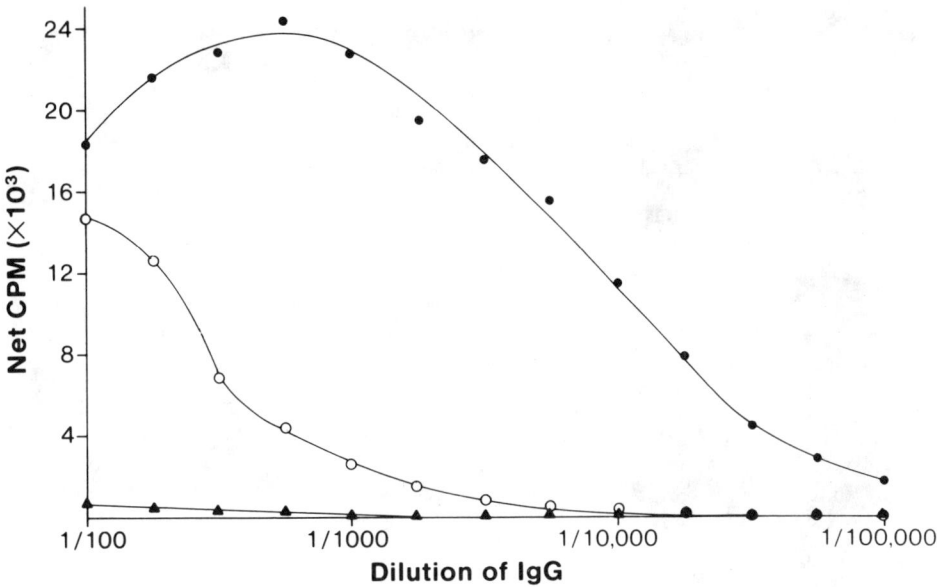

FIGURE 5. Reactivities of various dilutions of purified antidecapeptide antibodies in a solid-phase radioimmunoassay. Antigens were decapeptide (●), L-type pyruvate kinse (○), and M-type pyruvate kinase (▲). Antibodies reacting with each of the antigens were detected with [^{125}I]-labeled protein A.

RIA. These data further map the contribution of individual amino acids to recognition of the antigenic epitope and are presented in Table 3. The antibodies recognized the N^α-acetylated decapeptide, the decapeptide methyl ester, and the doubly blocked decapeptide about as well as the decapeptide itself, indicating that they were directed more to the internal portion of the peptide than to either the NH$_2$- or COOH-terminal amino acids. The N^α-acetylated decapeptide methyl ester most closely resembles the original immunogen used to raise the antibodies.

A series of short synthetic peptides corresponding to the COOH-terminal part of the decapeptide were used to evaluate the contribution of the arginyl residues as antigenic determinants. The peptide Ala–Ser–Val–Ala, which is missing the two basic amino acids, was not recognized by the antibodies. The pentapeptide Arg–Ala–Ser–Val–Ala cross-reacted, but a 1400-fold greater concentration as compared to the decapeptide was required for displacement of radiolabeled ligand. The hexapeptide Arg–Arg–Ala–Ser–Val–Ala, which contains both arginyl residues, cross-reacted to a much greater degree. The addition of leucine to this peptide gave the heptapeptide Leu–Arg–Arg–Ala–Ser–Val–Ala, which reacted nearly as well as the decapeptide. These results indicate that the antidecapeptide antibodies recognize the arginyl residues. Also, it is clear the the NH$_2$-terminal three amino acids in the decapeptide do not contribute greatly as antigenic determinants.

The contribution of the COOH-terminal –Val–Ala sequence to recognition by the antibodies was shown by the fact that the pentapeptide Leu–Arg–Arg–Ala–Ser did not cross-react in the assay. In addition, the synthetic peptide substrate Kemptide was examined. The sequence of Kemptide is Leu–Arg–Arg–Ala–Ser–Leu–Gly, which corresponds closely to the sequence in porcine L-type pyruvate kinase.[133,134] Kemptide differs in structure from the COOH-terminal heptapeptide analogue of the rat decapeptide sequence by a substitution of –Leu–Gly for the –Val–Ala at the COOH-terminus. Despite the fact that the heptapeptide and Kemptide both contain the required –Arg–Arg–Ala–Ser– antigenic determinant, Kemptide was 450-fold less reactive with the antibodies. This large difference indicates that the COOH-terminal –Val–Ala in the decapeptide is an important determinant of antigenicity that is recognized by the antidecapeptide antibodies.

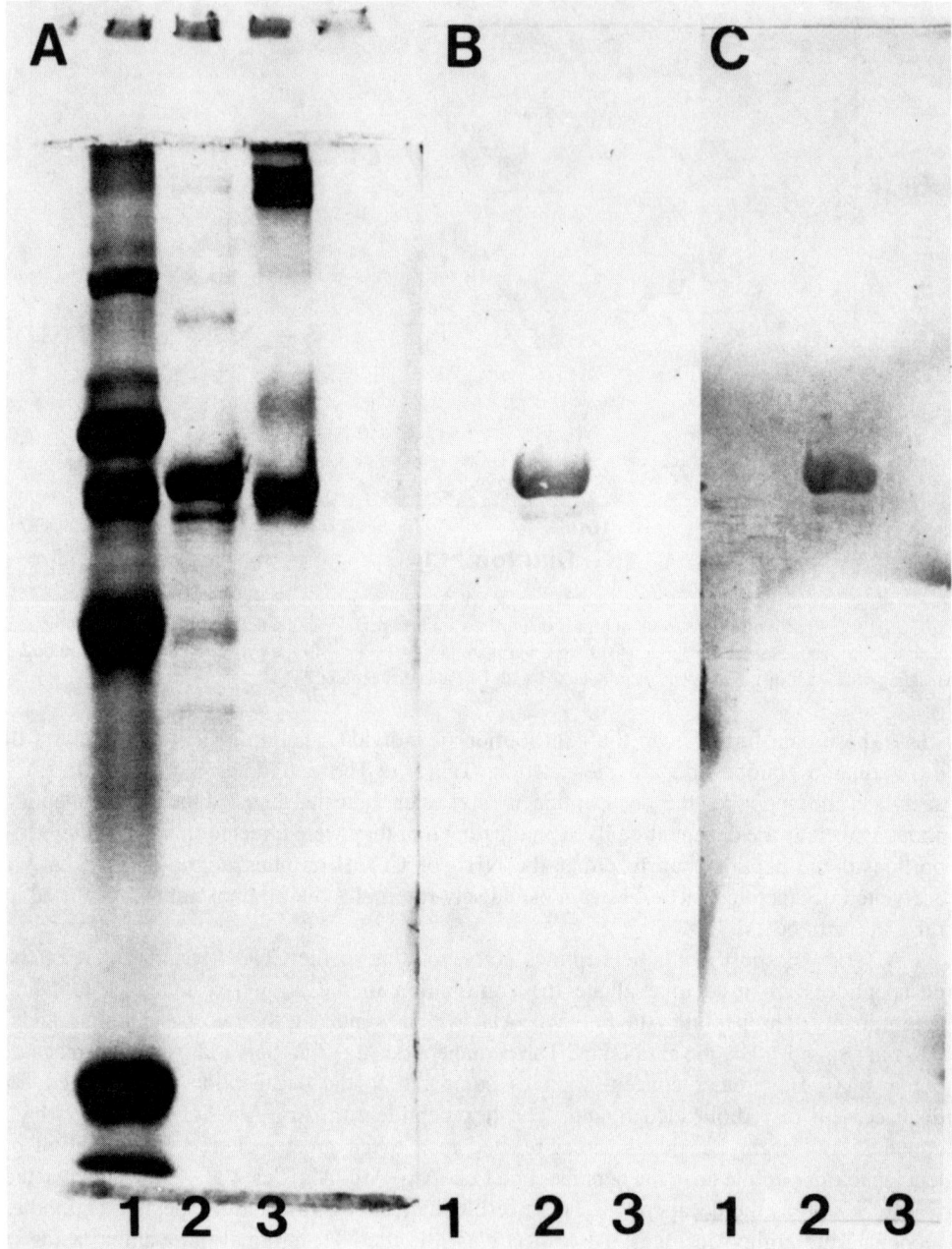

FIGURE 6. Immune overlay of protein blots of L- and M-type pyruvate kinases with purified anti-phosphorylation site antibodies and purified antidecapeptide antibodies. Panel A: Coomassie brilliant blue staining of blotted proteins. Panel B: immune overlay of blotted proteins with purified anti-phosphorylation site antibodies. The blot was probed with a 1:100 dilution of anti-phosphorylation site antibodies. Bound antibody was detected with an enzyme-linked second antibody, horseradish peroxidase-labeled goat anti-rabbit IgG, after incubation with the peroxidase substrates H_2O_2 and diaminobenzidine. Panel C: immune overlay of blotted proteins with purified antidecapeptide antibodies. The blotting was performed with a 1:200 dilution of antidecapeptide antibodies in a manner identical to that described for panel B. In all panels, protein molecular weight standards and L- and M-type pyruvate kinases were electroblotted onto Zeta-Probe membranes from an SDS-polyacrylamide gel electrophoretogram. Lane 1 contained the following protein standards: β-galactosidase (130 kDa), bovine serum albumin (68 kDa), catalase (60 kDa), ovalbumin (43 kDa), and trypsin inhibitor (21.5 kDa). Lane two contained 5 μg of L-type pyruvate kinase, and lane 3 contained 5 μg of M-type pyruvate kinase.

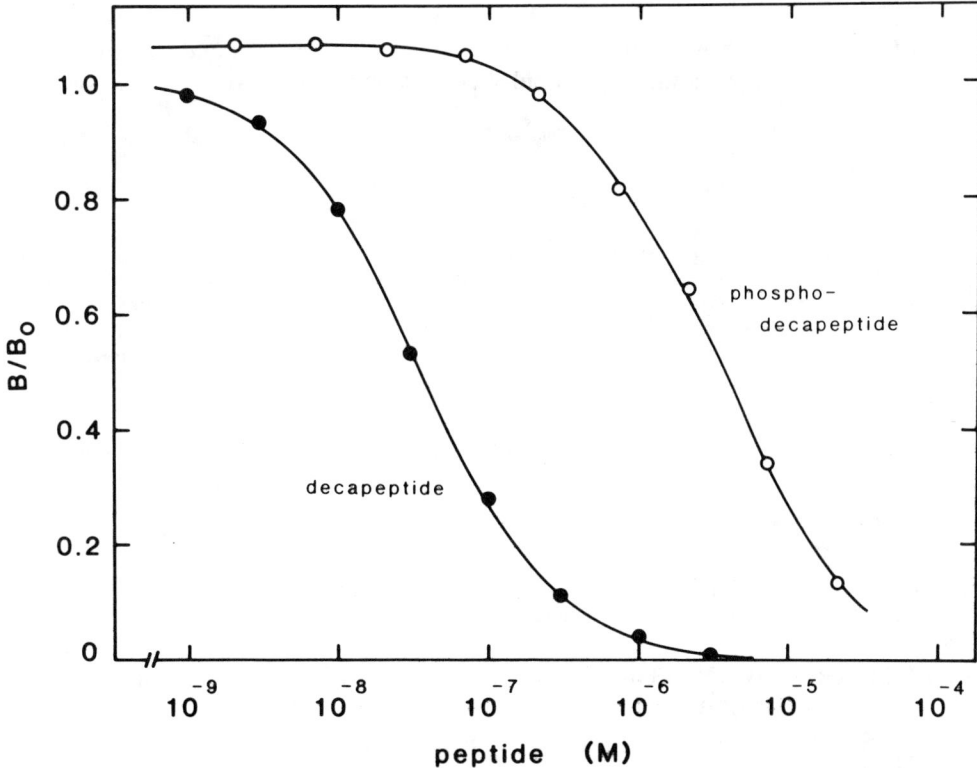

FIGURE 7. Competitive displacement of [^{125}I]-labeled decapeptide from purified antidecapeptide antibodies by peptides in a liquid-phase radioimmunoassay. Competing peptides were decapeptide in its dephosphorylated form (●) and stoichiometrically phosphorylated decapeptide (○).

These epitope mapping studies with synthetic peptides showed that the antibodies recognized the –Leu–Arg–Arg–Ala–Ser–Val–Ala sequence. These, presumably, are the antigenic determinants that the antibodies recognize in L-type pyruvate kinase. As shown in Table 3, the dephosphorylated form of intact L-type pyruvate kinase also reacted in the liquid-phase RIA, but the average affinity of the antidecapeptide antibodies for this antigen was appreciably less than for the decapeptide itself. The phosphorylated form of L-type pyruvate kinase was recognized less well than was dephosphorylated pyruvate kinase, but the concentration causing 50% inhibition (IC$_{50}$) could not be determined accurately due to limiting concentrations of this antigen.

The potential cross-reactivity of the antidecapeptide antibodies with other purified protein substrates of cAMP-dependent protein kinase was analyzed in three different immune assays. These substrates of cAMP-dependent protein kinase contain sites of phosphorylation of known sequence, all of which contain basic amino acids NH$_2$-terminal to the phosphorylatable seryl residue.[76-78] Many of these substrate proteins, including phosphorylase kinase, glycogen synthase, and 6-phosphofructo-2-kinase/fructose-2,6-bisphosphatase, contain the –X–Arg–Arg–X–Ser–X– sequence.[72,176] In solid- and liquid-phase RIAs there was a nearly complete lack of recognition of substrates of cAMP-dependent protein kinase other than L-type pyruvate kinase by the antidecapeptide antibodies (data not shown). As shown in Figure 8, the same result was obtained in a Western blot analysis in which the substrate proteins containing phosphorylation sites were denatured to maximize immune recognition of these linear sequences.

TABLE 3
Affinities of Purified Antidecapeptide Antibodies for Peptides and Proteins as Determined by Liquid-Phase Radioimmunoassay

Competing ligand	IC_{50}^a (μM)
Peptides	
Ala-Gly-Tyr-Leu-Arg-Arg-Ala-Ser-Val-Ala	0.035
Ala-Gly-Tyr-Leu-Arg-Arg-Ala-(PO_3)Ser-Val-Ala	3.6
Ala-Gly-Tyr-Leu-Arg-Arg-Ala-Ser-Val-Ala methyl ester	0.10
Acetyl-Ala-Gly-Tyr-Leu-Arg-Arg-Ala-Ser-Val-Ala	0.023
Acetyl-Ala-Gly-Tyr-Leu-Arg-Arg-Ala-Ser-Val-Ala methyl ester	0.038
Leu-Arg-Arg-Ala-Ser-Leu-Gly	100
Leu-Arg-Arg-Ala-Ala-Leu-Gly	4000
Ala-Ser-Val-Ala	No cross-reactivity
Arg-Ala-Ser-Val-Ala	50
Arg-Arg-Ala-Ser-Val-Ala	0.90
Leu-Arg-Arg-Ala-Ser-Val-Ala	0.22
Leu-Arg-Arg-Ala-Ser	No cross-reactivity
Proteins	
L-type pyruvate kinase (dephosphorylated)	3.3
Bovine serum albumin	No cross-reactivity

[a] This value is the concentration of competing peptide or protein that produces a 50% inhibition of binding of [^{125}I]decapeptide to purified antibody in a liquid-phase radioimmunoassay.

B. ANTI-PHOSPHORYLATION SITE ANTIBODIES FROM ANTI-PYRUVATE KINASE ANTISERA

All 12 rabbits immunized with L-type pyruvate kinase formed antibodies against this antigen. In solid-phase RIAs 6 out of 12 of these antisera recognized the decapeptide antigen, although the reactivities of 3 antisera were weak. Three of the anti-pyruvate kinase antisera reacted well with decapeptide, but at lower titers than against L-type pyruvate kinase itself. As described in Section VI.A, one of these antisera was subjected to immunoaffinity chromatography on an immobilized peptide to isolate a subpopulation of anti-pyruvate kinase antibodies that were directed against the phosphorylation site in the enzyme.

The reactivities of the purified anti-phosphorylation site antibodies to three antigens in the solid-phase RIA are illustrated in Figure 9. The purified IgG was concentrated 18-fold with respect to the volume of anti-pyruvate kinase antiserum from which it was isolated. The reactivities of the anti-phosphorylation site antibodies against L-type pyruvate kinase and decapeptide were of a similar magnitude. Also, like the antidecapeptide antibodies, these antibodies did not react with M-type pyruvate kinase. The lack of recognition of the M-type isozyme by the anti-phosphorylation site antibodies in a Western blot is depicted in Panel B of Figure 6. These results with the purified antibody were in distinct contrast to the reactivities displayed by the unfractionated anti-pyruvate kinase antiserum (data not shown). First, although the antiserum cross-reacted with decapeptide, the degree of reactivity toward the peptide was only about 10% of that against L-type pyruvate kinase. Second, the crude antiserum did recognize M-type pyruvate kinase, although its reactivity toward this antigen was somewhat less than that toward the original immunogen, L-type pyruvate kinase. Therefore, as expected, at least some of the antibodies in the heterogeneous anti-pyruvate kinase antiserum were specific for epitopes that occur on both the L- and M-type isozymes. The lack of recognition of M-type pyruvate kinase by the purified anti-phosphorylation site antibodies and their greatly enhanced reactivity toward decapeptide indicate that this subpopulation of anti-pyruvate kinase antibodies are indeed directed against determinants contained within the phosphorylation site of the enzyme. In addition, this study demonstrates that at least one immunogenic determinant on L-type pyruvate kinase is the site of phosphorylation.

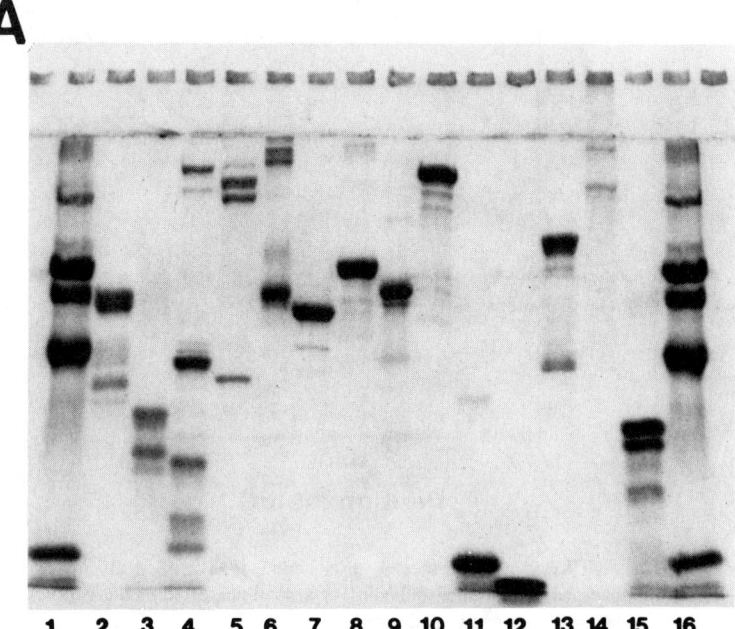

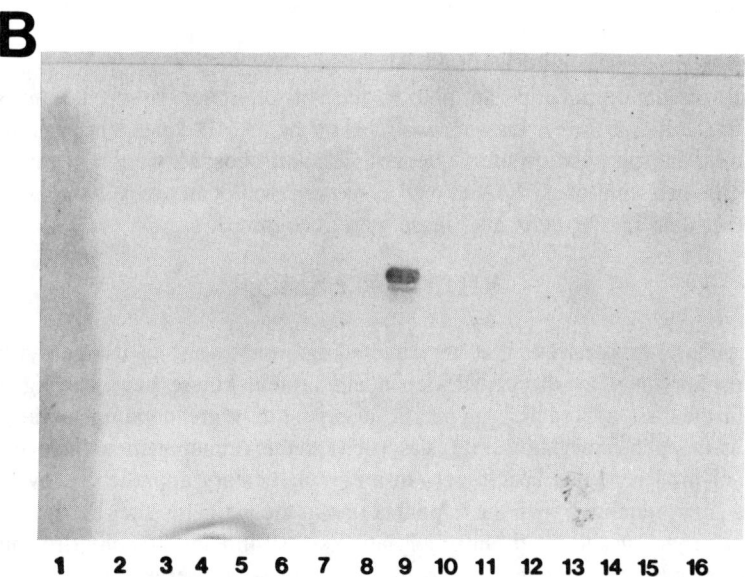

FIGURE 8. Immune overlay of protein blots of known substrates of cAMP-dependent protein kinase with purified antidecapeptide antibodies. Panel A: Coomassie brilliant blue staining of blotted proteins. Panel B: immune overlay of blotted proteins with purified antidecapeptide antibodies. The blot was probed with a 1:100 dilution of anti-decapeptide antibodies, and bound antibody was detected with horseradish peroxidase-labeled goat anti-rabbit IgG as described in the legend to Figure 6. In each panel, protein molecular weight standards and purified substrate proteins were electroblotted onto a Zeta-Probe membrane from an SDS-polyacrylamide gel electrophoretogram. Protein standards (lanes 1 and 16) were as indicated in the legend to Figure 6. Protein substrates of cAMP-dependent protein kinase (and several control proteins) were type II regulatory subunit of cAMP-dependent protein kinase (lane 2), casein (lane 3), cardiac troponin (lane 4), phosphorylase kinase (lane 5), M-type pyruvate kinase (lane 6), 6-phosphofructo-2-kinase/fructose-2,6-bisphosphatase (lane 7), bovine serum albumin (lane 8), L-type pyruvate kinase (lane 9), cardiac C-protein (lane 10), myelin basic protein (lane 11), histone H2B (lane 12), glycogen synthase (lane 13), type I collagen (lane 14), and mixed histone (lane 15). A total of 5 μg of protein were applied to each lane in the SDS-PAGE.

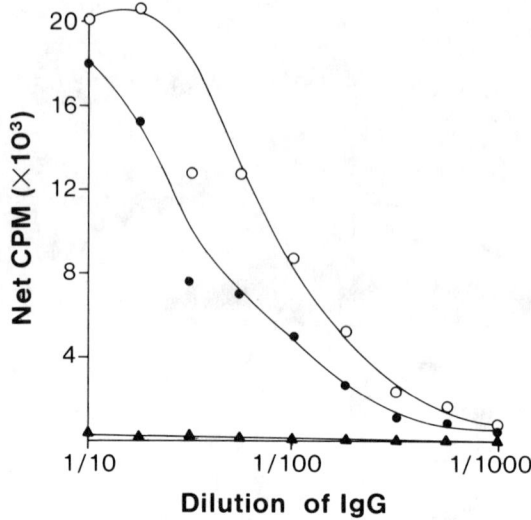

FIGURE 9. Reactivities of various dilutions of purified anti-phosphorylation site antibodies in a solid-phase radioimmunoassay. Antigens were decapeptide (●), L-type pyruvate kinase (○), and M-type pyruvate kinase (▲). The presence of antibodies reacting with each of the antigens was detected with [^{125}I]-labeled protein A.

As with the antidecapeptide antibodies, the anti-phosphorylation site antibodies were tested for recognition of other known substrates of the cAMP-dependent protein kinase that contain similar phosphorylation site sequences. The anti-phosphorylation site antibodies were highly specific in a solid-phase RIA as well as in immunoblot analyses. None of the substrate proteins other than L-type pyruvate kinase were recognized.

VIII. DISCUSSION

Two antibody preparations that are directed against the site in L-type pyruvate kinase that is phosphorylated by the cAMP-dependent protein kinase have been generated and purified. In the first approach, a synthetic decapeptide corresponding to the amino acid sequence at the phosphorylation site was used as the immunogen to induce antipeptide antibodies of predetermined specificity. In a complementary approach, L-pyruvate kinase was used as an immunogen with the hope that one of the antigenic sites on the protein would be the phosphorylation site. Anti-phosphorylation site antibodies did comprise approximately 10 to 15% of the antibodies in the anti-pyruvate kinase antiserum. This is typical of the immune response to a single epitope on a protein that contains multiple antigenic determinants.[137,177] To obtain antibodies of the desired specificity, the anti-pyruvate kinase antiserum was fractionated by immunoaffinity chromatography using an immobilized phosphorylation-site peptide.

Both the antidecapeptide antibodies and the anti-phosphorylation site antibodies are directed against the dephosphorylated form of the phosphorylation site. Epitope mapping of the antidecapeptide antibodies indicates that they recognize structures in the amino acid sequence -Leu-Arg-Arg-Ala-Ser-Val-Ala. The importance of the arginines as antigenic determinants in this epitope is demonstrated by the fact that one arginyl residue is essential for any recognition at all by the antidecapeptide antibodies and that two adjacent arginyl residues substantially increase immunoreactivity. These basic amino acids and the seryl residue are specificity determinants that also are recognized by the catalytic subunit of the cAMP-dependent protein kinase in its substrate proteins. This suggests that the antidecapeptide antibodies could be used to elicit anti-idiotype antibodies which would recognize the active site of the protein kinase if this interaction was not otherwise sterically hindered.

Although both antibodies clearly recognize L-type pyruvate kinase, as determined by solid- and liquid-phase RIAs and by immunoblotting, they are nonprecipitating toward this antigen. Precipitation of antigen-antibody complexes depends on the interaction of multivalent antigens with multivalent (and usually polyclonal) immunoglobulins to form a large, insoluble lattice of interconnected antigen and antibody molecules.[173] With respect to the purified antibodies in this study, L-type pyruvate kinase contains only four antigenic sites (i.e., phosphorylation sites) per tetrameric holoenzyme, one on each monomer. The affinity-purified antibodies are monospecific, reacting only with these four sites. Therefore, it is probable that the divalent IgG antibodies bind to two phosphorylation sites in the same pyruvate kinase molecule, thus preventing formation of a large antigen-antibody complex. Another possibility is that the antibodies cross-link different pyruvate kinase molecules, but only nonprecipitating dimers or trimers are formed. The fact that these antibodies are nonprecipitating toward L-type pyruvate kinase will better allow their use as probes of the structure and function of this highly regulated protein.

It was hoped that one or more of the antibodies would be a useful general reagent to probe sites of phosphorylation in numerous substrates of the cAMP-dependent protein kinase. If the antibodies cross-reacted among known substrate proteins, they could be used to detect yet uncharacterized substrate proteins in various cells and tissues. However, both the antidecapeptide and anti-phosphorylation site antibodies are extremely specific. Although these antibodies are directed against the sequence –Leu–Arg–Arg–Ala–Ser–Val–Ala, they do not recognize other known substrates of the protein kinase, many of which contain the sequence –X–Arg–Arg–X–Ser–X–. In many cases, investigators using the antipeptide antibody technique desire highly specific antibodies, but observe cross-reactivity with proteins unrelated to the target protein under study.[5] Why was greater cross-reactivity not observed in this study of anti-phosphorylation site antibodies?

One of two explanations is probable: either the phosphorylation sites in the different substrate proteins are in different conformations which cannot be recognized by the antibodies, or single amino acids in the X positions around the phosphorylatable seryl residues in substrate proteins are as critical for immune recognition as are the serine and arginines themselves. Reactivity of the antidecapeptide antibodies with pyruvate kinase was demonstrated in three different immunoassays. Because the extent of denaturation of the protein antigen differed greatly in each of these assays, it appears that recognition of L-type pyruvate kinase by the antibodies is not dependent on a particular structural conformation of the epitope in this protein. Also, the protein kinase can recognize different phosphorylation site sequences readily, and, if they do exist in highly variable conformations, the enzyme is able to induce their fit into the active site so that catalysis can occur. Most probably, the antidecapeptide and anti-phosphorylation site antibodies do not cross-react with other substrate proteins because of minor differences in the exact amino acids within the general concensus sequence at various phosphorylation sites.[76-79] The antidecapeptide antibodies lost two to three orders of magnitude in affinity when the COOH-terminal two residues in the peptide antigen underwent a conservative change from –Val–Ala to –Leu–Gly. It is well known that changes in a single amino acid in an antigenic site can have dramatic effects on the reactivity of antibodies toward that epitope.[6,45,156,158] Also, the frequency of cross-reactivity of antibodies with proteins containing similar sequences may depend on the total number of amino acids involved in a given epitope.[178] A high frequency of cross-reactivity is expected for antibodies recognizing four or fewer residues, but the expected frequency of cross-reactivity drops greatly for recognition of seven amino acids.[178] The antidecapeptide antibodies are directed against seven amino acids in the pyruvate kinase phosphorylation site sequence, and only three of these residues are constant in other substrate proteins.

The inability of the antidecapeptide and anti-phosphorylation site antibodies to serve as general immunological reagents for the study of substrates of the cAMP-dependent protein kinase certainly limits their usefulness. Nonetheless, these antibodies are being used to study the structure and function of L-type pyruvate kinase, including its state of phosphorylation, effect of phosphorylation on activity, and the conformational consequences of the binding of allosteric effectors of the enzyme.[131,179]

IX. CONCLUDING REMARKS

Synthetic peptides are invaluable tools in manipulating the immune system. The ability of peptides to act as immunological mimics of epitopes on cellular proteins allows the generation of antipeptide antibodies of defined specificity. These antibodies can then be used as immunological tools for the identification, purification, and investigation of the structures and functions of target proteins of interest. Among these proteins are certainly the numerous protein kinases and their substrates that undergo phosphorylation. Since sites of phosphorylation are on the surface of substrate proteins and the amino acid sequences at these sites of covalent modification are in many cases known or can be deduced from cloned cDNAs, antipeptide antibodies can be directed against these determinants. Techniques exist to develop antibodies that are specific for either the dephosphorylated or phosphorylated forms of phosphorylatable sites. These antibodies can be used to study substrates of protein kinases and also most protein kinases themselves, since these enzymes undergo autophosphorylation. Important topics for current and future research are the nature of the structural changes in a substrate protein that result from its phosphorylation and the mechanisms by which these covalent and conformational modifications alter the function of the substrate protein. Anti-phosphorylation site antibodies should prove useful in such investigations.

ACKNOWLEDGMENTS

The work from the author's laboratory was supported by U.S. Public Health Service Grant GM28144. We thank Drs. H. C. Hartzell for cardiac C-protein, J. F. Kuo for myelin basic protein, S. J. Pilkis for 6-phosphofructo-2-kinase/fructose-2,6-bisphosphatase, T. R. Soderling for glycogen synthase, T. D. Chrisman for phosphorylase kinase, and J. M. McPherson for type I collagen. We also thank Dr. R. B. Fritz for helpful advice on immune assays.

REFERENCES

1. **Lerner, R. A.**, Tapping the immunological repertoire to produce antibodies of predetermined specificity, *Nature (London)*, 299, 592, 1982.
2. **Green, N., Alexander, H., Olson, A., Alexander, S., Shinnick, T. M., Sutcliffe, J. G., and Lerner, R. A.**, Immunogenic structure of the influenza virus hemagglutinin, *Cell*, 28, 477, 1982.
3. **Niman, H. L., Houghten, R. A., Walker, L. E., Reisfeld, R. A., Wilson, I. A., Hogle, J. M., and Lerner, R. A.**, Generation of protein-reactive antibodies by short peptides is an event of high frequency: implications for the structural basis of immune recognition, *Proc. Natl. Acad. Sci. U.S.A.*, 80, 4949, 1983.
4. **Sutcliffe, J. G., Shinnick, T. M., Green, N., and Lerner, R. A.**, Antibodies that react with predetermined sites on proteins, *Science*, 219, 660, 1983.
5. **Walter, G. and Doolittle, R. F.**, Antibodies against synthetic peptides, *Genet. Eng.*, 5, 61, 1983.
6. **Alexander, H., Johnson, D. A., Rosen, J., Jerabek, L., Green, N., Weissman, I. L., and Lerner, R. A.**, Mimicking the alloantigenicity of proteins with chemically synthesized peptides differing in single amino acids, *Nature (London)*, 306, 697, 1983.
7. **Geysen, H. M., Barteling, S. J., and Meloen, R. H.**, Small peptides induce antibodies with a sequence and structural requirement for binding antigen comparable to antibodies raised against the native protein, *Proc. Natl. Acad. Sci. U.S.A.*, 82, 178, 1985.

8. **Ada, G. L.**, Ed., *Synthetic Peptides as Antigens,* Ciba Found. Symp. No. 119, John Wiley & Sons, New York, 1986.
9. **Walter, G.**, Production and use of antibodies against synthetic peptides, *J. Immunol. Methods,* 88, 149, 1986.
10. **Arnon, R.**, Ed., *Synthetic Vaccines,* Vol. I and II, CRC Press, Boca Raton, FL, 1987.
11. **Bittle, J. L., Houghten, R. A., Alexander, H., Shinnick, T. M., Sutcliffe, J. G., Lerner, R. A., Rowlands, D. J., and Brown, F.**, Protection against foot-and-mouth disease by immunization with a chemically synthesized peptide predicted from the viral nucleotide sequence, *Nature (London),* 298, 30, 1982.
12. **Gerlin, J. L., Alexander, H., Shih, J. W.-K., Purcell, R. H., Dapolito, G., Engle, R., Green, N., Sutcliffe, J. G., Shinnick, T. M., and Lerner, R. A.**, Chemically synthesized peptides of hepatitis B surface antigen duplicate the d/y specificities and induce subtype-specific antibodies in chimpanzees, *Proc. Natl. Acad. Sci. U.S.A.,* 80, 2365, 1983.
13. **Emini, E. A., Jameson, B. A., and Wimmer, E.**, Priming for and induction of anti-poliovirus neutralizing antibodies by synthetic peptides, *Nature (London),* 304, 699, 1983.
14. **Ballou, W. R., Rothbard, J., Wirtz, R. A., Gordon, D. M., Williams, J. S., Gore, R. W., Schneider, I., Hollingdale, M. R., Beaudoin, R. L., Maloy, W. L., Miller, L. H., and Hockmeyer, W. T.**, Immunogenicity of synthetic peptides from circumsporozoite protein of *Plasmodium falciparum, Science,* 228, 996, 1985.
15. **Zavala, F., Tam, J. P., Cochrane, A. H., Quakyi, I., Nussenzweig, R. S., and Nussenzweig, V.**, Rationale for development of a synthetic vaccine against *Plasmodium falciparum* malaria, *Science,* 228, 1436, 1985.
16. **Kennedy, R. C., Henkel, R. D., Pauletti, D., Allan, J. S., Lee, T. H., Essex, M., and Dreesman, G. R.**, Antiserum to a synthetic peptide recognizes the HTLV-III envelope glycoprotein, *Science,* 231, 1556, 1986.
17. **Robbins, K. C., Devare, S. G., Reddy, E. P., and Aaronson, S. A.**, In vivo identification of the transforming gene product of simian sarcoma virus, *Science,* 218, 1131, 1982.
18. **Giallongo, A., Appella, E., Ricciardi, R., Rovera, G., and Croce, C. M.**, Identification of the c-*myc* oncogene product in normal and malignant B cells, *Science,* 222, 430, 1983.
19. **Luka, J., Sternas, L., Jornvall, H., Klein, G., and Lerner, R.**, Antibodies of predetermined specificity for the NH_2 terminus of a cellular protein p53 react with the native molecule: evidence for the presence of different p53s, *Proc. Natl. Acad. Sci. U.S.A.,* 80, 1199, 1983.
20. **Suni, J., Narvanen, A., Wahlstrom, T., Aho, M., Pakkanen, R., Vaheri, A., Copeland, T., Cohen, M., and Oroszlan, S.**, Human placental syncytiotrophoblast Mr 75,000 polypeptide defined by antibodies to a synthetic peptide based on a cloned human endogenous retroviral DNA sequence, *Proc. Natl. Acad. Sci. U.S.A.,* 81, 6197, 1984.
21. **SenGupta, D. N., Kumar, P., Zmudzka, B. Z., Coughlin, S., Vishwanatha, J. K., Robey, F. A., Parrott, C., and Wilson, S. H.**, Mammalian alpha-polymerase: cloning of partial complementary DNA and immunoblotting of catalytic subunit in crude homogenate protein blots, *Biochemistry,* 26, 956, 1987.
22. **Sutcliffe, J. G., Milner, R. J., Shinnick, T. M., and Bloom, F. E.**, Identifying the protein products of brain-specific genes with antibodies to chemically synthesized peptides, *Cell,* 33, 671, 1983.
23. **Niman, H. L., Houghten, R. A., and Bowen-Pope, D. F.**, Detection of high molecular weight forms of platelet-derived growth factor by sequence-specific antisera, *Science,* 226, 701, 1984.
24. **Christie, D. L., Birch, N. P., Aitken, J. F., Harding, D. R. K., and Hancock, W. S.**, Antisera to synthetic peptide recognize high-molecular-weight enkephalin-containing proteins, *Biochem. Biophys. Res. Commun.,* 120, 650, 1984.
25. **Seckler, R., Wright, J. K., and Overath, P.**, Peptide-specific antibody locates the COOH terminus of the lactose carrier of *Escherichia coli* on the cytoplasmic side of the plasma membrane, *J. Biol. Chem.,* 258, 10817, 1983.
26. **Schneider, W. J., Slaughter, C. J., Goldstein, J. L., Anderson, R. G. W., Capra, J. D., and Brown, M. S.**, Use of antipeptide antibodies to demonstrate external orientation of the NH_2-terminus of the low density lipoprotein receptor in the plasma membrane of fibroblasts, *J. Cell Biol.,* 97, 1635, 1983.
27. **Barkas, T., Juillerat, M., Kistler, J., Schwendimann, B., and Moody, J.**, Antibodies to synthetic peptides as probes of acetylcholine receptor structure, *Eur. J. Biochem.,* 143, 309, 1984.
28. **Gordon, R. D., Fieles, W. E., Schotland, D. L., Hogue-Angeletti, R., and Barchi, R. L.**, Topographical localization of the C-terminal region of the voltage-dependent sodium channel from *Electrophorus electricus* using antibodies raised against a synthetic peptide, *Proc. Natl. Acad. Sci. U.S.A.,* 84, 308, 1987.
29. **Samelson, L. E., Weissman, A. M., Robey, F. A., Berkower, I., and Klausner, R. D.**, Characterization of an anti-peptide antibody that recognizes the murine analogue of the human T cell antigen receptor-T3 delta-chain, *J. Immunol.,* 137, 3254, 1986.
30. **Van Eldik, L. J., Fok, K. F., Erickson, B. W., and Watterson, D. M.**, Engineering of site-directed antisera against vertebrate calmodulin by using synthetic peptide immunogens containing an immunoreactive site, *Proc. Natl. Acad. Sci. U.S.A.,* 80, 6775, 1983.

31. Clarke, A. R., Carroll, A. R., Rowlands, D. J., Nicholson, B. H., Houghten, R. A., Lerner, R. A., and Brown, F., Synthetic peptides mimic subtype specificity of foot-and-mouth disease virus, *FEBS Lett.*, 157, 261, 1983.
32. DiMarchi, R., Brooke, G., Gale, C., Cracknell, V., Doel, T., and Mowat, N., Protection of cattle against foot-and-mouth disease by a synthetic peptide, *Science*, 232, 639, 1986.
33. Moriarty, A. M., Alexander, H., Lerner, R. A., and Thornton, G. B., Antibodies to peptides detect new hepatitis B antigen: serological correlation with hepatocellular carcinoma, *Science*, 227, 429, 1985.
34. Dillner, J., Sternas, L., Kallin, B., Alexander, H., Ehlin-Henriksson, B., Jorvall, H., Klein, G., and Lerner, R., Antibodies against a synthetic peptide identify the Epstein-Barr virus-determined nuclear antigen, *Proc. Natl. Acad. Sci. U.S.A.*, 81, 4652, 1984.
35. Dillner, J., Kallin, B., Klein, G., Jornvall, H., Alexander, H., and Lerner, R., Antibodies against synthetic peptides react with the second Epstein-Barr virus-associated nuclear antigen, *EMBO J.*, 4, 1813, 1985.
36. Chow, M., Yabrov, R., Bittle, J., Hogle, J., and Baltimore, D., Synthetic peptides from four separate regions of the poliovirus type 1 capsid protein VP1 induce neutralizing antibodies, *Proc. Natl. Acad. Sci. U.S.A.*, 82, 910, 1985.
37. Chanh, T. C., Dreesman, G. R., Kanda, P., Linette, G. P., Sparrow, J. T., Ho, D. T., and Kennedy, R. C., Induction of anti-HIV neutralizing antibodies by synthetic peptides, *EMBO J.*, 5, 3065, 1986.
38. Pert, C. B., Hill, J. M., Ruff, M. R., Berman, R. M., Robey, W. G., Arthur, L. O., Ruscetti, F. W., and Farrar, W. L., Octapeptides deduced from the neuropeptide receptor-like pattern of antigen T4 in brain potently inhibit human immunodeficiency virus receptor binding and T-cell infectivity, *Proc. Natl. Acad. Sci. U.S.A.*, 83, 9254, 1986.
39. Beachey, E. H., Tartar, A., Seyer, J. M., and Chedid, L., Epitope-specific protective immunogenicity of chemically synthesized 13-, 18-, and 23-residue peptide fragments of streptococcal M protein, *Proc. Natl. Acad. Sci. U.S.A.*, 81, 2203, 1984.
40. Rothbard, J. B., Fernandez, R., Wang, L., Teng, N. N. H., and Schoolnik, G. K., Antibodies to peptides corresponding to conserved sequence of gonococcal pilins block bacterial adhesion, *Proc. Natl. Acad. Sci. U.S.A.*, 82, 915, 1985.
41. Cheung, A., Leban, J., Shaw, A. R., Merkli, B., Stocker, J., Chizzolini, C., Sander, C., and Perrin, L. H., Immunization with synthetic peptides of a *Plasmodium falciparum* surface antigen induces antimerozoite antibodies, *Proc. Natl. Acad. Sci. U.S.A.*, 83, 8328, 1986.
42. Good, M. F., Maloy, W. L., Lunde, M. N., Margalit, H., Cornette, J. L., Smith, G. L., Moss, B., Miller, L. H., and Berzofsky, J. A., Construction of synthetic immunogen: use of new T-helper epitope on malaria circumsporozoite protein, *Science*, 235, 1059, 1987.
43. Giardina, S. L., Evans, S. W., Gandino, L., Robey, F. A., Bonvini, E., Longo, D. L., and Varesio, L., Generation of a murine monoclonal antibody that detects the *fos* oncogene product, *Anal. Biochem.*, 161, 109, 1987.
44. Niman, H. L., Antisera to a synthetic peptide of the *sis* viral oncogene product recognize human platelet-derived growth factor, *Nature (London)*, 307, 180, 1984.
45. Tanaka, T., Slamon, D. J., and Cline, M. J., Efficient generation of antibodies to oncoprotein by using synthetic peptide antigens, *Proc. Natl. Acad. Sci. U.S.A.*, 82, 3400, 1985.
46. Wong, T. W. and Goldberg, A. R., Synthetic peptide fragment of *src* gene product inhibits the *src* protein kinase and crossreacts immunologically with avian *onc* kinases and cellular phosphoproteins, *Proc. Natl. Acad. Sci. U.S.A.*, 78, 7412, 1981.
47. Nigg, E. A., Sefton, B. M., Hunter, T., Walter, G., and Singer, S. J., Immunofluorescent localization of the transforming protein of Rous sarcoma virus with antibodies against a synthetic *src* peptide, *Proc. Natl. Acad. Sci. U.S.A.*, 79, 5322, 1982.
48. Tamura, T., Bauer, H., Birr, C., and Pipkorn, R., Antibodies against synthetic peptides as a tool for functional analysis of the transforming protein pp60^{src}, *Cell*, 34, 587, 1983.
49. Gentry, L. E., Rohrschneider, L. R., Casnellie, J. E., and Krebs, E. G., Antibodies to a defined region of pp60^{src} neutralize the tyrosine-specific kinase activity, *J. Biol. Chem.*, 258, 11219, 1983.
50. Plumer, R., Fels, G., and Maelicke, A., Antibodies against preselected peptides to map functional sites on the acetylcholine receptor, *FEBS Lett.*, 178, 204, 1984.
51. Ratnam, M. and Lindstrom, J., Structural features of the nicotinic acetylcholine receptor revealed by antibodies to synthetic peptides, *Biochem. Biophys. Res. Commun.*, 122, 1225, 1984.
52. Neumann, D., Gershoni, J. M., Fridkin, M., and Fuchs, S., Antibodies to synthetic peptides as probes for the binding site on the α subunit of the acetylcholine receptor, *Proc. Natl. Acad. Sci. U.S.A.*, 82, 3490, 1985.
53. Takemoto, D. J., Spooner, B., and Takemoto, L. J., Antisera to synthetic peptides of bovine rhodopsin: use as site-specific probes of disc membrane changes in retinal dystrophic dogs, *Biochem. Biophys. Res. Commun.*, 132, 438, 1985.

54. **Weiss, E. R., Hadcock, J. R., Johnson, G. L., and Malbon, C. C.**, Antipeptide antibodies directed against cytoplasmic rhodopsin sequences recognize the β-adrenergic receptor, *J. Biol. Chem.*, 262, 4319, 1987.
55. **Dixon, R. A. F., Sigal, I. S., Rands, E., Register, R. B., Candelore, M. R., Blake, A. D., and Strader, C. D.**, Ligand binding to the β-adrenergic receptor involve its rhodopsin-like core, *Nature (London)*, 326, 73, 1987.
56. **Harris, B. A., Robishaw, J. D., Mumby, S. M., and Gilman, A. G.**, Molecular cloning of complementary DNA for the alpha subunit of the G protein that stimulates adenylate cyclase, *Science*, 229, 1274, 1985.
57. **Lerea, C. L., Somers, D. E., Hurley, J. B., Klock, I. B., and Bunt-Milam, A. H.**, Identification of specific transducin α subunits in retinal rod and cone photoreceptors, *Science*, 234, 77, 1986.
58. **Gariepy, J., Mietzner, T. A., and Schoolnik, G. K.**, Peptide antisera as sequence-specific probes of protein conformational transitions: calmodulin exhibits calcium-dependent changes in antigenicity, *Proc. Natl. Acad. Sci. U.S.A.*, 83, 8888, 1986.
59. **Langford, M. P., Gray, P. W., Stanton, G. J., Lakhchaura, B., Chan, T.-S., and Johnson, H. M.**, Antibodies to a synthetic peptide corresponding to the N-terminal end of mouse gamma interferon (IFN), *Biochem. Biophys. Res. Commun.*, 117, 866, 1983.
60. **Chow, T. P., DeGrado, W. F., and Knight, E., Jr.**, Antibodies to synthetic peptides of human interferon-β. Use in biosynthetic studies, *J. Biol. Chem.*, 259, 12220, 1984.
61. **Aguet, M., Salgam, P., Butte, B., and Arneiter, H.**, A crystalline synthetic peptide representing the epitope of a monoclonal antibody raised against synthetic interferon-α1 fragment 111-166, *Eur. J. Biochem.*, 146, 689, 1985.
62. **Jacob, C. O., Sela, M., and Arnon, R.**, Antibodies against synthetic peptides of the B subunit of cholera toxin: crossreaction and neutralization of the toxin, *Proc. Natl. Acad. Sci. U.S.A.*, 80, 7611, 1983.
63. **Jacob, C. O., Sela, M., Pines, M., Hurwitz, S., and Arnon, R.**, Both cholera toxin-induced adenylate cyclase activation and cholera toxin biological activity are inhibited by antibodies against related synthetic peptides, *Proc. Natl. Acad. Sci. U.S.A.*, 81, 7893, 1984.
64. **Roop, D. R., Cheng, C. K., Titterington, L., Meyers, C. A., Stanley, J. R., Steinert, P. M., and Yuspa, S. H.**, Synthetic peptides corresponding to keratin subunits elicit highly specific antibodies, *J. Biol. Chem.*, 259, 8037, 1984.
65. **Konomi, H., Seyer, J. M., Ninomiya, Y., and Olsem, B. R.**, Peptide-specific antibodies identify the α2 chain as the proteoglycan subunit of type IX collagen, *J. Biol. Chem.*, 261, 6742, 1986.
66. **Hui, K. Y., Haber, E., and Matsueda, G. R.**, Monoclonal antibodies to a synthetic fibrin-like peptide bind to human fibrin but not fibrinogen, *Science*, 222, 1129, 1983.
67. **Scheefers-Borchel, U., Muller-Berghaus, G., Fuhge, P., Eberle, R., and Heimburger, N.**, Discrimination between fibrin and fibrinogen by a monoclonal antibody against a synthetic peptide, *Proc. Natl. Acad. Sci. U.S.A.*, 82, 7091, 1985.
68. **Day, E. D. and Hashim, G. A.**, Format determinants of synthetic myelin basic protein peptide S82 mimicked by a mixture of synthetic peptides S8 and S79, *Neurochem. Res.*, 9, 1445, 1984.
69. **Day, E. D. and Hashim, G. A.**, Affinity purification of two populations of antibodies against format determinants of synthetic myelin basic protein peptide S82 from S82-AH and S82-CH-Sepharose columns, *Neurochem. Res.*, 9, 1453, 1984.
70. **Mannie, M. D., Paterson, P. Y., U'Prichard, D. C., and Fluoret, G.**, Induction of experimental allergic encephalomyelitis in Lewis rats with purified synthetic peptides: delineation of antigenic determinants for encephalitogenicity, *in vitro* activation of cellular transfer, and proliferation of lymphocytes, *Proc. Natl. Acad. Sci. U.S.A.*, 82, 5515, 1985.
71. **Fletterick, R. J., Sygusch, J., Semple, H., and Madsen, N. B.**, Structure of glycogen phosphorylase *a* at 3.0 Å resolution and its ligand binding sites at 6 Å, *J. Biol. Chem.*, 251, 6142, 1976.
72. **Yeaman, S. J., Cohen, P., Watson, D. C., and Dixon, G. H.**, The substrate specificity of adenosine 3':5'-cyclic monophosphate-dependent protein kinase of rabbit skeletal muscle, *Biochem. J.*, 162, 411, 1977.
73. **Janski, A. M. and Graves, D. J.**, An antibody probe for the amino-terminal region of glycogen phosphorylase, *FEBS Lett.*, 82, 232, 1977.
74. **Janski, A. M. and Graves, D. J.**, Use of an antibody probe to study the regulation of glycogen phosphorylase by its NH_2-terminal region, *J. Biol. Chem.*, 254, 1644, 1979.
75. **Campbell, D. G., Hardie, D. G., and Vulliet, P. R.**, Identification of four phosphorylation sites in the N-terminal region of tyrosine hydroxylase, *J. Biol. Chem.*, 261, 10489, 1986.
76. **Krebs, E. G. and Beavo, J. A.**, Phosphorylation-dephosphorylation of enzymes, *Annu. Rev. Biochem.*, 48, 923, 1979.
77. **Glass, D. B. and Krebs, E. G.**, Protein phosphorylation catalyzed by cyclic AMP-dependent and cyclic GMP-dependent protein kinases, *Annu. Rev. Pharmacol. Toxicol.*, 20, 363, 1980.
78. **Edelman, A. M., Blumenthal, D. K., and Krebs, E. G.**, Protein serine/threonine kinases, *Annu. Rev. Biochem.*, 56, 567, 1987.

79. **Glass, D. B.,** Substrate specificity of the cyclic GMP-dependent protein kinase, in *Peptides and Protein Phosphorylation,* Kemp, B. E., Ed., CRC Press, Boca Raton, FL, 1990, chap. 9.
80. **Kemp, B. E. and Stull, J. T.,** Myosin light chain kinases, in *Peptides and Protein Phosphorylation,* Kemp, B. E., Ed., CRC Press, Boca Raton, FL, 1990, chap. 4.
81. **Pinna, L. A., Meggio, F., and Marchiori, F.,** Type-2 casein kinases: general properties and substrate specificity, in *Peptides and Protein Phosphorylation,* Kemp, B. E., Ed., CRC Press, Boca Raton, FL, 1990, chap. 6.
82. **Cohen, P.,** The role of protein phosphorylation in the hormonal control of enzyme activity, *Eur. J. Biochem.,* 151, 439, 1985.
83. **Hunter, T.,** Synthetic peptide substrates for a tyrosine protein kinase, *J. Biol. Chem.,* 257, 4843, 1982.
84. **Kemp, B. E., Graves, D. J., Benjamini, E., and Krebs, E. G.,** Role of multiple basic residues in determining the substrate specificity of cyclic AMP-dependent protein kinase, *J. Biol. Chem.,* 252, 4888, 1977.
85. **Rosevear, P. R., Fry, D. C., Mildvan, A. S., Doughty, M., O'Brian, C., and Kaiser, E. T.,** NMR studies of the backbone protons and secondary structure of pentapeptide and heptapeptide substrates bound to bovine heart protein kinase, *Biochemistry,* 23, 3161, 1984.
86. **Berzofsky, J. A.,** Intrinsic and extrinsic factors in protein antigenic structure, *Science,* 229, 932, 1985.
87. **Chen, P. P., Houghten, R. A., Fong, S., Rhodes, G. H., Gilbertson, T. A., Vaughan, J. H., Lerner, R. A., and Carson, D. A.,** Anti-hypervariable region antibody induced by a defined peptide: an approach for studying the structural correlates of idiotypes, *Proc. Natl. Acad. Sci. U.S.A.,* 81, 1784, 1984.
88. **Dyson, H. J., Cross, K. L., Houghten, R. A., Wilson, I. A., Wright, P. E., and Lerner, R. A.,** The immunodominant site of a synthetic immunogen has a conformational preference in water for a type-II reverse turn, *Nature (London),* 318, 480, 1985.
89. **Hopp, T. P. and Woods, K. R.,** Prediction of protein antigenic determinants from amino acid sequence, *Proc. Natl. Acad. Sci. U.S.A.,* 78, 3824, 1981.
90. **Kyte, J. and Doolittle, R. F.,** A simple method for displaying the hydropathic character of a protein, *J. Mol. Biol.,* 157, 105, 1982.
91. **Parker, J. M. R., Guo, D., and Hodges, R. S.,** New hydrophilicity scale derived from high-performance liquid chromatography peptide retention data: correlation of predicted surface residues with antigenicity and X-ray-derived accessible sites, *Biochemistry,* 25, 5425, 1986.
92. **Tainer, J. A., Getzoff, E. D., Alexander, H., Houghten, R. A., Olson, A. J., Lerner, R. A., and Hendrickson, W. A.,** The reactivity of anti-peptide antibodies is a function of the atomic mobility of sites in a protein, *Nature (London),* 312, 127, 1984.
93. **Tainer, J. A., Getzoff, E. D., Paterson, Y., Olson, A. J., and Lerner, R. A.,** The atomic mobility component of protein antigenicity, *Annu. Rev. Immunol.,* 3, 501, 1985.
94. **Westhof, E., Altschuh, D., Moras, D., Bloomer, A. C., Mondragon, A., Klug, A., and Van Regenmortel, M. H. V.,** Correlation between segmental mobility and the location of antigenic determinants in proteins, *Nature (London),* 311, 123, 1984.
95. **Sefton, B. M. and Hunter, T.,** Tyrosine protein kinases, *Adv. Cyclic Nucleotide Protein Phosphorylation Res.,* 18, 195, 1984.
96. **Hunter, T. and Cooper, J. A.,** Protein-tyrosine kinases, *Annu. Rev. Biochem.,* 54, 897, 1985.
97. **Gershoni, J. M. and Palade, G. E.,** Protein blotting: principles and applications, *Anal. Biochem.,* 131, 1, 1983.
98. **Parikh, I. and Cuatrecasas, P.,** Affinity chromatography in immunology, in *Immunochemistry of Proteins,* Atassi, M. Z., Ed., Plenum Press, New York, 1977, 1.
99. **Rosen, O. M.,** After insulin binds, *Science,* 237, 1452, 1987.
100. **Feldman, R. A., Tam, J. P., and Hanafusa, H.,** Antipeptide antiserum identifies a widely distributed cellular tyrosine kinase related to but distinct from the c-*fps/fes*-encoded protein, *Mol. Cell. Biol.,* 6, 1065, 1986.
101. **Kris, R. M., Lax, I., Gullick, W., Waterfield, M. D., Ullrich, A., Fridkin, M., and Schlessinger, J.,** Antibodies against a synthetic peptide as a probe for the kinase activity of the avian EGF receptor and v-*erb* protein, *Cell,* 40, 619, 1985.
102. **Kris, R. M., Lax, I., Sasson, I., Copf, B., Werlin, S., Gullick, W., Waterfield, M. D., Ullrich, A., Fridkin, M., and Schlessinger, J.,** Synthetic peptide approach to the analysis of kinase activities of avian EGF receptor and v-*erb*B protein, *Biochemie,* 67, 1095, 1985.
103. **Akiyama, T., Sudo, C., Ogawara, H., Toyoshima, K., and Yamamoto, T.,** The product of the human c-*erb*B-2 gene: a 185-kilodalton glycoprotein with tyrosine kinase activity, *Science,* 232, 1644, 1986.
104. **Schaffhausen, B., Benjamin, T. L., Pike, L., Casnellie, J., and Krebs, E.,** Antibody to the nonapeptide Glu-Glu-Glu-Glu-Tyr-Met-Pro-Met-Glu is specific for polyoma middle T antigen and inhibits in vitro kinase activity, *J. Biol. Chem.,* 257, 12467, 1982.
105. **Koch, W., Carbone, A., and Walter, G.,** Purified polyoma virus medium T antigen has tyrosine-specific protein kinase activity but no significant phosphatidylinositol kinase activity, *Mol. Cell. Biol.,* 6, 1866, 1986.

106. **Casnellie, J. E., Gentry, L. E., Rohrschneider, L. R., and Krebs, E. G.**, Identification of the tyrosine protein kinase from LSTRA cells by use of site-specific antibodies, *Proc. Natl. Acad. Sci. U.S.A.*, 81, 6676, 1984.
107. **Trevillyan, J. M., Canna, C., Maley, D., Linna, T. J., and Phillips, C. A.**, Identification of the human T-lymphocyte protein-tyrosine kinase by peptide-specific antibodies, *Biochem. Biophys. Res. Commun.*, 140, 392, 1986.
108. **Herrera, R., Petruzzelli, L., Thomas, N., Bramson, H. N., Kaiser, E. T., and Rosen, O. M.**, An antipeptide antibody that specifically inhibits insulin receptor autophosphorylation and protein kinase activity, *Proc. Natl. Acad. Sci. U.S.A.*, 82, 7899, 1985.
109. **Stadtmauer, L. and Rosen, O. M.**, Phosphorylation of synthetic insulin receptor peptides by the insulin receptor kinase and evidence that the preferred sequence containing Tyr-1150 is phosphorylation *in vivo*, *J. Biol. Chem.*, 261, 10000, 1986.
110. **Herrera, R. and Rosen, O. M.**, Autophosphorylation of the insulin receptor *in vitro*. Designation of phosphorylation sites and correlation with receptor kinase activation, *J. Biol. Chem.*, 261, 11980, 1986.
111. **Herrera, R., Petruzzelli, L. M., and Rosen, O. M.**, Antibodies to deduced sequences of the insulin receptor distinguish conserved and nonconserved regions in the IGF-1 receptor, *J. Biol. Chem.*, 261, 2489, 1986.
112. **Petruzzelli, L., Herrera, R., Arenas-Garcia, R., Fernandez, R., Birnbaum, M. J., and Rosen, O. M.**, Isolation of a *Drosophila* genomic sequence homologous to the kinase domain of the human insulin receptor and detection of the phosphorylated *Drosophila* receptor with anti-peptide antibody, *Proc. Natl. Acad. Sci. U.S.A.*, 83, 4710, 1986.
113. **Ek, B. and Hekdin, C.-H.**, Use of an antiserum against phosphotyrosine for the identification of phosphorylated components in human fibroblasts stimulated by platelet-derived growth factor, *J. Biol. Chem.*, 259, 11145, 1984.
114. **Pang, D. T., Sharma, B. R., and Shafer, J. A.**, Purification of the catalytically active phosphorylated form of the insulin receptor kinase by affinity chromatography with *O*-phosphotyrosyl-binding antibodies, *Arch. Biochem. Biophys.*, 242, 176, 1985.
115. **Migliaccio, A., Rotondi, A., and Auricchio, F.**, Estradiol receptor: phosphorylation on tyrosine in uterus and interaction with anti-phosphotyrosine antibody, *EMBO J.*, 5, 2867, 1986.
116. **Daniel, T. O., Tremble, T. M., Frackelton, R. A., Jr., and Williams, L. T.**, Purification of the platelet-derived growth factor receptor by using an anti-phosphotyrosine antibody, *Proc. Natl. Acad. Sci. U.S.A.*, 82, 2684, 1985.
117. **Ross, A. H., Baltimore, D., and Eisen, H. N.**, Phosphotyrosine-containing proteins isolated by affinity chromatography with antibodies to a synthetic hapten, *Nature (London)*, 294, 654, 1981.
118. **Frackelton, A. R., Jr., Ross, A. H., and Eisen, H. N.**, Characterization and use of monoclonal antibodies for isolation of phosphotyrosyl proteins from retrovirus-transformed cells and growth factor-stimulated cells, *Mol. Cell. Biol.*, 3, 1343, 1983.
119. **Comoglio, P. M., Di Renzo, M. F., Tarone, G., Giancotti, F. G., Naldini, L., and Marchisio, P. C.**, Detection of phosphotyrosine-containing proteins in the detergent-insoluble fraction of RSV-transformed fibroblasts by azobenzene phosphonate antibodies, *EMBO J.*, 3, 483, 1984.
120. **Naldini, L., Stacchini, A., Cirillo, D. M., Aglietta, M., Gavosto, F., and Comoglio, P. M.**, Phosphotyrosine antibodies identify the $p210^{c-abl}$ tyrosine kinase and proteins phosphorylated on tyrosine in human chronic myelogenous leukemia cells, *Mol. Cell. Biol.*, 6, 1803, 1986.
121. **Zippel, R., Sturani, E., Toschi, L., Naldini, L., Alberghina, L., and Comoglio, P. M.**, In vivo phosphorylation and dephosphorylation of the platelet-derived growth factor studied by immunoblot analysis with phosphotyrosine antibodies, *Biochim. Biophys. Acta*, 881, 54, 1986.
122. **Frackelton, A. R., Jr., Tremble, P. M., and Williams, L. T.**, Evidence for the platelet-derived growth factor-stimulated tyrosine phosphorylation of the platelet-derived growth factor receptor *in vivo*, *J. Biol. Chem.*, 259, 7909, 1984.
123. **Nairn, A. C., Detre, J. A., Casnellie, J. E., and Greengard, P.**, Serum antibodies that distinguish between the phospho- and dephospho-forms of a phosphoprotein, *Nature (London)*, 299, 735, 1982.
124. **Nairn, A. C. and Greengard, P.**, Cyclic GMP-dependent protein phosphorylation in mammalian brain, *Fed. Proc.*, 42, 3107, 1983.
125. **Detre, J. A., Nairn, A. C., Aswad, D. W., and Greengard, P.**, Localization in mammalian brain of G-substrate, a specific substrate for guanosine 3′,5′-cyclic monophosphate-dependent protein kinase, *J. Neurosci.*, 4, 2843, 1984.
126. **Safran, A., Neumann, D., and Fuchs, S.**, Analysis of acetylcholine receptor phosphorylation sites using antibodies to synthetic peptides and monoclonal antibodies, *EMBO J.*, 5, 3175, 1986.
127. **Smith, S. C., McAdam, W. J., Kemp, B. E., Morgan, F. J., and Cotton, R. G. H.**, A monoclonal antibody to the phosphorylated form of phenylalanine hydroxylase, *Biochem. J.*, 244, 625, 1987.
128. **Shenolikar, S. and Steiner, A. L.**, Immunological approach to the study of hormone action: regulation of phosphorylation state of protein phosphatase inhibitor-1 by hormones, *Adv. Cyclic Nucleotide Protein Phosphorylation Res.*, 17, 405, 1984.

129. **Engstrom, L.,** The regulation of liver pyruvate kinase by phosphorylation-dephosphorylation, *Curr. Top. Cell. Regul.,* 13, 29, 1978.
130. **Ljungstrom, O., Berlund, L., and Engstrom, L.,** Studies on the kinetic effects of adenosine-3'5'-monophosphate-dependent phosphorylation of purified pig liver pyruvate kinase type L, *Eur. J. Biochem.,* 68, 497, 1976.
131. **El-Maghrabi, M. R., Haston, W. S., Flockhart, D. A., Claus, T. H., and Pilkis, S. J.,** Studies on the phosphorylation and dephosphorylation of L-type pyruvate kinase by the catalytic subunit of cyclic AMP-dependent protein kinase, *J. Biol. Chem.,* 255, 668, 1980.
132. **Pilkis, S. J., El-Maghrabi, M. R., Coven, B., Claus, T. H., Tager, H. S., Steiner, D. F., Keim, P. S., and Heinrikson, R. L.,** Phosphorylation of rat hepatic fructose-1,6-biphosphatase and pyruvate kinase, *J. Biol. Chem.,* 255, 2770, 1980.
133. **Hjelmquist, G., Andersson, J., Edlund, B., and Engstrom, L.,** Amino acid sequence of a (^{32}P)phosphopeptide from pig liver pyruvate kinase phosphorylated by cyclic 3',5'-AMP-stimulated protein kinase and (γ-^{32}P)ATP, *Biochem. Biophys. Res. Commun.,* 61, 559, 1974.
134. **Humble, E.,** Amino acid sequence around the phosphorylated sites of porcine and rat liver pyruvate kinase, type L, *Biochim. Biophys. Acta,* 626, 179, 1980.
135. **Hoar, C. G., Nicoll, G. W., Schiltz, E., Schmitt, W., Bloxham, D. P., Byford, M. F., Dunbar, B., and Fothergill, L. A.,** Muscle and liver pyruvate kinases are closely related: amino acid sequence comparisons, *FEBS Lett.,* 171, 293, 1984.
136. **Feramisco, J. R., Glass, D. B., and Krebs, E. G.,** Optimal spacial requirements for the location of basic residues in peptide substrates for the cyclic AMP-dependent protein kinase, *J. Biol. Chem.,* 255, 4240, 1980.
137. **Atassi, M. Z.,** Antigenic structures of proteins. Their determination has revealed important aspects of immune recognition and generated strategies for synthetic mimicking of protein binding sites, *Eur. J. Biochem,* 145, 1, 1984.
138. **Novotny, J., Handschumacher, M., and Haber, E.,** Location of antigenic epitopes on antibody molecules, *J. Mol. Biol.,* 189, 715, 1986.
139. **Novotny, J., Handschumacher, M., Haber, E., Burccoleri, R. E., Carlson, W. B., Fanning, D. W., Smith, J. A., and Rose, G. D.,** Antigenic determinants in proteins coincide with surface regions accessible to large probes (antibody domains), *Proc. Natl. Acad. Sci. U.S.A.,* 83, 226, 1986.
140. **Kauzmann, W.,** Some factors in the interpretation of protein denaturation, *Adv. Protein Chem.,* 14, 1, 1959.
141. **Chou, P. Y. and Fasman, G. D.,** Empirical predictions of protein conformation, *Annu. Rev. Biochem.,* 47, 251, 1978.
142. **Garnier, J., Osguthorpe, D. J., and Robson, B.,** Analysis of the accuracy and implications of simple methods for predicting the secondary structure of globular proteins, *J. Mol. Biol.,* 120, 97, 1978.
143. **Yada, R. Y., Jackman, R. L., and Nakai, S.,** Secondary structure prediction and determination of proteins — a review, *Int. J. Peptide Protein Res.,* 31, 98, 1988.
144. **Karplus, P. A. and Schulz, G. E.,** Prediction of chain flexibility in proteins, *Naturwissenschaften,* 72, 212, 1985.
145. **Amit, A. G., Mariuzza, R. A., Phillips, S. E. V., and Poljak, R. J.,** Three-dimensional structure of an antigen-antibody complex at 2.8 Å resolution, *Science,* 233, 747, 1986.
146. **Welling, G. W. and Fries, H.,** Choice of peptide and peptide length for the generation of antibodies reactive with the intact protein, *FEBS Lett.,* 182, 81, 1985.
147. **Lone, Y.-C., Simon, M.-P., Kahn, A., and Marie, J.,** Complete nucleotide and deduced amino acid sequences of rat L-type pyruvate kinase, *FEBS Lett.,* 195, 97, 1986.
148. **Riou, J. P., Claus, T. H., and Pilkis, S. J.,** Stimulation by glucagon of *in vivo* phosphorylation of rat hepatic pyruvate kinase, *J. Biol. Chem.,* 523, 656, 1978.
149. **Perich, J. W.,** Modern methods of O-phosphoserine- and O-phosphotyrosine-containing protein synthesis, in *Peptides and Protein Phosphorylation,* Kemp, B. E., Ed., CRC Press, Boca Raton, FL, 1990, chap. 12.
150. **Sun, I. Y.-C., Johnson, E. M., and Allfrey, V. G.,** Affinity purification of 8 newly phosphorylated protein molecules, *J. Biol. Chem.,* 255, 742, 1980.
151. **Davies, J. and Watkins, J. C.,** Actions of D and L forms of 2-amino-5-phosphonovalerate and 2-amino-4-phosphonobutyrate in the cat spinal cord, *Brain Res.,* 235, 378, 1982.
152. **Itoh, M., Hagiwara, D., and Kamiya, T.,** A new *tert*-butyloxycarbonylating reagent, 2-*tert*-butyloxycarbonyloxyimino-2-phenylacetonitrile, *Tetrahedron Lett.,* 49, 4393, 1975.
153. **Glass, D. B.,** Synthesis of oligopeptides for the study of cyclic nucleotide-dependent protein kinases, *Methods Enzymol.,* 99, 119, 1983.
154. **Kent, S. and Clark-Lewis, I.,** Modern methods for chemical synthesis of biologically active peptides, in *Synthetic Peptides in Biology and Medicine,* Alitalo, K., Partanen, P., and Vaheri, A., Eds., Elsevier, Amsterdam, 1985, 29.

155. **Barany, G., Kneib-Cordonier, N., and Mullen, D. G.**, Solid-phase peptide synthesis: a silver anniversary report, *Int. J. Peptide Protein Res.*, 30, 705, 1987.
156. **Houghten, R. A.**, General method for the rapid solid-phase synthesis of large numbers of peptides: specificity of antigen-antibody interaction at the level of individual amino acids, *Proc. Natl. Acad. Sci. U.S.A.*, 82, 5131, 1985.
157. **Houghten, R. A., Bray, M. K., DeGraw, S. T., and Kirby, C. J.**, A simplified procedure for carrying out simultaneous multiple hydrogen fluoride cleavage of protected resins, *Int. J. Peptide Protein Res.*, 27, 673, 1986.
158. **Geysen, H. M., Meloen, R. H., and Barteling, S. J.**, Use of peptide synthesis to probe viral antigens for epitopes to a resolution of a single amino acid, *Proc. Natl. Acad. Sci. U.S.A.*, 81, 3998, 1984.
159. **Smith, J. A., Hurrell, J. G. R., and Leach, S. J.**, A novel method for delineating antigenic determinants: peptide synthesis and radioimmunoassay using the same solid support, *Immunochemistry*, 14, 565, 1977.
160. **Wilcox, P. E.**, Esterification, *Methods Enzymol.*, 11, 605, 1967.
161. **Schmitz, H. E., Atassi, H., and Atassi, M. Z.**, Production of monoclonal antibodies to surface regions that are non-immunogenic in a protein using free synthetic peptides as immmunogens: demonstration with sperm-whale myoglobin, *Immunol. Commun.*, 12, 161, 1983.
162. **Arnon, R., Shapira, M., and Jacob, C. O.**, Synthetic vaccines, *J. Immunol. Methods*, 61, 261, 1983.
163. **Benoit, R., Ling, N., Brazeau, P., Lavielle, S., and Guillemin, R.**, Peptides: strategies for antibody production and radioimmunoassays, in *Neuromethods: Peptides*, Vol. 6, Boulton, A. A., Baker, G. B., and Pittman, Q. J., Eds., Humana Press, Clifton, NJ, 1987, 43.
164. **Carelli, C., Audibert, F., Gaillard, J., and Chedid, L.**, Immunological castration of male mice by a totally synthetic vaccine administered in saline, *Proc. Natl. Acad. Sci. U.S.A.*, 79, 5392, 1982.
165. **Klapper, M. H. and Klotz, I. M.**, Acylation with dicarboxylic acid anhydrides, *Methods Enzymol.*, 25, 531, 1972.
166. **Schick, A. F. and Singer, S. J.**, On the formation of covalent linkages between two protein molecules, *J. Biol. Chem.*, 236, 2477, 1961.
167. **Tager, H. S.**, Coupling of peptides to albumin with difluorodinitrobenzene, *Anal. Biochem.*, 71, 367, 1976.
168. **Samokhin, G. P. and Filimonov, I. N.**, Coupling of peptides to carrier proteins by mixed anhydride procedure, *Anal. Biochem.*, 145, 311, 1985.
169. **King, T. P., Zhao, S. W., and Lam, T.**, Preparation of protein conjugates via intermolecular hydrazone linkage, *Biochemistry*, 25, 5774, 1986.
170. **Kohler, G. and Milstein, C.**, Continuous cultures of fused cells secreting antibody of predefined specificity, *Nature (London)*, 256, 495, 1975.
171. **Engvall, E.**, Enzyme immunoassay ELISA and EMIT, *Methods Enzymol.*, 70, 419, 1980.
172. **Langone, J. J.**, [^{125}I]protein A: a tracer for general use in immunoassay, *J. Immunol. Methods*, 24, 269, 1978.
173. **Dawson, J. R. and Cresswell, P.**, Immunogens (antigens) and antibodies and their determination, in *Zinsser Microbiology*, 18th ed., Joklik, W. K., Willet, H. P., and Amos, D. B., Eds., Appleton-Century-Crofts, Norwalk, CT, 1984, 261.
174. **Jonsson, S. and Kronvall, G.**, The use of protein A-containing *Staphylococcus aureus* as a solid phase anti-IgG reagent in radioimmunoassays as exemplified in the quantitation of α-fetoprotein in normal human adult serum, *Eur. J. Immunol.*, 4, 29, 1974.
175. **Scouten, W. H.**, A survey of enzyme coupling techniques, *Methods Enzymol.*, 135, 30, 1987.
176. **Murray, K. J., El-Maghrabi, M. R., Kountz, P. D., Lukas, T. J., Soderling, T. R., and Pilkis, S. J.**, Amino acid sequence of the phosphorylation site of rat liver 6-phosphofructo-2-kinase/fructose-2,6-bisphosphatase, *J. Biol. Chem.*, 259, 7673, 1984.
177. **Kabat, E. A.**, Basic principles of antigen-antibody reactions, *Methods Enzymol.*, 70, 3, 1980.
178. **Crawford, L., Leppard, K., Lane, D., and Harlow, E.**, Cellular proteins reactive with monoclonal antibodies directed against simian virus 40 T-antigen, *J. Virol.*, 42, 612, 1982.
179. **El-Maghrabi, M. R., Claus, T. H., McGrane, M. M., and Pilkis, S. J.**, Influence of phosphorylation on the interaction of effectors with rat liver pyruvate kinase, *J. Biol. Chem.*, 257, 233, 1982.

Chapter 12

MODERN METHODS OF O-PHOSPHOSERINE- AND O-PHOSPHOTYROSINE-CONTAINING PEPTIDE SYNTHESIS

John W. Perich

TABLE OF CONTENTS

I. Synthesis of O-Phosphoserine-Containing Peptides 290
    A. Introduction 290
    B. Global "Phosphite-Triester" Phosphorylation of Ser-Peptides Using Dibenzyl and Di-t-Butyl N,N-Diethylphosphoramidite 292
    C. Solution-Phase Synthesis of PSer-Peptides by Synthon Incorporation 295
        1. Synthesis of Boc–Ser(PO_3R_2)–OH Derivatives 295
            a. General Phosphorylation Procedures 298
            b. Synthesis of Boc–Ser(PO_3Ph_2)–OH 298
        2. Synthesis of Protected Ser(PO_3R_2)-Tripeptides 299
            a. General Procedure for the Incorporation of Boc–Ser(PO_3R_2)–OH Derivatives into Peptide Synthesis 300
        3. Deprotection of Protected Ser(PO_3R_2)-Peptides 301
            a. Synthesis of PSer-Peptides by the Hydrogenation of Protected Ser(PO_3Bzl_2)- and Ser(PO_3Ph_2)-Peptides 301
        4. Synthesis of Casein-Related PSer-Peptides 302
            a. Synthesis of Ac–Glu–PSer–Leu–PSer–PSer–PSer–Glu–Glu–NHMe 303
    D. Solid-Phase Synthesis of PSer-Peptides by the "Synthon Incorporation Strategy" 305

II. Synthesis of O-Phosphotyrosine-Containing Peptides 305
    A. Introduction 305
    B. Solution- and Solid-Phase Synthesis of PTyr-Peptides by the "Synthon Incorporation Strategy" 306
        1. General Procedure for the Phosphorylation of Boc–Tyr–ONBzl 307
        2. Synthesis of Boc–Tyr–(PO_3Me_2)–OH 307
        3. Synthesis of H–Asn–Glu–Tyr(PO_3H_2)–Thr–Ala–OH 310

III. Conclusion 310

IV. Appendix: Abbreviations 311

References 312

I. SYNTHESIS OF O-PHOSPHOSERINE-CONTAINING PEPTIDES

A. INTRODUCTION

Investigations into the chemical synthesis of PSer-containing peptides commenced in the early 1950s, after it had been elucidated some 20 years earlier that the phosphate content of the bovine milk caseins was in the form of O-phosphoserine.[1] Since 1957, the chemical synthesis of small PSer-peptides has been investigated by several groups,[2] in particular by Folsch,[3] and has generally been accomplished by the "global" phosphorylation of protected seryl-containing peptides with either diphenyl or dibenzyl phosphorochloridate/pyridine followed by hydrogenolytic removal of the phenyl or benzyl groups (Figure 1). However, in these studies, the synthesis of PSer-peptides by a "global phosphorylation strategy" was complicated due to either (1) incomplete platinum-catalyzed hydrogenolytic cleavage of phenyl groups from Ser(PO_3Ph_2)-peptides[4] or (2) the low reactivity of dibenzyl phosphorochloridate for the efficient phosphorylation of seryl hydroxyl groups.[5]

While investigations into the preparation of PSer-peptides by the "global" phosphorylation of Ser-peptides were ostensibly ignored since 1964, two reports appeared in 1987 which detailed the synthesis of the phosphorylated forms of the cAMP-dependent protein kinase peptides H–Arg–Arg–Ala–PSer–Val–Ala–OH[6] and Ac–Leu–Arg–Arg–Ala–PSer–Leu–Gly–R[7] (R = OMe or NHMe). While the PSer-hexapeptide was prepared by the global diphenyl phosphochloridate/pyridine phosphorylation of Z–Arg(Adoc)$_2$–Arg(Adoc)$_2$–Ala–Ser–Val–Ala–OBu^t followed by acidolytic and hydrogenolytic (platinum) deprotection, the global phosphorylation of the protected Ser-heptapeptide Ac–Leu–Arg(NO_2)–Arg(NO_2)–Ala–Ser–Leu–Gly–OMe was reported to be unsuccessful. As a consequence, the PSer-heptapeptides were prepared by a convergent strategy which involved diphenyl phosphorochloridate/pyridine phosphorylation of Z–Ser–Leu–Gly–R (R = OMe or NHMe), hydrogenolytic removal of the Z group, and DCC/HOBt coupling of Ac–Leu–Arg(NO_2)–Arg(NO_2)–Ala–OH with HCl·H–Ser(PO_3Ph_2)–Leu–Gly–R (R = OMe or NHMe), followed by hydrogenolytic deprotection of the protected Ser(PO_3Ph_2)-heptapeptides.

An interesting facet of these two studies is the contrasting phosphorylation outcome for the sequence-similar protected Ser-peptides. Recent studies in our laboratory have indicated that the successful phosphorylation of Ser-peptides is sequence dependent, with a base-mediated β-elimination or rearrangement process being particularly prominent with the generation of Ser(PO_3Ph_2)-residues. This statement is supported by our findings that (1) the diphenyl phosphorochloridate/pyridine phosphorylation of Ac–Gly–Ser–Gly–NHMe failed to give the Ser(PO_3Ph_2)-tripeptide and (2) the phosphorylation of Ac–Ser–NHMe did not give Ac–Ser(PO_3Ph_2)–NHMe, but instead gave the diphenyl phosphate salt of serine N-methylamide (PhO)$_2$PO$_2$H·H–Ser–NHMe as the major product.[8] While the formation of this latter product is unclear, it is probable that the process involves initial phosphorylation of the seryl hydroxyl group, oxygen to nitrogen diphenyl phosphate shift, and hydrolytic cleavage of the acetyl group followed by acidolysis of the diphenyl phosphoramidate linkage.

The unsuccessful phosphorylation of Ac–Gly–Ser–Gly–NHMe and Ac–Ser–NHMe is particularly interesting considering that Folsch[3] has reported the successful diphenyl phosphorochloridate phosphorylation of Z–Gly–Ser–Gly–OBzl and that Alewood et al.[9] have reported that the low-temperature dibenzyl phosphorochloridate/pyridine phosphorylation of Ac–Ser–NHMe gives Ac–Ser(PO_3Bzl_2)–NHMe in 66% yield. In view of these anomalous results, it is clear that further work is needed to rationalize the outcome of successful and unsuccessful phosphorylations in respect to both amino acid sequence and phosphate-substituent effects.

In addition to the global phosphorylation of Ser-peptides using diaralkyl phosphorochloridates, Okawa et al.[10] prepared Gly–PSer–Gly and Ala–PSer–Gly by the electrophilic addition of dibenzyl phosphoric acid or orthophosphoric acid to two Azi-containing tripep-

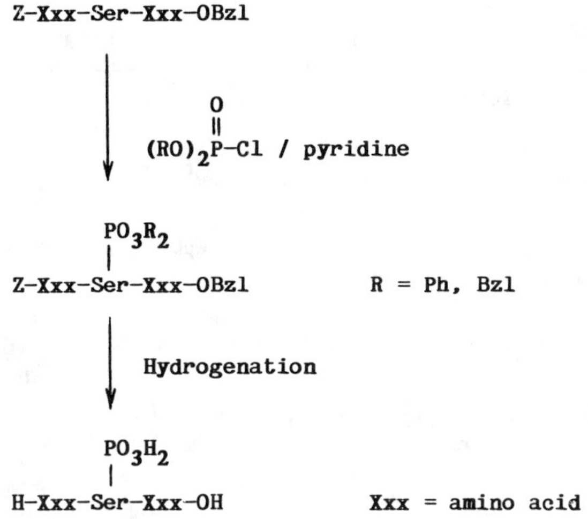

Z-Xxx-Ser-Xxx-OBzl

↓ (RO)$_2$P(=O)-Cl / pyridine

Z-Xxx-Ser(PO$_3$R$_2$)-Xxx-OBzl   R = Ph, Bzl

↓ Hydrogenation

H-Xxx-Ser(PO$_3$H$_2$)-Xxx-OH   Xxx = amino acid

FIGURE 1. Synthesis of PSer-peptides by the "global" phosphorylation of Ser-peptides using diaralkyl phosphorochloridates.

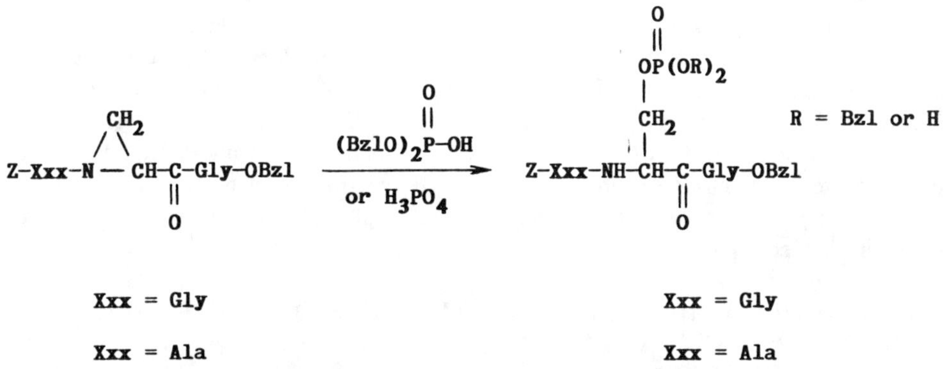

Z-Xxx-N(—CH$_2$—)CH-C(=O)-Gly-OBzl $\xrightarrow{\text{(BzlO)}_2\text{P(=O)-OH or H}_3\text{PO}_4}$ Z-Xxx-NH-CH(CH$_2$-OP(=O)(OR)$_2$)-C(=O)-Gly-OBzl   R = Bzl or H

Xxx = Gly   Xxx = Gly
Xxx = Ala   Xxx = Ala

FIGURE 2. Synthesis of PSer-peptides by the "global" phosphorylation of Ser-peptides using dibenzyl phosphoric acid or orthophosphoric acid.

tides, Z–Gly–Azi–Gly–OBzl and Z–Ala–Azi–Gly–OBzl (Figure 2). However, despite the high yields of the two PSer-tripeptides, this procedure has not been adopted for PSer-peptide synthesis due to the complicated synthesis of Azi-peptides[11] and a lack of general peptide versatility. The deficiencies of this strategy for PSer-peptide synthesis have been highlighted particularly by Johnson and Coward[7] in their attempted synthesis of the PSer-heptapeptide Ac–Leu–Arg–Arg–Ala–PSer–Leu–Gly–OH, the synthesis of the Azi-heptapeptide requiring several steps, there being extensive by-product formation during the electrophilic phosphorylation of the aziridine residue, and the isolation of polar PSer-peptides from solvent amounts of orthophosphoric acid necessitating barium precipitation.

In 1980 our laboratory became interested in the chemical properties of the bovine milk caseins and targeted the heavily phosphorylated sequence –Glu–PSer–Leu–PSer–PSer–PSer–Glu–Glu– for chemical and physical investigation. This peptide sequence is considered to be chemically significant since it occurs in both bovine α_{s1}- and β-casein as well as in human β-casein and is believed to be structurally responsible for the maintenance of the micellar integrity of the caseins via calcium bridging of the O-phosphoserine clusters.

Despite the good precedence for the "global" phosphorylation of Ser-peptides, the attempted phosphorylation of Ac–Glu–Ser–Leu–Ser–Ser–Ser–Glu(OBzl)–Glu(OBzl)–

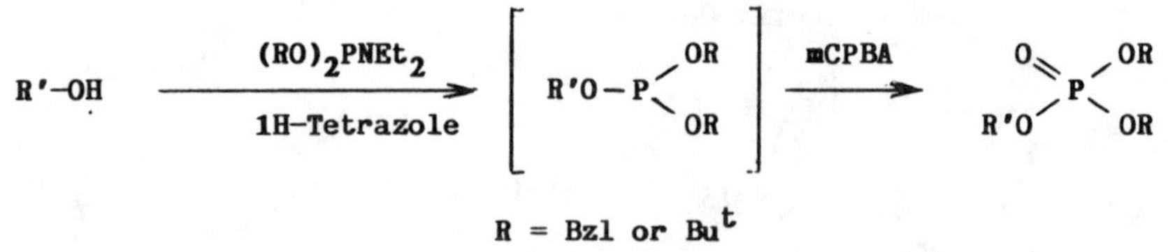

$$R'\text{-OH} \xrightarrow[\text{1H-Tetrazole}]{(RO)_2PNEt_2} \left[R'O-P\begin{smallmatrix}OR\\OR\end{smallmatrix} \right] \xrightarrow{mCPBA} R'O-P(=O)(OR)_2$$

R = Bzl or Bu^t

FIGURE 3. Synthesis of alkyl dibenzyl or di-*t*-butyl phosphates by the "global" benzyl- or *t*-butyl-phosphoramidite "phosphite-triester" phosphorylation of simple alkyl alcohols.

NHMe[12,13] with either diphenyl or dibenzyl phosphorochloridate proved unsuccessful, ^{13}C- and ^{31}P-nuclear magnetic resonance (NMR) spectroscopic analysis of the recovered products showing mixtures of partially phosphorylated peptides and serine degradation peptides. As amino acid analysis of the hydrogenated residue showed a marked deficiency of serine and high quantities of alanine and glycine, it was concluded that extensive degradation of the –Ser– or –Ser(PO$_3$R$_2$)– residues had occurred during phosphorylation and that such phosphorylation procedures would be unsuitable for the efficient "global" phosphorylation of multiple serine- or seryl cluster-containing peptides.

In order to overcome this synthetic difficulty, our laboratory initiated a research program in mid-1981 with the aim of developing new and general chemical procedures which would permit the efficient synthesis of complex and multiple *P*Ser-containing peptides. From these studies it was demonstrated that *P*Ser-peptides could be prepared by two different approaches: first, by the global "phosphite-triester" phosphorylation of Ser-peptides using the highly reactive dibenzyl or di-*t*-butyl *N,N*-diethylphosphoramidite; and second, by the direct incorporation of Boc–Ser(PO$_3$R$_2$)–OH derivatives into Boc/peptide synthesis. In addition, the combined use of ^{13}C-NMR spectroscopy, ^{31}P-NMR spectroscopy, and fast atom bombardment mass spectrometry (FAB-MS) were found to be powerful techniques for the characterization of *P*Ser-peptides.

B. GLOBAL "PHOSPHITE-TRIESTER" PHOSPHORYLATION OF SER-PEPTIDES USING DIBENZYL OR DI-*T*-BUTYL *N,N*-DIETHYLPHOSPHORAMIDITE

In order to overcome the low phosphorylative reactivity of dibenzyl phosphorochloridate, the highly reactive dibenzyl and di-*t*-butyl *N,N*-diethylphosphoramidite reagents were examined for the global "phosphite-triester" phosphorylation of Ser-peptides on account of the successful application of phosphoramidite chemistry to oligonucleotide synthesis. In a preliminary study, various alkyl and aryl dibenzyl or di-*t*-butyl phosphates were prepared in near-quantitative yields by the treatment of the alkyl or aryl alcohol with dibenzyl or di-*t*-butyl *N,N*-diethylphosphoramidite/1H-tetrazole followed by oxidation of the resultant alkyl dibenzyl or di-*t*-butyl phosphite triester with 3-chloroperoxybenzoic acid (mCPBA)[14,15] (Figure 3). Two major features of this phosphorylation procedure are that the phosphitylation and oxidation steps are very rapid (<10 min) and that the reaction is conducted under mildly acidic conditions, this being especially important with the phosphorylation of base-sensitive hydroxy-containing biomolecules.

The application of these two "phosphite-triester" phosphorylation procedures for *P*Ser-peptide synthesis was demonstrated by the high-yielding synthesis of Glu–*P*Ser–Leu by the "phosphite-triester" phosphorylation of Boc–Glu(OBu^t)–Ser–Leu–OBu^t with (BzlO)$_2$PNEt$_2$/ 1H-tetrazole and then mCPBA, or with (Bu^{t}O)$_2$PNEt$_2$/1H-tetrazole and then mCPBA, followed by hydrogenolytic or acidolytic deprotection of the respective protected Ser(PO$_3$Bzl$_2$)- and Ser(PO$_3$Bu$^t_2$)-tripeptides (Figure 4).[8] In addition, both the benzyl and *t*-butyl phosphoramidite "phosphite-triester" procedures also were extended to the efficient solid-phase phosphorylation of the polystyrene resin-bound Ser-tripeptide Z–Glu(OBzl)–Ser–Leu–O–

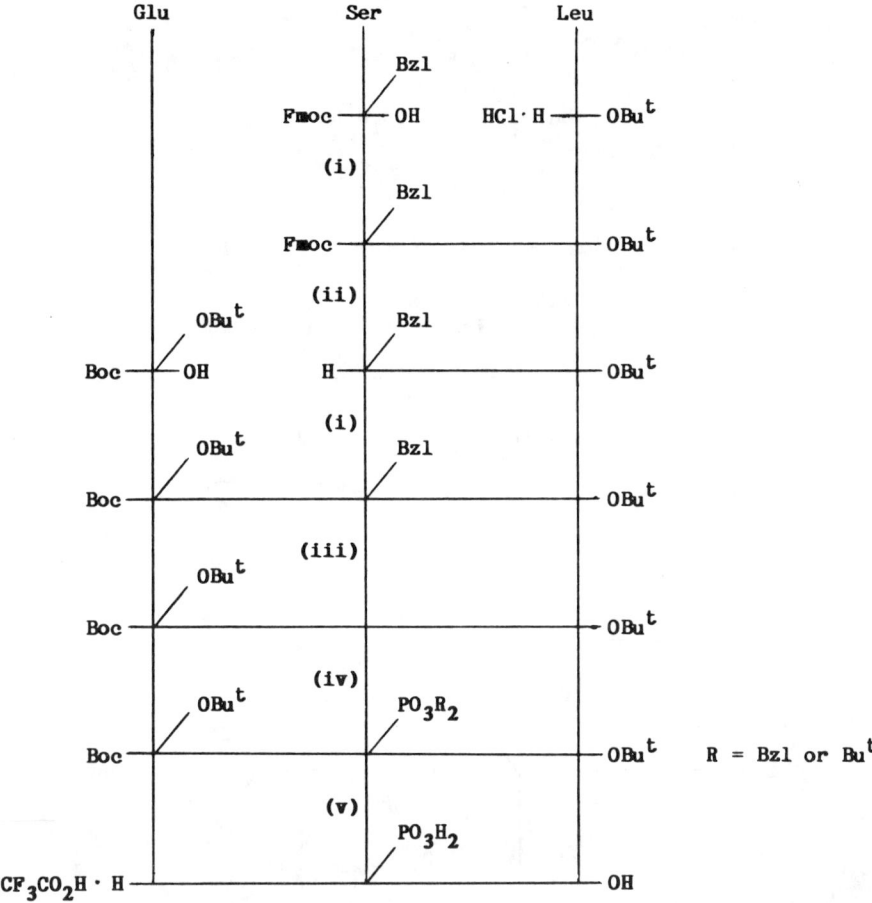

FIGURE 4. Synthesis of CF$_3$CO$_2$H·H–Glu–PSer–Leu–OH by the "global" benzyl- or *t*-butyl-phosphoramidite "phosphite-triester" phosphorylation approach. (i) NMM, IBCF, −20°C; (ii) 20% Et$_2$NH/THF (60 min, 20°C); (iii) H$_2$-Pd/C — 50% AcOH/MeOH; (iv-a) (BzlO)$_2$ PNEt$_2$/1H-tetrazole (10 min, 20°C), mCPBA (−40°C) or (iv-b) (Bu^{t}O)$_2$PNEt$_2$/1H-tetrazole (10 min, 20°C), mCPBA (−40°C); (v-a) H$_2$-Pd/C — 40% TFA/AcOH or (v-b) 40% TFA/AcOH.

resin[16] (Figure 5), quantitative seryl phosphorylation being confirmed by ^{13}C-NMR and ^{31}P-NMR spectroscopic examination of the phosphorylated resin-bound peptides.

The efficiency of the benzyl-based phosphoramidite "phosphite-triester" phosphorylation procedure for the preparation of large single PSer-containing peptides also was demonstrated by Bannwarth and Trzeciak[17] in 1987 by the synthesis of the delta sleep-inducing PSer-nonapeptide H–Trp–Ala–Gly–Gly–Asp–Ala–PSer–Gly–Glu–OH. The protected Ser(PO$_3$Bzl$_2$)-nonapeptide was prepared by a convergent strategy which involved "phosphite-triester" phosphorylation of the Ser-pentapeptide with (1) (BzlO)$_2$PN^{i}Pr$_2$/1H-tetrazole and (2) mCPBA, acidolytic cleavage of the N-terminal Boc group (1 M HCl/AcOH, 5 min), and subsequent mixed anhydride coupling of the Ser(PO$_3$Bzl$_2$)-pentapeptide hydrochloride with a protected tetrapeptide. Final hydrogenolytic deprotection of the protected Ser(PO$_3$Bzl$_2$)-nonapeptide followed by chromatographic purification (DEAE-Sephadex®) gave the PSer-nonapeptide in an overall yield of 22% (Figure 6).

Despite the successful synthesis of the PSer-nonapeptide, the synthesis of large PSer-peptides by a similar convergent strategy should be approached with caution in view that Perich[18] reported extensive acidolytic debenzylation of the seryl dibenzyl phosphorotriester functionality under various acidolytic conditions. In a ^{31}P-NMR study, approximately 5%

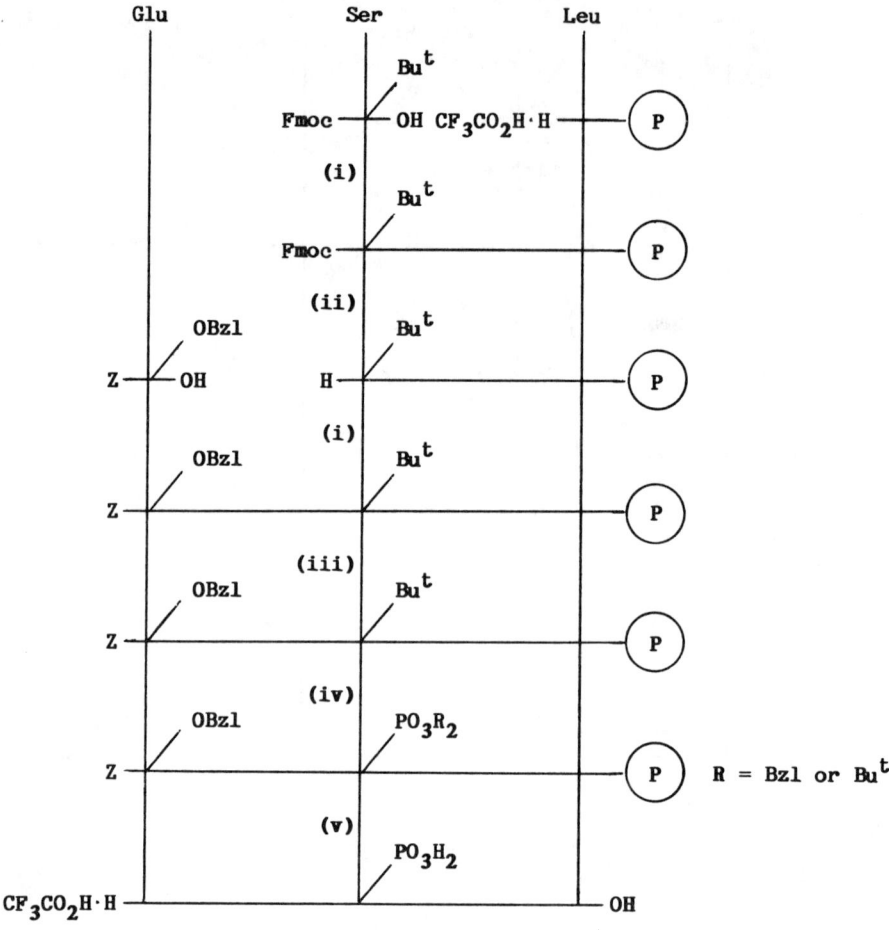

FIGURE 5. Synthesis of CF$_3$CO$_2$H·H–Glu–PSer–Leu–OH by the solid-phase "global" benzyl- or t-butyl-phosphoramidite "phosphite-triester" phosphorylation of a resin-bound Ser-containing peptide. (i) NMM, IBCF, −20°C; (ii) 20% piperidine/DMF (10 min, 30°C); (iii) 90% TFA/CH$_2$Cl$_2$ (60 min); (iv-a) (BzlO)$_2$PNEt$_2$/1H-tetrazole (20 min, 20°C), mCPBA (20°C) or (iv-b) (Bu^{t}O)$_2$PNEt$_2$/1H-tetrazole (20 min, 20°C), mCPBA (20°C); (v-a) H$_2$-Pd(OAc)$_2$/DMF, then 40% TFA/H$_2$O.

acidolytic loss of a benzyl group from the dibenzyl phosphorotriester functionality resulted from a 30-min treatment of isopropyl dibenzyl phosphate with 1 M HCl/AcOH (see Section I.2). Hence, it is probable that, under routine peptide synthesis conditions, low or moderate PSer-peptide yields (as with the PSer-nonapeptide) can be attributed to partial cleavage of the benzyl phosphate group during acidolytic removal of the Boc group.

In continuation of their studies, Perich and Johns[19] used dibenzyl N,N-diethylphosphoramidite for the preparation of Ac–PSer–NHMe and the multiple PSer-containing peptides Ac–PSer–PSer–NHMe and Ac–PSer–PSer–PSer–NHMe (compare with Figure 11). However, as yields of 95, 60, and 15% were obtained for the mono-, di-, and tri-Ser(PO$_3$Bzl$_2$)-peptides, respectively, the marked decrease in yields indicated that the "global phosphite-triester" phosphorylation procedure was not totally suitable for the phosphorylation of multiple-serine-containing peptides. Despite this limitation, the phosphoramidite-phosphorylation procedure nevertheless presents a practical and viable alternative to the use of diaralkyl phosphorochloridates for the phosphorylation of single-serine-containing peptides.

An interesting finding from the above study was the observation that the N-terminal Ac–Ser(PO$_3$Bzl$_2$)-residue of the three N-acetylated Ser(PO$_3$Bzl$_2$)-peptides underwent com-

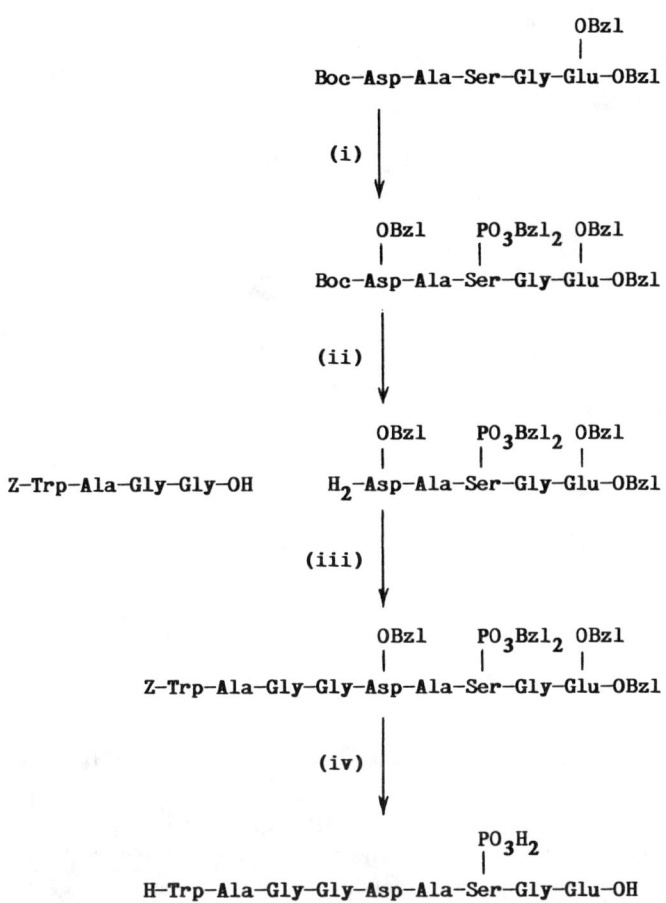

FIGURE 6. Convergent synthesis of the *P*Ser-nonapeptide using the "global" benzyl-based phosphoramidite "phosphite-triester" phosphorylation procedure. (i) (PhO)$_2$P(O)Cl/pyridine; (ii) 1 *M* HCl/AcOH (5 min, 20°C); (iii) NMM, IBCF (−15°C); (iv) H$_2$-Pd/C — AcOH/H$_2$O.

plete rearrangement to the (BzlO)$_2$PO$_2$H·H−Ser-residue after 12 months storage (10 to 20°C)[8,19] (Figure 7). In view of the slow rearrangement of these residues, it is recommended that peptides containing an N-terminal Ser(PO$_3$Bzl$_2$)-residue be hydrogenated as soon as possible and not stored for later use. It was also particularly interesting that while complete rearrangement of the Ac−Ser(PO$_3$Bzl$_2$)-residue of the mono-, di-, and tri-Ser(PO$_3$Bzl$_2$)-peptides occurred over 12 months, rearrangement of the N-terminal Ac−Ser(PO$_3$H$_2$)-residue of the mono-, di-, and tri-Ser(PO$_3$H$_2$)-peptides also occurred with prolonged storage. However, this process was much slower, with approximately 50% conversion of the Ac−Ser(PO$_3$H$_2$)-residue to the H$_3$PO$_4$·H−Ser-residue after a period of 5 years.[8,19]

C. SOLUTION-PHASE SYNTHESIS OF *P*SER-PEPTIDES BY SYNTHON INCORPORATION

1. Synthesis of Boc−Ser(PO$_3$R$_2$)−OH Derivatives

In order to overcome the inherent difficulties of a "global" phosphorylation procedure, it was recognized early that the direct incorporation of Boc−Ser(PO$_3$R$_2$)−OH derivatives into Boc/solution-phase peptide synthesis would lead to a general and versatile procedure for the efficient preparation of *P*Ser-peptides. However, while the protection of the amino, carboxyl, and side-chain functionalities of the majority of the amino acids was well established in

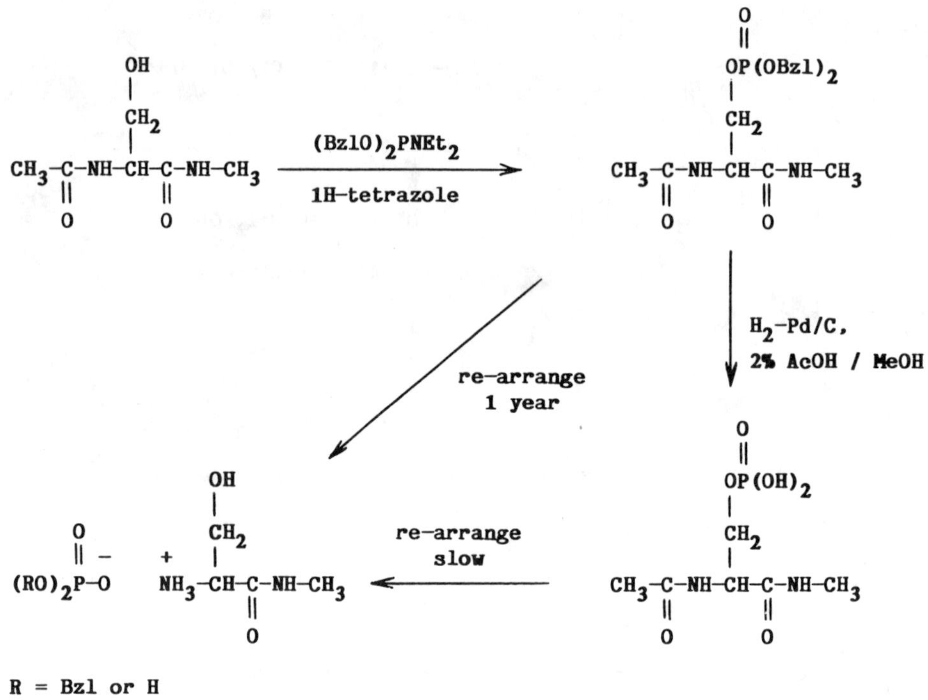

FIGURE 7. Synthesis of Ac–Ser(PO$_3$Bzl$_2$)–NHMe and Ac–Ser(PO$_3$H$_2$)–NHMe and its rearrangement products.

peptide chemistry,[20] the success of this synthetic approach lay in the compatibility of the phosphate protecting group with (1) the efficient phosphorylation of protected serine residues and (2) removal of the groups under conditions which do not cause rearrangement or β-elimination of the –Ser(PO$_3$R$_2$)– or –Ser(PO$_3$H$_2$)– residue. In his studies, Perich[18] examined the use of the phenyl, ethyl, methyl, benzyl, and t-butyl groups for temporary protection of the phosphate functionality because these groups permitted either acidolytic or hydrogenolytic deprotection.

The synthesis of the five Boc–Ser(PO$_3$R$_2$)–OH (R = Ph, Et, Me, Bzl, Bu^t) derivatives was accomplished by a three-step procedure which involved (1) initial protection of the carboxyl group of Boc–Ser–OH with the 4-nitrobenzyl group, (2) phosphorylation of the seryl hydroxyl by either dialkyl or diaryl phosphorochloridate, and (3) final removal of the carboxyl protecting group (Figure 8).[18,21-23] While the carboxyl 4-nitrobenzyl group generally was removed by catalytic hydrogenation, the presence of benzyl groups at the phosphate functionality necessitated sodium dithionite reduction[24] of the 4-nitrobenzyl group from Boc–Ser(PO$_3$Bzl$_2$)–ONBzl.[18,22]

The crucial step in this three-step synthetic procedure was determined to be dictated by the efficient and complete phosphorylation of Boc–Ser–ONBzl. While near-quantitative yields of Boc–Ser(PO$_3$Ph$_2$)–ONBzl and Boc–Ser(PO$_3$Et$_2$)–ONBzl were obtained through the use of diphenyl and diethyl phosphorochloridate, only moderate yields (40 to 60%) of Boc–Ser(PO$_3$Me$_2$)–ONBzl and Boc–Ser(PO$_3$Bzl$_2$)–ONBzl were initially obtained using literature-reported phosphorylation procedures. However, near-quantitative yields of these two products were subsequently obtained by working at low temperatures and through the use of second additions of dimethyl or dibenzyl phosphorochloridate.[22] From a ^{31}P-NMR spectroscopy study it was established that the initial moderate yields of these latter products were not due to the low reactivity of the dimethyl and dibenzyl phosphorochloridates, but were the result of pyridine-mediated dealkylation of the phosphorylating agent, presumed to be

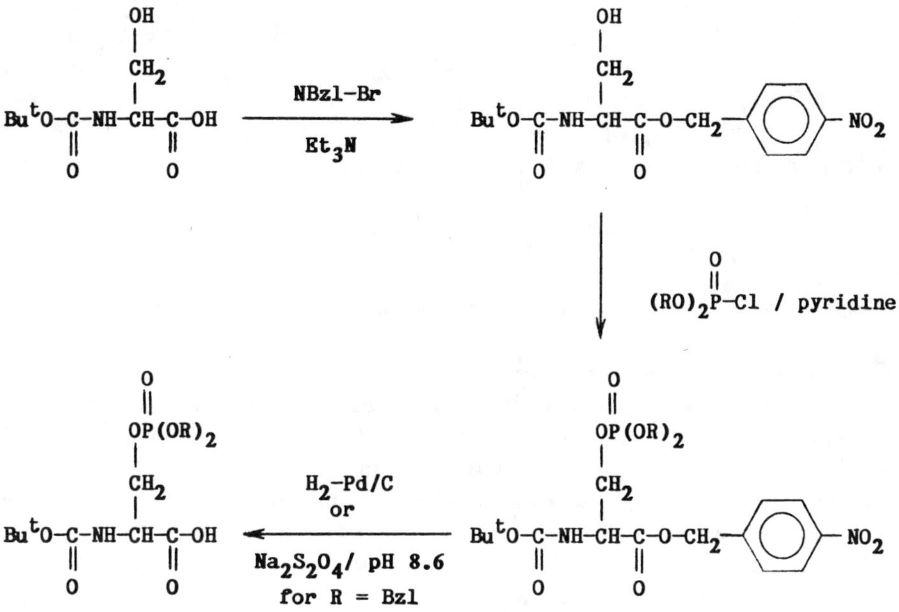

FIGURE 8. Synthetic route for the preparation of Boc–Ser(PO$_3$R$_2$)–OH derivatives.

the phosphoropyridinium chloride intermediate.[18,22] Contrary to a recent unqualified report,[17] the *modified* dibenzyl phosphorochloridate/pyridine phosphorylation procedure presents a particularly useful procedure for the phosphorylation of serine derivatives (but not Ser-peptides) and other hydroxyl-containing biomolecules since (1) dibenzyl phosphorochloridate is easily prepared from the treatment of commercially available dibenzyl hydrogen phosphonate with *N*-chlorosuccinimide,[25] (2) the low-temperature phosphorylation reaction is chemically simple, and (3) near-quantitative yields of pure products are consistently obtained under such conditions.

Apart from the above phosphorylation procedure, the efficient phosphorylation of Boc–Ser–ONBzl also was effected through the use of either the (1) diaralkyl phosphorochloridite (R = Ph, Et, Me)/pyridine[18,21] or (2) diaralkyl *N,N*-diethylphosphoramidite (R = Ph, Et, Me, Bzl, Bu^t)/1H-tetrazole[18,26] "phosphite-triester" phosphorylation procedure followed by an oxidation step. While the use of mCPBA was necessary for the nonnucleophilic oxidation of the benzyl and *t*-butyl phosphite triesters,[14,15,18] oxidation of the phenyl, ethyl, and methyl phosphite triesters could be effected by the use of either I$_2$/H$_2$O or mCPBA.[18,21] The use of I$_2$/H$_2$O for the oxidation of benzyl and *t*-butyl phosphites is not possible due to extensive iodide-mediated dealkylation during the nucleophilic oxidation process.

In addition to the preparation of the synthons by the three-step synthetic procedure, the phenyl, ethyl, and methyl derivatives also were prepared by a one-step procedure which involved treatment of Boc–Ser–OH with (RO)$_2$P–Cl (R = Ph, Et, Me) followed by I$_2$/H$_2$O oxidation of the resultant bis-phosphite intermediate (Figure 9).[27] While this one-step procedure is considerably timesaving, a drawback in its use is that product purity varies with each preparation and is a function of the purity of the diaralkyl phosphorochloridite. As the preparation of diaralkyl phosphorochloridites requires a high level of chemical experience, since their distillative isolation can be explosive, and because diaralkyl phosphorochloridites disproportionate with storage, the three-step synthetic procedure is recommended as the method of choice for use in biochemical laboratories.

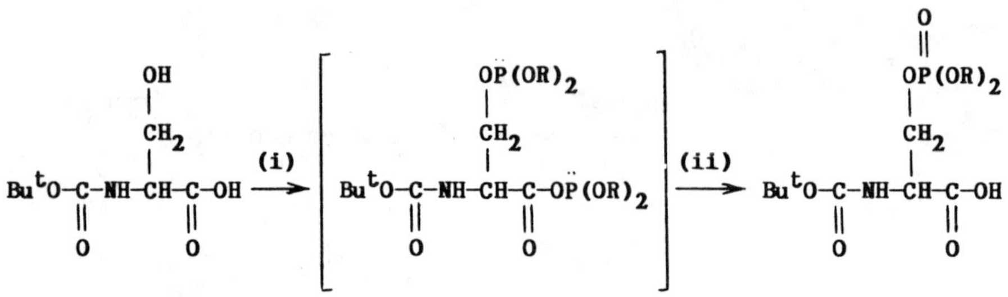

FIGURE 9. One-pot preparation of Boc–Ser(PO$_3$R$_2$)–OH derivatives using diaralkyl phosphorochloridites.

a. General Phosphorylation Procedures

Method 1a: Phosphorylation of Boc–Ser–ONBzl using diphenyl phosphorochloridate/pyridine — A solution of diphenyl phosphorochloridate (4.0 g, 5.0 mmol) in dry THF is added to Boc–Ser–ONBzl (3.40 g, 10.0 mmol) in dry pyridine (10 ml) at 20°C. After stirring for 4 h, water (5 ml) is added and the reaction mixture stirred for a further 10 min. Ethyl acetate (100 ml) is then added and the organic phase washed with 1 M HCl (5 × 30 ml), 5% NaHCO$_3$ (2 × 30 ml), dried (Na$_2$SO$_4$), and filtered. Solvent evaporation under reduced pressure gives a white solid which on crystallization from ethyl acetate/petroleum spirits gives Boc–Ser(PO$_3$Ph$_2$)–ONBzl as fine, off-white needles (5.60 g, 98%), mp = 113 to 114.5°C, [α]^{14} = +6.71° (c, 1 in CHCl$_3$).[18,23]

Method 1b: Phosphorylation of Boc–Ser–ONBzl using the modified dibenzyl phosphorochloridate/pyridine phosphorylation procedure — A solution of dibenzyl phosphorochloridate (4.45 g, 15.0 mmol) in dry THF is slowly added to a solution of Boc–Ser–ONBzl (3.40 g, 10.0 mmol) in pyridine/THF (5 ml/5 ml) such that the temperature is maintained at −40°C. On completion of addition, the reaction mixture is stirred for 3 h and then a second solution of dibenzyl phosphorochloridate (2.22 g, 7.5 mmol) in THF (2.5 ml) is slowly added at −40°C. After stirring for a further 3 h at −40°C, water (5 ml) is added and the reaction mixture stirred for 10 min. Diethyl ether (100 ml) is then added and the ethereal phase washed with 1 M HCl (5 × 30 ml), 5% NaHCO$_3$ (3 × 30 ml), dried (Na$_2$SO$_4$), and filtered. Solvent evaporation under reduced pressure gives a light yellow oil which on crystallization from diethyl ether/pentane gives Boc–Ser(PO$_3$Bzl$_2$)–ONBzl as an off-white solid (5.64 g, 94%), mp = 50 to 51°C, [α]^{14} = +3.04° (c, 1 in CHCl$_3$).[18,22]

Method 1c: Phosphorylation of Boc–Ser–ONBzl using dibenzyl or di-*t*-butyl *N,N*-diethylphosphoramidite — 1H-tetrazole (1.85 g, 25.0 mmol) is added in one portion to a solution of Boc–Ser–ONBzl (3.40 g, 10.0 mmol) in dry THF (10 ml) containing dibenzyl or di-*t*-butyl *N,N*-diethylphosphoramidite (3.80 or 3.00 g, 12.0 mmol) at +20°C and stirred for 10 min. The reaction mixture is then cooled to −40°C, and a solution of 85% mCPBA (2.44 g, 12.0 mmol) in CH$_2$Cl$_2$ (12 ml) is added such that the temperature is kept below 0°C. After stirring for 10 min at 20°C, 10% Na$_2$S$_2$O$_5$ (10 ml) is added, the aqueous phase discarded, and the ethereal phase washed with 10% Na$_2$S$_2$O$_5$ (1 × 30 ml), 5% NaHCO$_3$ (3 × 30 ml), 5% citric acid (pH 3.3; 1 × 30 ml), backwashed with 5% NaHCO$_3$ (1 × 30 ml), dried (Na$_2$SO$_4$), and filtered. Solvent evaporation under reduced pressure gives Boc–Ser(PO$_3$Bzl$_2$)–ONBzl and Boc–Ser(PO$_3$Bu$^t_2$)–ONBzl as clear oils (92 to 95%). Recrystallization of the latter oil from hexane gives Boc–Ser(PO$_3$Bu$^t_2$)–ONBzl as a white amorphous solid, mp = 70 to 72°C, [α]^{14} = −9.47° (c, 1 in CHCl$_3$).[18]

b. Synthesis of Boc–Ser(PO$_3$Ph$_2$)–OH

A solution of Boc–Ser(PO$_3$Ph$_2$)–ONBzl (5.60 g, 9.8 mmol) in ethyl acetate/acetic acid (49 ml:1 ml) containing 10% Pd/C (980 mg) is charged with hydrogen and left until the cessation of hydrogen uptake. The catalyst is then removed by gravity filtration and the

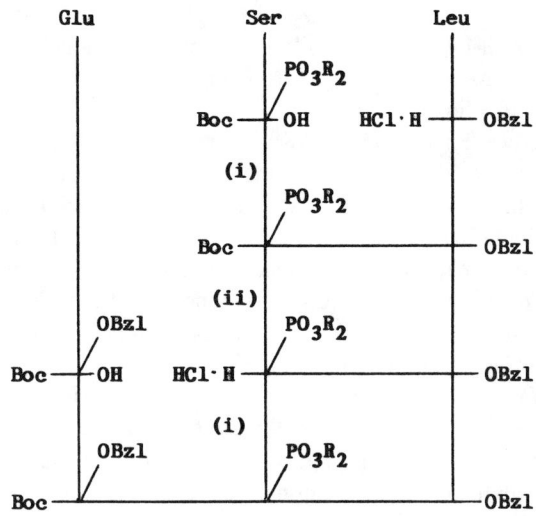

FIGURE 10. Synthesis of protected Ser(PO_3R_2)-tripeptides by the use of Boc–Ser(PO_3R_2)–OH derivatives. (i) NMM, IBCF ($-20°C$); (ii) 4 M HCl/dioxane (20 min, 20°C).

solvent evaporated under reduced pressure. The yellow oil is dissolved in diethyl ether (50 ml), washed with 1 M HCl (2 × 20 ml), and then extracted with 5% $NaHCO_3$ (3 × 20 ml). The combined base extracts are then washed with diethyl ether (2 × 20 ml), acidified to pH 1 (2 M HCl), extracted with dichloromethane (3 × 30 ml), dried (Na_2SO_4), and filtered. Solvent evaporation under reduced pressure gives Boc–Ser(PO_3Ph_2)–OH as a clear oil (4.31 g, 93%) which solidifies on storage, mp = 62 to 64°C, $[\alpha]^{14}$ = 40.0° (c, 1 in $CHCl_3$).[18,23]

2. Synthesis of Protected Ser(PO_3R_2)-Tripeptides

The suitability of the five Boc–Ser(PO_3R_2)–OH derivatives for use in Boc/solution-phase peptide synthesis was demonstrated by the synthesis of the protected Ser(PO_3R_2)-tripeptides Boc–Glu(OBzl)–Ser(PO_3R_2)–Leu–OBzl. While all five Ser(PO_3R_2)-dipeptides were isolated in 92 to 98% yields by their mixed anhydride condensation with HCl·H–Leu–OBzl, the four protected Ser(PO_3R_2)-tripeptides (R = Ph,[23] Et, Me, Bzl) were obtained in 94, 91, 95, and 33% yields, respectively, by initial acidolytic cleavage of the Boc group from the Ser(PO_3R_2)-dipeptides with 4 M hydrogen chloride/dioxane followed by the mixed anhydride coupling of Boc–Glu(OBzl)–OH with the resultant dipeptide hydrochlorides[13] (Figure 10). In the case of Boc–Ser($PO_3Bu^t_2$)–Leu–OBzl, N-terminal peptide extension was not possible due to the greater acidolytic susceptibility of the t-butyl phosphate groups over the Boc group (both t-butyl groups being observed in a ^{31}P-NMR spectroscopy study to be cleaved within 3 min by 4 M HCl/dioxane or 40% TFA/CH_2Cl_2).[15,18]

From a ^{31}P-NMR spectroscopy study using Z–Ser(PO_3Bzl_2)–Leu–OBzl as a model substrate, the low yield of the Z–Glu(OBzl)–Ser(PO_3Bzl_2)–Leu–OBzl was found to be due to extensive acidolytic debenzylation of the dibenzyl phosphorotriester functionality during acidolytic removal of the N-terminal Boc group, approximately 35 and 50% benzyl loss being observed from the Ser(PO_3Bzl_2)-dipeptide after a 30-min treatment with 4 M hydrogen chloride/dioxane and 1 M hydrogen chloride/acetic acid, respectively.[18] Under these conditions, acidolytic benzyl cleavage was presumed to proceed via initial protonation of the phosphate oxygen followed by chloride-mediated debenzylation of the resultant phosphonium chloride intermediate. Hydrogen chloride/dioxane was later established to be unsatisfactory for the deblocking of Boc-protected Ser(PO_3Bzl_2)-peptides, since the acid-base titration curve obtained from the ^{31}P-NMR chemical shift value/log [HCl] graph showed the phosphate oxygen to be totally protonated in 2 M hydrogen chloride/dioxane and 50% protonated in 1.2 M hydrogen chloride/dioxane.[18]

In a later study, Perich used 98% formic acid[28] for the deblocking of Boc–Ser(PO_3Bzl_2)–Leu–OBzl and found that the use of this nonnucleophilic Boc-deblocking procedure led to the isolation of Boc–Glu(OBzl)–Ser(PO_3Bzl_2)–Leu–OBzl and Boc–Glu(OBzl)–Ser(PO_3{Bzl}{H})–Leu–OBzl in 81[29] and 7% yields, respectively, the higher yield of the Ser(PO_3Bzl_2)-tripeptide being established to be due to minimal acidolytic debenzylation of the dibenzyl phosphorotriester functionality during Boc cleavage from Boc–Ser(PO_3Bzl_2)–Leu–OBzl ($t_{1/2}$ {Bzl_1} = 24 h for 98% HCOOH compared to $t_{1/2}$ {Bzl_1} = 0.5 h for 4 M hydrogen chloride/dioxane).[18]

To avoid excessive benzyl loss during Boc-peptide synthesis, Ppoc–Ser(PO_3Bzl_2)–OH (prepared by employing Ppoc–Ser–OH in the synthetic route outlined in Figure 8) was examined for possible application in peptide synthesis on account of the mild acidolytic cleavage of the 2-phenylisopropyloxycarbonyl (Ppoc) group.[30] By using 2% TFA/CH_2Cl_2 for cleavage of the Ppoc group from Ppoc–Ser(PO_3Bzl_2)–Leu–OBzl, Boc–Glu(OBzl)–Ser(PO_3Bzl_2)–Leu–OBzl was obtained in an overall yield of 89% (95 and 94% yields for the two coupling steps)[16] after chromatographic removal of Ppoc cleavage by-products. Despite the high yield of the Ser(PO_3Bzl_2)-tripeptide using Ppoc–Ser(PO_3Bzl_2)–OH, a Ppoc-based methodology was found to be inconvenient for the synthesis of larger, more complex Ser(PO_3Bzl_2)-peptides because (1) such peptide syntheses require a range of noncommercial Ppoc-amino acids, (2) the preparation of large quantities of Ppoc–Ser(PO_3Bzl_2)–OH is complicated during reductive carboxyl deprotection of the 4-nitrobenzyl group from Ppoc–Ser(PO_3Bzl_2)–ONBzl, and (3) the derivative decomposes during storage. Although not suitable for use in solution-phase syntheses, it is possible that Ppoc–Ser(PO_3Bzl_2)–OH may be of some future use in Bpoc/solid-phase peptide synthesis.

a. General Procedure for the Incorporation of Boc–Ser(PO_3R_2)–OH Derivatives into Peptide Synthesis

N-Methylmorpholine (0.30 g, 3.0 mmol) in THF (2 ml) and isobutyl chloroformate (0.38 g, 2.79 mmol) in THF (2 ml) are added successively to a solution of Boc–Ser(PO_3R_2)–OH (3.00 mmol) in THF (10 ml) at −20°C. After an activation period of 3 min, a solution of HCl·H–Leu–OBzl (0.55 g, 2.14 mmol) and N-methylmorpholine (0.216 g, 2.14 mmol) in THF (3 ml) is added at −20°C and the reaction mixture stirred at −20°C for 2 h prior to the addition of 5% $NaHCO_2$ (2 ml). After stirring at 0°C for 30 min, diethyl ether (70 ml) is added and the organic phase washed with 5% $NaHCO_3$ (2 × 30 ml), 1 M HCl (5% citric acid, pH 4, for Boc–Ser($PO_3Bu^t_2$)–Leu–OBzl) (2 × 30 ml), dried (Na_2SO_4), and filtered. Solvent evaporation under reduced pressure gives Boc–Ser(PO_3R_2)–Leu–OBzl as a clear, colorless oil (93 to 98%).[18]

Boc–Ser(PO_3R_2)–Leu–OBzl (2.0 mmol) is dissolved in 4 M hydrogen chloride/dioxane (5 ml) and allowed to stand at 20°C for 30 min [or 98% formic acid for Boc–Ser(PO_3Bzl_2)–Leu–OBzl for 60 min]. The solvent is then evaporated under reduced pressure, the gummy solid triturated with diethyl ether (3 × 30 ml) and dried under high vacuum.

N-Methylmorpholine (0.283 g, 2.8 mmol) in THF (2 ml) and isobutyl chloroformate (0.335 g, 2.60 mmol) in THF (2 ml) are added successively to a solution of Boc–Glu(OBzl)–OH (0.944 g, 2.80 mmol) in THF (10 ml) at −20°C. After an activation period of 3 min, a solution of HCl·H–Ser(PO_3R_2)–Leu–OBzl (2.00 mmol) and N-methylmorpholine (0.202 g, 2.00 mmol) in THF (3 ml) is added at −20°C and the reaction mixture stirred at −20°C for 2 h period to the addition of 5% $NaHCO_3$ (2 ml). After stirring at 0°C for 30 min, diethyl ether (70 ml) is added and the organic phase washed with 5% $NaHCO_3$ (2×30 ml), 1 M HCl (2×30 ml), dried (Na_2SO_4), and filtered. Solvent evaporation under reduced pressure gives Boc–Glu(OBzl)–Ser(PO_3R_2)–Leu–OBzl as a clear, colorless oil (81% for R = Bzl and 93 to 98% for R = Ph, Et, Me).[18]

3. Deprotection of Protected Ser(PO_3R_2)-Peptides

The final and most crucial step in the synthesis of *P*Ser-peptides requires the efficient removal of protecting groups — in particular, the phosphate protecting groups — from the protected Ser(PO_3R_2)-peptides. In the case of Boc–Ser($PO_3Bu^t_2$)–Leu–OBzl, the *t*-butyl phosphate groups were found to be particularly sensitive to acidolysis, being observed from a ^{31}P-NMR study to be cleaved within 3 min using 4 M hydrogen chloride/dioxane and within 4 h using 1 M hydrogen chloride/dioxane.[18] Subsequent hydrogenolysis of Boc–Ser–(PO_3H_2)–Leu–OBzl in 40% TFA/AcOH gave $CF_3CO_2H\cdot H$–Ser(PO_3H_2)–Leu–OH in quantitative yield.

In contrast to the ready acidolytic cleavage of *t*-butyl groups, the attempted acidolytic deprotection of $CF_3CO_2H\cdot H$–Glu–Ser(PO_3Et_2)–Leu–OH or $CF_3CO_2H\cdot H$–Glu–Ser–(PO_3Me_2)–Leu–OH using either liquid hydrogen fluoride, methanesulfonic acid/diethyl sulfide,[31] or 33% hydrogen bromide/acetic acid[32] were unsuccessful, since alkyl phosphate cleavage proceeded with simultaneous decomposition of the Ser(PO_3Et_2)- or Ser(PO_3Me_2)-residue. Similarly, silylative deprotection of the Ser(PO_3Et_2)- or Ser(PO_3Me_2)-tripeptides with either trimethylsilyl iodide,[33] trimethylsilyl bromide,[33] or phenylthiotrimethylsilane/potassium thiophenolate (catalytic)[34] were also unsuccessful due to complete silylative decomposition of the Ser(PO_3Et_2)- and Ser(PO_3Me_2)-residues under the reaction conditions. The unsuccessful deprotection of the Ser(PO_3Et_2)- and Ser(PO_3Me_2)-tripeptides was synthetically disappointing due to the straightforward and high-yielding synthesis of the protected Ser(PO_3Et_2)- and Ser(PO_3Me_2)-tripeptides using Boc–Ser(PO_3Et_2)–OH and Boc–Ser(PO_3Me_2)–OH.

In accordance with previous literature reports, hydrogenolytic deprotection of the Ser(PO_3Bzl_2)- and Ser(PO_3Ph_2)-tripeptides was found to be a more successful procedure for the preparation of *P*Ser-peptides.[3] For example, palladium-catalyzed hydrogenolysis of Boc–Ser(PO_3Bzl_2)–Leu–OBzl and Boc–Glu(OBzl)–Ser(PO_3Bzl_2)–Leu–OBzl in formic acid gave zwitterionic *P*Ser–Leu and Glu–*P*Ser–Leu, respectively, in near-quantitative yields.[18,29] The structural confirmation of both these *P*Ser-peptides was readily established by their ^{13}C-NMR spectra which contained phosphorus-coupled doublet signals for the Cα and Cβ carbons of the *P*Ser-residue. In addition, positive ion FAB-MS (argon) was demonstrated to be an excellent technique for molecular weight determination, sequence elucidation, structural characterization, and purity evaluation of *P*Ser-peptides.[35] In particular, a loss of 98 mass units was observed to be a diagnostic test for the identification of an *O*-phosphoseryl residue in a peptide.

As with benzyl phosphate deprotection, phenyl phosphate deprotection also proved to be a convenient procedure for the preparation of *P*Ser-peptides. While initial platinum-catalyzed hydrogenolytic deprotections of $CF_3CO_2H\cdot H$–Glu–Ser(PO_3Ph_2)–Leu–OH using procedures from the literature were frustrated by incomplete phenyl cleavage, hydrogenations conducted using 1.0 meq PtO_2/mmol phenyl group and 40 to 50% TFA/AcOH as the reaction medium proceeded with rapid phenyl cleavage and afforded $CF_3CO_2H\cdot H$–Glu–*P*Ser–Leu–OH in quantitative yield.[18,23] As in the above benzyl deprotection, FAB-MS proved to be a powerful and highly sensitive technique for confirming that complete phenyl cleavage had occurred under the modified reaction conditions.

a. *Synthesis of PSer-Peptides by the Hydrogenation of Protected Ser(PO₃Bzl₂)- and Ser(PO₃Ph₂)-Peptides*

Method 1a — A solution of Boc–Ser(PO_3Bzl_2)–Leu–OBzl (0.668 g, 1.0 mmol) in formic acid (5 ml) containing 10% palladium on charcoal (100 mg) is charged with hydrogen and stirred until the cessation of hydrogen uptake. The catalyst is then removed by gravity filtration and the solvent evaporated under reduced pressure. Repeated trituration of the gummy solid with diethyl ether (3 × 30 ml) and high-vacuum drying gives zwitterionic PSer–Leu (0.294 g, 99%) as light, white flakes, mp = 100 to 110°C (dec.), $[\alpha]^{21} = -15.0°$ (c, 1 in 1 M HCl);[29] lit.[4] $[\alpha]^{21} = -16.0°$; ^{13}C-NMR (D$_2$O) δ 21.6 and 22.6 (Leu δ-C), 24.9 (Leu γ-C), 39.7 (Leu β-C), 52.4 (Leu α-C), 53.9 (d, J_{PC} = 8.8 Hz, Ser α-C), 63.8 (d, J_{PC} = 4.4 Hz, Ser β-C), 167.3 (Ser CO), 176.5 (Leu CO); ^{31}P-NMR (D$_2$O) δ 0.8 ppm; FAB-MS (Ar, +ve mode) m/z 299 (MH$^+$), 201, and 132.[35]

Method 1b — A solution of Boc–Glu(OBzl)–Ser(PO$_3$Bzl$_2$)–Leu–OBzl (0.887 g, 1.0 mmol) in formic acid (5 ml) containing 10% palladium on charcoal (100 mg) is charged with hydrogen and stirred until the cessation of hydrogen uptake. The catalyst is then removed by gravity filtration and the solvent evaporated under reduced pressure. Repeated trituration of the gummy solid with diethyl ether (3 × 30 ml) and high-vacuum drying gives zwitterionic Glu–PSer–Leu[29] (0.423 g, 99%) as light, white flakes, mp = 150 to 155°C (dec.), $[\alpha]^{21} = -14.0°C$ (c, 1 in 1 M HCl); ^{13}C-NMR (D$_2$O) δ 21.1 and 22.7 (Leu δ-C), 24.9 (Leu γ-C), 26.3 (Glu β-C), 29.6 (Glu γ-C), 39.9 (Leu β-C), 52.1 (Leu α-C), 52.7 (Glu α-C), 54.6 (d, J_{PC} = 5.9 Hz, Ser α-C), 64.5 (d, J_{PC} = 5.9 Hz, Ser β-C), 169.7 (Ser CO), 170.6 (Glu CO), 176.6 (Leu CO), and 176.6 (Glu δ-CO); ^{31}P-NMR (D$_2$O) δ +0.3 ppm; FAB-MS (Ar, +ve mode) m/z 428 (MH$^+$), 299, 201, and 132.[35]

Method 1c — A solution of CF$_3$CO$_2$H·H–Glu–Ser(PO$_3$Ph$_2$)–Leu–OH (0.68 g, 1.0 mmol) in 40% TFA/acetic acid (10 ml) containing 83% platinum oxide (0.59 g, 2.16 mmol) is charged with hydrogen and stirred until the cessation of hydrogen uptake. The catalyst is then removed by gravity filtration and the solvent evaporated under reduced pressure. Repeated trituration of the gummy solid with diethyl ether (3 × 30 ml) and high-vacuum drying gives CF$_3$CO$_2$H·H–Glu–PSer–Leu–OH[18] (0.527 g, 99%) as white hygroscopic flakes.

4. Synthesis of Casein-Related PSer-Peptides

As a consequence of the above evaluation studies, Perich and Johns[36] prepared Ac–PSer–NHMe and the di- and tri-PSer-peptides Ac–PSer–PSer–NHMe and Ac–PSer–PSer–PSer–NHMe by (1) the solution-phase synthesis of the protected Ser(PO$_3$Ph$_2$)-peptides using Boc–Ser(PO$_3$Ph$_2$)–OH, (2) the isobutoxycarbonyl mixed anhydride of acetic acid for peptide N-acetylation, and (3) hydrogenolytic cleavage of the phenyl phosphate protecting groups. The syntheses of Ac–PSer–NHMe, Ac–PSer-PSer–NHMe, and Ac–PSer–PSer–PSer–NHMe (shown in Figure 11) were achieved in overall yields of 93, 90, and 88%, respectively [calculated from Boc–Ser(PO$_3$Ph$_2$)–NHMe], with their purity and structure being established by FAB-MS, ^{13}C-NMR spectroscopy, and ^{31}P-NMR spectroscopy.

The viability of preparing complex PSer-peptides by using Boc–Ser(PO$_3$Ph$_2$)–OH in the "synthon incorporation strategy" was demonstrated by the synthesis of the tetra-PSer-octapeptide Ac–Glu–PSer–Leu–PSer–PSer–PSer–Glu–Glu–NHMe (Figure 12),[18,37] this heavily phosphorylated peptide sequence corresponding to region 14 to 21 of bovine β-casein. The protected tetra-Ser(PO$_3$Ph$_2$)-octapeptide was prepared in 61% yield [calculated from Boc–Glu(OBzl)–NHMe], with all mixed anhydride couplings proceeding with 90 to 99% efficiency, the purity of the intermediate peptides being assessed at each stage by ^{13}C- and ^{31}P-NMR spectroscopy. The incorporation of the four Ser(PO$_3$Ph$_2$)-residues in the protected tetra-Ser(PO$_3$Ph$_2$)-octapeptide was established from its ^{31}P-NMR spectra, which contained four resolved peaks at −11.0, −12.9, −13.0, and −13.2 ppm.[18] Phenyl phosphate protection was found to facilitate peptide synthesis due to the good solubility of the intermediate Ser(PO$_3$Ph$_2$)-peptides in chlorinated solvents, this being in contrast to the poor solubility of the corresponding Ser(PO$_3$Et$_2$)- and Ser(PO$_3$Me$_2$)-peptides. The tetra-Ser(PO$_3$H$_2$)-octapeptide was obtained in 32% by a two-step deprotection procedure which involved initial

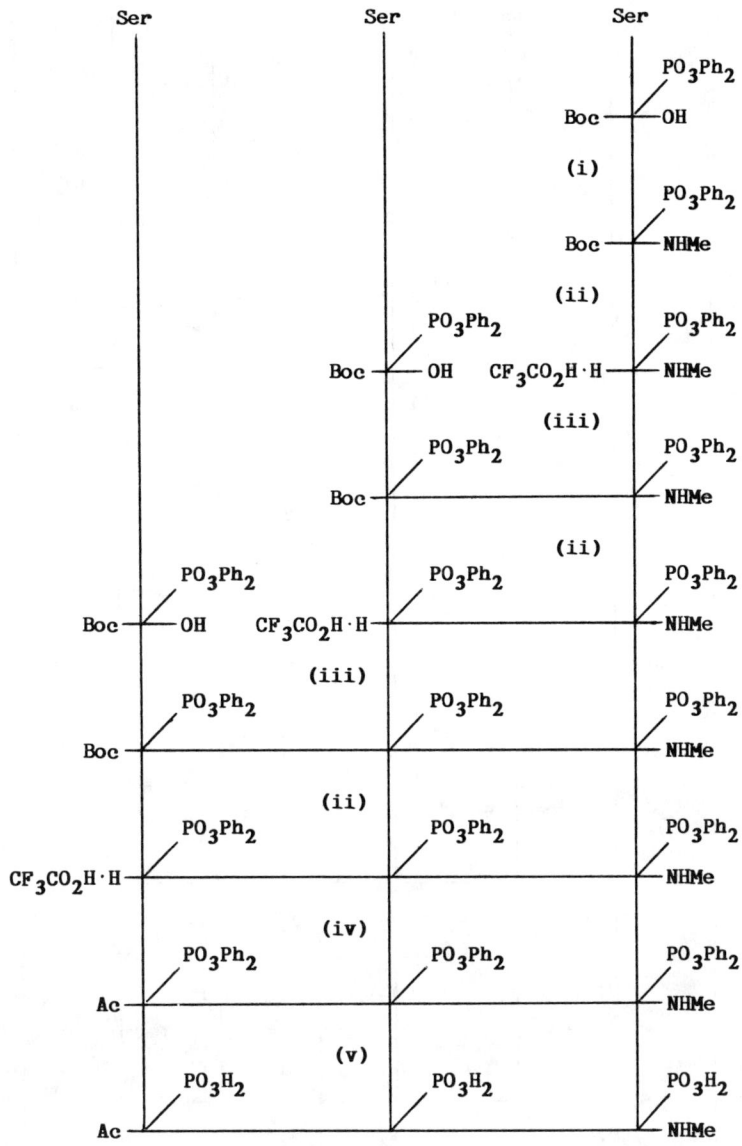

FIGURE 11. Synthesis of Ac–PSer–PSer–PSer–NHMe by the use of Boc–Ser(PO₃R₂)–OH in the "synthon incorporation strategy". (i) NMM, IBCF (−20°C), CH₃NH₂; (ii) 40% TFA/CH₂Cl₂ (60 min, 20°C); (iii) NMM, IBCF (−20°C); (iv) NMM IBCF (−20°C), CH₃CO₂H; (v) H₂-Pd/C — 40% TFA/AcOH.

palladium-catalyzed hydrogenolytic removal of the benzyl ester groups, platinum-catalyzed hydrogenolytic cleavage of the phenyl phosphate groups, and final C_{18} reversed-phase high-performance liquid chromatography (RP-HPLC) purification. Structural confirmation of the PSer-octapeptide was established by ^{13}C-NMR spectroscopy and FAB-MS, its FAB mass spectrum containing an intense [MH]⁺ peak at m/e 1242 and fragment ions consistent with four incorporated PSer-residues in the correct peptide sequence.

a. Synthesis of Ac–Glu–PSer–Leu–PSer–PSer–PSer–Glu–Glu–NHMe

A solution of Ac–Glu(OBzl)–Ser(PO₃Ph₂)–Leu–Ser(PO₃Ph₂)–Ser(PO₃Ph₂)–Ser(PO₃Ph₂)–Glu(OBzl)–Glu(OBzl)–NHMe (0.37 g, 0.2 mmol) [prepared in 61% yield from

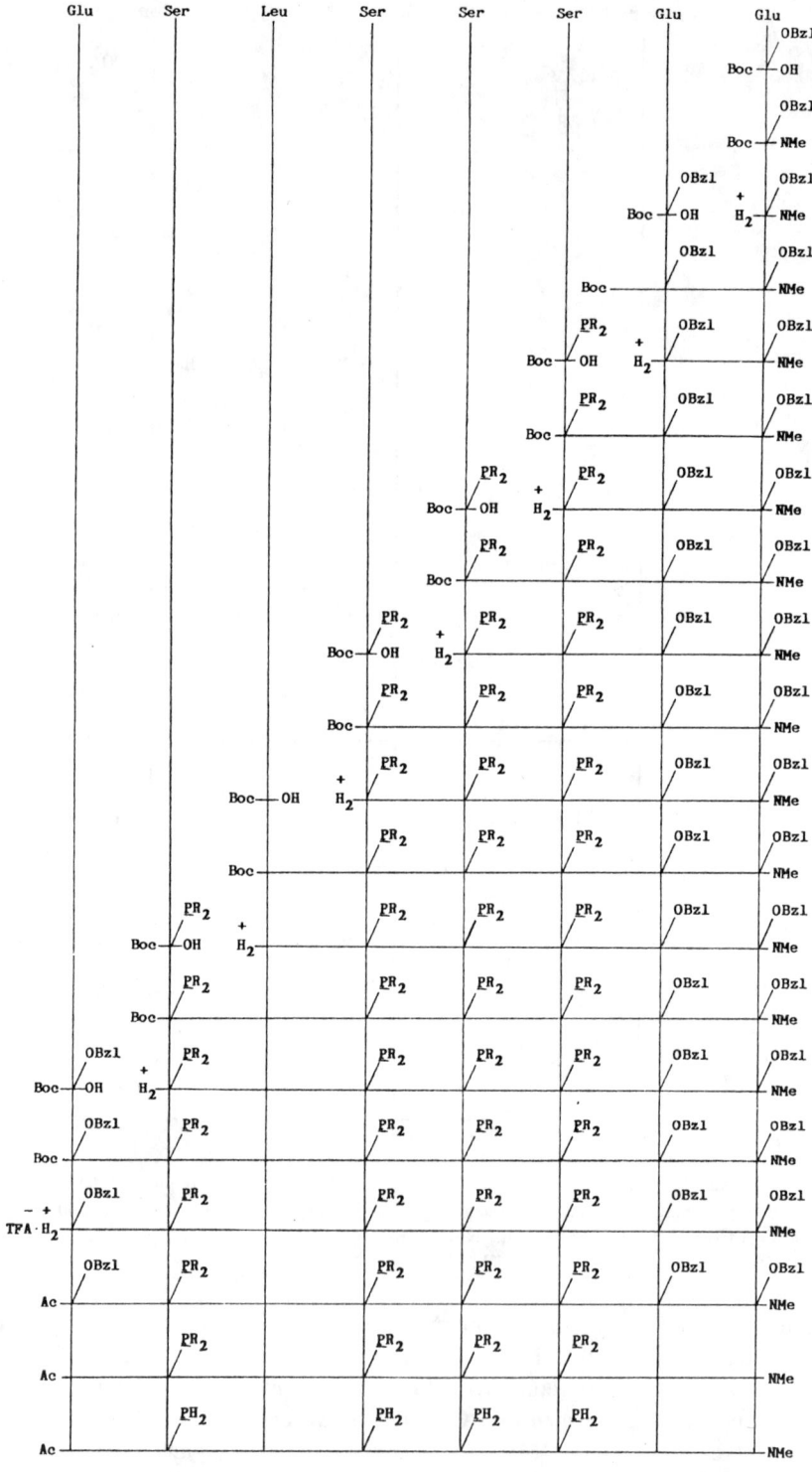

FIGURE 12. Synthesis of Ac–Glu–*P*Ser–Leu–*P*Ser–*P*Ser*P*Ser–Glu–Glu–NHMe using Boc–Ser(PO₃Ph₂)–OH in the "synthon incorporation strategy".

Boc–Glu(OBzl)–OH] in 40% TFA/AcOH (5 ml) containing 83% platinum oxide (0.438 g, 1.6 mmol) is charged with hydrogen and stirred until cessation of hydrogen uptake. The catalyst is removed by gravity filtration, the solvents evaporated under reduced pressure, and the oily residue triturated with diethyl ether (3 × 30 ml). The crude white solid is purified by C_{18} RP-HPLC chromatography (0.1% TFA/8% MeCN), the first of four fractions combined, and the solvents evaporated under reduced pressure. Lyophilization of the colorless oil gives Ac–Glu–PSer–Leu–PSer–PSer–PSer–Glu–Glu–NHMe as a light white, fluffy solid (0.13 g, 52%), ^{31}P-NMR δ (D_2O) = −0.42, FAB-MS (Ar, +ve mode) 1242 (MH^+), 1099, 970, 883, 829, 737, 687, 645, 595, 553, 503, and 457.[18,38]

D. SOLID-PHASE SYNTHESIS OF PSER-PEPTIDES BY THE "SYNTHON INCORPORATION STRATEGY"

In view of the ready β-elimination of Ser(PO_3R_2)-residues under basic conditions, the synthesis of PSer-peptides by solid-phase methodology is currently limited to a Boc-synthesis approach. In a preliminary study evaluating the potential of a Boc/solid-phase approach, the PSer-tripeptide HBr·H–Glu–Ser(PO_3Et_2)–Leu–OH[39] was obtained by the use of (1) Merrifield polystyrene resin (1% cross-linked) as the peptide support, (2) the incorporation of Boc–Ser(PO_3Et_2)–OH into the synthetic procedure, and (3) HBr/TFA for cleavage of the assembled peptide from the resin support. Hydrogen fluoride was unsuitable for this latter step due to decomposition of the Ser(PO_3Et_2)-residue under the strongly acidic conditions.

In a later study, the PSer-tripeptide CF_3CO_2H·H–Glu–PSer–Leu–OH was prepared by the use of (1) Boc–Ser(PO_3Ph_2)–OH, (2) hydrogenolytic cleavage of the assembled peptide from the resin support,[40] and (3) final platinum hydrogenolytic cleavage of the phenyl phosphate groups (Figure 13).[41] Successful incorporation of the Boc–Ser(PO_3Ph_2)–OH into the peptide sequence was established by ^{13}C- and ^{31}P-NMR spectroscopic analysis of the resin-bound peptide, its ^{13}C-NMR spectrum displaying the characteristic phosphorus-coupled doublet signals for the Cα- and Cβ-carbons of Ser(PO_3Ph_2)-residue. The use of HBr/TFA or HBr/AcOH for peptide-resin cleavage was not considered in this synthesis, since a ^{31}P-NMR-based spectroscopic study showed that slow decomposition of the Ser(PO_3Ph_2)-residue occurred under the strongly acidic conditions.[18]

In an independent study, Arendt et al.[42] prepared the PSer-tripeptide Val–PSer–Lys by employing Boc–Ser(PO_3Ph_2)–OH in a Merrifield Boc/solid-phase methodology. In this synthesis, the assembled peptide was cleaved from the resin support by HF treatment and fluoride cleavage of the phenyl phosphate protecting groups. Although the PSer-tripeptide was obtained using these latter treatments, we[39] and others[6] have discouraged the use of HF in PSer-peptide synthesis because it has been observed that such strong acid conditions cause extensive decomposition of Ser(PO_3R_2)-residues (R = Bzl, Et, Me).

II. SYNTHESIS OF O-PHOSPHOTYROSINE-CONTAINING PEPTIDES

A. INTRODUCTION

The chemical synthesis of PTyr-containing peptides is not as well developed in the literature as PSer-containing peptides; this has been due predominantly to the infrequent observation of such peptides in biochemical systems. Indeed, prior to 1983 there had only been two studies aimed at the synthesis of simple PTyr-containing peptides, both of these "global" phosphorylation procedures being tedious and lacking general applicability.[43,44] The resurgence of interest in the efficient synthesis of PTyr-peptides in 1983 was a consequence of the hypothesis that protein tyrosine phosphorylation possibly played important regulatory roles in several cellular processes. In view of the major limitations of the two synthetic approaches in the literature, our laboratory decided to approach the synthesis of

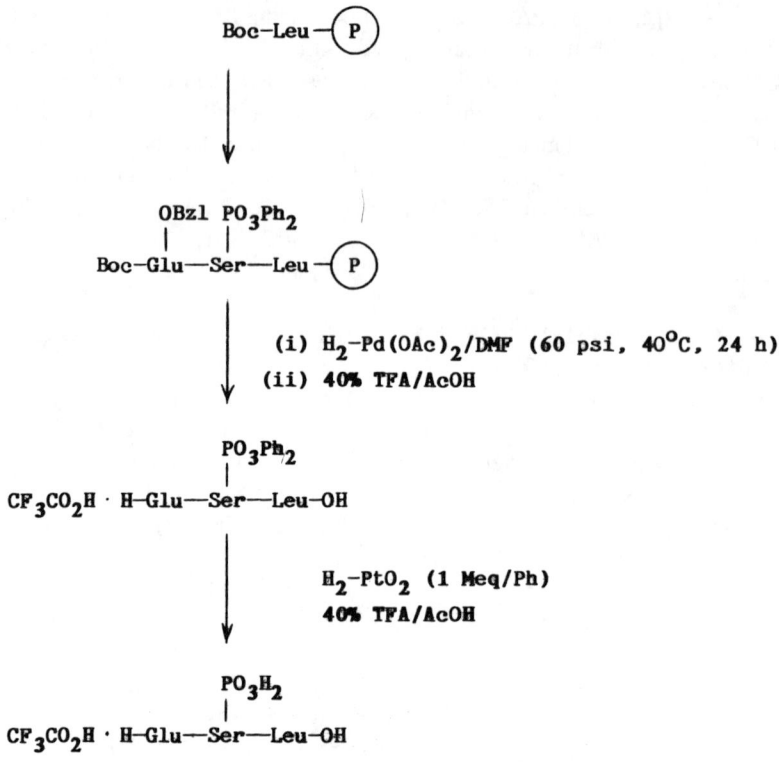

FIGURE 13. Solid-phase synthesis of CF$_3$CO$_2$H·H–Glu–PSer–Leu–OH by the use of Boc–Ser(PO$_3$Ph$_2$)–OH in the "synthon incorporation strategy".

PTyr-peptides by an alternative synthetic procedure which involved the direct incorporation of Boc–Tyr(PO$_3$R$_2$)–OH derivatives into Boc/solution or solid-phase peptide synthesis.

B. SOLUTION- AND SOLID-PHASE SYNTHESIS OF PTYR-PEPTIDES BY THE "SYNTHON INCORPORATION STRATEGY"

In 1983, Valerio[45] prepared Boc–Tyr(PO$_3$Et$_2$)–OH, Boc–Tyr(PO$_3$Me$_2$)–OH, and Boc–Tyr(PO$_3$Bzl$_2$)–OH from Boc–Tyr–OH in overall yields of 60 to 70% by a three-step procedure which involved initial carboxyl protection, dialkyl phosphorochloridate phosphorylation of the sodium phenoxide of tyrosine, and final removal of the carboxyl protecting group (Figure 14). While the 4-nitrobenzyl carboxyl protecting group was cleaved from the ethyl and methyl derivatives by catalytic hydrogenolysis, the presence of benzyl groups at the phosphate moiety necessitated saponific cleavage of the 4-nitrobenzyl group.

As an alternative to the use of dialkyl phosphorochloridates, Perich and Johns[8] phosphorylated the hydroxyl group of Boc–Tyr–ONBzl by the use of the (1) dialkyl N,N-diethylphosphoramidite (R = Et, Me, Bzl)/1H-tetrazole and (2) I$_2$/H$_2$O (R ≠ Bzl) or mCPBA "phosphite-triester" phosphorylation procedure. Likewise, Bannwarth and Trzeciak[17] effected the "phosphite-triester" phosphorylation of Bzl–Tyr–OBzl using dibenzyl N,N-diisopropylphosphoramidite.

Of the three Boc–Tyr(PO$_3$R$_2$)–OH (R = Et, Me, Bzl) derivatives, Boc–Tyr(PO$_3$Me$_2$)–OH was determined to be the most suitable for use in Boc/peptide synthesis, since ^{31}P-NMR studies using Ac–Tyr(PO$_3$Me$_2$)–NHMe as a model substrate established that the methyl phosphate groups were readily cleaved within 1 h by 10% TMSBr/CH$_3$CN or 33% HBr/AcOH. In contrast, ethyl cleavage from Ac–Tyr(PO$_3$Et$_2$)–NHMe using 10% TMSBr/MeCN was complete after 2 h, while ethyl cleavage using 33% HBr/AcOH required 22 h and gave

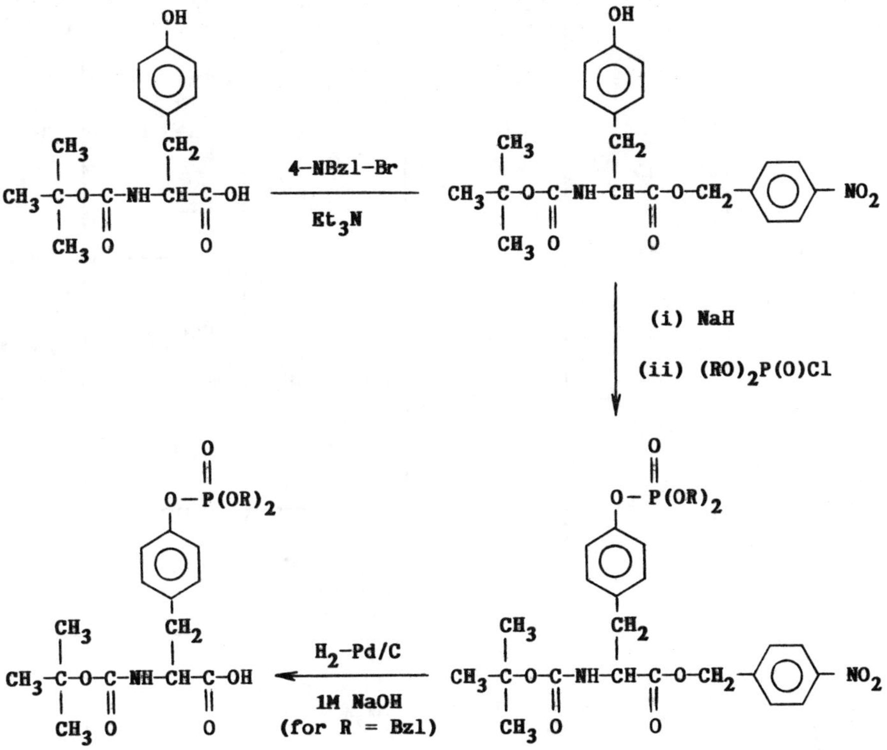

FIGURE 14. Synthetic route for the preparation of Boc–Tyr(PO$_3$R$_2$)–OH derivatives.

rise to extensive by-product formation. Boc–Tyr(PO$_3$Bzl$_2$)–OH was not compatible with Boc/Tyr(PO$_3$Bzl$_2$)-peptide synthesis due to concomitant benzyl loss during cleavage of the Boc group from Tyr(PO$_3$Bzl$_2$)-peptides by 4 M HCl/dioxane or 40% TFA/CH$_2$Cl$_2$ treatments.[45]

1. General Procedure for the Phosphorylation of Boc–Tyr–ONBzl

A solution of Boc–Tyr–ONBzl (1.67 g, 4.0 mmol) in dry dioxane (4 ml) is added dropwise to a suspension of oil-free sodium hydride (0.2 g) in dioxane (3 ml) at 10°C. After stirring for 30 min at 20°C, the reaction mixture is cooled to 10°C and dimethyl phosphorochloridate (0.87 g, 6.0 mmol) added in one portion to the bright orange solution (the solution immediately becomes colorless). After stirring for 30 min at 20°C, water (5 ml) is added and the solvents evaporated under reduced pressure. The oily residue is dissolved in ethyl acetate (30 ml), washed with 1 M HCl (1 × 10 ml), 5% NaHCO$_3$ (1 × 10 ml), H$_2$O (1 × 10 ml), dried (Na$_2$SO$_4$), and filtered. Evaporation of the solvent under reduced pressure gives Boc–Tyr(PO$_3$Me$_2$)–ONBzl as a light yellow oil (1.89 g, 90%), [α]^{20} = −10° (c, 1 in MeOH), ^{31}P-NMR δ (CDCl$_3$) = −3.8.[45,46]

2. Synthesis of Boc–Tyr(PO$_3$Me$_2$)–OH

A solution of Boc–Tyr(PO$_3$Me$_2$)–ONBzl (5.5 g, 10.5 mmol) in 10% acetic acid/methanol (30 ml) containing 5% palladium on charcoal (1.05 g) is charged with hydrogen and stirred until cessation of hydrogen uptake. The catalyst is then removed by gravity filtration and the solvent evaporated under reduced pressure. The oily residue is then dissolved in ethyl acetate (30 ml), washed with 1 M HCl (1 × 10 ml), and extracted with NaHCO$_3$ (3 × 10 ml). The combined base extracts are washed with diethyl ether (1 × 10 ml) and then acidified to pH 1 (3 M HCl). The aqueous phase is then extracted with ethyl acetate (3 ×

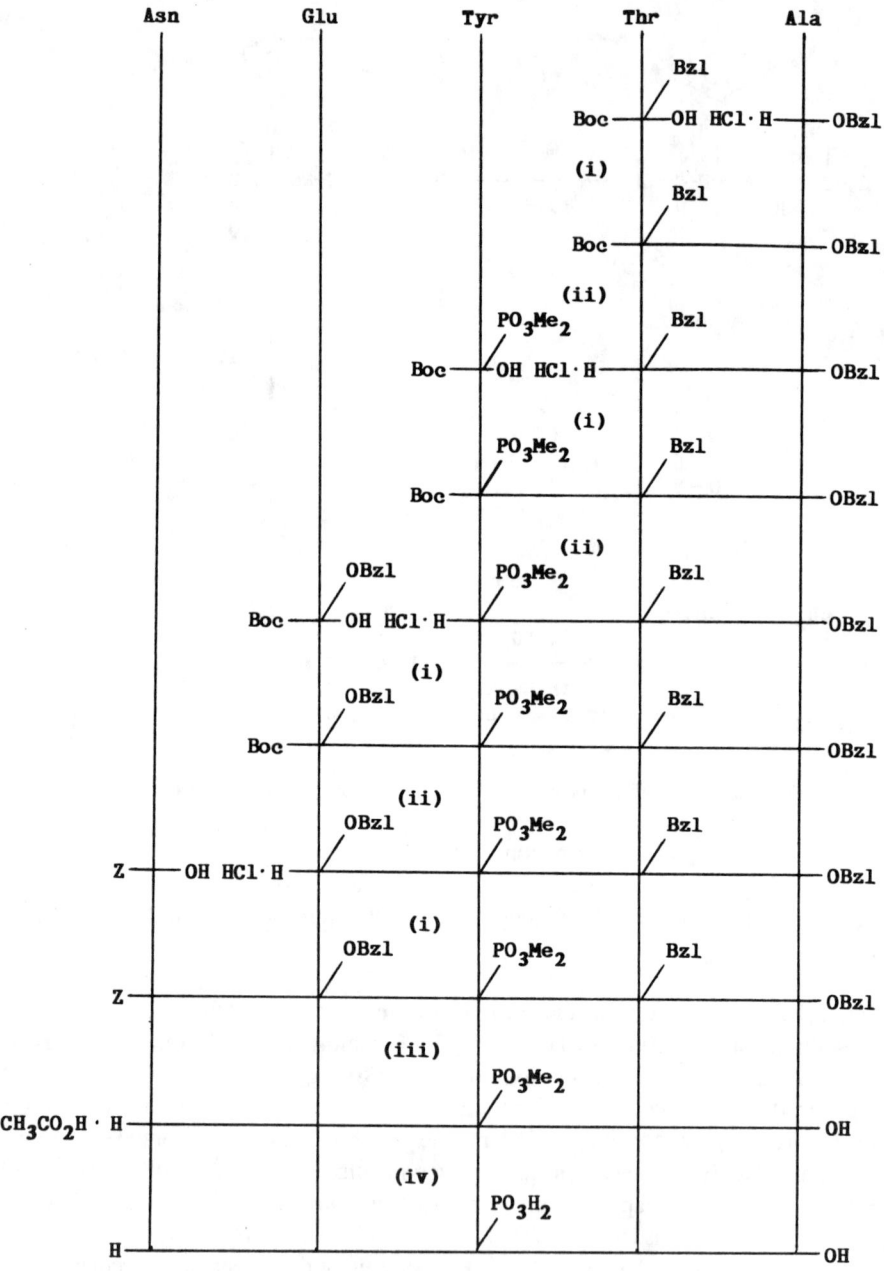

FIGURE 15. Solution-phase synthesis of H–Asn–Glu–*P*Tyr–Thr–Ala–OH by the use of Boc–Tyr(PO$_3$Me$_2$)–OH in the "synthon incorporation strategy". (i) NMM, IBCF, $-20°C$; (ii) 4 M HCl/dioxane; (iii) H$_2$-Pd(OAc)$_2$ — AcOH; (iv) 10% TMSBr/MeCN 95 h, 20°C.

20 ml), dried (Na$_2$SO$_4$), and filtered. Solvent removal under reduced pressure gives Boc–Tyr(PO$_3$Me$_2$)–OH as a clear oil (4.0 g, 98%), [α] = $-0.1°$ (c, 1 in MeOH), ^{31}P-NMR δ (CDCl$_3$) = -3.7.

The efficient preparation of *P*Tyr-peptides by the "synthon incorporation strategy" was demonstrated by the Boc/solution-phase synthesis of the *P*Tyr-pentapeptide H–Asn–Glu–*P*Tyr–Thr–Ala–OH (Figure 15),[45] this nonphosphorylated peptide sequence corresponding to region 414 to 418 of the Rous sarcoma virus (RSV) protein kinase. The

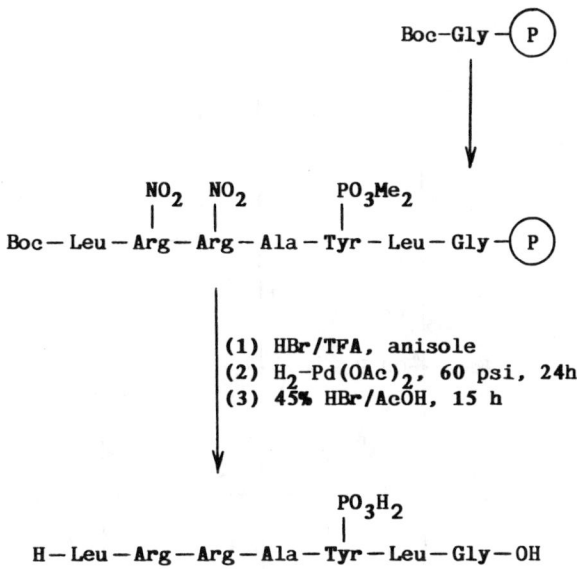

FIGURE 16. Solid-phase synthesis of $CH_3CO_2H \cdot H$–Leu–Arg–Arg–Ala–PTyr–Leu–Gly–OH by the use of Boc–Tyr-(PO_3Me_2)–OH in the "synthon incorporation strategy".

synthesis of the PTyr-pentapeptide was accomplished by the adoption of a global benzyl protection strategy, the use of the mixed anhydride coupling procedure for the formation of peptide linkages, and 4 M HCl/dioxane for repetitive removal of the Boc group from intermediate peptides. Subsequent deprotection of the protected Tyr(PO_3Me_2)-pentapeptide was accomplished by a two-step procedure which involved initial palladium acetate-catalyzed hydrogenolytic cleavage of the benzyl protecting groups followed by removal of the methyl phosphate groups using 10% TMSBr/MeCN. Final purification of the crude O-phosphopeptide using SP-Sephadex® afforded the PTyr-pentapeptide a 6% overall yield. Structural confirmation of the PTyr-heptapeptide was established by ^{31}P-NMR spectroscopy (single peak, -4.9 ppm), phosphate analysis, and FAB-MS (m/z 677, MH$^+$). The use of 33% HBr/AcOH for methyl phosphate cleavage was not considered in the deprotection procedure due to potential acetylation of the threonine hydroxyl group.

In addition to the above solution-phase synthesis, Valerio also prepared the PTyr-heptapeptide H–Leu–Arg–Arg–Ala–PTyr–Leu–Gly–OH[45,46] by the application of Boc–Tyr(PO_3Me_2)–OH in Boc/solid-phase methodology (Figure 16), the nonphosphorylated sequence corresponding to the cAMP-dependent protein kinase substrate with tyrosine substituted for serine. The synthesis of the PTyr-heptapeptide was accomplished by the use of (1) Merrifield polystyrene resin (1% divinylbenzene cross-linked) as peptide support, (2) Boc–Tyr(PO_3Me_2)–OH, and (3) the DCC/HOBt coupling procedure. After cleavage of the peptide from the resin support by HBr/TFA, the crude peptide was deprotected by initial palladium acetate-catalyzed hydrogenolytic reduction of the nitroarginine residues followed by acidolytic cleavage of the methyl phosphate groups with 33% HBr/AcOH. Final SP-Sephadex® purification of the crude peptide gave the PTyr-heptapeptide in an overall yield of 22%.

Although Valerio showed that Boc–Tyr(PO_3Bzl_2)–OH was unsuitable for use in Boc/solution or solid-phase synthesis, Perich et al.[47] demonstrated that this derivative has particular application in the synthesis of PTyr-peptides in which the PTyr-residue is located at the N-terminus. To illustrate its use, the PTyr-tripeptide $CF_3CO_2H \cdot H$–PTyr–Leu–Gly–OH was prepared in 96% yield by the mixed anhydride coupling of Boc–Tyr(PO_3Bzl_2)–OH with

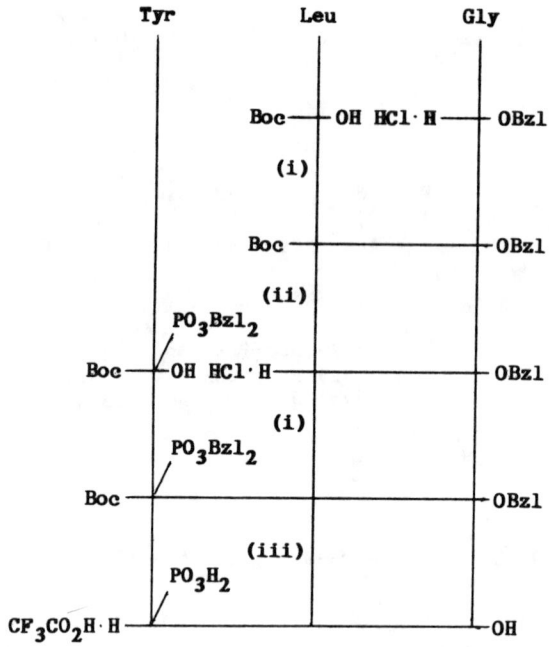

FIGURE 17. Solution-phase synthesis of zwitterionic H–PTyr–Leu–Gly–OH by the use of Boc–Tyr (PO$_3$Bzl$_2$)–OH in the "synthon incorporation strategy". (i) NMM, IBCF (−20°C); (ii) 4 M HCl/dioxane; (iii) H$_2$-Pd/C — MeOH.

HCl·H–Leu–Gly–OBzl followed by a one-step hydrogenolytic deprotection of Boc–Tyr(PO$_3$Bzl$_2$)–Leu–Gly–OBzl (Figure 17). Two positive features in the use of benzyl phosphate protection are that (1) benzyl cleavage is rapid and quantitative and (2) the final hydrogenolytic deprotection step permits ready isolation of the PTyr-peptide in a metal-free environment. However, an obvious limitation of this approach is that it cannot be used for the preparation of Met- or Cys-containing peptides because these residues cause catalyst poisoning.

3. Synthesis of H–Asn–Glu–Tyr(PO$_3$H$_2$)–Thr–Ala–OH

A solution of Z–Asn–Glu(OBzl)–Tyr(PO$_3$Me$_2$)–Thr(Bzl)–Ala–OBzl (0.2 g, 0.18 mmol) in acetic acid containing palladium acetate (100 mg) is charged with hydrogen and stirred until cessation of hydrogen uptake. The catalyst is then removed by gravity filtration and the solvent removed under reduced pressure. The crude solid is then suspended in 10% TMSBr/acetonitrile (10 ml) and stirred for 4 h at 20°C after complete dissolution. The solvent is evaporated under reduced pressure, and methanol (2 ml) is added and then evaporated under reduced pressure. The oily residue is triturated with diethyl ether (2 × 20 ml) and then purified by SP-Sephadex® cation exchange chromatography using a linear pyridinium acetate elution gradient (0.05 M [pH 2.5] to 2 M [pH 5.0]). The fractions containing the PTyr-pentapeptide are combined and the solvent removed under reduced pressure. Lyophilization of the oil gives H–Asn–Glu–Tyr(PO$_3$H$_2$)–Thr–Ala–OH as a white solid (0.015 g, 12%); ^{31}P-NMR δ (D$_2$O) = −4.9, FAB-MS (Ar, +ve mode) m/z 677 (MH$^+$).

III. CONCLUSION

The solution-phase synthesis of the casein-related phosphopeptide Ac–Glu–PSer–Leu–PSer–PSer–PSer–Glu–Glu–NHMe and the PTyr-pentapeptide

H–Asn–Glu–PTyr–Thr–Ala–OH clearly indicates that the "synthon incorporation strategy" is ideally suited for the synthesis of complex PSer- and PTyr-peptides and is a far superior approach to the traditional "global phosphorylation strategy". While only Boc–Ser(PO_3Ph_2)–OH and Boc–Tyr(PO_3Me_2)–OH were used successfully in the synthesis of the above peptides, it is probable that the development of alternative phosphate protecting groups will lead to greater synthetic flexibility and permit the preparation of large, complex PSer- and PTyr-peptides. In view of the straightforward incorporation of Boc–Ser(PO_3Et_2)–OH and Boc–Ser(PO_3Me_2)–OH into peptide synthesis it is likely that the development of substituted alkyl groups, which can be removed without destruction of the PSer residue, may permit the efficient synthesis of PSer-peptides. Furthermore, the application of these derivatives to solid-phase peptide synthesis methodology indicates that the "synthon incorporation strategy" will be a useful technique for the synthesis of many important PSer- and PTyr-peptides.

IV. APPENDIX: ABBREVIATIONS

Protecting groups

Ac	—acetyl
Adoc	—adamantyloxycarbonyl
Boc	—t-butyloxycarbonyl
Bpop	—2-biphenylisopropyloxycarbonyl
Bu^t	—t-butyl
Bzl	—benzyl
Et	—ethyl
Me	—methyl
NEt_2	—diethylamine
NHMe	—N-methylamide
N^iPr_2	—diisopropylamine
NO_2	—nitro
OBu^t	—t-butyl ester
OBzl	—benzyl ester
OMe	—methyl ester
ONBzl	—4-nitrobenzyl ester
Ph	—phenyl
Ppoc	—2-phenylisopropyloxycarbonyl

Reagents

AcOH	—acetic acid
DCC	—dicyclohexylcarbodiimide
DMF	—dimethylformamide
HOBt	—hydroxybenzotriazole
IBCF	—isobutyl chloroformate
mCPBA	—m-chloroperoxybenzoic acid
MeCN	—acetonitrile
NMM	—N-methylmorpholine
Pd/C	—palladium on charcoal
TFA	—trifluoroacetic acid
THF	—tetrahydrofuran
TMSBr	—trimethylsilyl bromide

Amino acids

Ala	—alanine
Arg	—arginine
Asn	—asparagine
Asp	—aspartic acid
Azi	—aziridine
Glu	—glutamic acid
Gly	—glycine
Leu	—leucine
Lys	—lysine
PSer	—O-phosphoserine
PTyr	—O-phosphotyrosine
Ser	—serine
Thr	—threonine
Trp	—tryptophan
Tyr	—tyrosine
Val	—valine

REFERENCES

1. **Folsch, G.**, Synthesis of phosphopeptides. I. Peptides of D,L-serine and glycine, *Acta Chem. Scand.*, 12, 561, 1958.
2. **Frank, A. W.**, Synthesis and properties of N-, O-, and S-phospho derivatives of amino acids, peptides, and proteins, *Crit. Rev. Biochem.*, 16, 51, 1984.
3. **Folsch, G.**, The synthesis of peptides of O-phosphorylserine, *Sven. Kem. Tidskr.*, 79, 38, 1967.
4. **Folsch, G.**, Synthesis of phosphopeptides. II. O-Phosphorylated dipeptides of L-serine, *Acta Chem. Scand.*, 13, 1407, 1959.
5. **Folsch, G.**, O-Phosphorylserine, *Acta Chem. Scand.*, 11, 1232, 1957.
6. **Grehn, L., Fransson, B., and Ragnarsson, U.**, Synthesis of substrates of cyclic AMP-dependent protein kinase and use of their protected precursors for the convenient preparation of phosphoserine peptides, *J. Chem. Soc. Perkin Trans. 1*, p. 529, 1987.
7. **Johnson, T. B. and Coward, J. K.**, Synthesis of oligophosphopeptides and related ATP γ-peptide esters as probes for cAMP-dependent protein kinase, *J. Org. Chem.*, 52, 1771, 1987.
8. **Perich, J. W. and Johns, R. B.**, unpublished data, 1986.
9. **Alewood, P.F., Palma, S., and Johns, R. B.**, Synthesis of a model phosphopeptide, *Aust. J. Chem.*, 37, 425, 1984.
10. **Okawa, K., Yuki, M., and Tanaka, T.**, A convenient method of the synthesis of phosphopeptide via aziridine ring opening reaction, *Chem. Lett.*, p. 1085, 1979.
11. **Okawa, K., Nakijima, K., Tanaka, T., and Kawana, Y.**, Hydroxyamino acids. V. Synthesis and N-acylation of 3-methyl-L-azylylglycine benzyl ester, *Chem. Lett.*, p. 591, 1975.
12. **Alewood, P. F., Johns, R. B., and Perich, J. W.**, Synthesis of casein related peptides, in *Peptides — Proc. 7th Am. Peptide Symp.*, Pierce Chemical Co., Rockford, Illinois, 1981, 65.
13. **Perich, J. W., Alewood, P. F., and Johns, R. B.**, Synthesis of casein related peptides and phosphopeptides. I. Solution-phase synthesis and ^{13}C N.M.R. spectroscopy of the Nα-acetyl octapeptide N-methylamide corresponding to region 14—21 of bovine β-casein A², *Aust. J. Chem.*, 40, 257, 1987.
14. **Perich, J. W. and Johns, R. B.**, A new, convenient and efficient general procedure for the conversion of alcohols into their dibenzyl phosphorotriesters using dibenzyl N,N-diethylphosphoramidite, *Tetrahedron Lett.*, 28, 101, 1987.
15. **Perich, J. W. and Johns, R. B.**, Di-t-butyl N,N-diethylphosphoramidite. A new phosphitylating agent for the efficient phosphorylation of alcohols, *Synthesis*, p. 142, 1988.
16. **Perich, J. W., Guillaume, H. A., Tregear, G. W., and Johns, R. B.**, The synthesis of O-phosphoamino acid-containing peptides, Paper 28, 12th Annu. Conf. Protein Structure and Function, Lorne, Victoria, Australia, 1987.
17. **Bannwarth, W. and Trzeciak, A.**, A simple and effective chemical phosphorylation procedure for biomolecules, *Helv. Chim. Acta*, 70, 175, 1987.

18. **Perich, J. W.,** Synthesis of Phosphopeptides, Ph.D. thesis, University of Melbourne, Victoria, Australia, 1986.
19. **Perich, J. W. and Johns, R. B.,** unpublished data, 1982.
20a. **Geiger, R. and Konig, W.,** Amine protecting groups, in *The Peptides: Analysis, Synthesis, Biology,* Vol. 3, Gross, E. and Meienhofer, J., Eds., Academic Press, New York, 1981, chap. 1.
20b. **Roeske, R. W.,** Carboxyl protecting groups, in *The Peptides: Analysis, Synthesis, Biology,* Vol. 3, Gross, E. and Meienhofer, J., Eds., Academic Press, New York, 1981, chap. 2
20c. **Stewart, J. M.,** Protection of the hydroxyl group in peptide synthesis, in *The Peptides: Analysis, Synthesis, Biology,* Gross, E. and Meienhofer, J., Eds., Academic Press, New York, 1981, chap. 4.
21. **Alewood, P. F., Perich, J. W., and Johns, R. B.,** Synthesis of a protected phosphoamino acid, N^α-tert-butyloxycarbonyl-*O*-diethylphosphoro-L-serine, *Synth. Commun.,* 12, 821, 1982.
22. **Alewood, P. F., Perich, J. W., and Johns, R. B.,** Preparation of $N\alpha$-(*tert*-butyloxycarbonyl)-*O*-(dibenzylphosphono)-L-serine, *Aust. J. Chem.,* 37, 429, 1984.
23. **Perich, J. W., Alewood, P. F., and Johns, R. B.,** Solution-phase synthesis of an *O*-phosphoseryl-containing peptide using phenyl phosphorotriester protection, *Tetrahedron Lett.,* 27, 1373, 1986.
24. **Guibe-Jampel, E. and Wakselman, M.,** Selective cleavage of *p*-nitrobenzyl esters with sodium dithionite, *Synth. Commun.,* 12, 219, 1982.
25. **Atherton, F. R.,** Dibenzyl phosphorochloridate, *Biochem. Prep.,* 5, 1, 1957.
26. **Alewood, P. F., Johns, R. B., Kemp, B. E., Perich, J. W., and Valerio, R. M.,** Phosphopeptide synthesis, presented at the 7th Natl. Royal Australian Chemical Institute Convention, Canberra, A.C.T., Australia, August 22 to 27, 1982, 26.
27. **Perich, J. W., Alewood, P. F., and Johns, R. B.,** An efficient one-pot synthesis of *N*-protected α-amino-β-dialkoxyphosphinyloxy(diphenylphosphinyloxy)-carboxylic acids (phosphate-group esters of *O*-phosphorylated *N*-protected α-amino-β-hydroxyamino acids), *Synthesis,* p. 572, 1986.
28. **Halpern, B. and Nitecki, D.E.,** The deblocking of *t*-butyloxycarbonyl-peptides with formic acid, *Tetrahedron Lett.,* p. 3031, 1967.
29. **Alewood, P. F., Perich, J. W., and Johns, R. B.,** A novel approach to phosphopeptide synthesis — preparation of Glu-PSer-Leu, *Tetrahedron Lett.,* 25, 987, 1984.
30. **Sandberg, B. E. B. and Ragnarsson, U.,** Preparation of some 2-phenylisopropyloxycarbonyl amino acids. Evaluation of their properties with particular reference to application in solid phase peptide synthesis, *Int. J. Peptide Protein Res.,* 6, 119, 1974.
31. **Jacob, L., Julia, M., Pfeiffer, B., and Rolando, C.,** Selective demethylation of mixed alkyl phosphates, *Synthesis,* p. 451, 1983.
32. **Zygmut, J., Kafarski, P., and Masterlerz, P.,** Preparation of oxoalkanephosphonic acids, *Synthesis,* p. 609, 1978.
33. **Olah, G.A. and Narang, S. G.,** Iodotrimethylsilane — a versatile synthetic reagent, *Tetrahedron,* 38, 2225, 1982.
34. **Takeuchi, Y., Demachi, Y., and Yoshii, E.,** Stepwise and selective dealkylation of phosphorotriesters with phenylthiotrimethylsilane, *Tetrahedron Lett.,* p. 1231, 1979.
35. **Johns, R. B., Alewood, P. F., Perich, J. W., Chaffee, A. L., and MacLeod, J. K.,** Fast atom bombardment mass spectrometry of seryl- and *O*-phosphoseryl-containing peptides, *Tetrahedron Lett.,* 27, 4791, 1986.
36. **Perich, J. W. and Johns, R. B.,** unpublished data, 1987.
37. **Perich, J. W., Valerio, R. M., Langford, N. J., Johns, R. B., and Alewood, P. F.,** Solution- and solid-phase synthesis of casein-related phosphopeptides, presented at the 11th Annu. Conf. Protein Structure and Function, Lorne, Victoria, Australia, February 2 to 6, 1986, 3.
38. **Perich, J. W., Chaffee, A. L., and Johns, R. B.,** unpublished data, 1985.
39. **Perich, J. W., Alewood, P. F., Valerio, R. M., and Johns, R. B.,** unpublished data, 1982.
40. **Schlatter, J. M., Mazur, R. H., and Goodmonson, O.,** Hydrogenation in solid phase peptide synthesis. I. Removal of product from the resin, *Tetrahedron Lett.,* p. 2851, 1977.
41. **Perich, J. W., Valerio, R. M., and Johns, R. B.,** Solid-phase synthesis of an *O*-phosphoseryl-containing peptide using phenyl phosphorotriester protection, *Tetrahedron Lett.,* 1377, 1986.
42. **Arendt, A., McDowell, J. H., and Hargrave, P. A.,** Solid-phase synthesis of phosphopeptides: synthesis of peptide analogs from the carboxyl-terminus of rhodopsin, Paper P-Wth-58, 9th Am. Peptide Symp., Toronto, Ontario, Canada, June 23 to 28, 1985.
43. **Posternak, T. and Grafl, S.,** Protection against enzymic hydrolysis afforded by phosphorylated groups. II. Preparation of some peptides derived from phosphotyrosine and their enzymic degradation, *Helvetica,* 28, 1258, 1945.
44. **Anastasi, A., Bernadi, L., Bertaccini, G., Bosisio, G., DeCastiglione, R., Erspamer, V., Goffredo, O., and Impicciatore, M.,** Synthetic peptides related to caerulein I, *Experientia,* p. 771, 1968.

45. **Valerio, R. M.,** Synthesis and Biological Studies of O- and C-Phosphonopeptides, Ph.D. thesis, University of Melbourne, Victoria, Australia, 1986.
46. **Valerio, R. M., Alewood, P. F., Johns, R. B., and Kemp, B. E.,** Synthesis of protected derivatives of O-phosphotyrosine — incorporation in a heptapeptide, *Tetrahedron Lett.*, 25, 2609, 1984.
47. **Perich, J. W. and Johns, R. B.,** *J. Org. Chem.*, 54, 1750, 1989.

INDEX

A

abl, 100
Acetyl CoA carboxylase, 162
Acidic cluster, 156
Acidic-residue-requiring protein kinases, 160, see also specific types
Acidic residues, 155, 156
Acid phosphatase, 140, 141
Acid precipitation, 50
ACTH, see Adrenocorticotrophic hormone
Actin, 116, 117
Activator protein 2 (AP-2), 4
Adenosine, 55
Adenosine diphosphate (ADP), 17, 55, 123
Adenosine 5'-(β,γ-imido)-triphosphate (AMPPNP), 27
Adenosine 5'-(β,γ-methylene)-triphosphate (AMPPCP), 8
Adenosine monophosphate (AMP), 55
Adenosine triphosphatase (ATPase), 117
Adenosine triphosphate (ATP), 17, 55
 AMPPNP vs., 27
 binding of, 24, 27
 binding sites of, 8—13, 32
 casein kinases and, 146
 hydrolysis of, 116
 magnesium-, see Magnesium-ATP
 mapping with analogues of, 8—10
 as phosphate donor, 147
 phosphates of, 30
 protein phosphotransferases ratio to, 11
Adenosine triphosphate (ATP)-binding proteins, 31, see also specific types
Adenosine triphosphate (ATP)-phosphotransferases, 22
Adenylate cyclase, 2
Adenylate kinase, 29, 32
5'-Adenylimidodiphosphate (AMP-PNP), 46, 55
ADP, see Adenosine diphosphate
Adrenocorticotrophic hormone (ACTH), 71, 160, 161
Affinity labeling, 11—13, 15, 16, 32
Alanine, 62, 229
Alkaline phosphatases, 138, 140
Alloxan-induced diabetes, 72
Altered gene products, 99
L-2-Amino-4-phosphonobutyric acid, 263
Amino-terminal region roles, 99—101
Amino-terminal tyrosine phosphorylation, 97, 98
AMP, see Adenosine monophosphate
AMPPCP, see Adenosine 5'-(β,γ-methylene)-triphosphate
AMPPNP, see Adenosine 5'-(β,γ-imido)-triphosphate
AMP-PNP, see 5'-Adenylimidodiphosphate
Angiotensin, 140, 240, 241, 247
Antibodies, 255—280, see also specific types
 antisera screening and, 267, 268
applications of, 257—260
AP2, see Activator protein 2
ASV, see Avian sarcoma virus
Atrial natriuretic factor, 210, 218
Autophosphorylation, 7, 95, 96, 148, 211, 257
 of cGMP-dependent protein kinase, 215—217
 functional significance of, 217
 inhibition of, 229
 of insulin receptor, 258
 kinetics of, 215, 216
 modeling of with synthetic peptides, 216, 217
Avian sarcoma virus (ASV), 86, 92, 93

B

Bacteria, 256, see also specific types
Basic-residue-requiring protein kinases, 159
Benzamidine, 198
p-Benzoyl-phenylalanine, 16
Bowmann-Birk antiproteases, 153

C

Calcineurin, 139
Calcium, 7, 119
Calcium-calmodulin, 6
Calcium-dependent myosin light chain kinase, 116
Calcium-dependent myosin phosphorylation, 118
Calcium regulatory system, 116
Caldesmon, 117
Calmodulin, 6, 101, 122, 153, 157
 activation of, 125—127
 binding of, 121, 127
Calmodulin-dependent enzymes, 118
Calmodulin-dependent myosin light chain kinase, 116
Calmodulin-dependent protein kinase, 123
Calmodulin-dependent protein kinase II, 124
Calmodulin-dependent protein phosphatase, 139, 140
cAMP, see Cyclic AMP
CII antiproteases, 153
Capsid proteins, 256
Carboxyl groups, 17, 22, see also specific types
Cascade mechanism, 2, 4
Casein, 157, 172
Casein kinase-1 (type-1 casein kinases), 158—161
Casein kinase-2, see Type-2 casein kinases
Casein kinases, see also specific types
 classification of, 146, 147
 history of, 146, 147
 mammary gland, 158—161
 type-1, 158—161
 type-2, see Type-2 casein kinases
Catalysis, 122—125
Catalytic subunit, see C-subunit
CD, see Circular dichroism
cGMP, see Cyclic GMP
Chorionic gonadotropin, 173

Chou and Fasman analysis, 67, 69
Chromatography, 200, see also specific types
   DEAE, 51, 72
   high pressure liquid (HPLC), 50, 303, 305
   immunoaffinity, 257, 260, 267—270, 278
   ion-exchange, 50
Chromogenic substrates, 136
Circular dichroism (CD), 26, 27, 66, 67, 69
CK, see Casein kinases
Cleavage, 180
Complementary DNA, 256, 262
Contractile elements, 116, 117
Corticotrophin-stimulated ACTH release, 72
Critical acidic residues, 155, 156
crk, 100
Cross bridge (latch bridge) of myosin, 119
C-subunit complex of inhibitor protein, 54—56
C-subunit of cAMP-dependent protein kinase, 18—20
   conformational changes associated with, 26, 27
   crystal structure of, 32
C-terminal phosphorylation, 94—96
Cyclic adenosine monophosphate, see Cyclic AMP
Cyclic AMP, 4, 55
   assay of, 44
   binding sites of, 19
   gene expression regulated by, 71
   inhibitor protein as tool for study of action of, 70, 71
   physiological modulator of functions mediated by, 44
Cyclic AMP-dependent protein kinase, 1—32
   active site residues and, 22, 23
   affinity labeling and, 11—13, 16, 32
   carboxyl groups and, 17, 22
   C-subunit of, see C-subunit
   description of, 44—49
   domain structure of, 27—32
   enzyme mechanism and, 23—26
   fluorescent analogues and, 10, 11
   inhibitor protein of, see Inhibitor protein
   inhibitors of C-subunit of, 18—20
   kinetics of catalytic mechanism of, 46
   model substrate for, 257
   as modular structure, 5—8
   nucleotide fold and, 13
   peptide analogues and, 14—15
   peptide binding site and, 14—20
   peptide substrates and, 17, 18
   phosphorylation site for, 122, 124, 137
   PKI peptides as model for substrates of, 64—66
   R-subunit of, 45, 139
   smooth muscle myosin light chain phosphorylation by, 123
   substrate selectivity for, 217, 218
   substrates of, 17, 18, 64—66, 260
   substrate specificity of, see under Substrate specificity
   sulfhydryl groups and, 16, 17
   synthetic peptides and, 191
Cyclic GMP, 227, 228
Cyclic GMP-dependent protein kinase, 5, 11, 70, 71
   autophosphorylation of, 215—217
   catalytic domains of, 213, 215
   peptide inhibitors of, 228, 229
   regulatory domains of, 211—213
   substrate selectivity for, 218—221
   substrate specificity of, see under Substrate specificity
Cyclic guanosine monophosphate, see Cyclic GMP
Cyclic nucleotide-dependent protein kinases, 211—215, see also specific types
Cys-199, 16, 17, 19, 23, 24, 27
Cys-343, 16, 27
Cytochemical localization, 72
Cytochrome P-450, 173
Cytoplasmic calcium, 119
Cytoskeleton, 101
Cytosolic phosphotyrosine phosphatases, 140

D

DCCD, see Dicyclohexylcarbodiimide
DEAE-cellulose, 146, 147
DEAE chromatography, 51, 72
Deamidation, 51
Denaturation, 72
Denaturing gel electrophoresis, 52
Dephosphorylation, 2, 161
Deprotection, 301, 302
Desaturation, 50
Diabetes, 73
Diacylglycerol, 7
Dibenzyl N,N-diethylphosphoramidite, 292—295
Dibenzyl phosphorochloridate/pyridine phosphorylation, 290
Di-*t*-butyl N,N-diethylphosphoramidite, 292—295
Dicyclohexylcarbodiimide (DCCD), 22—24, 28
Diphenyl phosphorochloridate/pyridine phosphorylation, 290
5,5′-Dithio-bis(-2-nitrobenzoic acid) (DTNB), 16
DNA, 256, 262
DNase, 50
DNA topoisomerases, 152
Dodecapeptides, 178
Domain structure, 27—32
Drosophila spp., 88
DTNB, see 5,5′-Dithio-bis(-2-nitrobenzoic acid)

E

EDC, see 1-Ethyl-3(3-dimethylaminopropyl)carbodiimide
EDTA, 198
EF-Tu, see Elongation factor Tu
EGF, see Epidermal growth factor
Electron microscopy, 118
Electrophoresis, 52, 122, 148, 198, 200, 268, 270, see also specific types
ELISA, see Enzyme-linked immunosorbent assay
Elongation factor Tu (EF-Tu), 28
Endogenous substrates, 150—152, 226—228, see also specific types

Enolase, 101, 245
Enzyme-ATP-metal complex, 23
Enzyme-linked immunosorbent assay (ELISA), 257, 267
Enzymes, see also specific types
 calmodulin-dependent, 118
 mechanism of, 23—26
 preparation of, 198—200
 proteolytic, 121
Enzyme-specific substrates, 136, see also specific types
Epidermal growth factor (EGF), 104
Epidermal growth factor (EGF) receptor, 2, 7, 13, 22, 76, 99, 103, 258
Epidermal growth factor (EGF) receptor kinase, 123, 241, 243
Epithelial cells, 103
Epitopes, 260, 262, 264, 265, 267, 272—275, see also specific types
Epstein-Barr virus, 256
Erythrodehydroneopterin triphosphate synthetase, 173
Escherichia coli, 212
1-Ethyl-3(3-dimethylaminopropyl)carbodiimide (EDC), 17
Evolution, 88

F

FAB-MS, see Fast atom bombardment mass spectrometry
Fast atom bombardment mass spectrometry (FAB-MS), 292, 301
Feline sarcoma virus (FeSV), 86
FeSV, see Feline sarcoma virus
fgr, 86, 101
Fibrinogen, 160, 161
Fibroblasts, 103, 104
Flavonoids, 150, see also specific types
Fluorescence energy transfer, 27
Fluorescent analogues, 10, 11
Fluorescent substrates, 138
Fluorogenic substrates, 136
p-Fluorosulfonylbenzoyladenosine, 214
5-Fluorosulfonylbenzoyladenosine, 200
p-Fluorosulfonylbenzoyl-5'-adenosine (FSBA), 12, 31
Foot-and-mouth virus, 256
Fourier transform infrared (FTIR) spectroscopy, 67
fps, 100
Free phosphotyrosine, 140, 141
Fructose-2,6-bisphosphatase/fructose-6-phosphate-2-kinase, 173
Fructose-2,6-diphosphate, 4
Fructose-6-phosphate, 4
FSBA, see *p*-Fluorosulfonylbenzoyl-5'-adenosine
FTIR, see Fourier transform infrared
fyn, 86, 88

G

GAP, 100

Gap-junctional conductance, 72
Gardner-Rasheed strain feline sarcoma virus (GR-FeSV), 86
GDP, see Guanosine diphosphate
Gel electrophoresis, 52, 122, 148, 198, 200, 268, see also specific types
Gel filtration, 50, 52, 72, 146
Gene expression, 72
Global phosphorylation, 290, 292—295
Glu-91, 23
Glucagon, 2
Glutamic acid, 262
Glycine, 13, 63—65, 69, 70, 176, see also specific types
Glycogen metabolism, 2
Glycogen synthase, 2, 150, 151, 173, 197
Glycogen synthase kinase, 140, 151
GMP, cyclic, see Cyclic GMP
Gonadotropin, 173
Gramylate cyclase, 210
GR-FeSV, see Gardner-Rasheed strain feline sarcoma virus
Group-specific labeling, 32
Growth factor receptors, 2, 4, 86, 247, see also specific types
Growth factors, see also specific types
 epidermal, see Epidermal growth factor (EGF)
 insulin-like (IGF), 259
 platelet-derived, see Platelet-derived growth factor (PDGF)
 polypeptide, 247
 transforming (EGF), 104
GTP, see Guanosine triphosphate
GTPase, see Guanosine triphosphatase
Guanethidine, 56
Guanosine diphosphate (GDP), 30
Guanosine monophosphate (GMP), cyclic, see Cyclic GMP
Guanosine triphosphatase (GTPase), 30, 100
Guanosine triphosphate (GTP), 55, 146, 149, 150
Guanosine triphosphate (GTP)-binding proteins, 28

H

hck, 88
Heat desaturation, 50
Heat-stable inhibitor protein, 218, 226, 229
Heat-stable protein kinase inhibitor, 20, 27
Heparin, 149, 150
Hepatitis B, 256
Hepatocytes, 71
High order structure, 90—92, 139, 140
High pressure liquid chromatography (HPLC), 50, 303, 305
Histone H4, 190
Histones, 150, 172, 200, 203, 219, 222, 229, see also specific types
HIV, see Human immunodeficiency virus
H4 kinase, 190
 extended specificity determinants for, 203—205
 protein kinase C and derivation of, 200—203

purification of, 198
substrate specificity of, 190—196
HMG protein, 151, 153, 157
Holoenzyme, 19, 211
Homology, 211—215
Hormones, 72, 160, 161, see also specific types
HPLC, see High pressure liquid chromatography
Human chorionic gonadotropin, 173
Human immunodeficiency virus (HIV), 256
Hydrogenolytic deprotection, 301
Hydroxyproline, 178

I

IGF, see Insulin-like growth factor
IgG, see Immunoglobulin G
Immune assays, 267, 268, see also specific types
Immunization, 263, 265—267
Immunoaffinity chromatography, 257, 260, 267—270, 278
Immunocytochemistry, 257
Immunodiffusion, 268, 270
Immunoglobulin G (IgG), 270
Immunologic reagents, 256, see also specific types
Immunoprecipitation, 268
Inhibitor protein, 43—77, see also Protein kinase inhibitors (PKI)
  C-subunit complex of, 54—56
  heat-stable, 218, 226, 229
  multiple-charge isomers of, 50
  multiple molecular forms of, 51, 52*
  nucleotide-dependent PKI and, 53, 54
  physicochemical properties of, 49—51
  physiological function of, 72—76
  PKI structure and, 66—71
  protein kinase modulator and, 53
  purification of, 49, 50, 76, 77
  recognition determinants of, 56—63
  from testis, 52, 53
  as tool for study of cAMP action, 71, 72
Inhibitors, see also specific types
  of cGMP-dependent protein kinase, 228, 229
  of C-subunit of cAMP-dependent protein kinase, 18—20
  protein kinase, see Protein kinase inhibitors (PKI)
Insulin binding, 7
Insulin-like growth factor (IGF), 259
Insulin receptor, 2, 7, 22, 162, 258
Insulin receptor kinase, 248
In vitro substrate specificity, 217—222, 225, 226
Ion-exchange chromatography, 50
Isoleucine, 63, 64, 68—70
ITP, 55

K

Kinases, see Protein kinases
Kinetics, 218
  of autophosphorylation, 215, 216
  of cAMP-dependent protein kinase catalytic mechanism, 46

of histone H4, 190
of myosin phosphorylation, 117

L

Lactate dehydrogenase (LDH), 29, 31, 101
Latch bridge (cross bridge) of myosin, 119
lck, 86
LDH, see Lactate dehydrogenase
Lectin, 103
Leukemia, 248
Leupeptin, 198
Lipocortin, 101
Localization, 72, 100, 101, 257
LSTRA, 86, 88, 95, 96, 102, 248
L-type pyruvate kinase, 260, 262—265, 267, 269
  function of, 280
  immunization with, 270
  reactivity against, 272, 276
  recognition of, 278
  structure of, 280
Lymphoma, 86, 88, 95, 96, 102, 248
Lys-27, 32
Lys-47, 31
Lys-72, 23, 24, 28, 30, 31, 32
Lys-78, 31
Lys-81, 31
Lys-83, 31
Lys-721, 13
Lysozyme, 172

M

Madin-Darby canine kidney (MDCK) epithelial cells, 103
Magnesium, 22, 48, 55
Magnesium-ADP, 8, 10, 16, 24
Magnesium-ATP, 2, 8, 10, 12, 16—18
  binding of, 8, 148
  conformational changes and, 26
  inhibitor protein and, 54, 56
  ligands of magnesium and, 22, 23
  protection by, 27, 28, 30, 31
Magnesium-ATPase, 117, 118, 123, 124
  increased activity of, 120
  myosin activated, 197
Magnesium-ATP-dependent protein phosphatase, 138, 140
Malaria, 256
Mammary gland casein kinases, 158—161
Manganese, 55
Mapping
  with ATP analogues, 8—10
  of MgATP binding site features, 8
Mass spectrometry (MS), 292, 301
MDCK, see Madin-Darby canine kidney
2-Mercaptoethanol, 198
MES, see 2-(N-Morpholino)ethanesulfonic acid
Metal-ATP complexes, 26, 28, 31, see also specific types
Mitochondrial protein phosphatase, 139

Mitosis, 71, 95, 99
MLCK, see Myosin light chain kinases
Monoclonal antibodies, 91, 259, 266, 267
2-(N-Morpholino)ethanesulfonic acid (MES), 198
MS, see Mass spectrometry
M-type pyruvate kinase, 276
Multiple-charge isomers of inhibitor protein, 50
Multiple molecular forms of inhibitor protein, 51, 52
Multiple phosphate-accepting residues, 137
Murine lymphoma, 86, 88, 95, 96, 102, 248
Murine lymphosarcoma protein kinase, 190
Muscle
  contraction of, 118
  skeletal, 116, 120, 121, 125
  smooth, see Smooth muscle
Myelin, 172, 173
Myosin, 116, 194
  actin interaction with, 116
  cross bridge of, 119
  phosphorylation of, 117—121
Myosin ATPase, 117
Myosin light chain kinases (MLCK), 6, 22, 94, 115—129, 194, 196—198, 201, see also specific types
  calcium-dependent, 116
  calmodulin activation and, 125—127
  calmodulin-dependent, 116
  catalysis and, 122—125
  contractile elements of, 116, 117
  defined, 121
  H4 kinase and, 196
  myosin phosphorylation and, 117—121
  physiochemical properties of, 121, 122
  protein kinase C vs., 124
  pseudosubstrate control and, 125—127
  serine phosphorylation and, 201
  specificity of, 124
Myristylation, 89, 101

N

NAD, 31
Natriuretic factor, 210, 218
Neuroaminidase, 50
p-Nitrophenyl phosphate, 140
NMR, see Nuclear magnetic resonance
NOESY, see Nuclear Overhauser effect spectroscopy
Noncatalytic modules, 5
Nondenaturing gel electrophoresis, 52
Nonspecific phosphatases, 138
N-terminal residues, 156
N-terminal variable region, 100
Nuclear magnetic resonance (NMR), 14, 26, 31, 67
  atomic coordinates and, 46
  ^{13}C, 292, 302
  ^{31}P, 292, 309
  of peptide substrates, 17, 18
  two-dimensional, 32
Nuclear Overhauser effect spectroscopy (NOESY), 68
Nucleolar phosphoproteins, 153

Nucleotide-dependent protein kinase inhibitors, 53, 54
Nucleotide fold, 13

O

Octapeptide, 178
Oligomeric structure of type-2 casein kinases, 147—149
Oligonucleotides, 20, see also specific types
Oncogenicity, 92—94
Oocyte maturation, 72
Ornithine decarboxylase, 152, 153
Ouchterlony immunodiffusion, 268, 270
Ovalbumin, 160, 161

P

PAK, see Protease-activated kinase
PCS, see Polycation-stimulated
PDGF, see Platelet-derived growth factor
Pentapeptide, 176
PEPCK, see Phosphoenolpyruvate carboxykinase
Pepsin, 160, 161
Peptide analogues, 14, 15, see also specific types
Peptide binding site, 14—20
Peptide recognition site, 14
Phenylalanine hydroxylase, 259
Phenylmethylsulfonyl fluoride (PMSF), 198
Phorbol diester, 103
Phorbol esters, 4, see also specific types
Phosphatase inhibitor-1, 138, 139, 173
Phosphatase inhibitor-2, 138, 152
Phosphatases, 4, see also specific types
  acid, 140, 141
  alkaline, 140
  calmodulin-dependent, 139, 140
  MgATP-dependent, 140
  nonspecific, 138
  substrate specificity of, 135—141
Phosphate-accepting residues, 137
Phosphate-accepting site, 173
Phosphate acceptors, 2, see also specific types
Phosphates, 30, see also specific types
Phosphatidyl serine, 7
Phosphite-triester phosphorylation, 292—295
Phosphoamino acids, 137
Phosphocellulose paper, 241
Phosphoenolpyruvate carboxykinase (PEPCK), 4
6-Phosphofructo-2-kinase/fructose-2,6-bisphosphatase, 2—4, 218, 222, 275
Phosphoglycerate mutase, 101
Phospholipase C, 100
Phosphorylase, 137, 140
Phosphorylase a, 138
Phosphorylase kinase, 2, 6, 139, 140, 172, 173
  discovery of, 2
  smooth muscle myosin light chain phosphorylation by, 123
Phosphorylase kinase kinase, see Cyclic AMP-dependent protein kinase

Phosphorylase phosphatase, 138, 140
Phosphoserine, 137
O-Phosphoserine-containing peptide synthesis, 290—305
  global phosphite-triester and, 292—295
  solid-phase, 305
  solution-phase, 295—305
    of Boc-Ser(PO$_3$R$_2$)-OH derivatives, 295—299
    of casein-related PSer-peptides, 302—305
    of protected Ser(PO$_3$R$_2$)-tripeptides, 299—301
Phosphothreonine, 137
Phosphotransferase, 146
Phosphotransfer assay, 8
Phosphotyrosine, 140, 141, 258
Phosphotyrosine-angiotensin, 140
O-Phosphotyrosine-containing peptides, 305—310
Phosphotyrosine phosphatases, 140
Phosphotyrosyl, 140
Phosphotyrosyl phosphatase, 140
Phosvitin, 146, 153, 157, 163
Phosvitin kinase, 146
o-Phthalaldehyde, 17
PKC, see Protein kinase C
PKI, see Protein kinase inhibitor
Plasma membrane, 100
Plasmodium falciparum, 256
Platelet-derived growth factor (PDGF), 95, 97, 99, 103, 256
Platelet-derived growth faetor (PDGF) receptor kinase, 248
PMSF, see Phenylmethylsulfonyl fluoride
Poliovirus, 256
Polyamines, 150, see also specific types
Polycation-stimulated (PCS) phosphorylase phosphatase, 138
Polyclonal antibodies, 266
Polyglutamate, 150
Polyoma virus, 93, 94
Polypeptide growth factors, 247
Potassium conductance, 72
Potato acid phosphatase, 138
Preimmune serum, 267, 270
Proline, 178, 204
Prolyl residue, 225
Protamine, 203
Protease-activated kinase I (PAK-I), 123, 190
Protease-activated kinase II (PAK-II), 190
Protease-generated phosphopeptides, 137
Proteases, 180, see also specific types
Protein kinase C (PKC), 4, 6, 7, 22, 54, 75, 99
  H4 kinase and, 190, 196 197, 200—203
  HMM phosphorylation by, 124
  myosin light chain kinase vs., 124
  preparation of, 200
  smooth muscle myosin light chain phosphorylation by, 123
Protein kinase inhibitor (PKI), see also Inhibitor protein; specific types
  heat-stable, 20, 27
  inhibitor protein relationship to, 53, 54
  as model of substrates of cAMP-dependent protein kinase, 64—66
  nucleotide-dependent, 53, 54
  three-dimensional structure of, 66—71
Protein kinase modulator, 53
Protein kinase, see also specific types
  acidic-residue-requiring, 160
  activation of, 2
  antibodies to, see Antibodies
  basic-residue-requiring, 159
  basic subsite sequences of, 63
  cAMP-dependent, see Cyclic AMP-dependent protein kinase
  casein kinases compared to, 158—161
  cGMP-dependent, see Cyclic GMP-dependent protein kinase
  depression of activity of, 93—95
  as modular structures, 5—8
  myosin light chain, see Myosin light chain kinases (MLCK)
  nucleotide-dependent, 211—215, see also specific types
  serine/threonine, 259, 260
  src family of, see *src* kinases
  tyrosine, see Tyrosine protein kinases
Protein phosphatases, see Phosphatases
Protein-tyrosine kinases, see Tyrosine protein kinases
Proteolysis, 122
Proteolytic cleavage, 180
Proteolytic digestion, 91
Proteolytic enzymes, 121, see also specific types
Proto-oncogenes, 86—88, see also specific types
Protozoans, 256
Pseudosubstrates, 125—127, 218, 229, see also specific types
Pyruvate dehydrogenase, 139
Pyruvate dehydrogenase kinase, 139
Pyruvate dehydrogenase phosphatase, 139
Pyruvate kinase, 2, 137, 138, 172, 173, 176, 177
  L-type, see L-type pyruvate kinase
  M-type, 276

Q

Quercetin, 150

R

Rabbit muscle adenylate kinase, 32
Radioimmunoassay (RIA), 73, 257, 259, 268, 270, 272
  liquid-phase, 275
  solid-phase, 275, 278
Recognition determinants of inhibitor protein, 56—64
Regulatory, noncatalytic modules, 5
Regulatory subunit, see R-subunit
Rephosphorylation, 152
Reporter groups, 136
Reversed-phase HPLC, 303, 305
RIA, see Radioimmunoassay

Ribosomal proteins, 221
RNA polymerase II, 151
RNase, 50
Rous sarcoma virus (RSV), 86, 92, 256
R-subunit, 18—20
  of cAMP-dependent protein kinase, 45, 139
  of inhibitor protein, 54
RSV, see Rous sarcoma virus

S

Saccharomyces cerevisiae, 19
Sarcoma virus, 86, 92, 256
SDS, see Sodium dodecyl sulfate
Sequence homology, 211—215
Serine, 2, 56, 98, 99, 153, 172, 178, 179, 201
Serine/threonine protein kinases, 259, 260
Ser-peptides, 292—295, see also specific types
Skeletal muscle, 116, 120, 121, 125
Smooth muscle, 116—119, 124—227
Sodium dodecyl sulfate-polyacrylamide gel
    electrophoresis (SDS-PAGE), 52, 122, 148,
    198, 200, 268, 270
Solid-phase synthesis, 305—310
Solution-phase synthesis, 295—305
  of Boc-Ser(PO_3R_2)-OH derivatives, 295—299
  of casein-related *P*Ser-peptides, 302—305
  of protected Ser(PO_3R_2)-tripeptides, 299—301
Specificity
  of antibodies, 272—278
  of substrates, see Substrate specificity
Spectrin, 100
Spectroscopy, 136, see also specific types
  fast atom bombardment mass (FAB-MS), 292, 301
  Fourier transform infrared (FTIR), 67
  NMR, see Nuclear magnetic resonance
  nuclear Overhauser effect (NOESY), 68
src kinases, 7, 85—104, 256, 258, see also specific
    types
  altered gene products and, 99
  amino-terminal phosphorylation and, 97, 98
  amino-terminal region roles, 99—101
  C-terminal phosphorylation and, 94—96
  evolution and, 88
  functions of, 103, 104
  higher order structure of, 90—92
  mutant forms of, 93
  N-terminal variable region of, 100
  oncogenic activation and, 92, 93
  oncogenicity and, 93, 94
  physiological regulation of activity of, 95
  polyoma virus and, 93, 94
  primary sequences of, 89, 90
  proto-oncogene detection and, 86—88
  serine phosphorylations and, 98, 99
  structure of, 88—92
  subcellular localization signals and, 100, 101
  substrates for, 101, 102
  three-dimensional structure of, 88—90
Subcellular localization, 100, 101, 257

Substrates, see also specific types
  assay of, 162
  of cAMP-dependent protein kinase, 17, 18, 64—66,
    260
  chromogenic, 136
  endogenous, 150—152, 226—228, see also specific
    types
  enzyme-specific, 136
  fluorescent, 138
  fluorogenic, 136
  model, 156
  NMR of, 17, 18
  pseudo-, 125—127, 218, 229
  specificity of, see Substrate specificity
  for *src* kinases, 101, 102
  synthesis of, 162
  synthetic peptides as, 154—158, 221—226
  for type-2 casein kinases, 154—158
Substrate specificity, 93, 135—141, 229
  of cAMP-dependent protein kinase, 171—182, 260
    phosphorylation sites and, 174—180
    synthetic peptides as, 173—178
  of cGMP-dependent protein kinase, 209—230
    autophosphorylation and, 215—217
    endogenous substrates and, 226—228
    homology and, 211—215
    in vitro, 217—222, 225, 226
    peptide inhibitors and, 228, 229
  of H4 kinase, 190—196
  in vitro, 217—222, 225—226
  of tyrosine protein kinases, 240—247
Sulfhydryl groups, 16, 17
Sulfhydryl-specific reagents, 16
Synthetic peptides, 229, see also specific types
  antibodies against, see Antibodies
  antibodies made from, 260, 262—265, 268
  autophosphorylation modeled with, 216, 217
  cAMP-dependent protein kinase and, 192
  design of, 196—198
  as model substrates, 154—158, 221—226
  O-phosphoserine-containing peptides and, 154—
    158
  type-2 casein kinases and, 154—158
  for tyrosine protein kinases, 247—249
Synthon incorporation
  O-phosphoserine-containing peptides and, 295—
    305
  O-phosphotyrosine-containing peptides and, 306—
    310

T

TCA, see Trichloroacetic acid
T-cells, 103, 104, 248
Temperature-sensitive mutations, 92
Testicular inhibitor protein, 52, 53
TGF, see Transforming growth factor
Thermolysis, 91
Thionitropyridine, 16
Three-dimensional structure, 172

of PKI peptides, 66—71
of *src* kinases, 88—90
Threonine, 2, 19, 152, 173, 178, 196—198, 203, see also specific types
tkl, 86
Transforming growth factor (TGF), 104
Trichloroacetic acid (TCA), 50
Tropomyosin, 116
Troponin, 116
Troponin-T, 151
Trypsin, 72, 126
Trypsin-labile factor, 44
Tumor antigens, 93
Two-dimensional NMR, 32
Type-1 casein kinases, 158—161, see also specific types
Type-2 casein kinases, 145—163, see also specific types
 classification of, 146, 147
 endogenous substrates and, 150—152
 history of, 146, 147
 oligomeric structure of, 147—149
 phosphorylation sites for, 153
 physiological implications of, 150—152
 regulation of, 149, 150
 smooth muscle myosin light chain phosphorylation by, 123
 specificity of, 152—161
  other kinases compared to, 158—161
  phosphorylation site structure and, 152, 153
  synthetic peptides and, 154—158
 substrates for, 154—158
 subunits of, 161
 synthetic peptides and, 154—158
 type-1 casein kinases vs., 158—161
Tyr-92, 98
Tyr-394, 96
Tyr-416, 96, 97, 99
Tyr-505, 95, 96
Tyr-527, 95— 97
Tyrosine, 2, 68, 70, 104
Tyrosine hydroxylase, 173, 218

Tyrosine protein kinases, 15, 159, 239, 240, see also specific types
 antibodies to, 258, 259
 characterization of, 247—249
 isolation of, 247
 normal cellular, 247, 248
 purification of, 247—249
 src family of, see *src* kinases
 substrate specificity of, 240—247
 synthetic peptides for, 247—249

V

ortho-Vanadate, 97, 102
Vascular smooth muscle, 226, 227
Viruses, 256, see also specific types
 avian sarcoma (ASV), 86, 92, 93
 Epstein-Barr, 256
 foot-and-mouth, 256
 Gardner-Rasheed strain feline sarcoma (GR-FeSV), 86
 hepatitis B, 256
 human immunodeficiency (HIV), 256
 polio, 256
 polyoma, 93, 94
 Rous sarcoma (RSV), 86, 92, 256
 Yamaguchi 73 avian sarcoma (Y73-ASV), 86
Vitamin D, 73

W

Western blotting, 257, 268

X

X-ray crystallography, 32

Y

Yamaguchi 73 avian sarcoma virus (Y73-ASV), 86
Y73-ASV, see Yamaguchi 73 avian sarcoma virus
Yeast cells, 95
yes, 86, 88, 101